Structural
Analysis

Classical
Structural
Analysis

A Modern Approach

Anthony E. Armenàkas

Professor of Aerospace Engineering
Polytechnic University of New York

McGraw-Hill Book Company

New York   St. Louis   San Francisco   Auckland   Bogotá
Caracas   Colorado Springs   Hamburg   Lisbon
London   Madrid   Mexico   Milan   Montreal
New Delhi   Oklahoma City   Panama   Paris
San Juan   São Paulo   Singapore
Sydney   Tokyo   Toronto

Library of Congress Cataloging-in-Publication Data

Armenàkas, Anthony E., date.
  Classical structural analysis.

  Includes index.
  1. Structures, Theory of.   I. Title.
TA645.A76   1988   624.1'71   87-3253
ISBN 0-07-002349-2

1234567890   HAL/HAL   8954321098

ISBN 0-07-002349-2

*The editors for this book were Nadine Post and David E. Fogarty, the
designer was Naomi Auerbach, and the production supervisor was Richard
A. Ausburn. It was set in Times Roman by University Graphics, Inc. The
illustrations were drafted by Stella Graphic Arts.*

Printed and bound by Halliday Lithograph.

*To my wife, Stella D. Petroutsa
for filling my life with love and alacrity*

Contents

Part 3   Analysis of Statically Indeterminate Structures

Preface

In the past thirty years, new methods have been developed for analyzing structures with the aid of electronic computers. These methods are referred to either as *modern methods* or as *matrix methods* of structural analysis inasmuch as the pertinent relations are expressed in matrix form. The methods which were employed in analyzing structures prior to the development of matrix methods are referred to as *classical methods*.

In the author's judgment, the reader would acquire a better comprehension of the fundamental concepts of structural analysis by first studying a selection of classical methods before being introduced to the more general modern methods. Moreover, the structural engineer should be able to analyze simple structures by hand calculations, and the classical methods are best suited for this purpose. The modern matrix methods are best suited for writing programs for analyzing groups of structures with the aid of a computer. They constitute a powerful tool for analyzing complex structures requiring the solution of large numbers of simultaneous equations.

This is the first of two books which cover some of the classical and modern methods of structural analysis in a unified manner. This book is devoted entirely to classical methods while the second book is devoted entirely to modern methods.

This book is divided into three parts. Part 1 consists of Chaps. 1 and 2. In Chap. 1, certain preliminary concepts are presented, including the idealizations and assumptions of structural analysis. In Chap. 2, the response of symmetric structures, subjected to symmetric or antisymmetric loads, is discussed qualitatively. It is considered desirable to introduce the reader to the response of symmetric structures early in the study of structural analysis.

Part 2 consists of Chaps. 3 to 6. These chapters cover the analysis of statically determinate structures. In Chaps. 3 and 4, methods are presented for computing the reactions and internal actions of structures subjected to nonmoving and to moving loads, respectively. In Chap. 5, the principle of "virtual" work is formulated and the method of virtual work (virtual forces) is applied in computing the components of displacement of points of structures. The principle of virtual work is of paramount importance in structural analysis. It is valid for structures made of any material and forms the basis for the method of virtual work (virtual forces) and the method of virtual displacements. For this reason, we present a detailed derivation of the principle of virtual work. In Chap. 6, the conjugate beam method for computing the displacements of beams is presented.

Part 3 consists of Chaps. 7, 8, and 9. These chapters cover the analysis of statically indeterminate structures. In Chap. 7, the classical approach (system approach) to the basic force or flexibility method for analyzing statically indeterminate structures is presented, as well as a number of other versions of the flexibility method. The relations of the basic flexibility method are presented in matrix form, resulting in a more compact and systematic presentation. This approach facilitates the reader's ability to concentrate on the fundamental concepts without the distraction of lengthy algebraic details. Moreover, it introduces the reader to some of the characteristics of the matrix methods. In Chap. 8, two classical displacement methods for analyzing statically indeterminate beams and frames are presented: the slope deflection method and the displacement method with moment distribution. The derivation of the slope deflection equation is presented in a way that emphasizes the similarities of this method to the direct stiffness method presented in the second book. In Chap. 9, the Müller-Breslau principle is employed in plotting influence lines for statically determinate and indeterminate structures.

In Appendix A, the fundamental relations of mechanics of materials which are pertinent to structural analysis are derived and discussed. This appendix has been included for those interested in a better comprehension of the limitations of these relations.

The second book is entitled *Modern Structural Analysis—The Matrix Method Approach*. In the matrix methods of structural analysis, a structure is considered as an assemblage of elements, and its properties are determined by assembling the properties of its elements (element approach). Every relation is presented in matrix form. Moreover, the matrices can be generated by the computer, using as input the geometric parameters of the structure, its type, the constants of the materials from which its members are made, as well as the global coordinates and the type of its nodes.

These two books have been written as reference texts for practicing engineers and could be used as texts for undergraduate and graduate courses in structural analysis. The knowledge required for studying them is contained in basic books in statics, strength of materials, calculus, and elements of matrix algebra.

In each section, the presentation of the pertinent theory is followed by a number of solved examples which contribute to a better understanding of the theory and illustrate its application. Moreover, a number of photographs of interesting structures with a brief description and sketches of important details are presented in the two books. They bring to the attention of the reader some of the interesting structural or aesthetic features of these structures, as well as ingenious aspects of their construction.

The author is deeply appreciative and forever grateful to his valued colleague, his wife, Stella, who has provided inestimable assistance and indefatigable support during the preparation of this book. Moreover, the author wishes to express his appreciation and thanks to his students, Professor J. T. Katsikadelis and Dr. V. C. Koumousis for reading parts of the manuscript and making constructive comments, and to Dr. K. V. Spiliopoulos and Mr. E. Sapountzakis for checking some of the answers to the problems presented at the end of each chapter.

This book was written during the time the author was professor and director of the Institute of Structural Analysis of the National Technical University of Athens, and his secretary Miss Dia Troullinou did all the typing of the manuscript. Her superior ability and patience has been greatly appreciated.

Anthony E. Armenàkas

Partial List of Symbols

A	Area.
$\{A\}$	Column matrix of the internal actions at both ends of all members of a structure.
$\{A^{PL}\}$	Column matrix of the internal actions at both ends of all the members of the primary structure subjected to the given loading, except the movement of the supports corresponding to chosen redundants.
$[A^{PX}]$	The ith column of this matrix represents the internal actions at both ends of all the members of the primary structure subjected to a unit value of the ith redundant ($X_i = 1$).
C^{ip}	Carry-over factor of end p ($p = j$ or k) of the ith member.
E	Modulus of elasticity.
e_{ij} $(i, j = 1, 2, 3)$	Components of strain.
F_{ij}	Flexibility coefficient of a structure.
$[F]$	Flexibility matrix of a structure corresponding to the redundants $[X]$.
F_k^{ip}	Component of internal force in the x_k direction acting at end p ($p = j$ or k) of the ith member.
G	Shear modulus.
$\mathbf{i}_1, \mathbf{i}_2, \mathbf{i}_3$	Unit vectors directed along the positive x_1, x_2, x_3 axes, respectively.
I_i $(i = 2, 3)$	Moment of inertia of the cross section of a member about the x_i centroidal axis.
$h^{(i)}$ $(i = 2, 3)$	Distance between the points of intersection of the x_i axis with the boundary of a cross section of a member.
$k(x_1)$	Curvature of a member.
$k^I(x_1)$	Initial curvature of a member.
k^{ip}	Rotational stiffness of end p ($p = j$ or k) of the ith member that is rigidly connected at both ends.
$\hat{k}^{ip}$	Rotational stiffness of the rigidly connected end p ($p = j$ or k) of the ith member that is rigidly connected at one end and pinned at the other.
K_{ij}^t	Translational stiffness coefficient of a structure.
$[K^t]$	Translational stiffness matrix of a structure.
$M_i^{(m)}$	Component of external moment in the x_i direction, applied at point m of a structure.

M_k^{ip}	Component of the internal moment in the x_k direction acting at end p ($p = j$ or k) of the ith member.
$N^{(i)}$	Axial force in the ith member.
$P_i^{(m)}$	Component of external force in the x_i direction, applied at point m of a structure.
q	External distributed force per unit length.
Q_k^{ip}	Component of the shearing force in the x_k direction acting at end p ($p = j$ or k) of the ith member.
$R_k^{(i)}$	Component of the reaction in the x_k direction at support i.
$\hat{u}_k$	Component of displacement in the x_k direction of a particle of a structure.
u_k	Component of displacement in the x_k direction of a particle of the axis of a member. It is also referred to as the component of *translation*.
u_k^{ip}	Component of translation in the x_k direction of the end p ($p = j$ or k) of the ith member.
U_s	Strain energy density.
U_c	Complementary energy density.
ΔT_c	Change of temperature at the centroid of a cross section.
δT_k	$T_k^{(+)} - T_k^{(-)}$ ($k = 2, 3$)
$T_k^{(+)}, T_k^{(-)}$	Temperature of the points of a cross section of a member where the positive and negative x_k ($k = 2, 3$) axis, respectively, intersects its perimeter.
T_0	Uniform temperature at which the structure was constructed.
x_i ($i = 1, 2, 3$)	Local cartesian coordinates.
$\bar{x}_i$ ($i = 1, 2, 3$)	Global cartesian coordinates.
X_i	ith redundant (reaction or internal action).
$\{X\}$	Matrix of the redundants of a structure.
α	Coefficient of linear thermal expansion.
θ_k^{ip}	Component of rotation, about the x_k axis, of the end p ($p = j$ or k) of the ith member.
Δ_i	Global component of displacement (translation or rotation) of a joint of a structure.
Δ^s ($s = 1, 2, \ldots, S$)	Given displacements of the supports of a structure.
Δ_i^{PL}	Displacement of the primary structure, corresponding to the redundant X_i, resulting from the given loading of the structure except the given movements of its supports which correspond to the redundants.
Δ_i^t	Component of translation of a joint of a structure, corresponding to the holding force H_i.
ν	Poisson's ratio.
ρ	Radius of curvature at a point of a member.
ρ^I	Initial radius of curvature at a point of a member.
τ_{ij} ($i, j = 1, 2, 3$)	Components of stress.
$(\bar{\ })$	A bar at the top of a component of internal action or displacement indicates that it is referred to global axes.

Photograph and Brief Description of the Twin Towers of New York

Figure a   The Twin Towers of New York City. *(Courtesy of the Port Authority of New York and New Jersey.)*

The Twin Towers (Fig. a) were completed in 1970. Their foundation rests on rock at a depth of 21.5 m below ground. They rise 411.5 m above ground. Each tower has 110 stories above ground and 20 m of underground floors used for train stations, parking areas, etc. Each tower acts as a vertical cantilever beam of rectangular cross section. Its exterior walls consist of a skeleton of steel columns which are connected at the levels of the floors by horizontal steel beams (see Fig. b). The columns and the beams of the exterior wall have been constructed from prefabricated, welded panels (see Fig. c). All the interior columns of each tower are located in a rectangular core. The floors are supported on trusses (see Fig. e) which extend from the core to the exterior wall without any intermediate support.

Above the floor trusses there is a corrugated metal top on which the concrete floor slab and tile surface are placed. The floors were constructed by prefabricated sections of 18 m in length and 4 m in width (see Fig. e).

In each tower some 10,000 viscoelastic damping units (100 per floor) were employed to decrease wind-induced sway. These damping units are shock absorbers which dissipate energy in the form of heat or friction.

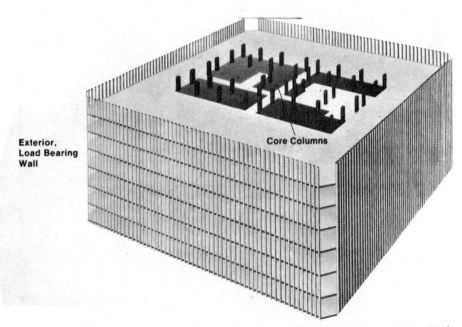

Exterior, Load Bearing Wall

Core Columns

Figure b Model of a section of the Tower. *(Courtesy of the Port Authority of New York and New Jersey.)*

Figure c   Raising tower walls. *(Courtesy of the Port Authority of New York and New Jersey.)*

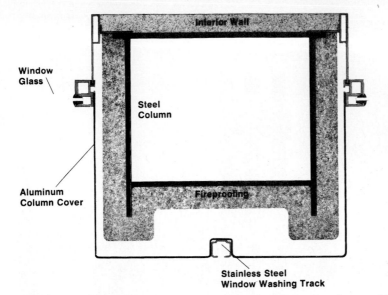

Figure d   Exterior wall column section. *(Courtesy of the Port Authority of New York and New Jersey.)*

Figure e   A prefabricated section of the floor. *(Courtesy of the Port Authority of New York and New Jersey.)*

Preliminary Considerations

1

Idealization, Loading, and Classification of Framed Structures

1.1   Introduction

From ancient times humans have been preoccupied with the planning, design, and construction of structures.

Planning a structure involves the selection of the most suitable type of structure and the choice of its general layout and overall dimensions on the basis of economic, aesthetic, functional, and other criteria.

Designing a structure entails determining the disturbances (external forces, change of temperature, etc.) to which it is expected to be exposed during its lifetime and then choosing the dimensions of its members as well as the details of their connections. The structure is then analyzed, that is, the internal forces and moments in its members and the displacements of some of its cross sections are computed. The components of stress acting at any point on a cross section of a member of a structure are established from the internal forces and moments acting on this cross section. The members of a structure must have sufficient strength and rigidity so that when the structure is subjected to the disturbances to which it is expected to be exposed, the components of stress and displacement at any of its points do not exceed the maximum allowable values given in the appropriate design codes. Moreover, the members of a structure must be such that the structure, or any of its parts, do not reach a state of instability (buckling) when subjected to the disturbances which are expected to act on it. If the results of the analysis show that the members of a structure do not have sufficient strength and rigidity to satisfy the aforementioned requirements, the structure is redesigned. That is, new dimensions of the cross section are chosen for some of its members, and the resulting structure is reanalyzed. The process is repeated until a structure is obtained which satisfies all the aforementioned requirements. Moreover, when the analysis of a structure indicates that some of its members are not stressed sufficiently, the structure is redesigned and reanalyzed.

On the basis of the foregoing, it is apparent that structural analysis is an integral part of the design of structures; consequently, every competent structural designer must be well versed in structural analysis.

1.2 Idealization of Structures

Many structures of engineering interest may be considered as an assemblage of one-dimensional (line) and two-dimensional (surface) members (see Fig. 1.1). Generally, the length of a one-dimensional member of a structure is large compared to its other dimensions (see Fig. 1.1*a*). The axis (locus of the centroids of its cross sections) of a line member could be a straight line or a curve. Moreover, the dimensions and orientation of the cross sections of a line member could change along its length. A line member is referred to as *prismatic* or *cylindrical* when the dimensions and the orientation of its cross sections do not change along its length.

A two-dimensional member has two, usually parallel surfaces, referred to as its faces, and a lateral surface which is generally normal to the two faces. The thickness of a two-dimensional (surface) member is smaller than its other two dimensions (see Fig. 1.1*b*). Its middle surface can be a plane surface or a curved surface. In the first case, the two-dimensional member is referred to as a *plate,* while in the second case, it is referred to as a *shell.*

A structure made of line members joined together is referred to as a *framed structure.* In this text, we limit our attention to framed structures.

As a result of the geometry of line members, it is possible to make certain assumptions as to their deformed configuration and as to the distribution of the

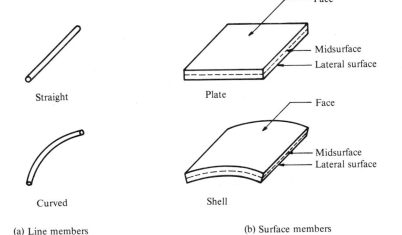

(a) Line members   (b) Surface members

Figure 1.1 One-dimensional (line) and two-dimensional (surface) members.

 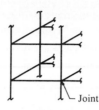 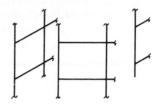

(a) Actual structure   (b) Line diagram   (c) Planar structures

Figure 1.2 Idealization of a structure.

components of stress. On the basis of these assumptions, the components of stress at any point of a line member are expressed in terms of the components of the resultant force and moment acting on this cross section by the well-known relations of the theory of strength of materials (see Sec. A.13). Moreover, the deformed configuration of a line member is determined by the deformed configuration of its axis. For these reasons, in structural analysis the configuration of framed structures is described by a *line diagram* (see Fig. 1.2*b*). Therein, the line members of the framed structure are represented by single lines which, generally, are the loci of the centroids of their cross sections.† A cross section of a member of a framed structure is represented by a point on its line diagram. Moreover, in a line diagram, a connection of two or more members is represented by a point and is referred to as a *joint* (see Fig. 1.2*b*).

In the analysis of framed structures, we are interested in establishing the components of displacement of the points of the axis of their members and the components of internal force and moment acting on any cross section of their members. That is, we are dealing with quantities which, in each member, are functions only of its axial coordinate. The components of stress at any point of a cross section of a member can be computed from the internal force and moment acting on this cross section.

---

† The continuous haunched beam shown in Fig. 1.3 is an example of a framed structure whose members are not represented by lines which are the loci of the centroids of their cross sections. The effect of the haunch at the middle support on the line diagram of this beam is disregarded. That is, the line diagram is taken as a straight line. However, the effect of the haunch on the moment of inertia of the cross sections of the beam cannot be disregarded.

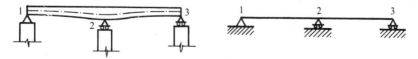

(a) Actual structure   (b) Line diagram

Figure 1.3 Idealization of a haunched beam.

In this text, we limit our attention to straight-line members which either have constant cross section or cross sections whose geometry changes so that the direction of their principal centroidal axes remains constant throughout their length. Moreover, in this text, we limit our attention to curved members whose axis lies in one plane, whereas one of the principal centroidal axes of their cross sections is normal to this plane.

In general, framed structures have a three-dimensional configuration. Often, however, for purposes of analysis and design, a framed structure may be broken down into planar parts whose response can be considered as two-dimensional (see Fig. 1-2c). That is, these parts may be considered as lying in a plane and subjected to forces whose line of action is in the same plane. These parts are called planar framed structures. Thus, framed structures may be classified as

1. Planar

2. Space

The emphasis in this text is on planar framed structures. The analysis of space framed structures generally entails tedious, lengthy calculations while it does not differ conceptually from the analysis of planar framed structures.

1.2.1 Idealization of the joints of framed structures

The joints of a planar framed structure are usually idealized as *rigid joints* or *pinned joints*. Rigid joints develop full continuity of the connected members (see Fig. 1.4). That is, the ends of each member connected to a rigid joint undergo the same translation and rotation. Moreover, the connected members can transfer forces and moments to the rigid joint. That is, generally, on a cross section of a member of a planar structure adjacent to a rigid joint, it is possible to have

1. An internal force in the plane of the structure. In Fig. 1.4c, the two components of the internal force are shown.

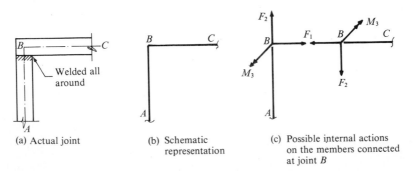

(a) Actual joint

(b) Schematic representation

(c) Possible internal actions on the members connected at joint B

Figure 1.4 Rigid joint of a planar framed structure.

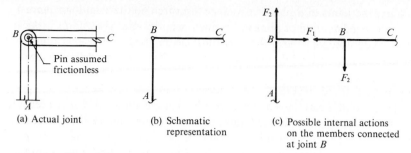

Figure 1.5 Pinned joint of a planar framed structure.

2. An internal moment whose vector is perpendicular to the plane of the structure.

Pinned joints permit rotation of the connected ends of the members about the axis of the connecting pin (see Fig. 1.5) and, consequently, the components of translation of the ends of the members of the structure connected to a pinned joint are the same while their components of rotation are not. Moreover, a member cannot transfer a moment to a pinned joint. That is, generally, on a cross section of a member adjacent to the pinned joint, it is possible to have only an internal force which lies in the plane of the structure. In Fig. 1.5c, the two components of the internal force are shown.

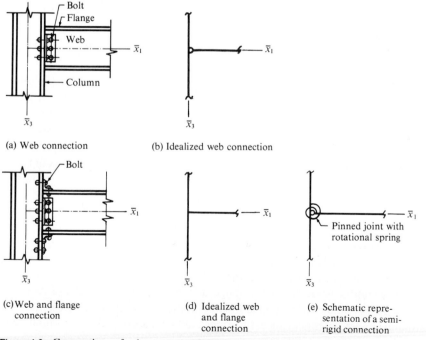

Figure 1.6 Connections of a beam to a column.

In reality, the joints of a planar structure are often neither rigid nor pinned. They are idealized as rigid or pinned. For instance, the joint shown in Fig. 1.6a permits considerable rotation of the horizontal beam about the $\bar{x}_2$ axis and may be considered as pinned, while the joint shown in Fig. 1.6c permits very small rotation of the horizontal beam about the $\bar{x}_2$ axis and may be assumed as rigid. In analyzing a framed structure, if it is necessary to take into account that one of its joints is neither rigid nor pinned, the idealization shown in Fig. 1.6e may be used. In the joint shown in this figure, the relative rotation of the end of the horizontal member is restrained by a rotational spring.

The joints of a space structure are idealized as *rigid*, as *ball-and-socket*, or as *pinned* joints. Rigid joints develop full continuity of the connected members. In general, on a cross section of a member adjacent to a rigid joint, it is possible to have an internal force and an internal moment, each of which could have three components with respect to a given system of axes. Ball-and-socket joints permit rotation of the ends of the connected members about any axis. Thus, a member does not transfer a moment to a ball-and-socket joint. Pinned joints permit rotation of the ends of connected members only about the axis of the pin. Thus, a member does not transfer to a pinned joint component of moment in the direction of the axis of the pin.

1.2.2 Idealization of the supports of framed structures

The connections of a structure to the body supporting it are referred to as its *supports*. At a support of a framed structure, one or more of its members are connected to the body supporting the structure. The supports of a planar framed structure are idealized as:

1. *Roller support.* This support permits the supported ends of the members of the structure to rotate about an axis normal to the plane of the structure and to move only in one direction, referred to as the direction of rolling. It is equivalent to one link, normal to the direction of rolling. A roller support can exert a reacting force on the structure acting in the direction normal to the direction of rolling and of magnitude equal to that required to counteract the applied loads. The schematic representation of this support is shown in Fig. 1.7a.

2. *Hinged support.* This support restrains the supported ends of the members of the structure from translating. However, it permits them to rotate about an axis normal to the plane of the structure. It is equivalent to two nonparallel links connected to the same point of the axis of the supported members of the structure. A hinged support can exert on the structure a reacting force R which passes through the center of the hinge and has the magnitude and direction required to counteract the applied loads. That is, the reacting force can have two independent components with respect to two axes of reference which lie in

the plane of the structure. The schematic representation of this support is shown in Fig. 1.7*b*.

3. *Fixed support.* This support restrains the supported end of the members of the structure from translating and rotating (see Fig. 1.7*c*). It is equivalent to three nonparallel links which do not pass through the same point. It can exert on the structure a reacting force acting in any required direction in the plane of the structure and a moment whose vector is normal to the plane of the structure. That is, the reacting force can have two independent components with respect to a set of two axes of reference which lie in the plane of the structure.

4. *Helical spring support.* This support partially restrains the supported ends of the members of the structure from moving in the direction of the axis of the spring (see Fig. 1.7*d*). However, it permits them to translate in the direction normal to the axis of the spring and to rotate about an axis normal to the

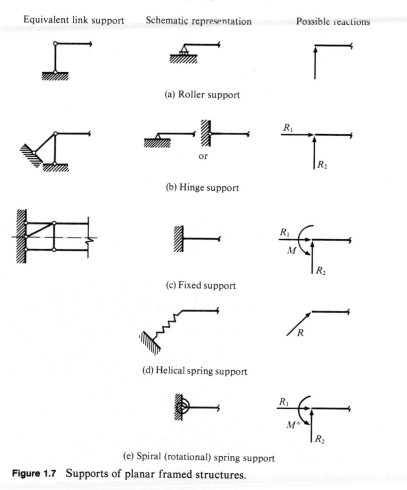

Equivalent link support   Schematic representation   Possible reactions

(a) Roller support

(b) Hinge support

(c) Fixed support

(d) Helical spring support

(e) Spiral (rotational) spring support

Figure 1.7 Supports of planar framed structures.

plane of the structure. This support can exert a reacting force on the structure in the direction of the axis of the spring whose magnitude in a known function (usually a linear function) of the deformation of the spring.

5. *Spiral spring support.* This support restrains the supported end of the members of the structure from translating and partially from rotating (see Fig. 1.7*e*). It can exert a reacting moment on the structure whose magnitude is a

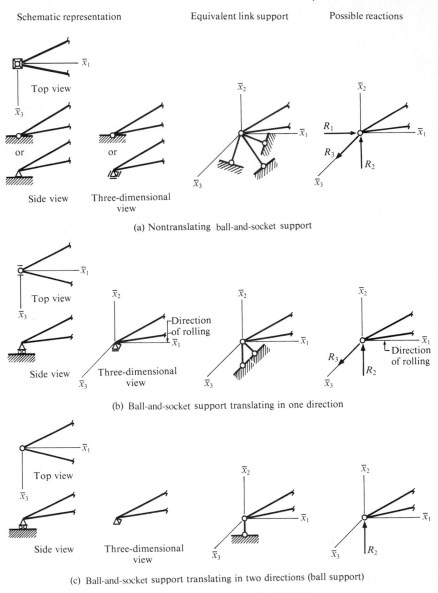

(a) Nontranslating ball-and-socket support

(b) Ball-and-socket support translating in one direction

(c) Ball-and-socket support translating in two directions (ball support)

Figure 1.8 Ball-and-socket supports of space framed structures.

known function (usually linear) of the rotation of the connected ends of the members.

The supports of space framed structures restrain one or more of the components of translation and/or rotation of the supported ends of the members of the structures. Supports which permit rotation of the supported ends of the members of the structure about any axis are referred to as *ball-and-socket* supports. Supports which permit rotation of the supported ends of the members of the structure about only one axis are referred to as *cylindrical or pin supports*. Supports which do not permit rotation of the supported ends of the members of the structure are referred to as *fixed-against-rotation supports*. Each of the aforementioned types of supports can be either nontranslating or translating in one or two directions. Usually, however, supports fixed against rotation are nontranslating and are referred to as *fixed supports*. These supports can exert a reacting force and a reacting moment on the structure, both acting in any required direction. That is, both the reacting force and moment may have three independent components with respect to a chosen set of axes of reference. Generally, the supports of space trusses are ball-and-socket supports. These supports have the following properties:

1. *Nontranslating ball-and-socket support.*   This support is equivalent to three noncoplanar links connected to the supported members of the structure at one point (see Fig. 1.8a). It can exert a reacting force on the structure which may have three independent components with respect to a set of axes of reference (see support 5 of the frame or Fig. 1.9a).

2. *Ball-and-socket support translating (rolling) in one direction.*   This support is equivalent to two nonparallel links located in the plane normal to the direction of rolling (see Fig. 1.8b). It can exert a reacting force on the structure acting in the plane normal to the direction of rolling. That is, the reacting force may have two independent components with respect to a set of two axes lying in the plane normal to the direction of rolling (see Fig. 1.8b).

3. *Ball-and-socket support translating in two directions.*   This support is also referred to as *ball support*. It is equivalent to one link normal to the plane on which the support translates and can exert a reacting force on the structure acting normal to this plane (see Fig. 1.8c).

1.3   Loads on Structures

We refer to the disturbances which cause internal forces and moments in the members of a structure and/or displacements of its points as *the loads* on the structure. These disturbances may be classified as

1. External actions

2. Displacements of the supports

3. Change of environmental conditions (usually change of temperature)

4. Initial stresses

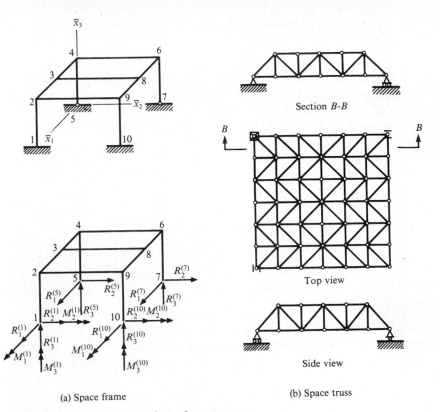

(a) Space frame

(b) Space truss

Figure 1.9 Supports of space framed structures.

The establishment of the loads acting on a structure is one of the most important steps in the process of designing this structure because the accuracy of the results of its analysis depends on the accuracy of the loads used.

A structure must be designed to withstand the most critical combinations of loads to which it is expected to be exposed during its lifetime. However, it is unlikely that the maximum effect of each loading condition will occur simultaneously. For instance, it is unlikely that a structure will be subjected simultaneously to full snow load and to full wind load inasmuch as the wind will blow away the snow. The load combinations which must be considered in designing a particular type of structure are usually mentioned in the appropriate design codes.

1.3.1 External actions

External forces may act on each material particle of a body and are referred to as *body forces,* or they may be exerted on the surface of the body through direct contact with other bodies and are referred to as *surface forces.*

The body forces are the result of the presence of a body in a force field and are distributed throughout its mass. Examples of body forces are the weight of the body resulting from its presence in the earth's gravitational field and the forces exerted on an electrically charged body located in an electromagnetic field.

In certain cases involving forces of very high intensity applied over a very small portion of the volume, or over a very small portion of the surface of a body, it is mathematically convenient to replace these forces, depending upon the nature of their distribution, either by equivalent forces or moments distributed along a line (see Fig. 1.10), by an equivalent concentrated force or moment, or by a combination of these forces and moments. Moreover, this replacement of forces is desirable since it is difficult to establish the distribution of loads of high intensity, whereas the equivalent forces or moments distributed along a line, or the equivalent concentrated force or moment are readily obtainable. The difference in the magnitude of the components of stress established using the equivalent forces and moments, or the actual forces, is negligible at points of the body which are distant from the region of the application of these forces and moments.

In reality, it is not possible to have forces or moments distributed along a line or to have concentrated forces or moments acting on a deformable body inasmuch as the stress along the line or at the point of the application of these forces and moments will be infinite.

In the analysis of framed structures, we consider the resultant body force as acting along the axis of the members of the structure. Moreover, the surface forces acting on a member are converted into equivalent forces or moments acting on its axis. Thus, line members may be subjected to external concentrated forces and moments and external distributed forces and moments acting on their axis. The distributed external forces or moments are given in units of force or moment, respectively, per unit length of the axis of the member (see Fig. 1.11a), or per unit length along some other direction as, for instance, the horizontal direction (see Fig. 1.11b).

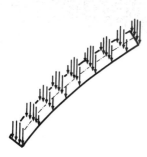

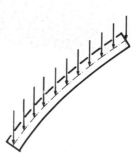

(a) Actual distribution of forces   (b) Equivalent distribution of forces

Figure 1.10 Forces distributed along a line.

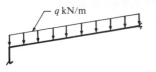

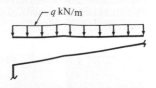

(a) Distributed force in units of force
   per unit length of member axis

(b) Distributed force in units of force
   per unit length in horizontal direction

Figure 1.11 Uniformly distributed forces.

The external forces and moments are called the *external actions*. They include

1. *Dead loads.*   These loads are vertical (gravitational) forces of constant magnitude and direction. They include the weight of the members of the structure and the weight of other immovable items.

2. *Live loads.*   Usually, these loads are vertical forces and are divided into *movable* and *moving* loads. Movable loads are those which may be moved from one place to another. Examples of movable loads are equipment, furniture, or goods on the floor of a building. Moving loads are those which move on their own power. Examples of moving loads are vehicles on a bridge or people walking in a building.

Live loads may be obtained directly from building codes or design specifications (see Table 1.1) and may be distributed or concentrated forces. (See Table 1.2 for examples of recommended live loads.) The effect of these forces depends not only on their magnitude but also on their location on the structure. For example, in order to compute the maximum internal bending moment at a cross section of a girder of a bridge, it is necessary to determine the position

---

TABLE 1.1 U.S. Design Specifications

1. AASHTO Standard Specifications for Highway Bridges, American Association of State Highway and Transportation Officials, Washington, D.C., 1973.

2. ACI Building Code Requirements for Reinforced Concrete (ACI 318-71), American Concrete Institute, Detroit, 1971.

3. AISC Specification for the Design, Fabrication, and Erection of Structural Steel for Buildings, American Institute of Steel Construction, New York, 1971.

4. AISI Specification for the Design of Light Cage Cold-Formed Steel Structural Members, American Iron and Steel Institute, New York, 1962.

5. AREA Specifications for Steel Railway Bridges, American Railway Engineering Association, Chicago, 1965.

6. ASTM Standards, American Society for Testing and Materials, Philadelphia, 1967.

7. UBC Uniform Building Code, International Conference of Building Officials (subsidiary of Pacific Coast Conference of Building Officials), Pasadena, Calif., 1967

TABLE 1.2   Minimum Uniform Live Loads for Buildings, American National
Standards Institute (ANSI) Code 1972

Type of building	lb/ft^2	kN/m^2†
Apartment and hotel rooms	40	1.92
Assembly halls		
Fixed seats	60	2.87
Movable seats	100	4.79
Balconies	100	4.79
Bleachers	100	4.79
Corridors in general	100	4.79
Corridors of apartment houses	80	3.83
Dance halls, gymnasiums	100	4.79
Driveways, sidewalks	250	11.98
Restaurants	100	4.79
Garages (passenger cars)	50	2.40
Libraries		
Reading room	60	2.87
Stacks	150	7.19
Manufacturing	125	5.99
Offices	50	2.46
Recreational areas (bowling alleys, poolrooms, etc)	75	3.60
Skating rinks	100	4.79
Stairs	100	4.79
Stores		
First floor	100	4.79
Upper floors	75	3.60
Theatres		
Orchestra floor and balconies	60	2.87
Stage	150	7.19
Warehouses		
Light storage	125	5.99
Heavy storage	250	11.98

† Conversion factors from the international system of units (SI) to the U.S. Customary System
(USCS) are given on the inside of the front cover.

of the live loads for which the moment at the cross section in question is maximum. As discussed in Chap. 4, this is accomplished by the use of influence lines.

3. *Wind loads.*   These loads are of paramount importance in the design of many structures, as for instance, tall buildings, long span bridges, etc. The Task Committee on Wind Forces of the Structural Division of the American Society of Civil Engineers (ASCE) published a report on the wind forces acting on structures.[1]† The Committee recommended the following approximate formula for computing the pressure p (lb/ft^2) acting on the surfaces of tall buildings which are exposed to the wind and are normal to its direction:

$$p = (2.56 \times 10^{-3})C_s V^2 \qquad (1.1)$$

† Superscript numbers indicate references listed at the end of the chapter.

where C_s is a coefficient which depends on the shape of the structure. For box-like structures, C_s is taken as 0.8 for pressure on the windward side and as 0.5 for suction on the leeward side (see Fig. 1.12); V is the maximum wind velocity in miles per hour, measured 30 ft (9.1 m) above the ground at the location of the building. These velocities are assembled from weather stations and are expected to occur once in 50 years. They are provided in codes and design specifications. For instance, the wind velocity for New York City is 80 mi/h while it is 120 mi/h for Key West, Florida.

4. *Snow and ice loads.* Snow loads are taken into account in the design of roofs but are not usually considered in the design of bridges because they are very small compared to the other loads acting on them. Ice loads are taken into account in the design of both roofs and bridges. For bridges, in addition to the effect of the weight of ice, the increase of the area of their members due to ice is taken into account in computing the wind loads. Snow and ice loads are considered movable loads; thus, they must be placed on the structure so as to produce a maximum effect. A layer of snow 10 mm thick weighs approximately 120 N/m^2. Values of snow and ice loads are provided in design codes and specifications. They are based on past records of snowfall in each geographic area. For instance, the recommended snow load for New York City is 30 lb/ft^2 (1437 Pa) while for San Francisco it is only 5 lb/ft^2 (239.5 Pa).

5. *Soil and hydrostatic pressure.* Structures such as retaining walls, buildings, culverts, etc., which are in contact with soil are subjected to soil pressure. This pressure depends upon a number of factors whose effect is discussed in texts on soil mechanics.

Structures such as dams, tanks, etc., which are in contact with a fluid are subjected to hydrostatic pressure. This pressure is normal to the surface of the structure in contact with the fluid, and its change of magnitude Δp between two points is proportional to the change of depth Δh between these two points. That is,

$$\Delta p = \gamma \, \Delta h \qquad (1.2)$$

where γ is the specific weight of the fluid.

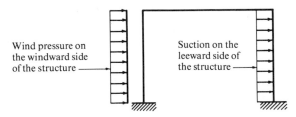

Wind pressure on the windward side of the structure →

Suction on the leeward side of the structure →

Figure 1.12   Wind load.

1.3.2 Displacements of the supports

1. *Forces due to translation or rotation of a support.*   As shown in Chap. 7, when a support of a structure settles or rotates, internal forces and moments may be produced in the members of the structure.

2. *Earthquake forces.*   During an earthquake, the foundation of a structure moves with an acceleration which is primarily horizontal, if the structure is sufficiently distant from the epicenter. The acceleration of some earthquakes may reach values of the order of $0.50g$ ($g = 9.81$ m/s^2). The ground exerts a horizontal force on the structure at the level of the foundation which accelerates the structure linearly and angularly. A gross approximation for establishing the order of magnitude of the inertia force due to an earthquake may be obtained by assuming that the structure responds as a rigid body, and by disregarding the effect of its angular acceleration. On this basis, the resultant inertia force acting on any part of the structure of mass W/g, for a ground acceleration $a = 0.5g$, is equal to

$$F = \frac{W}{g} a = 0.5W \qquad (1.3)$$

1.3.3 Change of environmental conditions—thermal forces

When relative movement of some points of a structure cannot occur freely, changes of temperature induce internal forces and moments in its members. For instance, consider a member made of aluminum, subjected to a temperature increase of $\Delta T = 30°$C. If the member is free to expand, it will not be stressed because of the increase of temperature. The coefficient of linear expansion of aluminum is $\alpha = 23 \times 10^{-6}$ $°$C^{-1}. Consequently, the resulting axial component of strain (change of length per unit length) in the member, when it is free to expand, will be equal to

$$e_{11} = \alpha \, \Delta T = (23 \times 10^{-6})30 = 69 \times 10^{-5} \qquad (1.4)$$

If the member is fixed at both ends, as its temperature increases its supports will exert a compressive force on it which will prevent it from elongating. Consequently, a compressive stress will act at the particles of the member when its ends are fixed, which corresponds to a value of the axial component of strain equal to 69×10^{-5}. The response of aluminum to external loads may be approximated by that of a linearly elastic material having a modulus of elasticity $E = 69,000$ MPa. Consequently, the stress induced in the member under consideration is

$$\tau_{11} = Ee_{11} = 69,000(69 \times 10^{-5}) = 47.61 \text{ MPa} \qquad (\text{compression}) \qquad (1.5)$$

Notice that this stress is independent of both the length and the cross-sectional area of the member.

1.3.4   Initial stresses

In certain instances a member of a structure is manufactured with its length or curvature slightly different than that required to fit. In order for such a member to be connected to the structure, it must be initially stressed. That is it must be subjected to the external actions required to make it fit. When these actions are subsequently removed the member may or may not assume its undeformed geometry. It depends on whether the other members of the structure can restrain it. If the member assumes its undeformed geometry, the members of the structure will neither be subjected to internal actions nor will they deform but will move as rigid bodies. If, however, the member is restrained from assuming its undeformed geometry, the members of the structure will be subjected to a distribution of internal forces and moments, and they will deform. These initial forces and moments could be added to those induced by other loads and, consequently, the capacity of the members of the structure to carry the other loads decreases. For instance, as shown in Fig. 1.13, when an initially slightly bent member is used to make a beam fixed at both ends, it must be subjected to an initial distribution of internal forces and moments in order to be forced into place. This becomes apparent if we suppose that initially this bent member is used to form a simply supported beam. For sufficiently small values of the angle ω, referring to Fig. 1.13b from geometric considerations, we have

$$\omega = \theta^{(1)} + \theta^{(2)} = \theta^{(2)} \left(\frac{L - a}{a} + 1 \right)$$

Thus,
$$\theta^{(2)} = \frac{\omega a}{L} \tag{1.6}$$

and
$$\theta^{(1)} = \left(\frac{L - a}{L} \right) \omega \tag{1.7}$$

It is apparent, that in order to convert the simply supported beam of Fig. 1.13b into a fixed-at-both-ends beam, we must subject it to end moments which will eliminate its end rotations $\theta^{(1)}$ and $\theta^{(2)}$. These moments are added to those induced by the weight of the beam or other transverse forces.

Members of structures are often prestressed in order to induce in them initial internal forces and moments which are opposite to those induced by other loads (usually dead loads). Consequently, these initial actions increase the capacity of the structure to carry the other loads. An interesting example is the prestressing of concrete structures. Concrete has considerable strength in compression but very little strength in tension. For this reason, concrete structures are

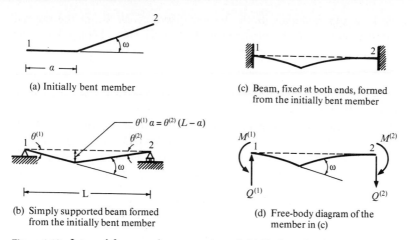

(a) Initially bent member

(c) Beam, fixed at both ends, formed
from the initially bent member

(b) Simply supported beam formed
from the initially bent member

(d) Free-body diagram of the
member in (c)

Figure 1.13 Internal forces and moments in an initially bent fixed-at-both-ends member.

usually reinforced with steel rods which carry the tensile stress acting on their cross sections. Instead of reinforcing a concrete structure, one could prestress it. A simple example of prestressing is illustrated in Fig. 1.14. The concrete beam shown in this figure has been cast with a hole extending all through its length, located at a distance e from its middle plane. After the concrete hardens, a steel rod is passed through the hole. The rod is then prestressed by pulling its ends by hydraulic jacks until it elongates by a prescribed amount. Subsequently, a pair of nuts are tightened at the two ends of the rod (see Fig. 1.14a), and the hydraulic jacks are removed. The beam prevents the rod from assuming its undeformed length, and the rod exerts, through the nuts, an eccentric, compressive force (prestressing force) at the two ends of the beam (see Fig. 1.14b). The magnitude of this force depends on the initial elongation

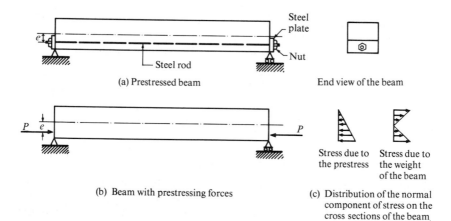

(a) Prestressed beam

End view of the beam

(b) Beam with prestressing forces

(c) Distribution of the normal
component of stress on the
cross sections of the beam

Figure 1.14 Prestressed beam.

of the rod. It is apparent that at the bottom fibers of the beam the normal component of stress due to the prestressing force (see Fig. 1.14c) reduces the tensile stress due to the weight of the beam or other transverse forces.

1.3.5 Static, quasi-static, and dynamic loads

The loads acting on a structure may be classified as *static, quasi-static,* or *dynamic.* Static loads are the dead loads. Quasi-static loads are either movable loads which are slowly applied to the structure or very slowly moving loads. When analyzing a structure these loads must be placed at locations that produce maximum effect. The structure may then be analyzed by disregarding the acceleration of its particles (static analysis). Dynamic loads are fast moving loads or suddenly applied movable loads whose magnitude changes rapidly. Examples of dynamic loads are those due to an unbalanced mass of a rotating rotor or due to an earthquake, a strong wind, an explosion, or a fast moving vehicle. The effects of dynamic loads are established by considering the structure as a moving body whose particles accelerate (dynamic analysis). The methods of analyzing a structure as a moving body are not discussed in this text. In certain cases, in order to avoid analyzing a structure as a moving body, the dynamic loads acting on it are replaced by equivalent static loads. For instance, during an earthquake, the inertia forces acting on a building that is not very tall could be replaced by equivalent static forces acting at each floor, which are proportional to the mass of each floor. The factor of proportionality is given in codes and depends on the seismicity of the region where the building is located. As a second example, we may mention that the loads of vehicles moving on bridges are multiplied by a *dynamic magnification factor,* and the bridges are then analyzed as being in equilibrium. For highway bridges, the dynamic magnification factor I given in the specifications of AASHTO (see Table 1.1) is

$$I = 1 + \frac{50}{L + 125} \quad \text{and} \quad I < 1.30 \tag{1.8}$$

where L is the length in feet of the portion of the span which, when loaded, produces the maximum value of the quantity (moment, shearing force, axial force) which we are interested in computing.

1.4 Classification of Framed Structures

Framed structures may be classified on the basis of how their members resist the applied loads as:

1. Structures whose members are primarily subjected to an axial force, which induces mainly a uniformly distributed axial tensile or compressive com-

ponent of stress. These structures include cables, cable structures, and trusses (planar or space).

2. Structures whose members are subjected to shearing forces, axial forces, bending moments, and possibly torsional moments, inducing a nonuniform distribution of the axial component of stress on their cross sections. These structures include beams, arches, planar frames, grids, and space frames.

3. Structures with some members subjected only to uniform distribution of axial stress and others subjected to a nonuniform distribution of axial stress.

In what follows, we present a brief description of the different types of structures which are considered in this text.

1.4.1 Cable structures

Cables used in structures are made either of wire strands, wire ropes, or parallel wires. A wire strand consists of a number of wires twisted together. The wire rope is made of a number of strands twisted around a strand core. The wires of cables are high-strength drawn steel wires, usually protected with a uniform coating of zinc. They have a fracture strength of 203,000 lb/in^2 (1400 MPa) and a modulus of elasticity of 20×10^6 to 24×10^6 lb/in^2 (138,000 to 166,000 MPa).

Cables are almost perfectly flexible, and thus can carry only a tensile force. As a result, their shape depends on the applied loads. For instance, as shown in Fig. 1.15a, when a cable is subjected only to its own weight, it assumes the shape of a catenary, while as shown in Fig. 1.15b, when a cable is subjected to a large concentrated force, its shape is a broken line consisting of two almost straight line segments.

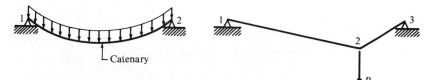

(a) Cable under its own weight   (b) Cable under a large concentrated force

Figure 1.15 Cables subjected to external forces.

Cables are often used as structural components to support loads over long spans in suspension bridges (see Figs. 1.16 and 1.58 to 1.61) in cable-stayed bridges (see Figs. 1.17 and 2.18 to 2.20) and in cable roofs (see Figs. 1.18 to 1.20). A brief description of cable structures is presented in the sequel.

Suspension bridges. Today, it is well recognized that for very long spans (over 600 m), the suspension bridge is superior to any other type of bridge. The superstructure of suspension bridges is made of steel, and it is either of the

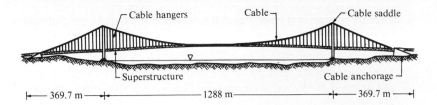

(a) Elevation of the Verrazano-Narrows Bridge in New York City

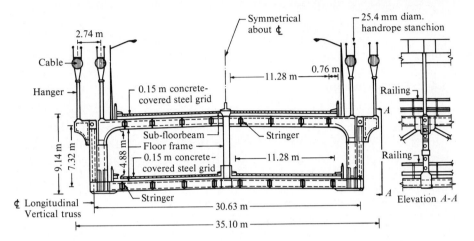

(b) Cross section and side view of the Verrazano-Narrows Bridge in New York City

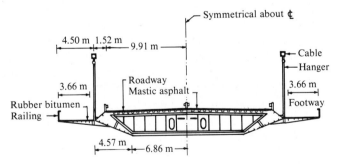

(c) Cross section of the Severn River Bridge in Great Britain

Figure 1.16 Elevation of a suspension bridge and types of superstructures of suspension bridges.

laced-wall tubular type (see Fig. 1.16b) or of the winged-box section type (see Fig. 1.16c). In the laced-wall tubular-type superstructure, the roadway is usually a concrete-filled steel grid supported on transverse subfloor beams (see Fig. 1.16b) which, in turn, are supported on longitudinal beams (stringers). In modern suspension bridges (e.g., the Verrazano-Narrows Bridge; see Figs. 1.58 to 1.61), the stringers are placed in the planes of the upper and lower chords of the two stiffening, longitudinal, vertical trusses and are rigidly connected to the floor beams which are the horizontal members of transverse floor frames. The vertical members of these frames are rigidly connected to the longitudinal trusses (see Figs. 1.16b and 1.59 at the end of this chapter). Moreover, the stringers and the horizontal members of the floor frames and the top and bottom chords of the longitudinal trusses are interconnected by horizontal diagonal members located below the stringers. Thus, the superstructure resists the vertical and horizontal forces to which it is subjected as a continuous, laced-wall tubular structure in which the floor beam frames act as heavy diaphragms.

The superstructure of a suspension bridge is supported at relatively short intervals by vertical cable hangers which are suspended from the main cables. The superstructure of a suspension bridge is discontinuous at the towers. The main span of the bridge and its approaches are connected to the towers by specially designed metal hangers which do not transmit moment to the towers.

The cables are the main carrying members of a suspension bridge. They consist either of wire strands, wire rope, or parallel bridge wire. Parallel bridge wire cables are used for long-span suspension bridges. In several recently built bridges, parallel wire cables have been built from parallel wire bundles, also called strands. A machine carrying steel wire, shuttles back and forth between the two anchorages of the bridge, unreeling the wires of each strand, one at a time. When each strand is finished, it is adjusted for sagging and placed into the body of the cable. After placing all the strands, the cable is compressed into a round shape, and it is given some form of protection from the elements. For example, the cables of the Verrazano-Narrows Bridge† are protected by a coating of lead paste and by a soft, galvanized wire wrapped around them.

The cables of a suspension bridge are supported at the top of its towers on saddles (usually cast steel) which provide for a smooth change in the direction of the cables. The cables are anchored at each end of the bridge into massive concrete anchorages which resist their enormous pull. The cables pass through the anchorage saddles and each strand is supported on a chain of eye bars (see Fig. 1.61b); alternatively the ends of each strand are socketed, and the sockets are held in place by anchor bolts.

The towers of suspension bridges are usually made of steel and are subjected to large vertical and horizontal forces. The vertical forces consist of the weight of the towers and the reactions of the cables resulting from the dead load of

---

† For a detailed description of the Verrazano-Narrows Bridge see the series of articles in the March 1966 issue of the *Journal of the Construction Division of ASCE.*

the superstructure and the live loads acting on it. For example, the vertical forces acting on each tower of the Verrazano-Narrows Bridge consist of 845,100 kN (190,000 kips) dead-load cable reaction, 97,869 kN (22,000 kips) live-load cable reaction, and 240,200 kN (54,000 kips) weight of the tower. The longitudinal, horizontal forces are due primarily to wind and to the difference in length and loading between the main span and the side span. The towers are fixed on a pedestal resting either on a caisson or a footing supported on piles.

In designing suspension bridges, the aerodynamic response of the bridge to the wind must be investigated. There is no available method for predicting accurately the aerodynamic response of a bridge. For this reason, models of the bridge are tested in wind tunnels in order to check their aerodynamic response.

Cable-stayed bridges.   During the last 25 years, a considerable number of cable-stayed bridges have been constructed, and a great number are presently in the planning and design stages. Cable-stayed bridges are considered competitive in the span-length range of 600 to 2000 ft (180 to 600 m). They are preferred over truss bridges for both aesthetic and economic reasons. A cable-stayed bridge[†] consists of a girder or truss superstructure supported at a number of points along its length by cables emanating from tall towers. The superstructure of cable-stayed bridges usually is also supported at the towers by bearing shoes. However, in some cases, the superstructure is not supported directly on the towers. The superstructure may be constructed of steel or reinforced concrete (see Fig. 1.17a) or of a combination of steel and reinforced concrete (hybrid) (see Fig. 1.17b). The development of long-span cable-stayed bridges with a concrete superstructure is relatively recent; thus fewer of these bridges exist than cable-stayed bridges with a steel superstructure. Cable-stayed bridges with concrete superstructure have shorter lengths of panel between stays and more stays per span than cable-stayed bridges with steel superstructure (see Fig. 2.18). This is necessary in order to avoid very heavy concrete superstructures.

The cables of cable-stayed bridges may be located in one vertical plane or in two planes which may be vertical or sloping. There are three main arrangements for the cables of cable-stayed bridges: (a) the harp type shown in Fig. 1.17c in which the cables are parallel, (b) the radiating type shown in Fig. 1.17a in which the cables emanate from one point of the tower (see also Fig. 2.18a), and (c) the fan type shown in Fig. 1.17b which is similar to the radiating type except that the cables emanate from different points located on a finite length of the tower (see also Figs. 2.19 and 2.20). The towers of cable-stayed bridges may consist of one or two columns (see Fig. 1.17 and Figs. 2.18 to 2.20) and may be constructed of steel or concrete.

---

† For further reading see Podolny and Scalzi[2] and Subcommittee on Cable-stayed Bridges.[3]

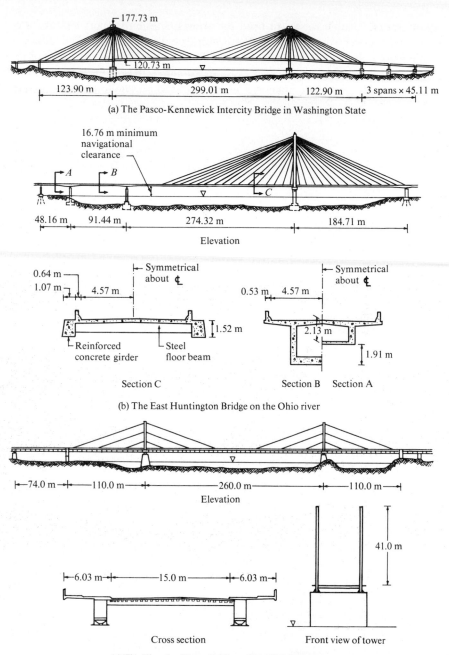

(a) The Pasco-Kennewick Intercity Bridge in Washington State

Elevation

(b) The East Huntington Bridge on the Ohio river

Elevation

Cross section   Front view of tower

(c) The Theodor-Heuss Bridge at Dusseldorf, Germany

Figure 1.17 Types of cable-stayed bridges. *(Courtesy Arvid Grant and Associates, Inc., Consulting Engineers, Olympia, Wash., and Mr. Gerard Fox, Partner, Howard, Needles, Tammen, and Bergendoff, Architects, Engineers, Planners, New York, N.Y.)*

Cable roofs.   Cable roofs are used for arenas, stadiums, sport centers, etc., where they cover a large area without interior column support. The cables of cable roofs may be arranged in a variety of ways. For example, the roof of the arena in Raleigh, N.C., shown in Figs. 1.18a and 1.20 is supported on a network of cables which are anchored in two inclined parabolic arches. The cir-

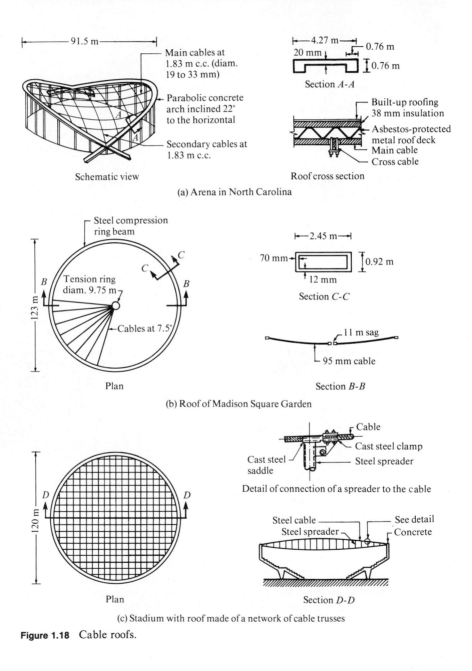

Schematic view

Main cables at 1.83 m c.c. (diam. 19 to 33 mm)

Parabolic concrete arch inclined 22° to the horizontal

Secondary cables at 1.83 m c.c.

91.5 m

4.27 m
20 mm
0.76 m
0.76 m
Section A-A

Built-up roofing
38 mm insulation
Asbestos-protected metal roof deck
Main cable
Cross cable

Roof cross section

(a) Arena in North Carolina

Steel compression ring beam

Tension ring diam. 9.75 m

Cables at 7.5°

123 m

Plan

2.45 m
70 mm
12 mm
0.92 m
Section C-C

11 m sag
95 mm cable
Section B-B

(b) Roof of Madison Square Garden

120 m

Plan

Cable
Cast steel clamp
Steel spreader
Cast steel saddle

Detail of connection of a spreader to the cable

Steel cable
Steel spreader
See detail
Concrete

Section D-D

(c) Stadium with roof made of a network of cable trusses

Figure 1.18   Cable roofs.

(a)

(b)

Figure 1.19 The Madison Square Garden Sports and Entertainment Center of New York City. The 404 ft (123.14 m) diameter roof of the center is supported on the 48 radial cables shown in the photograph of Fig. 1.19*b* which was taken during construction. There are no interior columns in the Sports Center, and thus, its 20,234 seats have an unobstructed view. *(Courtesy New York Convention and Visitors Bureau and Bethlehem Steel Corporation, Bethlehem, Pa.)*

Figure 1.20 The Dorton Arena in Raleigh, N.C. The saddle shaped roof of this arena is supported on a network of cables which are anchored in two inclined 14 ft (4.28 m) wide arches. The maximum height of the arches is 90 ft (27.04 m). The shape of the arena is elliptical. The concrete floor has a maximum length of 221 ft (67.36 m) and a maximum width of 127 ft (38.71 m). *(Courtesy North Carolina State Fair.)*

cular roof of Madison Square Garden in New York City, shown in Figs. 1.18*b* and 1.19, is supported on a single layer of radial cables anchored in an outer steel compression ring and in an inner steel tension ring. The circular roof of the stadium in Fig. 1.18*c* is supported on a network of cable trusses.

The cables of cable roofs are usually pretensioned to ensure that they remain in tension under the most adverse conditions of loading. The cables of networks are usually pretensioned by pulling their ends with hydraulic jacks. Cable trusses are pretensioned by the insertion of vertical members (spreaders) placed between their two cables (see Fig. 1.18*c*). The detail of the connection of these vertical members to the cables is shown in Fig. 1.18*c*. A number of long-span cable roofs have been built without pretensioning the cables. In this case, in order to provide sufficient stiffness to the roof system, the roofing is either made out of concrete which is a relatively heavy material or weights are hung from the roof cables. The circular roof of Madison Square Garden in New York City (Figs. 1.18*b* and 1.19) is the most outstanding example of a suspended roof built without pretensioning the cables. The roofing of cable roofs is made of sheet metal, glass, plastics, concrete, etc. (see Fig. 1.18*a*).

1.4.2 Trusses

Trusses are framed structures whose members are straight and assumed connected by frictionless pins; moreover, the axes of their members which frame at a joint are assumed to intersect at a point. Trusses are loaded by concentrated forces acting at their joints (see Fig. 1.22). The weight of the members of trusses is usually neglected or considered as acting at their joints. Thus, it is assumed that the members of trusses are not subjected to end moments or to intermediate external actions. Consequently, they are subjected only to axial forces, inducing a uniform state of axial tension or compression. However, for economy and facility of fabrication and erection, the members of trusses are not connected with pins but are bolted or welded together, as shown in Fig. 1.21. Thus, the joints of trusses are relatively rigid and, consequently, some bending moment could develop in their members. The reason for this becomes apparent when one realizes that, as a result of the application of the loads, the angles between the members of an idealized, pin-connected truss change by a small amount, in order to accommodate the changes of the lengths of the members. For a truss with relatively rigid joints, the angle between its members cannot change freely, as in the case of an idealized pin-connected truss, and, thus, its members are slightly bent. However, if a truss is properly designed and constructed, in any member the axial component of stress associated with this bending is very small compared to that induced by the axial force. For this reason, the axial component of stress in the members of a truss associated with bending is referred to as *secondary stress,* and it is usually not taken into account in its design.

Examples of space trusses are shown in Fig. 1.22. The tower shown in Fig. 1.22*a* has legs of constant slope. This tower can be analyzed by treating each of its sides as a planar truss (see Fig. 1.22*b*). The tower shown in Fig. 1.22*c* has legs of varying slope. This tower cannot be analyzed by treating each of its sides as a planar truss. Unless the legs of a tower have a constant slope throughout their length, a tower should be analyzed as a space structure. A hexagonal Schwedler-type dome is shown in Fig. 1.22*d*. Each of the supports of this dome can supply a vertical and a horizontal reaction. The horizontal reaction acts normal to the radius of the circle circumscribing the polygon of the base of the dome.

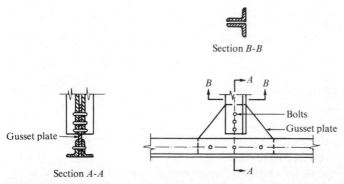

Section *B-B*

Section *A-A*

Figure 1.21   Detail of a joint of a truss.

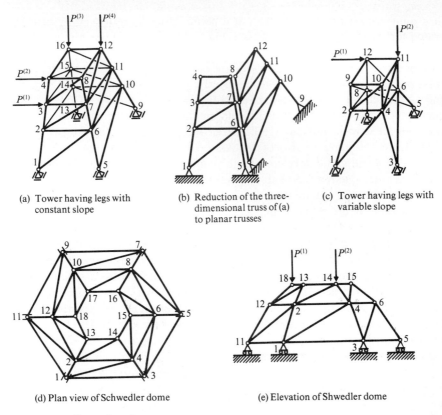

(a) Tower having legs with constant slope

(b) Reduction of the three-dimensional truss of (a) to planar trusses

(c) Tower having legs with variable slope

(d) Plan view of Schwedler dome

(e) Elevation of Shwedler dome

Figure 1.22 Examples of space trusses.

A planar simple truss is constructed by starting from a pin-connected tri-angle and adding pairs of new members. One end of each member of a pair of new members is connected to one of two adjacent joints of the truss, while their other ends are connected together forming a new joint. A planar, simple truss is not a mechanism; that is its joints cannot move without its members deform-ing. Moreover, on the basis of its definition, a planar simple truss with NJ joints must have the following number of members

$$NM = 3 + 2(NJ - 3) = 2NJ - 3 \qquad (1.9)$$

In Fig. 1.23, different types of planar trusses used for roofs and bridges are shown. Those of Fig. 1.23a, b, c, f, and h are simple.

The loads are transferred to the joints of roof trusses by purlins, as shown in Fig. 1.24a, and to the joints of bridge trusses by a floor system composed of stringers and floor beams, as shown in Fig. 1.24b. The stringers are parallel to the trusses, while the floor beams are transverse to them. Thus, the floor slab of a bridge rests on stringers that are supported by floor beams which are sup-ported by the two trusses.

(a) Warren truss

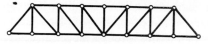

(b) Warren truss with verticals

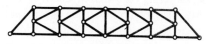

(c) Pratt truss

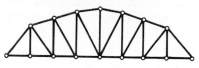

(d) Subdivided Pratt truss

(e) K truss

(f) Parker truss

(g) Fink truss

(h) Howe truss

Figure 1.23 Types of planar trusses.

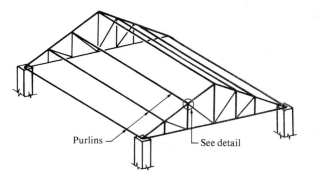

Purlins

See detail

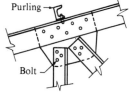

Purling

Bolt

(a) Roof

Detail of joint

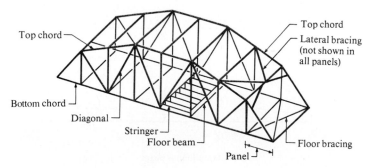

Top chord

Top chord

Lateral bracing (not shown in all panels)

Bottom chord

Diagonal

Stringer

Floor beam

Panel

Floor bracing

(b) Bridge

Figure 1.24 Examples of structures with trusses.

31

1.4.3 Beams

Beams are loaded so that they carry the load primarily in bending. We can have either *planar or space beams*. The external forces acting on a planar beam lie in one plane (the plane of the beam) which passes through the shear center of the cross section of the beam (see Sec. A.10), and it is parallel to the plane which contains its axis and a principal centroidal axis of its cross sections. Moreover, the vector of the external moments acting on a planar beam is normal to its plane (see Fig. 1.25). Consequently, on each cross section of a planar beam we could have only an axial and a shearing component of internal force lying in the plane of the beam as well as a component of internal bending moment whose vector is normal to the plane of the beam. Moreover, every cross section of a planar beam rotates only about the axis normal to the plane of the beam. When the external forces and moments applied on a beam do not meet one or more of the requirements described previously, then the beam is a space beam.

Beams are classified according to the manner in which they are supported. Some types of beams are shown in Fig. 1.25. Simply supported beams with or without one or two overhangs are supported by a hinged support at one point and by a roller support at another point (see Fig. 1.25a). They are used to span distances of up to about 200 ft (70 m) in buildings, bridges, and houses. Cantilever beams are completely fixed at one end and free at the other end (see Fig. 1.25b). Continuous beams are supported on three or more supports (see Fig. 1.25d).

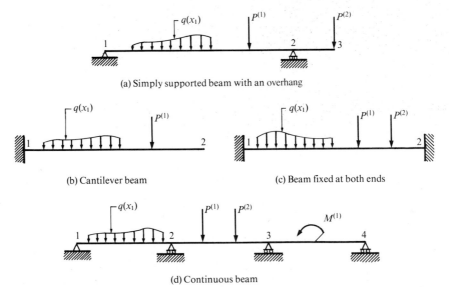

(a) Simply supported beam with an overhang

(b) Cantilever beam   (c) Beam fixed at both ends

(d) Continuous beam

Figure 1.25 Types of planar beams.

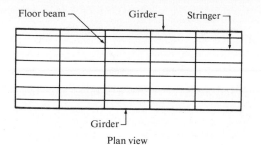

Plan view

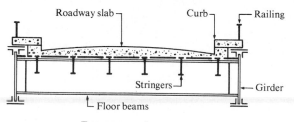

Transverse section

Figure 1.26   Girder bridge.

In long girder bridges, the loads are often transferred to the primary beams (girders) by a floor system of stringers and floor beams (see Fig. 1.26) analogous to that used in truss bridges. The roadway slab rests on the stringers. The floor system could be at the top, bottom, or intermediate level of the girders.

In steel-frame buildings the loads are transferred to the girders by a floor system similar to that employed in bridges. However, in this case, the top flanges of all beams are usually at the same level, and thus the floor slab rests directly on the stringers, the floor beams, and the girders.

1.4.4   Arches

An arch could be designed so that it would be in a state of uniform axial compression (no bending) when subjected to a given loading. However, an arch possesses flexural stiffness, and thus it cannot assume the configuration required to carry another loading without bending. Generally, therefore, the distribution of the normal component of stress on the cross sections of an arch is nonuniform. However, in arches, the axial component of stress resulting from the axial force is considerably larger than that resulting from the bending moment. For this reason, arches are often classified as structures whose members are primarily subjected to an axial force, inducing primarily a uniform axial component of stress. The supports of arches may be either hinged or fixed (see Fig. 1.27). Moreover, a hinge could exist at some point of the arch (see

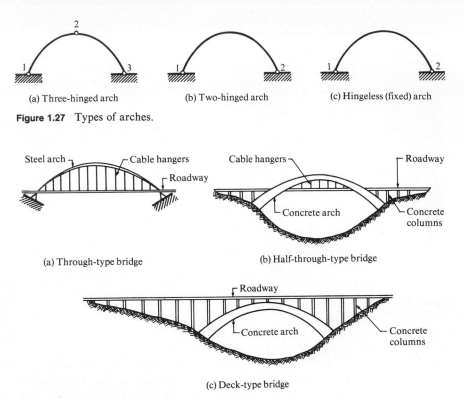

(a) Three-hinged arch   (b) Two-hinged arch   (c) Hingeless (fixed) arch

Figure 1.27 Types of arches.

(a) Through-type bridge   (b) Half-through-type bridge

(c) Deck-type bridge

Figure 1.28 Types of arch bridges.

Fig. 1.27*a*). Arches are used as supporting members in bridges and buildings. In Fig. 1.28 different types of arch bridges are shown. Photographs of arch bridges are presented in Fig. 3.17, 3.18, and 3.20.

1.4.5 Frames

Frames are the most general type of framed structures. Their members are subjected to axial and shearing forces, bending moments, and, possibly, torsional moments. They can have both rigid and nonrigid joints and can be loaded in any way. Usually, frames are space structures. Frequently, however, they can be broken down into parts which can be considered as planar frames or grids (see Fig. 1.29*b* and 1.29*c*). The members of a planar frame lie in one plane. Moreover, the external forces act in the same plane, while the vector of the external moments is normal to this plane. The members of a grid also lie in one plane. However, the external forces are normal to the plane of the grid and the vector of the external moments lies in this plane. Examples of frames are shown in Fig. 1.29.

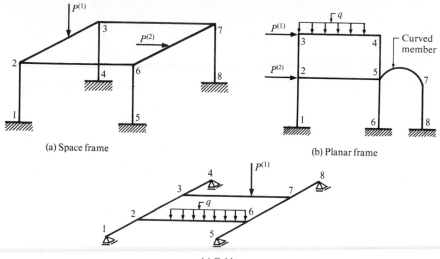

(a) Space frame

(b) Planar frame

(c) Grid

Figure 1.29 Types of frames.

1.5   Numbering of the Members and Joints of Framed Structures—Global and Local Axes

We number the joints of a structure consecutively. Moreover, whenever necessary we number the members of a structure by placing a number in a circle,

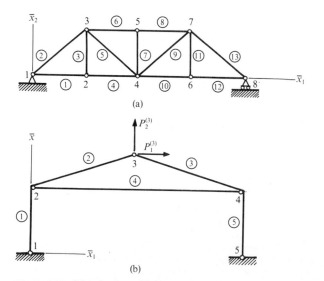

(a)

(b)

Figure 1.30 Numbering of joints and members of planar framed structures.

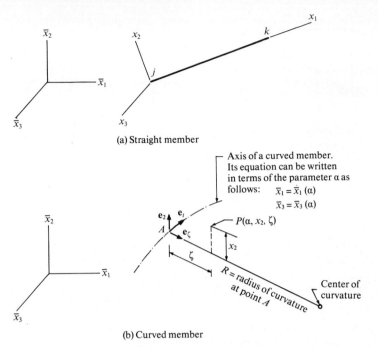

(a) Straight member

(b) Curved member

Figure 1.31 Local and global axes.

as shown in Fig. 1.30. Furthermore, we denote the ends of each member by j and k; where j is the end of the member connected to the joint having the smallest number (see Fig. 1.31).

Whenever necessary we refer each framed structure to a set of right handed† rectangular system of axes (cartesian axes) $\bar{x}_1$, $\bar{x}_2$, and $\bar{x}_3$, termed the *global system of axes* of the structure. Moreover, we refer each straight-line member of the structure to a right-handed cartesian set of axes x_1, x_2, and x_3 (see Fig. 1.31a). This set of axes has its origin at the centroid of the cross section at end j of the member; the x_1 axis coincides with the axis of the member and is directed from its end j to its end k, while the x_2 and x_3 axes are principal‡ centroidal axes of the cross section of the member, except if stated otherwise. This set of axes is referred to as the *local axes* of the member

Furthermore, we associate a set of orthogonal curvilinear coordinates (α, x_2, ζ) with each curved member. The coordinate α of a point P (α, x_2, ζ) specifies

---

† A set of axes $\bar{x}_1$, $\bar{x}_2$, and $\bar{x}_3$ is right-handed if a right-handed screw placed parallel to the $\bar{x}_3$ axis moves in the direction of the positive $\bar{x}_3$ axis when it is turned from the $\bar{x}_1$ to the $\bar{x}_2$ axis.

‡ The rectangular axes x_2 and x_3 are principal axes of a plane surface of area A if the

the point where the axis of the member intersects the plane normal to it which contains point P. The coordinate x_2 is measured in the direction normal to the plane of the member, while the coordinate ζ is measured from the axis of the member along the direction normal to it (see Fig. 1.31b). With the set of orthogonal curvilinear coordinates (α, x_2, ζ), we associate a set of right-hand orthogonal unit vectors $\mathbf{e}_t$, $\mathbf{e}_2$, and $\mathbf{e}_\zeta$. As shown in Fig. 1.31b, the unit vector $\mathbf{e}_t$ is tangent to the axis of the member, while the unit vectors $\mathbf{e}_\zeta$ and $\mathbf{e}_2$ are directed along the positive ζ and x_2 directions, respectively. It is apparent that, generally, the direction of the unit vectors $\mathbf{e}_t$ and $\mathbf{e}_\zeta$ changes with α.

1.6 Components of Internal Actions—Sign Convention

When a structure is subjected to external loads, a distribution of normal and shearing components of stress could exist on any cross section of its members. This distribution of stress on a cross section is statically equivalent to a force acting at its centroid and to a moment about its centroid. We refer to this force and moment as the *internal actions* acting on the cross section of the member. The components of the internal actions acting on a cross section of a member of a structure are referred either to the local axes of the member or to the global axes of the structure. The local component of the internal force acting on a cross section of a member in the direction of its axis is called *axial*, while the local components of internal force acting along the other two local axes of

---

product of inertia I_{23} of the surface, with respect to these axes, vanishes. That is,

$$I_{23} = \iint\limits_A x_2 x_3 \, dA = 0$$

Every point of the plane of a planar surface is the origin of at least one pair of principal axes of this surface. Moreover, for every point of an axis of symmetry of a planar surface, this axis and the perpendicular to it form a pair of principal axes (see Fig. 1.32).

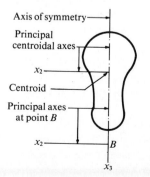

Axis of symmetry

Principal centroidal axes

x_2

Centroid

Principal axes at point B

x_2

B

x_3

Figure 1.32 Plane surface with an axis of symmetry.

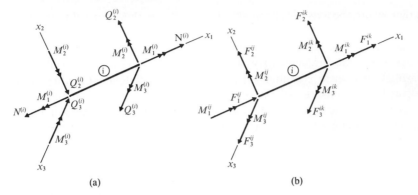

Figure 1.33 Sign conventions for internal actions.

the member are called *shearing*. The local component of internal moment acting in the direction of the axis of a member is called *torsional,* while the local components of internal moment acting along the other two local axes of the member are called *bending*. We represent a moment either by a vector with two arrow heads (see Fig. 1.33) or by a curl (see Fig. 1.34).

We denote the axial component of force by N, the shearing components of force acting in the direction of the x_m axis by Q_m ($m = 2, 3$), and the components of the moment about the x_m axis (torsional and bending) by M_m($m = 1, 2, 3$). If it is necessary to specify the member on which the components of a force or a moment act, we add the number of the member as a superscript to their symbol.

We consider as positive the components of force and moment acting on a positive† cross section if their sense coincides with the positive sense of the local axes x_1, x_2, x_3 (see Fig. 1.33a). Moreover, we consider as positive the components of force and moment acting on a negative cross section, if their sense coincides with the negative sense of the local axes x_1, x_2, x_3 (see Fig. 1.33a). Thus, a tensile axial force is considered positive while a compressive axial force is considered negative.

We choose the local axis x_2 of the members of planar structures normal to their plane and directed toward the observer. Thus, the direction of the positive shearing force and bending moment acting on a planar structure is as shown in Fig. 1.34. That is, a positive bending moment induces a tensile stress on the interior fibers of a planar frame or on the bottom fibers of a planar horizontal beam. Furthermore, for simplicity, we denote the shearing force and bending moment acting on a cross section of a planar structure by Q and M, respectively; that is, we omit their subscripts (Q_3, M_2).

---

† We call a cross section of a member positive or negative if the unit vector normal to it is directed along the positive or negative x_1 axes, respectively.

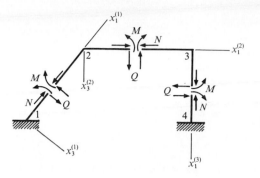

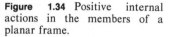

Figure 1.34 Positive internal actions in the members of a planar frame.

When we analyze structures using the slope deflection or the displacement method with moment distribution (see Chap. 8), we denote the components of the internal force and moment acting on the end p ($p = j, k$) of the ith member by F_m^{ip} and M_m^{ip} ($m = 1, 2, 3$), respectively, when referred to local axes and by $\bar{F}_m^{ip}$ and $\bar{M}_m^{ip}$, respectively, when referred to global axes (see Fig. 1.35). Moreover, we chose the x_3 axis normal to the plane of the structure, and we consider as positive the components of force and moment acting on the ends of a member if their sense coincides with the positive sense of the corresponding local axes (see Fig. 1.33b).

When analyzing a framed structure, we compute the components of internal actions acting on the cross sections of its members. The components of stress acting on a cross section of a member can be established from the components of internal actions acting on this cross section using relations (A.116).

1.7 Components of Displacements

When framed structures are subjected to external loads, their members translate and rotate as rigid bodies and they deform. In structural analysis the warping of the cross sections of the members of a structure is disregarded when their deformed configuration is specified. Consequently, the deformed config-

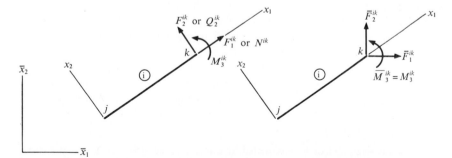

Figure 1.35 Local and global components of the internal action at the end k of the ith member of a planar structure.

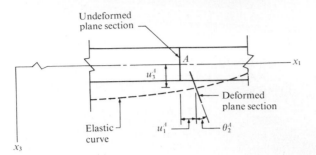

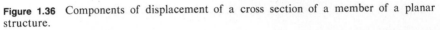

Figure 1.36 Components of displacement of a cross section of a member of a planar structure.

uration of a cross section of a member of a space framed structure is specified by the three components of the displacement vector of its centroid, referred to a set of rectangular axes, and by the three components of its rotation vector,† referred to the same set of axes. Moreover, the deformed configuration of a cross section of a member of a planar framed structure, due to its deformation, is specified by the two components (in the plane of the structure) of the displacement vector of its centroid and by the component of its rotation about the axis normal to the plane of the structure (see Fig. 1.36). The components of the displacement vector of the centroid of a cross section of a member of a structure are referred to as the *components of translation* of this cross section. Inasmuch as a cross section of a structure is represented by a point on the line diagram of the structure, we refer to the components of translation and rotation of the cross section represented by point A on the line diagram of the member of the structure as the components of translation and rotation of point A of the member.

The deformation of members of framed structures subjected to internal axial forces, shearing forces, and bending moments is such that plane sections normal to their axis before deformation can be considered plane after deformation. Actually this is true only for members subjected to internal axial forces or to bending moments. Plane sections normal to the axis of members subjected to internal shearing forces do not remain plane after deformation but they warp. However, the warping is very small, and it is neglected. Moreover, plane sections normal to the axis of members of noncircular cross section subjected to torsional moments do not remain plane after deformation but they warp. However, plane sections normal to the axis of members of circular (hollow or full) cross section subjected to torsional moments remain plane after deformation.

---

† A rotation about an axis is represented by a vector acting along this axis and pointing in the direction in which a right-hand screw moves when subjected to this rotation. Small rotations are vector quantities while large rotations are not. In this text we consider structures whose deformation involves only small rotations.

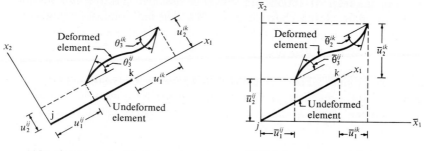

(a) Local components of displacement (b) Global components of displacement

Figure 1.37 Local and global components of displacement of the ends of the ith member of a planar structure.

We denote the components of translation and rotation of a cross section of a member by u_i and θ_i ($i = 1, 2, 3$), respectively, when referred to local axes, and by $\bar{u}_i$ and $\bar{\theta}_i$ ($i = 1, 2, 3$), respectively, when referred to global axes. If it is necessary to specify the member with which a component of translation or rotation is associated, we add the number of the member as a superscript to its symbol. Moreover, if we want to specify the point or the joint with which a component of translation or rotation is associated, we add the number or the letter specifying the point or joint as superscript to its symbol. Thus, u_2^{ij} is the component of translation of the end j or the ith member along the x_2 local axes (see Fig. 1.37), while $\bar{u}_1^{(3)}$ is the global component of translation of joint 3 in the direction of the global axis $\bar{x}_1$.

In Chap. 8, we denote the components of translation and rotation of the joints of a structure by Δ_i ($i = 1, 2, \ldots, n$), where n is the total number of unknown components of translation and rotation of the joints of the structure.

The components of translation (u_1, u_2, u_3) and the components of rotation ($\theta_1, \theta_2, \theta_3$) of a cross section of a member of a structure are called its *components of displacement*. Moreover, the deformed axis of a member is called its *elastic curve*.

Throughout this text *the local or global components of displacement (translation and rotation) of a point or of a joint of a member are assumed positive when their sense coincides with the positive sense of the corresponding local or global axes.* The positive components of displacement of the ends of a member of a planar structure are shown in Fig. 1.37.

1.8 Free-Body Diagrams

A free-body diagram is a sketch of a part of a structure showing all external actions and all possible reactions acting on it, as well as all internal actions acting on the cross sections where the part under consideration is cut from the structure. Usually, in a free-body diagram either the local or the global components of the internal actions are shown as assumed positive. For instance, consider the planar frame shown in Fig. 1.38a. The uniform load acting on this frame is given in kilonewtons (kN) per meter of length along the horizontal direction. We consider as positive the global components of the internal actions acting in the positive direction of the global axes. At support 1, the frame is fixed to the ground, while at support 4 it is pinned. Thus, at support 1, it is possible to have a reacting force acting in any direction required, as well as a reacting moment, while at support 4, it is possible to have only a reacting force acting in any direction required.

The free-body diagram of the whole structure is shown in Fig. 1.38b. It includes the given external actions (forces and moments) and the components

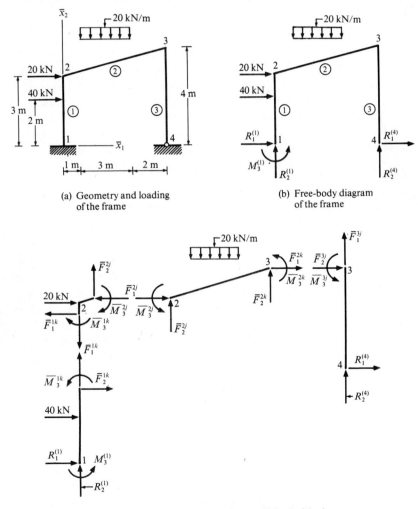

(a) Geometry and loading of the frame

(b) Free-body diagram of the frame

(c) Free-body diagram of the members and joint 2 of the frame

Figure 1.38 Free-body diagrams of parts of a frame.

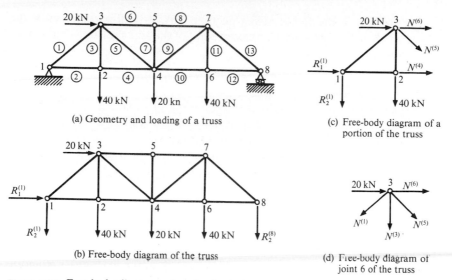

(a) Geometry and loading of a truss

(b) Free-body diagram of the truss

(c) Free-body diagram of a portion of the truss

(d) Free-body diagram of joint 6 of the truss

Figure 1.39 Free-body diagrams of parts of a truss.

of the possible unknown reactions acting at the points of support of the structure. The free-body diagrams of members 1, 2, and 3 and of joint 2 are shown in Fig. 1.38c. In the free-body diagrams, for members 1, 2, and 3 the known external actions acting on them and the unknown global components of the reactions at the supports of the structure (points 1 and 4) are shown. Moreover, the global components of the internal actions are shown which act on the cross sections where the members are cut. In the free-body diagram of the joint 2, in addition to the external actions acting on this joint, the global components of the internal actions transmitted to it by the members framing on the joint are also shown. The latter must be opposite to the components of the internal actions acting at the ends of the members framing in the joint.

As a second example, consider the simple truss shown in Fig. 1.39a. The free-body diagram of the whole truss is shown in Fig. 1.39b. It includes the given external actions acting on the truss as well as the possible unknown reactions. The free-body diagram of a part of a truss is shown in Fig. 1.39c, while the free-body diagram of joint 3 is shown in Fig. 1.39d. The unknown internal forces are assumed positive as shown in these diagrams, that is, when they subject the members on which they act to tension.

1.9 Restrictions in the Analysis of Framed Structures

In this text, we only consider framed structures which have the following attributes:

1. *They are in equilibrium under the influence of the external loads.* This implies that if we isolate any part of the structure and draw its free-body dia-

gram, the sum of the forces acting on that part, as well as the sum of the moments about any conveniently chosen point of the actions acting on the part, must vanish. Thus

$$\Sigma \mathbf{F} = 0 \qquad \Sigma \mathbf{M} = 0 \qquad (1.10)$$

Equations (1.10) are referred to as the *equations of statics*.

A vector in a three-dimensional space may be resolved into three components with respect to any rectangular system of axes x_1, x_2, x_3. A necessary and sufficient condition that the sum of a number of vectors is equal to zero is that the sums of their components vanish. Thus, Eqs. (1.10) may be rewritten as

$$\Sigma F_1 = 0 \qquad \Sigma F_2 = 0 \qquad \Sigma F_3 = 0$$
$$\Sigma M_1 = 0 \qquad \Sigma M_2 = 0 \qquad \Sigma M_3 = 0 \qquad (1.11)$$

In the above equations, ΣF_i ($i = 1, 2, 3$) represents the algebraic sum of the components along the x_i axis of the forces acting on the part of the structure under consideration. Moreover, ΣM_i ($i = 1, 2, 3$) represents the algebraic sum of the components, along the x_i axis, of the moments of the external actions acting on the part of the structure under consideration, about any conveniently chosen point.

In planar structures, the external and internal forces do not have a component in the direction normal to the plane of the structure and the moments do not have components in the plane of the structure. Thus, for planar structures, in the $x_1 x_3$ plane, the equations of statics (1.11) reduce to:

$$\Sigma F_1 = 0, \qquad \Sigma F_3 = 0, \qquad \Sigma M_2 = 0 \qquad (1.12)$$

2. *The ratio of the thickness to the radius of curvature of curved line members is very small compared to unity.* In general, the distribution of the normal component of stress on the cross section of curved members† subjected to forces acting in their plane is hyperbolic. However, this distribution can be approximated by a linear distribution when the ratio of the thickness to the radius of curvature of the member is small compared to unity.

3. *The structures or any group of their members cannot move without deforming, under the influence of any loading.* For instance, the structures of Fig. 1.40 would collapse if we would apply a horizontal force. Thus, we do not consider these structures.

4. *The deformation of the members of the structures is within the range of validity of the theory of small deformation.* This theory is based on the assumption that the deformation of a structure is such that

    a. The change of length per unit length of any infinitesimal material line

---

† We consider only curved members whose axes lie in one plane, whereas one of the principal centroidal axes of their cross sections is normal to this plane.

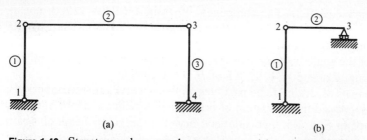

Figure 1.40 Structures whose members can move without deforming.

of the structure (unit elongation) (see Sec. A.2) is negligible compared to unity

b. The change of angle between any two mutually perpendicular infinitesimal material lines of the structure (unit shear) is negligible compared to unity

c. The angles of rotation of the particles of the structure are negligible compared to unity and not of a higher order of magnitude than the unit elongations and the unit shears

As a result of these assumptions, the following approximations are valid for structures whose deformation is within the range of validity of the theory of small deformation:

1. *The deformation of a particle is completely specified by its six components of strain* which referred to the rectangular system of axes x_1, x_2, x_3 are denoted by e_{11}, e_{22}, e_{33}, $e_{12} = e_{21}$, $e_{13} = e_{31}$, and $e_{23} = e_{32}$. The components of strain of a particle are related to its components of displacement $\hat{u}_1$, $\hat{u}_2$, and $\hat{u}_3$ by the following linear relations:

$$e_{11} = \frac{\partial \hat{u}_1}{\partial x_1} \qquad e_{22} = \frac{\partial \hat{u}_2}{\partial x_2} \qquad e_{33} = \frac{\partial \hat{u}_3}{\partial x_3}$$

$$e_{21} = e_{12} = \frac{1}{2}\left(\frac{\partial \hat{u}_2}{\partial x_1} + \frac{\partial \hat{u}_1}{\partial x_2}\right)$$

$$e_{31} = e_{13} = \frac{1}{2}\left(\frac{\partial \hat{u}_3}{\partial x_1} + \frac{\partial \hat{u}_1}{\partial x_3}\right) \qquad (1.13)$$

$$e_{32} = e_{23} = \frac{1}{2}\left(\frac{\partial \hat{u}_2}{\partial x_3} + \frac{\partial \hat{u}_3}{\partial x_2}\right)$$

The component of strain e_{ii} ($i = 1, 2, 3$) of a particle is referred to as the *normal component of strain* at the particle in the direction x_i. It represents the change of length per unit length of an infinitesimal line segment of length dx_i

($i = 1, 2, 3$) at the particle. The component of strain e_{ij} ($i, j = 1, 2, 3; i \neq j$) of a particle is referred to as the *shearing component of strain* at the particle in the directions of the x_i and x_j axes. It represents half the change due to the deformation of the angle between two mutually perpendicular infinitesimal line segments dx_i and dx_j ($i, j = 1, 2, 3; i \neq j$) at the particle.

The components of strain of the particles of planar curved-line members are conveniently expressed in the system of orthogonal curvilinear coordinates (α, x_2, ζ) described in Sec. 1.5. The relations between the components of strain (normal e_{tt} and shearing e_{t2} and $e_{t\zeta}$) of a particle of a planar curved member and the components of displacement of this particle (axial $\hat{u}_t$, normal to the plane of the element u_2 and radial u_ζ) are[4]

$$e_{tt} = \frac{\partial \hat{u}_t}{\partial s} - \frac{\hat{u}_\zeta}{R}$$

$$2e_{t2} = \frac{\partial \hat{u}_t}{\partial x_2} + \frac{\partial \hat{u}_2}{\partial s} \qquad (1.14)$$

$$2e_{t\zeta} = \frac{\partial \hat{u}_t}{\partial \zeta} + \frac{\partial \hat{u}_\zeta}{\partial x_2} + \frac{\hat{u}_t}{R}$$

where R is the radius of curvature of the axis of the member and ds is the length of an infinitesimal segment of the axis of the member. In relations (1.14) the ratio ζ/R has been disregarded as compared to unity.

2. *The effect of the change due to the deformation of the dimensions of the cross sections of the members of the structure is disregarded when computing the components of stress from the internal actions (forces and moments).*

3. The effect of the change of the geometry of the structure on the internal actions of its members is negligible. Consequently, *the undeformed configuration of a structure is used when considering its equilibrium*; that is, the change of the geometry of the structure due to its deformation is disregarded in drawing the free-body diagrams of the structure and its parts.

4. The angle between the tangent line of the deformed axis of a member and its undeformed axis is so small that its cosine is approximately equal to unity, while its sine is approximately equal to the angle in radians.

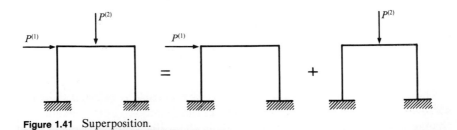

Figure 1.41 Superposition.

5. When the members of a structure are made of linearly elastic materials (see Sec. A.5.3), the relations between the loading acting on the structure and the resulting internal actions in its members and the relations between the loading acting on the structure and the displacements of its cross sections are linear. In this case, we say that *the response of the structure to the external loading is linear*. This implies that a component of an internal action or a component of displacement of a cross section of a member of a structure produced by a number of loads acting together on it is equal to the sum of the corresponding component of internal actions or component of displacement produced by each load acting alone on the structure (see Fig. 1.41). That is, the *results can be superimposed*.

Notice that we can distinguish two types of nonlinear behavior of structures:

1. Material nonlinearity. That is, the relations between the components of stress and strain are nonlinear.

2. Geometric nonlinearity. That is, the deformation of the structure is not in the range of validity of the theory of small deformation.

The theory of small deformation cannot be used in analyzing certain structures of interest to the structural designer when they are subjected to certain types of loading. For example, the theory of small deformation cannot be employed in the following cases:

1. In analyzing beams subjected to transverse and axial forces when the effect of the axial forces on their bending moment cannot be neglected

2. In establishing the loading under which a structure or a group of its members reaches a state of unstable equilibrium

3. In analyzing kinematically unstable structure.

In the following section, we discuss these three cases briefly

1.10 Elastic Structures Whose Analysis Requires a Nonlinear Theory

In this section we present a brief discussion of three cases of interest to the structural engineer which involve elastic structures whose analysis requires a nonlinear theory.

1.10.1 Effect of an axial force on the bending moment of a beam

Consider the simply supported planar beam shown in Fig. 1.42 which is subjected to an external transverse force, as well as to an axial force. By intuition, one can deduce that for any given value of the external transverse force, the

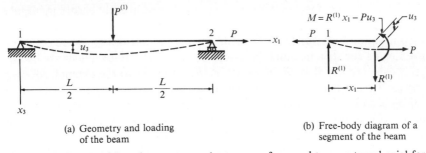

(a) Geometry and loading
of the beam

(b) Free-body diagram of a
segment of the beam

Figure 1.42 Beam subjected to an external transverse force and to an external axial force.

translation u_3 of a point of the axis of the beam depends on the value of the axial force P. If the beam is subjected to an axial tensile force, its translation u_3 decreases as P increases, whereas if the beam is subjected to an axial compressive force, its translation u_3 increases as P increases. That is, an axial compressive force decreases the ability of the beam to carry external moments and transverse forces. In most cases of practical interest, the effect of the axial force on the ability of a member to carry external moments and transverse forces is very small and can be neglected. In these cases, the member can be analyzed using the theory of small deformation. However, for members whose length is very long compared to the dimensions of their cross section, the effect of an axial compressive force on their ability to carry external moments and transverse forces is not negligible; thus these members must be analyzed using a nonlinear theory.

In order to take into account the effect of the axial force on the translation u_3, we must consider the deformed configuration of the beam when we examine the equilibrium of its parts. Thus, referring to Fig. 1.42*b*, the moment at a cross section $(x_1 < L/2)$ of the beam is equal to:

$$M = R^{(1)}x_1 - Pu_3 \qquad x_1 < \frac{L}{2} \qquad (1.15)$$

Notice that in the above relation the translation u_3 is a function of the force P; consequently, the moment is not a linear function of the load P.

1.10.2 Buckling of structures

Consider a perfectly straight, slender column of constant cross section made of a homogeneous, isotopic, linearly elastic material. Assume that one end of this column is fixed to the ground, while its other end is free. Suppose that an axial compressive force P is applied at the centroid of the cross section of the column at its free end (see Fig. 1.43*a*). As the load P increases, the column deforms axially but remains straight. The column is in a state of stable equilibrium until the axial force P reaches a certain value, referred to as the *critical value* P_{cr}

(see Example 1 of App. B). That is, when the value of the axial force is less than the critical value, if the column is subjected to a small transverse force, it will experience a small transverse displacement. However, when the transverse force is removed, the column will return to its straight configuration. When the axial force reaches its critical value, the column attains a state of unstable equilibrium. That is, for this value of the axial force, the column can be in equilibrium in two configurations (straight and bent). When the column is in the bent configuration, we say that it has *buckled.* At the critical load, the two equilibrium configurations of the column differ by an infinitesimal amount. That is, up to the critical load, the force displacement curve is a straight line (the $u = 0$ axis), while at the critical load, this curve has a bifurcation point and is divided into two curves (see Fig. 1.43c); the continuation of the original straight line represents positions of unstable equilibrium while the curve starting perpendicular to the original straight line represents positions of stable equilibrium. As the load increases above its critical value, the column could remain in its straight configuration or could assume its bent configuration. However, the straight configuration is unstable; thus, the smallest disturbance will force the column to its stable, bent (buckled) configuration.

As shown in Fig. 1.43b and $c,$ a small increase of the value of the force above its critical value results in an appreciable increase of the lateral displacement of the column, and, consequently, a large increase of its bending moment. For this reason, the critical load is an important design parameter of the column. It is considered as the load at which the column fails in buckling.

Consider a symmetric frame subjected to symmetric compressive forces, as shown in Fig. 1.44a. When the forces reach a certain critical value, depending on its geometry, the frame will buckle to one of the configurations shown in Fig. 1.44b or c for which the value of the critical load is the lowest.

Consider a symmetric truss consisting of two slender members subjected to a force $P,$ as shown in Fig. 1.45a. The deformed configuration of the truss is specified by the displacement of joint 2, considered positive, downward. If the

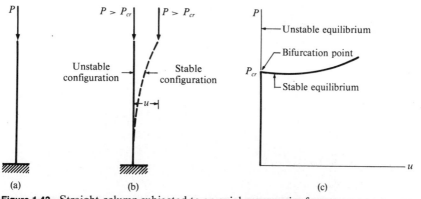

Figure 1.43 Straight column subjected to an axial compressive force.

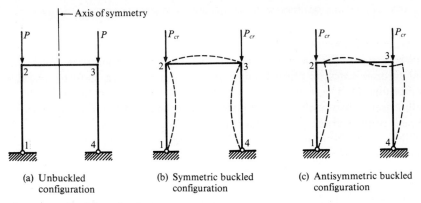

(a) Unbuckled
configuration

(b) Symmetric buckled
configuration

(c) Antisymmetric buckled
configuration

Figure 1.44 Buckling of a frame.

h/L ratio of the truss is not small, then for a certain value of the force P, the members of the truss reach a state of unstable equilibrium and buckle as columns. If the h/L ratio of the truss is sufficiently small, the truss reaches a state of unstable equilibrium and jumps (buckles) to a configuration of stable equilibrium which, as shown in Fig. 1.45b, differs considerably from the unstable configuration (see Example 2 of App. B). This type of buckling is known as

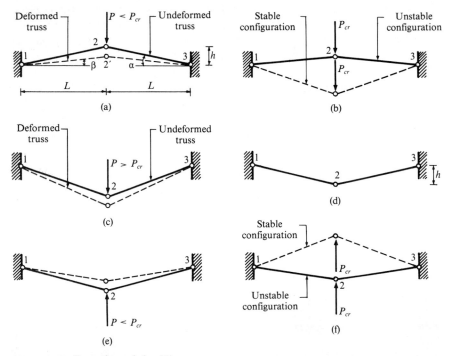

Figure 1.45 Snap-through buckling.

snap-through buckling. As the load increases, the truss deforms without reaching another state of unstable equilibrium until its members yield. If the load is slowly reduced before yielding occurs in the members of the truss, its deformation will decrease monotonically. When the load becomes zero, the truss assumes the configuration shown in Fig. 1.45*d*. If the truss is subsequently subjected to an upward force *P*, it will reach a state of unstable equilibrium and, as shown in Fig. 1.45*f*, it will jump (buckle) to a stable equilibrium configuration when the force *P* reaches its critical value (see Example 2 of App. B). Low arches pinned at both ends could experience snap-through buckling when subjected to lateral forces.

1.10.3 Kinematically unstable structures

Consider the structure of Fig. 1.46*a* subjected to a transverse force. Referring to Fig. 1.46*b*, it can be seen that when the transverse force is applied, the members of the structure are not subjected to any internal force and, consequently, they do not deform. Thus, the members of the structure rotate instantaneously without deforming until they reach a configuration which allows them to carry the applied force in tension. Thus, in order to establish the internal force in the members of this structure, their rotation must be taken into account (see Example 2 of App. B).

The geometry of some structures can be such that for certain types of loading, a number of their members can move instantaneously without deforming. These structures are termed *kinematically unstable*. The internal actions acting on the members of kinematically unstable structures cannot be computed if the change of their geometry due to their deformation is not taken into account. That is the internal actions in the members of kinematically unstable

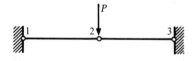

(a) Geometry and loading of the structure

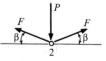

(b) Free-body diagram of joint 2 not taking into account the change of direction of the members of the structure due to deformation

(c) Free-body diagram of joint 2 taking into account the change of direction of the members of the structure due to deformation

Figure 1.46 Kinematically unstable structure.

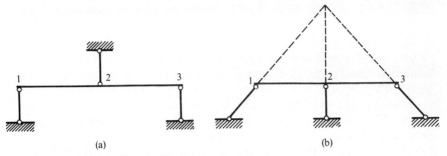

Figure 1.47 Kinematically unstable simple structures.

structures cannot be established using the theory of small deformation. For this reason, the analysis of these structures is usually laborious (see Example 2 of App. B). Examples of kinematically unstable structures are shown in Fig. 1.47. When the structure of Fig. 1.47*a* is subjected to a horizontal force, member 1,2,3 translates instantaneously in a horizontal direction without any members of the structure deforming. The reason for this becomes apparent when one notes that only when member 1,2,3 moves horizontally by an infinitesimal amount can the vertical members support a horizontal force and consequently deform. Similarly, the structure of Fig. 1.47*b* rotates instantaneously about point *A* when subjected to a force whose line of action does not pass through this point without its members deforming.

Notice that when a structure approaches a kinematically unstable configuration certain of its members are subjected to very large internal actions even for relatively small values of the external actions. Thus, kinematically, unstable structures should be avoided. Therefore, *the structural designer is not con-*

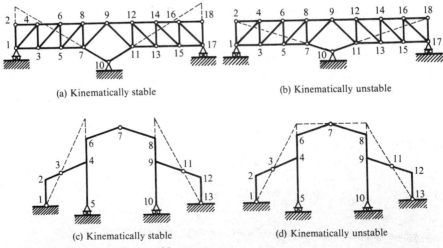

Figure 1.48 Kinematically unstable structures.

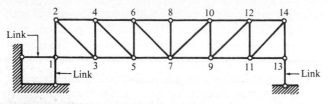

Figure 1.49 Kinematically stable truss.

cerned with the analysis of kinematically unstable structures but rather with the detection of kinematically unstable structures or structures whose configuration approaches a kinematically unstable configuration.

Kinematically unstable simple structures, such as those shown in Fig. 1.47 can be detected by inspection. However, kinematically unstable, complex structures (see Fig. 1.48) cannot be detected by inspection. When analyzing structures with the aid of a computer, kinematically unstable structures are detected automatically because their stiffness matrix is singular.

Simple trusses whose supports are equivalent to three nonparallel and nonconcurrent links are kinematically stable (see Fig. 1.49).

When a structure or a group of its members can move without deforming (see Fig. 1.40), or when a structure is kinematically unstable, we say that the structure constitutes a *mechanism*.

In App. B we analyze, using a nonlinear theory, two structures which, under certain conditions, cannot be analyzed on the basis of the theory of small deformation. The first is a beam-column subjected to an eccentric force; the second is the structure of Fig. 1.45a, which becomes kinematically unstable when $h/L = 0$.

1.11 Statically Determinate and Indeterminate Structures

If the reactions (forces and moments) of a structure and the internal actions in its members can be computed using the equations of equilibrium alone, then we say that the structure is *statically determinate*. That is, the number of unknown reactions and internal actions of a statically determinate structure is equal to the independent equations of statics which can be written for this structure. For example, a beam whose one end is hinged to a rigid support and whose other end is connected to a roller support is statistically determinate (see Fig. 1.50b). The first support can exert on the beam a reacting force that has two independent components, while the second support can exert on the beam a reacting force that is normal to the direction of rolling. These reacting forces can be computed by considering the free-body diagram of the beam and using the equations of equilibrium ($\Sigma F_1 = 0$, $\Sigma F_2 = 0$, $\Sigma M_3 = 0$). Moreover, the internal actions acting at any cross section of the beam can be computed by considering the equilibrium of a part of the beam. Additional examples of statistically determinate structures are shown in Fig. 1.50. The reactions and internal actions of the members of these structures are obtained by using only the equations of equilibrium.

If the sum of the independent components of the reactions of a structure and of the internal actions of its members exceeds the number of independent equations of equilibrium, we say that the structure is *statically indeterminate*. The degree of static indeterminancy IND of a structure is equal to the number by which the sum of the independent components of the reactions of the structure and the independent components of the internal actions in its members exceeds the number of independent equations of equilibrium.

The number of independent components of internal actions in a member of a structure is equal to the number of components of internal actions which must be specified in order to be able to establish uniquely the internal actions acting at any cross section of the member by considering the equilibrium of appropriate parts of the member. Thus, referring to Fig. 1.51*a,* there is only one independent action in a member of a planar or space truss (the axial force acting at one end of the member). Moreover, referring to Fig. 1.51*b,* there are three independent components of internal actions in a member of a planar

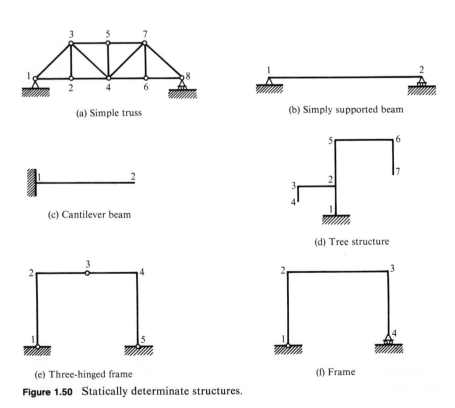

(a) Simple truss

(b) Simply supported beam

(c) Cantilever beam

(d) Tree structure

(e) Three-hinged frame

(f) Frame

Figure 1.50 Statically determinate structures.

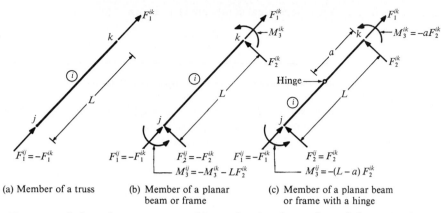

(a) Member of a truss (b) Member of a planar   (c) Member of a planar beam
                           beam or frame   or frame with a hinge

Figure 1.51 Independent components of internal actions in members of planar structures.

frame, if internal release mechanisms do not exist in this member. Further-more, there are six independent components of internal actions in a member of a space beam or frame if internal release mechanisms do not exist in this mem-ber. If internal release mechanisms exist in a member, the number of indepen-dent components of internal actions in this member decreases by the number of actions released by these mechanisms. For example, an internal hinge at a point of a member releases (renders it equal to zero) the moment about the axis of the hinge at that point of the member. Thus, the member of a planar structure shown in Fig. 1.51c has two independent components of internal actions because it has a hinge at some point along its length which renders the internal moment at that point equal to zero.

Inasmuch as the structures that we are considering are in equilibrium, the forces acting on their joints must satisfy the equations of equilibrium. The forces acting on a joint of a planar truss must satisfy two equations of equilib-rium ($\Sigma \overline{F}_i = 0$, $i = 1, 2$), while the forces acting on a joint of a space truss must satisfy three equations of equilibrium ($\Sigma \overline{F}_i = 0$, $i = 1, 2, 3$). Moreover, the forces acting on a joint of a planar beam or frame must satisfy three equa-tions of equilibrium ($\Sigma \overline{F}_i = 0$, $i = 1, 2$, $\Sigma M_3 = 0$), while the forces acting on a joint of a space beam or frame must satisfy six equations of equilibrium ($\Sigma \overline{F}_i = 0$, $\Sigma \overline{M}_i = 0$, $i = 1, 2, 3$). The number of independent equations of equilibrium that can be written for any structure is equal to the sum of inde-pendent equations of equilibrium that can be written for the joints of the struc-ture. Thus, for a planar truss with NJ joints, we have 2NJ independent equa-tions of equilibrium, while for a space truss with NJ joints we have 3NJ independent equations of equilibrium. Moreover, for a planar beam or frame with NJ joints, we have 3NJ independent equations of equilibrium, while for a space beam or frame with NJ joints we have 6NJ independent equations of equilibrium. Thus, the degree of static indeterminacy IND of a planar or a space truss having NR independent components of reactions and NM members is equal to

$$IND = NR + NM - 2NJ \quad \text{for a planar truss} \quad (1.16a)$$

$$IND = NR + NM - 3NJ \quad \text{for a space truss} \quad (1.16b)$$

Moreover, the degree of static indeterminacy IND of a planar or a space beam or frame having NR independent components of reactions, NM members, and NAR actions released by release mechanisms is equal to

$$IND = NR + 3NM - 3NJ - NAR \quad \text{for a planar beam or frame} \quad (1.17a)$$

$$IND = NR + 6NM - 6NJ - NAR \quad \text{for a space beam or frame} \quad (1.17b)$$

Notice, that

1. If IND $\leq$ 0, the structure is either kinematically unstable or under certain types of loading some of its members move without deforming. That is, the structure is a mechanism.

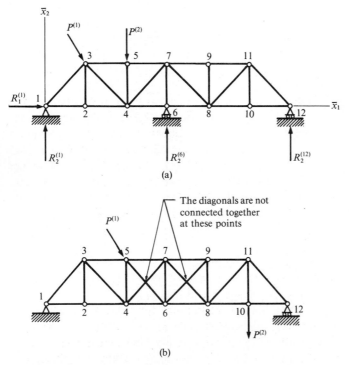

(a)

(b)

Figure 1.52 Statically indeterminate planar trusses.

2. If IND = 0 or if IND $\geq$ 0, then we cannot say if the structure is or is not a mechanism. However, if the structure is not a mechanism and IND = 0, then it is statically determinate, while if the structure is not a mechanism and IND $\geq$ 0, it is statically indeterminate to the IND degree.

Relations (1.16) and (1.17) may be employed to establish the degree of static indeterminacy of a structure. For example, for the planar truss of Fig. 1.52a we have

$$\text{NR} = 4 \qquad \text{NJ} = 12 \qquad \text{NM} = 21 \qquad (1.18)$$

Substituting the above values into relation (1.16a), we obtain

$$\text{IND} = 4 + 21 - 2(12) = 1 \qquad (1.19)$$

Consequently, the truss of Fig. 1.52a is statically indeterminate to the first degree.

As a second example, consider the space truss shown in Fig. 1.53. This truss is not a mechanism. It has a nontranslating ball-and-socket support at point 1, ball supports at points 9 and 11, and a translating, along the x_3 axis, ball-and-socket support at point 5. Thus,

$$\text{NR} = 7 \qquad \text{NJ} = 12 \qquad \text{NM} = 30 \qquad (1.20)$$

Substituting the above values into relation (1.16b), we obtain

$$\text{IND} = 7 + 30 - 3(12) = 1 \qquad (1.21)$$

Consequently, since the truss is not a mechanism, it is statically indeterminate to the first degree.

As a third example, consider the planar frame shown in Fig. 1.55d. This frame is not a mechanism. Referring to Fig. 1.55d, we have

$$\text{NR} = 12 \qquad \text{NJ} = 11 \qquad \text{NM} = 12 \qquad \text{NRL} = 4 \qquad (1.22)$$

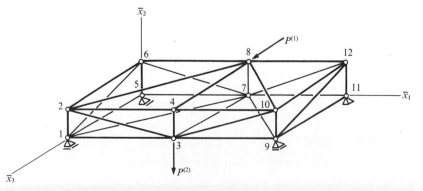

Figure 1.53 Statically indeterminate space truss.

(a)   (b)

Figure 1.54 Statically indeterminate beams.

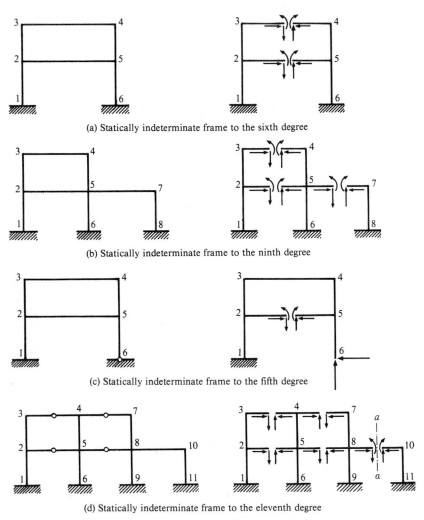

(a) Statically indeterminate frame to the sixth degree

(b) Statically indeterminate frame to the ninth degree

(c) Statically indeterminate frame to the fifth degree

(d) Statically indeterminate frame to the eleventh degree

Figure 1.55 Statically indeterminate frames.

Substituting the above values into relation (1.17a) we have

$$IND = 12 - 3(11) + 3(12) - 4 = 11 \qquad (1.23)$$

Consequently, the frame of Fig. 1.55d is statically indeterminate to the eleventh degree.

Another method for determining the degree of static indeterminacy of a structure is to reduce it to one or more statically determinate parts which are not mechanisms by releasing some of its reactions and internal actions. This is accomplished by removing some of the supports of the structure, by cutting some of its members, or by inserting some action release mechanisms in some of its members. The reactions and the internal actions released are called *the redundants*. The number of redundants of a structure is equal to the degree of static indeterminacy of the structure. For example, the truss of Fig. 1.52a can be reduced to a simple truss (see Fig. 1.50a) by removing its support at point 6. Moreover, the truss of Fig. 1.52b can be reduced to a simple truss by removing two of its members (members 5, 6 and 6, 9). A simple truss is a statically determinate structure, and it is not a mechanism. Consequently, the truss of Fig. 1.52a is statically indeterminate to the first degree, while the truss of Fig. 1.52b is statically indeterminate to the second degree.

The beam of Fig. 1.54a can be reduced to a simply supported beam (see Fig. 1.50b) by removing supports 2 and 3. A simply supported beam is statically determinate, and it is not a mechanism. Consequently, the beam of Fig. 1.54a is statically indeterminate to the second degree.

The frame of Fig. 1.55a can be reduced to two tree structures (see Fig. 1.50d) by cutting it at two points. A tree structure is statically determinate, and it is not a mechanism. At every cut, three internal actions, i.e., an axial force, a shearing force, and a bending moment, have been released. Thus, six unknown actions have been released by cutting the structure of Fig. 1.55a and reducing it to two tree structures. Consequently, the structure is statically indeterminate to the sixth degree.

Similarly, the structure shown in Fig. 1.55b is statically indeterminate to the ninth degree, while the structure of Fig. 1.55c is statically indeterminate to the fifth degree. The structure of Fig. 1.55d has four internal action release mechanisms (hinges). As shown in Fig. 1.55d, when this structure is cut into statically determinate parts, two internal actions are released at each of the internal hinges and three internal actions at cut a-a. Thus, this structure is statically indeterminate to the eleventh degree.

1.12 Compound Structures

Statically determinate structures may be classified as *simple* or *compound*. Compound trusses are made of two or more simple trusses connected to each other or to the supporting body by hinges or links (see Fig. 1.56). Compound

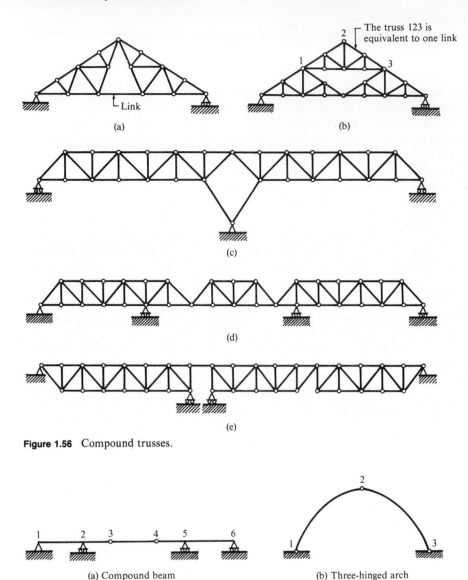

Figure 1.56 Compound trusses.

(a) Compound beam   (b) Three-hinged arch

Figure 1.57 Compound structures.

beams, frames, and arches are built with internal release mechanisms, such as hinges, links, rollers, etc. (see Fig. 1.57).

1.13 Problems

1. Determine the degrees of static indeterminacy of the structures of Fig. 1.P1*a* to *w*.

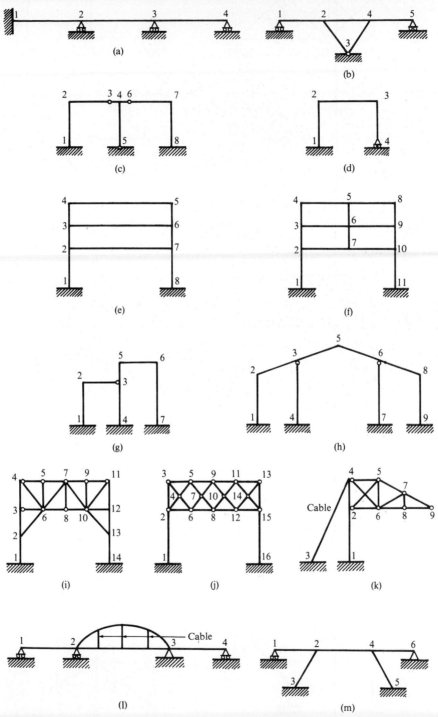

Figure 1.P1

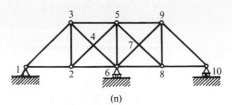

(n)

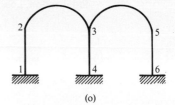

(o)

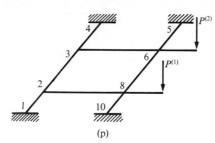

(p)

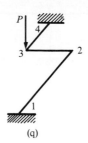

(q)

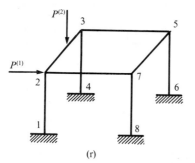

(r)

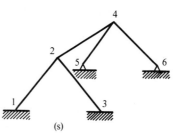

(s)

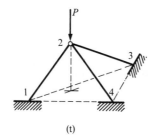

(t)

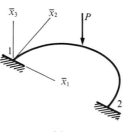

(u)

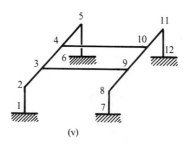

(v)

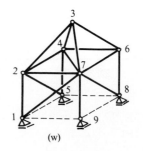

(w)

Figure 1.P1 (*Continued*)

REFERENCES

1. *ASCE Transactions,* vol. V, no. 126, 1961, pp. 1124–1198.
2. W. Podolny, Jr. and J. B. Scalzi, *Construction and Design of Cable-Stayed Bridges,* John Wiley & Sons, Inc., New York, 1976.
3. Subcommittee on Cable-Stayed Bridges, "Bibliography and Data on Cable-Stayed Bridges," *Journal of the Structural Division of ASCE,* Oct. 1977, pp. 1971–2004.
4. H. Kraus, *Thin Elastic Shells,* John Wiley & Sons, Inc., New York, 1967, p. 30.

Photographs and Brief Description of a Suspension Bridge

The Verrazano-Narrows Bridge† shown in Fig. 1.58 connects Long Island and Staten Island of New York City. It has the longest span between towers, 4260 ft (1288 m), of all the suspension bridges constructed up to now. Its two side spans are each 1213 ft (369.7 m). Its deck rises at a slope of 4 percent toward the center of the main span where a curve connects the two tangents. Its upper and lower roadways are concrete-filled steel grids supported on transverse subfloor beams which in turn are supported on longitudinal stringers. The latter are rigidly connected (see Fig. 1.59) to the horizontal members of transverse floor frames. The two vertical members of these frames are rigidly connected to two stiffening longitudinal vertical trusses forming a continuous laced-wall tubular structure. The cross section of the superstructure is shown in Fig. 1.16b. The superstructure is suspended from four cables. Each cable consists of 26,108 parallel, galvanized steel wires of 0.196 in (5 mm) diameter grouped in 61 strands. The cables are supported at the top of the two towers and are anchored at each end of the bridge into massive concrete anchorages 230 ft (70.1 m) wide and 345 ft (105.16 m) long. The steel towers rise 629'-9¾" (192 m) above the top of their pedestals. Each tower consists of two shafts connected at the top with an arched portal strut and below the roadway with a second strut. All faces of the shafts are tapered. The pedestals are built on a rectangular, cellular caisson. Each caisson

Figure 1.58 The Verrazano-Narrows Bridge in New York City. *(Courtesy of the Triborough Bridge and Tunnel Authority, New York, N.Y.)*

---

† For a detailed description of the Verrazano-Narrows Bridge, see the series of articles in the March 1966 issue of the *Journal of the Construction Division of ASCE.*

Figure 1.59 The Verrazano-Narrows Bridge under construction. A prefabricated segment of the superstructure is lifted into place. The segment consists of two floor frames, portions of the two vertical trusses, and the stringers on which the upper and lower roadways are supported. *(Courtesy of the Triborough Bridge and Tunnel Authority, New York, N.Y.)*

was sunk by its own weight to the required elevation by removing the soil underlying it with clam-shell buckets through its 17 ft (5.18 m) diameter cylindrical dredging wells (see Figs. 1.60 and 1.61*a*). At the bottom, the cylindrical dredging wells become 5.18 m $\times$ 5.18 m square wells. The bottom 2.13 m of the outer and interior walls of the caisson are heavily armored with steel plates and form cutting edges (see Fig. 1.60*c*) which, as the caisson was sinking, broke the

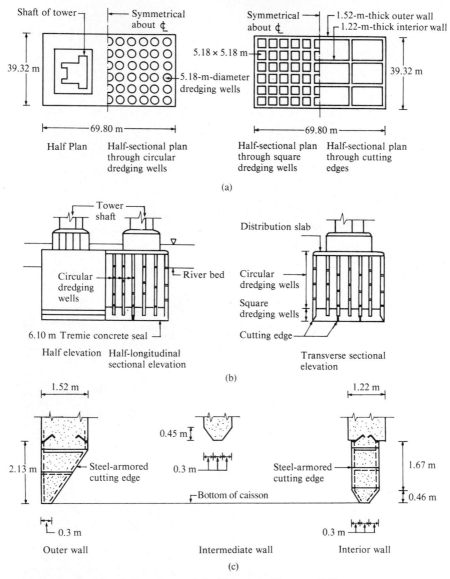

Figure 1.60 Details of the caissons of the Verrazano-Narrows Bridge.

boulders and displaced them into the dredging wells from which they were excavated.

After a caisson was sunk to the proper elevation, its bottom was cleared out and sealed with tremie-concrete. The dredging wells were left filled with water and covered with precast covers. Prior to the construction of the pedestals a distribution slab was cast over the entire top of each caisson.

Figure 1.61 View of the caisson and cable anchorage of the Verrazano-Narrows Bridge.

(a) View of a caisson during construction, showing the 66 dredging wells of 17-ft (5.18-m) diameter.

(b) View of the eyebars during anchoring of the strands. Three eyebars are pinned in series to form a chain. Each cable strand is anchored at the upper head of the first eyebar of a chain. The lower head of the first eyebar and the upper head of the second eyebar have elongated holes to permit shimming required for cable adjustment. The last eyebar of each chain is pinned to a supporting steel frame. *(Courtesy of the Triborough Bridge and Tunnel Authority, New York, N.Y.)*

(a)

(b)

2

Symmetric Structures

2.1 Symmetric and Antisymmetric Functions

A function $x_3 = f_s(x_1)$ is symmetric if

$$f_s(x_1) = f_s(-x_1) \tag{2.1}$$

while a function $x_3 = f_a(x_1)$ is antisymmetric if

$$f_a(x_1) = -f_a(-x_1) \tag{2.2}$$

Referring to Fig. 2.1a, it can be seen that the slope of a symmetric function at x_1 is equal to the negative of its slope at $-x_1$. Thus,

$$\left.\frac{df_s}{dx_1}\right|_{x_1} = -\left.\frac{df_s}{dx_1}\right|_{-x_1} \tag{2.3}$$

Referring to Fig. 2.1b, it can be seen that an antisymmetric function is either discontinuous or equal to zero at the origin. Moreover, the slope of an antisymmetric function at x_1 is equal to its slope at $-x_1$. Thus,

$$\left.\frac{df_a}{dx_1}\right|_{x_1} = \left.\frac{df_a}{dx_1}\right|_{-x_1} \tag{2.4}$$

These findings may be extended to higher derivatives as follows:

1. The odd derivatives of a symmetric function are antisymmetric functions.
2. The even derivatives of a symmetric function are symmetric functions.
3. The odd derivatives of an antisymmetric function are symmetric functions.
4. The even derivatives of an antisymmetric function are antisymmetric functions.

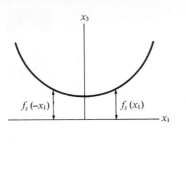

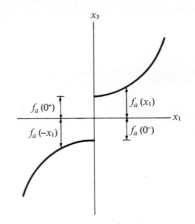

(a) Symmetric function

(b) Antisymmetric function

Figure 2.1   Symmetric and antisymmetric functions.

Any nonsymmetric function can be expressed as the sum of a symmetric and an antisymmetric function, as follows:

$$f(x_1) = f_s(x_1) + f_a(x_1)$$

Moreover,

$$f(-x_1) = f_s(-x_1) + f_a(-x_1) = f_s(x_1) - f_a(x_1)$$

From the above relations, we get

$$f_s(x_1) = \frac{f(x_1) + f(-x_1)}{2} \qquad f_a(x_1) = \frac{f(x_1) - f(-x_1)}{2} \qquad (2.5)$$

2.2   Symmetric Planar Structures

Many planar framed structures have either an axis of symmetry (see Fig. 2.2) or a point (center) of symmetry (see Fig. 2.3).

A planar structure (in the $\bar{x}_1\bar{x}_3$ plane) is symmetric with respect to an axis, say the $\bar{x}_3$ axis, if its geometry as well as the material properties of its members are the same with respect to two sets of right-hand orthogonal axes $\bar{x}_1, \bar{x}_2, \bar{x}_3$ and $\bar{x}'_1, \bar{x}'_2, \bar{x}'_3$, such that $\bar{x}_1 = -\bar{x}'_1, \bar{x}_2 = -\bar{x}'_2$, and $\bar{x}_3 = \bar{x}'_3$. It is apparent that the system of axes $\bar{x}'_1, \bar{x}'_2$, and $\bar{x}'_3$ is obtained by rotating the system of axes $\bar{x}_1, \bar{x}_2$, and $\bar{x}_3$ by 180° about the $\bar{x}_3$ axis.

A planar structure (in the x_1x_3 plane) is symmetric with respect to a point (center of symmetry) if its geometry and the material properties of its members are the same with respect to two sets of orthogonal axes $\bar{x}_1, \bar{x}_2, \bar{x}_3$ and $\bar{x}'_1, \bar{x}'_2, \bar{x}'_3$, such that, $\bar{x}_1 = -\bar{x}_1, \bar{x}_2 = -\bar{x}_2$, and $\bar{x}_3 = -\bar{x}_3$. Notice, that the $\bar{x}_1, \bar{x}_2,$

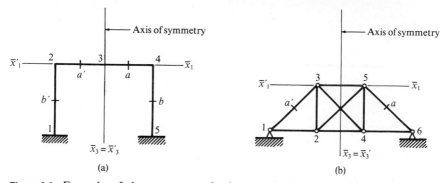

Figure 2.2 Examples of planar structures having an axis of symmetry lying in their plane.

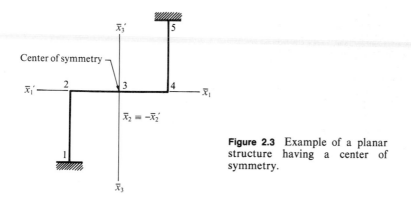

Figure 2.3 Example of a planar structure having a center of symmetry.

$\bar{x}_3$ system of axes is a right-hand system, while the $\bar{x}'_1$, $\bar{x}'_2$, $\bar{x}'_3$ system of axes is a left-hand system.

2.3 Symmetric and Antisymmetric Actions

Actions acting in a plane may be symmetric or antisymmetric with respect to an axis in that plane, or with respect to a point of that plane. Examples of symmetric and antisymmetric pairs of actions with respect to an axis in their plane are shown in Fig. 2.4. Examples of symmetric and antisymmetric pairs of actions with respect to a point in their plane are shown in Fig. 2.5. Notice, that the components of an action with respect to the system of axes $\bar{x}_1$, $\bar{x}_2$, and $\bar{x}_3$ are identical to the components of its symmetric counterpart referred to the system of axes $\bar{x}'_1$, $\bar{x}'_2$, and $\bar{x}'_3$. Moreover, notice that in order to establish the curl of a moment or of a rotation, the right-hand rule must be used for right-hand systems of axes and the left-hand rule for left-hand systems of axes.

Any loading may be replaced by the sum of a symmetric and an antisymmetric set. Examples of replacing a set of actions by the sum of their symmetric and antisymmetric components are illustrated in Fig. 2.6.

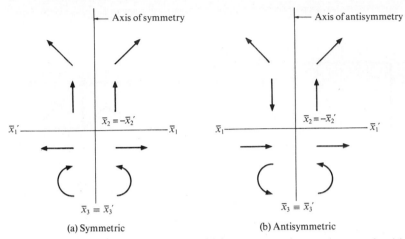

(a) Symmetric   (b) Antisymmetric

Figure 2.4 Pairs of forces and moments which are symmetric or antisymmetric with respect to an axis lying in their plane.

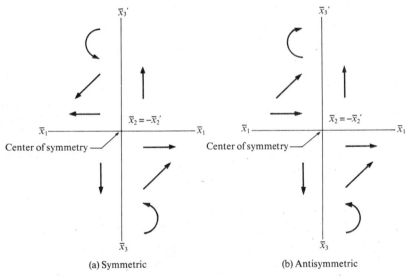

(a) Symmetric   (b) Antisymmetric

Figure 2.5 Pairs of forces and moments which are symmetric or antisymmetric with respect to a point in their plane.

2.4 Symmetric Planar Structures Subjected to Symmetric or Antisymmetric Loading

Symmetric structures subjected to symmetric loading have symmetric reactions, internal actions, and displacements. Symmetric structures subjected to antisymmetric loading have antisymmetric reactions, internal actions, and displacements. For instance, as shown in Fig. 2.7b, the internal actions in a symmetric beam subjected to symmetric actions are symmetric, while as shown in

Nonsymmetric load = symmetric load + antisymmetric load

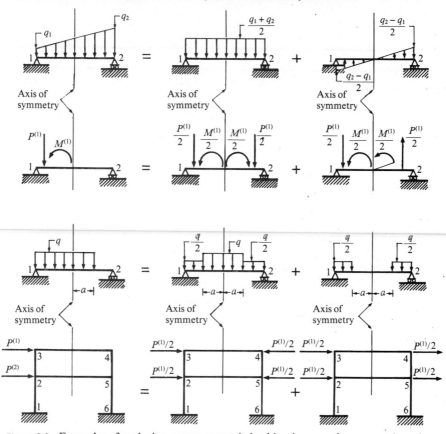

Figure 2.6 Examples of replacing a nonsymmetric load by the sum of a symmetric and an antisymmetric load.

Fig. 2.8*b,* the internal actions in a symmetric beam subjected to antisymmetric loading are antisymmetric. However, notice that, as shown in Fig. 2.7*c,* the shear diagram in a symmetric beam subjected to symmetric actions is antisymmetric, while, as shown in Fig. 2.8*c* the shear diagram of a symmetric beam subjected to an antisymmetric loading is symmetric. This is a result of the sign convention adopted for the shearing force. Notice that we may arrive at the same conclusion regarding the distribution of internal actions in symmetric beams subjected to symmetric or antisymmetric loading by considering the well-known relations between the applied force q and the shearing force (A.56), the shearing force Q and the moment (A.59), and the moment M and the curvature (A.74*b*) derived in App. A. That is,

$$q = -\frac{dQ}{dx_1} \qquad q = -\frac{d^2M}{dx_1^2} \qquad \frac{M}{EI} = -\frac{d^2u}{dx_1^2} \qquad (2.6)$$

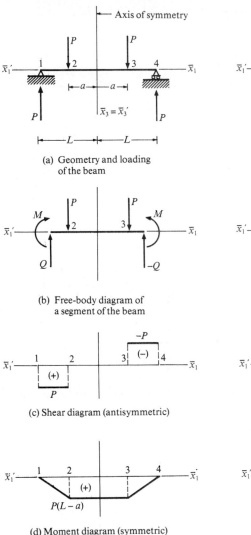

(a) Geometry and loading of the beam

(b) Free-body diagram of a segment of the beam

(c) Shear diagram (antisymmetric)

(d) Moment diagram (symmetric)

Figure 2.7 Simply supported beam subjected to symmetric loading.

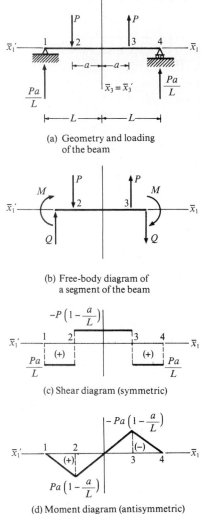

(a) Geometry and loading of the beam

(b) Free-body diagram of a segment of the beam

(c) Shear diagram (symmetric)

(d) Moment diagram (antisymmetric)

Figure 2.8 Simply supported beam subjected to antisymmetric loading.

It is apparent, that if q is symmetric, dQ/dx_1 and d^2M/dx_1^2 are symmetric, and as discussed in Sec. 2.1, Q must be antisymmetric and M must be symmetric. Moreover, the displacement u must be symmetric and the derivative du/dx_1 must be antisymmetric.

In what follows, we further consider the response of symmetric frames subjected to symmetric or antisymmetric external actions, as shown in Figs. 2.10 to 2.14. As mentioned previously the effect of axial deformation of the mem-

bers of a frame on their internal moments, generally, is small and can be disregarded. Thus, referring to Fig. 2.10 or 2.11, the vertical movement of joints
2 and 3 of the structure is disregarded.

The structure of Fig. 2.10a or b has an axis of symmetry. When it is sub-
jected to a symmetric loading, the translation $\bar{u}_1^{(c)}$ and the rotation $\bar{\theta}_2^{(c)}$ at point
C vanish. Moreover, from the equilibrium of a segment of the structure con-
taining point C (see Fig. 2.9a) it is apparent that if a concentrated force parallel to the axis of symmetry of the structure does not act at point C, the shearing force $Q^{(c)}$ vanishes. However, if a concentrated force parallel to the axis of
symmetry acts at point C, the shearing force on each side of point C is equal
to half this force. The free-body diagram of the left half of the structure is
shown in Fig. 2.10c. The force $N^{(c)}$ is required to restrain point C from moving
horizontally, while the moment $M^{(c)}$ is required to maintain the rotation at
point C equal to zero. Consider the auxiliary structure of Fig. 2.10d. Its members are identical to the corresponding members on the left half of the structure
of Fig. 2.10a. Its loading is identical to that of the left half of the structure of
Fig. 2.10a. Moreover, both structures are fixed at point 1. The support of the
auxiliary structure at point C permits movement of this point only along the
$\bar{x}_3$ axis. Consequently, it does not supply a vertical force. However, it prevents
the rotation of member 2 at point C as well as the movement of point C along

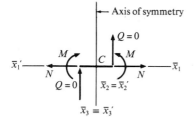

(a) Segment of a symmetric structure
subjected to symmetric loads
(see Fig. 2.10) with symmetric
pairs of internal actions

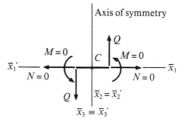

(c) Segment of a symmetric structure
subjected to symmetric loads
(see Fig. 2.11) with symmetric
pairs of internal actions

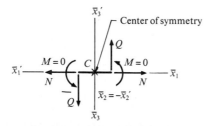

(b) Segment of a symmetric structure
subjected to antisymmetric loads
(see Fig. 2.13) with antisymmetric
pairs of internal actions

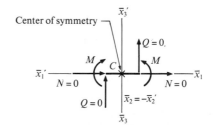

(d) Segment of a symmetric structure
subjected to antisymmetric loads
(see Fig. 2.14) with antisymmetric
pairs of internal actions

Figure 2.9 Segments of symmetric structures including their axis or center of symmetry.

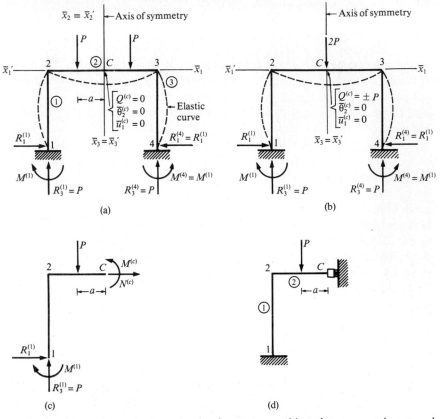

Figure 2.10 Planar frame having an axis of symmetry subjected to symmetric external actions.

the $\bar{x}_1$ axis. Thus it is apparent that the three boundary conditions $[\bar{u}_1^{(c)} = 0,$ $\bar{\theta}_2^{(c)} = 0, R_3^{(c)} = 0]$ at point C of the auxiliary structure of Fig. 2.10d are equal to the corresponding conditions $[\bar{u}_1^{(c)} = 0, \bar{\theta}_2^{(c)} = 0, Q^{(c)} = 0]$ at point C of the structure of Fig. 2.10a. Consequently, the internal actions in the members of the left half of the structure of Fig. 2.10a, as well as its deformed configuration, are identical to those of the auxiliary structure of Fig. 2.10d. Thus, we can analyze the structure of Fig. 2.10d instead of the structure of Fig. 2.10a. Notice, that the structure of Fig. 2.10a is statically indeterminate to the third degree, while the structure of Fig. 2.10d is statically indeterminate to the second degree.

The structure of Fig. 2.11a has an axis of symmetry; consequently, when it is subjected to an antisymmetric loading the internal axial force $N^{(c)}$ and the component of translation $\bar{u}_3^{(c)}$ at point C vanish. Moreover, from the equilibrium of a segment of the structure containing point C (see Fig. 2.9b), it is apparent that if a concentrated moment does not act at point C, the internal moment $M^{(c)}$ vanishes. If a concentrated moment acts at point C, the internal moment on each side of point C is equal to half the applied moment. The free-body diagram of the left half of the structure is shown in Fig. 2.11c. The force $Q^{(c)}$ is required to keep the vertical component of translation of point C equal to zero. Consider the auxiliary structure of Fig. 2.11d. Its members are identical to the corresponding members of the left half of the structure of Fig. 2.11a. Its loading is identical to that of the left half of the structure of Fig. 2.11a. Moreover, both structures are fixed at point 1. The support of the auxiliary structure at point C does not permit translation of this point along the $\bar{x}_3$ axis, and, moreover, it cannot supply an axial force or a bending moment. Thus, it is apparent that the three boundary conditions ($\bar{u}_3^{(c)} = 0$, $R_1^{(c)} = 0$, $M^{(c)} = 0$) at point C of the auxiliary structure of Fig. 2.11d are equal to the corresponding conditions ($\bar{u}_3^{(c)} = 0$, $N^{(c)} = 0$, $M^{(c)} = 0$) at point C of the structure of Fig. 2.11a. Consequently, the internal actions in the members of the left half of the structure of Fig. 2.11a as well as its deformed configuration

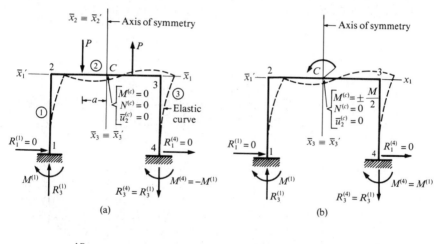

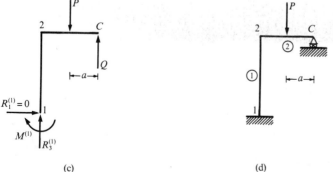

Figure 2.11 Planar frame having an axis of symmetry subjected to antisymmetric external actions.

are identical to those of the structure of Fig. 2.11d. Thus, we can analyze the structure of Fig. 2.11d instead of the structure of Fig. 2.11a. Notice, that the structure of Fig. 2.11a is statically indeterminate to the third degree, while the structure of Fig. 2.11d is statically indeterminate to the first degree.

The axis of symmetry of the structure in Fig. 2.12a passes through point 3. When this structure is subjected to a symmetric loading, the component of translation $\bar{u}_1^{(3)}$ and the component of rotation $\bar{\theta}_2^{(3)}$ at point 3 vanish. Moreover as mentioned previously, since the effect of axial deformation of the members of the frame on their internal moments is negligible, the component of displacement $\bar{u}_3^{(3)}$ may be assumed equal to zero. On the basis of the preceding discussion, it is apparent that point 3 of the structure of Fig. 2.12a can be considered fixed; consequently, the internal actions in the members of the left half of this structure, as well as its deformed configuration, are identical to those of the auxiliary structure of Fig. 2.12b. Thus, we can analyze the auxiliary structure of Fig. 2.12b instead of the structure of Fig. 2.12a. Notice that the structure of Fig. 2.12a is statically indeterminate to the sixth degree, while the structure of Fig. 2.12b is statically indeterminate to the third degree.

The structure of Fig. 2.13a has a center of symmetry and, consequently, when it is subjected to a symmetric loading, the components of translation $\bar{u}_1^{(c)}$ and $\bar{u}_2^{(c)}$ at the center of symmetry vanish. Moreover, from the equilibrium of a segment of the structure containing point C (see Fig. 2.9c), it is apparent that if a concentrated moment does not act at point C, the internal moment $M^{(c)}$ vanishes. If a concentrated moment acts at point C, the internal moment on each side of point C is equal to half this moment. The free-body diagram of the left half of the structure is shown in Fig. 2.13c. The forces $N^{(c)}$ and $Q^{(c)}$ are those required to keep point C from translating. Consider the auxiliary structure of Fig. 2.13d. Its members are identical to the corresponding mem-

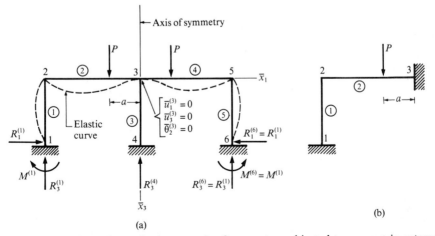

Figure 2.12 Planar frame having an axis of symmetry subjected to symmetric external actions.

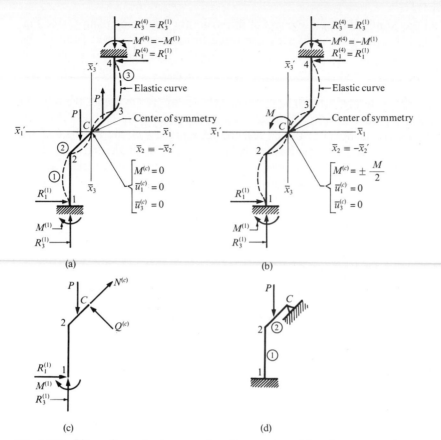

Figure 2.13 Planar frame having a center of symmetry subjected to symmetric external actions.

bers of the left half of the structure of Fig. 2.13a. Its loading is identical to that of the left half of the structure of Fig. 2.13a. Moreover, both structures are fixed at point 1. The support of the auxiliary structure at point C does not permit translation of this point in any direction. However, it does not restrain the rotation of member 2. Thus, it is apparent that the three boundary conditions ($\bar{u}_1^{(c)} = 0$, $\bar{u}_3^{(c)} = 0$, $M^{(c)} = 0$) at point C of the auxiliary structure of Fig. 2.13d are equal to the corresponding conditions at point C of the structure in Fig. 2.13a. Consequently, the internal actions in the members of the left half of the structure of Fig. 2.13a, as well as its deformed configuration, are identical to those of the auxiliary structure of Fig. 2.13d. Thus, we can analyze the structure of Fig. 2.13d instead of the structure of Fig. 2.13a. Notice that the structure of Fig. 2.13a is statically indeterminate to the third degree, while the structure of Fig. 2.13d is statically indeterminate to the second degree.

The structure of Fig. 2.14a has a center of symmetry; consequently, when it is subjected to an antisymmetric loading, the rotation $\bar{\theta}_2^{(c)}$ vanishes. Moreover, from the equilibrium of a segment of the structure containing point C (see Fig. 2.9d), it is apparent that if a concentrated force does not act at point C, the internal forces $N^{(c)}$ and $Q^{(c)}$ vanish. However, if a concentrated force acts at point C, the internal forces $N^{(c)}$ and $Q^{(c)}$ are equal to half the component of this force in their direction. The free-body diagram of the left half of the structure is shown in Fig. 2.14c. The moment $M^{(c)}$ is required to render the rotation at point C equal to zero. Consider the auxiliary structure of Fig. 2.14d. Its members are identical to the corresponding members of the left half of the structure of Fig. 2.14a. Its loading is identical to that of the left half of the structure of Fig. 2.14a. Moreover both structures are fixed at point 1. The support of the auxiliary structure at point C permits translation of this point in any direction. However, it prevents the rotation of member 2 at point C. Thus, it is apparent that the boundary conditions ($N^{(c)} = 0$, $R_3^{(c)} = 0$, $\bar{\theta}_2^{(c)} = 0$) at point C of the auxiliary structure of Fig. 2.14d are equal to the corre-

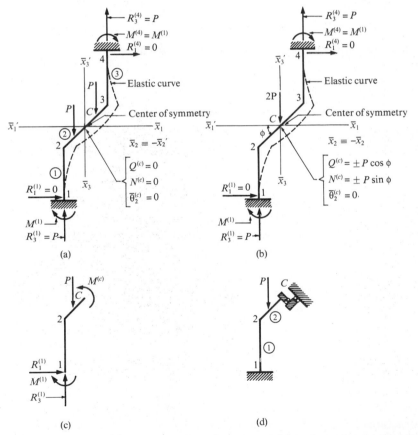

Figure 2.14 Planar frame having a center of symmetry subjected to antisymmetric external actions.

sponding conditions ($N^{(c)} = 0$, $Q^{(c)} = 0$, $\bar{\theta}_2^{(c)} = 0$) at point C of the structure of Fig. 2.14a. Consequently, the internal actions in the members of the left half of the structure of Fig. 2.14a, as well as its deformed configuration, are identical to those of the auxiliary structure of Fig. 2.14d. Notice that the structure of Fig. 2.14a is statically indeterminate to the third degree, while the structure of Fig. 2.14d is statically indeterminate to the first degree.

2.5 Symmetric Space Structures

A structure can be defined by specifying its geometry and its mechanical properties with reference to a rectangular, cartesian system of axes $\bar{x}_1$, $\bar{x}_2$, and $\bar{x}_3$. Certain structures have identical geometric and mechanical properties when described in either of two systems of axes $\bar{x}_1$, $\bar{x}_2$, $\bar{x}_3$ and $\bar{x}_1'$, $\bar{x}_2'$, $\bar{x}_3'$. Depending on the relation of these two systems, a structure may have one or more of the following three types of symmetry:

1. *Axial symmetry.* The structure of Fig. 2.15a is symmetric with respect to the $\bar{x}_2$ axis. In this case, the relation between the two systems of axes is

$$\bar{x}_1' = -\bar{x}_1 \qquad \bar{x}_2' = \bar{x}_2 \qquad \bar{x}_3' = -\bar{x}_3 \qquad (2.7)$$

Notice that in this case both systems of axes $\bar{x}_1$, $\bar{x}_2$, $\bar{x}_3$ and $\bar{x}_1'$, $\bar{x}_2'$, $\bar{x}_3'$ are right-hand systems. Moreover, notice that when a structure possesses axial symmetry about the $\bar{x}_2$ axis, an observer attached to the system of axes $\bar{x}_1$, $\bar{x}_2$, and $\bar{x}_3$ has the same view of the structure after the structure is rotated by 180° about the $\bar{x}_2$ axis.

2. *Planar symmetry.* The structure of Fig. 2.16a is symmetric with respect to the plane $\bar{x}_1 = 0$. In this case, the relationship between the two systems of axes is

$$\bar{x}_1' = -\bar{x}_1 \qquad \bar{x}_2' = \bar{x}_2 \qquad \bar{x}_3' = \bar{x}_3 \qquad (2.8)$$

Notice that in this case the axes $\bar{x}_1$, $\bar{x}_2$, and $\bar{x}_3$ form a right-hand system while the axes $\bar{x}_1'$, $\bar{x}_2'$, and $\bar{x}_3'$ form a left-hand system. Moreover, notice that when a

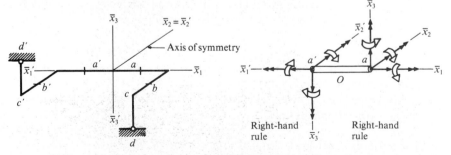

(a) Geometry of the structure   (b) Free-body diagram of a segment aOa' of the structure

Figure 2.15 Space frame having an axis of symmetry.

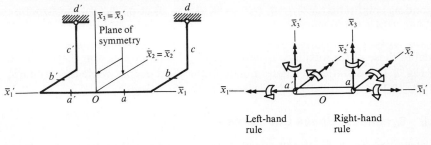

(a) Geometry of the structure   (b) Free-body diagram of segment aOa' of the structure

Figure 2.16 Space frame having a plane of symmetry.

structure possesses symmetry with respect to the $\bar{x}_2\bar{x}_3$ plane, the reflection of the image obtained from a mirror at the plane $\bar{x}_1 = 0$ is identical to the part of the actual structure which is behind the mirror.

Planar structures which possess planar symmetry also have an axis of symmetry. For instance, the frame in Fig. 2.2a has a plane of symmetry (the $\bar{x}_2\bar{x}_3$ plane) as well as an axis of symmetry (the x_3 axis).

3. *Center of symmetry.* The structure of Fig. 2.17a is symmetric with respect to point O. In this case, the relation between the two systems of axes is

$$\bar{x}_1' = -\bar{x}_2 \qquad \bar{x}_2' = -\bar{x}_2 \qquad \bar{x}_3' = -\bar{x}_3 \qquad (2.9)$$

Notice, that the axes $\bar{x}_1$, $\bar{x}_2$, and $\bar{x}_3$ form a right-hand system, while the axes $\bar{x}_1'$, $\bar{x}_2'$, and $\bar{x}_3'$ form a left-hand system.

Any given loading may be replaced by the sum of a symmetric and an antisymmetric loading, obtained in the same fashion as in the case of planar structures. That is:

1. The symmetric loading is obtained by dividing the given set of actions (forces and moments) by 2 and adding to the resulting set of actions the set which is symmetric to it.

2. The antisymmetric loading is obtained by dividing the given set of actions (forces and moments) by 2 and adding to the resulting set of actions the set which is antisymmetric to it.

For each of the three types of structural symmetry—axial, planar, and center—the following rules are valid provided that in each case, the proper definitions are employed for symmetric and antisymmetric loads, reactions, displacements, and internal actions:

1. *Symmetric* external loads applied to a *symmetric* structure produce *symmetric* reactions, internal actions, and displacements.

2. *Antisymmetric* external loads applied to a *symmetric* structure produce *antisymmetric* reactions, internal actions, and displacements.

3. In order to establish the curl of a moment or a rotation, the right-hand rule must be used for right-hand systems of axes and the left-hand rule must be employed for left-hand systems of axes.

In Figs. 2.15*b*, 2.16*b*, and 2.17*b*, the free-body diagram of a segment of the structures of Figs. 2.15*a*, 2.16*a*, and 2.17*a*, respectively, is shown, which includes the origin O of the two systems of axes $\bar{x}_1, \bar{x}_2, \bar{x}_3$ and $\bar{x}_1', \bar{x}_2', \bar{x}_3'$. The internal actions acting on the right-hand face of each segment are positive, while the internal actions acting on its left-hand face have the sense required, in order that the aforementioned rules of symmetry and antisymmetry be satisfied. As can be seen from Figs. 2.15*b*, 2.16*b*, 2.17*b*, as the length of the segments under consideration shrinks to zero, certain of their end actions do not satisfy the conditions for equilibrium; consequently, they must be zero. For instance, in order that the sum of the components in the x_2 direction of the forces acting on the segment shown in Fig. 2.16*b* vanish, the shearing force Q_2 must vanish.

A similar argument may be employed to establish that certain components of displacement vanish at point O. The components of displacement of the two end points of the segments of Fig. 2.15*b*, 2.16*b*, and 2.17*b* must satisfy the rules of symmetry and antisymmetry. Wherever this requirement is not satisfied by a particular displacement component of a segment, this component must vanish.

In Table 2.1, the conditions existing at point O for each type of structural symmetry under symmetric and antisymmetric loading are summarized. A (0) denotes that the quantity is zero at point O, while (?) indicates that the quan-

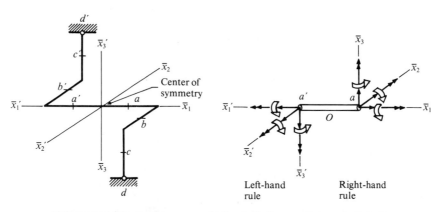

(a) Geometry of the structure   (b) Free-body diagram of segment aOa' of the structure

Figure 2.17 Space frame having a center of symmetry.

TABLE 2.1   Internal Actions and Displacements, at Point _0_, of Symmetric Structures Subjected to Symmetric Loading SL or to Antisymmetric Loading AL

Int. actions	Axial symmetry		Planar symmetry		Point symmetry		Ext. disp.	Axial symmetry		Planar symmetry		Point symmetry	
	SL	AL	SL	AL	SL	AL		SL	AL	SL	AL	SL	AL
N_1	?	0	?	0	?	0	u_1	0	?	0	?	0	?
Q_2	0	?	0	?	?	0	u_2	?	0	?	0	0	?
Q_3	?	0	0	?	?	0	u_3	0	?	?	0	0	?
M_1	?	0	0	?	0	?	θ_1	0	?	?	0	?	0
M_2	0	?	?	0	0	?	θ_2	?	0	0	?	?	0
M_3	?	0	?	0	0	?	θ_3	0	?	0	?	?	0

tity remains unknown. The following comments may be made from a study of Table 2.1.

1. Quantities which are zero (0) at point O under symmetric loading are unknown (?) at this point under antisymmetric loading and vice versa.

2. If an internal action is zero (0) at point O, the corresponding displacement is unknown (?) at this point and vice versa.

2.6   Problems

1 to 21.   Sketch the elastic curve of the structure loaded as shown in Fig. 2.P1. Explain the details of the sketch. The members of the structure are made of the same material and have the same cross section. Do not show the effect of the axial deformation of the members of the structure. Repeat for Figs. 2.P2 to 2.P21.

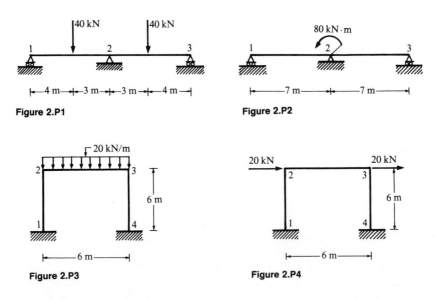

Figure 2.P1

Figure 2.P2

Figure 2.P3

Figure 2.P4

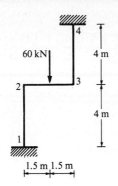

Figure 2.P5

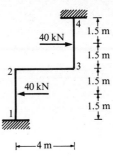

Figure 2.P6

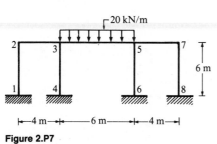

Figure 2.P7

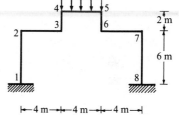

Figure 2.P8

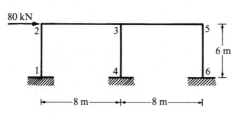

Figure 2.P9

Figure 2.P10

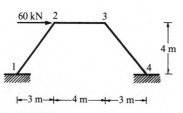

Figure 2.P11

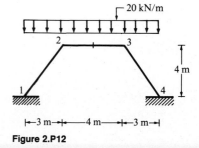

Figure 2.P12

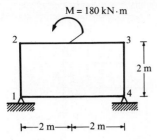

Figure 2.P13

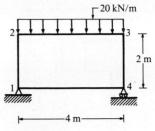

Figure 2.P14

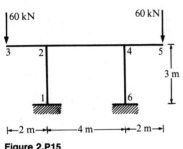

Figure 2.P15

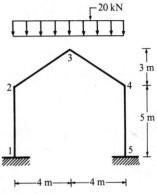

Figure 2.P16

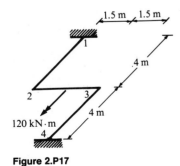

Figure 2.P17

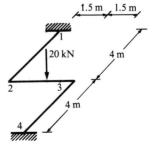

Figure 2.P18

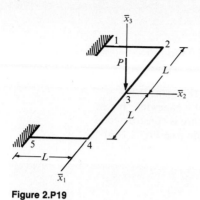

Figure 2.P19

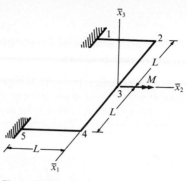

Figure 2.P20

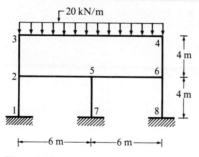

Figure 2.P21

Photographs and Brief Descriptions of Two Cable-Stayed Bridges

The Pasco-Kennewick Intercity Bridge (Fig. 2.18) spans the Columbia River between the cities of Pasco and Kennewick, Wash. Its main span is 981 ft (299 m). Its superstructure is constructed of reinforced concrete and has an open cross section (see Fig. 2.18b). A

(a)

(b)

Figure 2.18 The Pasco-Kennewick Intercity Bridge in Washington State. *(Courtesy of Arvid Grant and Associates Inc., Consulting Engineers, Olympia, Washington.)*

superstructure having this type of cross section does not offer high resistance to bending and, moreover, offers small resistance to torsion (see Sec. A.11). For this reason, as can be seen in Fig. 2.18a, the superstructure is supported by many cable stays at close intervals. Thus, the length of the unsupported spans of the superstructure is small; consequently, the bending and torsional moments resulting from the loads acting on it are small. The superstructure is not supported directly on the towers in order to avoid the increase of bending moment which would have been induced by the rigid tower support. In Fig. 2.18b, a prefabricated section of the concrete superstructure is lifted in place.

The East Huntington Bridge (Fig. 2.19) spans the Ohio River between East Huntington, W. Va., and Proctorville, Ohio. Its two cable-supported spans are 900 ft (274.32 m) and 608 ft (185.32 m) long. These spans have shallow concrete edge girders with transverse steel beams between them and there is a concrete roadway (see section C in Fig. 1.17b). The two longer approach spans (nonsupported by cables) have relatively shallow concrete box girders (see sections A and B in Fig. 1.17b). The rest of the approaches have steel-plate girders with a concrete deck. The superstructure of the two cable-supported spans was constructed as a double cantilever starting from the tower and proceeded, using prefabricated sections, toward the cantilevers which were constructed starting from the neighboring piers (Fig. 2.20).

Figure 2.19 The East Huntington Bridge on the Ohio River. *(Courtesy of West Virginia Department of Highways and Arvid Grant and Associates, Inc., Consulting Engineers, Olympia, Wash.)*

(a) A double cantilever started from the tower.

(b) View of the anchorage of the stays at the top of the tower.

Figure 2.20 The East Huntington Bridge on the Ohio River during construction. *(Courtesy of West Virginia Department of Highways and Arvid Grant and Associates, Inc., Consulting Engineers, Olympia, Wash.)*

Part 2

Analysis of Statically Determinate Structures

Chapter 3

Computation of the Internal Actions in the Members of Statically Determinate Framed Structures

3.1 Introduction

The internal actions in the members of statically determinate structures result from the application of external actions on the structures. Since the members of statically determinate structures are free to deform, their internal actions are not affected by changes in temperature, lack of fit of a member, or movement of a support. However, these disturbances change the geometry of the structure, and, concomitantly, the components of displacements of its joints. This is illustrated in Fig. 3.1.

The internal actions of the members of a statically determinate structure are obtained by considering the equilibrium of appropriate parts of the structure. For any part of a space structure, we can write the following set of six equations of equilibrium:

$$\Sigma F_1 = 0 \qquad \Sigma F_2 = 0 \qquad \Sigma F_3 = 0$$
$$\Sigma M_1^{(i)} = 0 \qquad \Sigma M_1^{(i)} = 0 \qquad \Sigma M_3^{(i)} = 0 \tag{3.1}$$

The first three relations of Eqs. (3.1) express the requirement that the sum of the components along the x_1, x_2, and x_3 axes of all the forces acting on the part of the structure under consideration must vanish. The last three relations of Eqs. (3.1) express the requirement that the sum of the components along the x_1, x_2, and x_3 axes of the moments of all the actions acting on the part of the structure under consideration must vanish. The moments are taken about a conveniently chosen point i.

When we consider the equilibrium of a member of a structure, we usually write Eqs. (3.1) in local form. That is, the components of the forces and

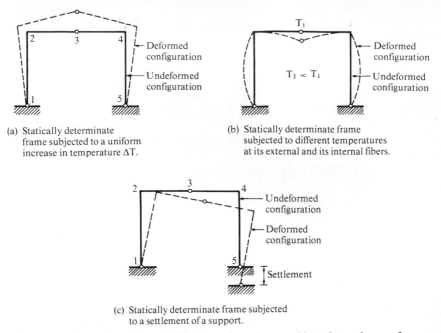

(a) Statically determinate frame subjected to a uniform increase in temperature ΔT.

(b) Statically determinate frame subjected to different temperatures at its external and its internal fibers.

(c) Statically determinate frame subjected to a settlement of a support.

Figure 3.1 Deformation of a statically determinate frame subjected to a change of temperature and to a settlement of a support.

moments in Eqs. (3.1) are referred to the local axes of the member. When we consider the equilibrium of a joint of a structure or a portion of a structure consisting of more than one member, we usually write the equations of equilibrium in global form. That is, the components of the forces and moments in Eqs. (3.1) are referred to the global axes of the structure. To indicate this, we use a bar over the letter F or M ($\Sigma \bar{F}_1 = 0$, $\Sigma \bar{M}_1^{(i)} = 0$).

For any part of a planar structure in the $\bar{x}_1 \bar{x}_3$ plane, the equations of equilibrium (3.1) in global form reduce to the following set of three equations.

$$\Sigma \bar{F}_1 = 0 \qquad \Sigma \bar{F}_3 = 0 \qquad \Sigma \bar{M}_2^{(i)} = 0 \qquad (3.2)$$

When the $\bar{x}_1$ and $\bar{x}_3$ axes are the horizontal and vertical axes, respectively, we write the equations of equilibrium as

$$\Sigma \bar{F}_h = 0 \qquad \Sigma \bar{F}_v = 0 \qquad \Sigma \bar{M}^{(i)} = 0 \qquad (3.3)$$

The last equation expresses the requirement that the sum of the moments of all the actions acting on the part of the structure under consideration about the axis normal to the plane of the structure at a conveniently chosen point i must vanish. For brevity, we say that the sum of the moments of all actions acting on the part of the structure under consideration about point i must vanish.

Often, it is convenient to employ a different set of equilibrium equations that are equivalent to Eqs. (3.1) or (3.2). For example, for any part of a planar structure in the $\bar{x}_1\bar{x}_3$ plane, we can use one of the following sets of equilibrium equations:

$$\Sigma \bar{F}_1 = 0 \qquad \Sigma \bar{M}^{(i)} = 0 \qquad \Sigma \bar{M}^{(j)} = 0 \qquad (3.4)$$

or

$$\Sigma \bar{M}^{(i)} = 0 \qquad \Sigma \bar{M}^{(j)} = 0 \qquad \Sigma \bar{M}^{(k)} = 0 \qquad (3.5)$$

where in Eqs. (3.4), the moments are taken about two different points i and j which are not located on the same line normal to the $\bar{x}_1$ axis. Moreover, in Eqs. (3.5), the moments are taken about three different points i, j, and k which are not located on the same line.

Usually, the first step when manually computing the internal actions in the members of statically determinate structures is to compute their reactions. For simple structures, this is accomplished by considering the equilibrium of the whole structure. That is, the free-body diagram of the whole structure is drawn, and the equations of equilibrium involving the unknown reactions are written and solved. In order to compute the reactions of compound structures, in addition to considering the equilibrium of the whole structure, we may have to consider the equilibrium of appropriate parts of the structure so as to take into account the conditions imposed by the internal hinges (see Examples 4 and 5 of Sec. 3.2 and Examples 3 and 4 of Sec. 3.4).

In Chaps. 3 to 7, we use the sign convention described in Fig. 1.33a for the internal actions in the members of a structure. Thus, a tensile axial force is considered positive while a compressive axial force is considered negative. Moreover, we choose the local axis x_2 of the members of planar structures normal to their plane and directed toward the observer. Thus, the direction of the positive shearing forces and bending moments acting on a planar structure is as shown in Fig. 3.2. A positive bending moment induces a tensile stress on the interior fibers of a planar frame or on the bottom fibers of a planar horizontal beam. Furthermore, for simplicity, we denote the shearing force and bending moment acting on a cross section of a planar structure by Q and M, respectively; that is, we omit their subscripts (Q_3, M_2).

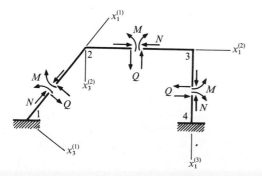

Figure 3.2 Positive internal actions in the members of a planar frame.

3.2   Analysis of Statically Determinate Planar Trusses

As discussed in Sec. 1.4.2, a member of a truss is subjected only to an axial, tensile, or compressive force. Thus, in general the internal forces exerted by the members of a planar truss on a joint and the external forces acting on this joint must satisfy two equations of equilibrium. For a joint of a planar truss these equations are written in global form as

$$\Sigma \bar{F}_h = 0 \qquad \Sigma \bar{F}_v = 0 \tag{3.6}$$

That is, for a planar truss with NJ joints, we have a system of 2NJ equations of equilibrium. For trusses which do not form mechanisms (see Sec. 1.9) these equations are independent. Moreover, as discussed in Sec. 1.11, for statically determinate trusses, the number of independent equations of equilibrium is equal to the number of unknown reactions plus the number of unknown internal axial forces in their members. Consequently, we can establish the reactions of statically determinate trusses and the internal forces in their members by drawing the free-body diagrams of their joints, writing the equations of equilibrium for each joint, and solving the resulting system of linear, algebraic equations. When we have a digital computer at our disposal, the solution of the resulting system of linear algebraic equations even for trusses with many joints can be easily obtained. However, when we perform the analysis of a truss by hand or with a desk calculator, we use methods which avoid the solution of simultaneous equations. For example, the reactions of a simple truss can be computed by drawing its free-body diagram and considering its equilibrium. Once the reactions of a simple truss are established a joint can be found which is subjected to two or less unknown internal forces. Thus, these forces can be computed by writing and solving the equations of equilibrium for the joint. The remaining unknown forces in the members of the truss can be established by considering the equilibrium of its joints one after the other. Care must be taken to select joints for which the forces of all but two or less of the members framing into them have previously been computed. This approach is referred to as

(a) Joint m

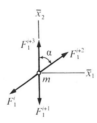

(b) Free-body diagram of joint m.

Figure 3.3 Joint connecting four members of a truss.

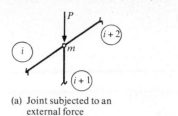

(a) Joint subjected to an
   external force

(b) Joint not subjected
   to an external force

Figure 3.4 Joint connecting three members of a truss.

the method of joints. We can also establish internal forces in the members of a truss by first computing its reactions and then considering the equilibrium of suitable portions of the truss. This approach is referred to as *the method of sections,* and it is employed when the internal forces in only a few members of a truss are required.

The analysis of a truss can be simplified if one is able to establish by inspection which of its members are not subjected to an internal force and which of its members are subjected to equal internal forces. For this purpose consider the following cases.

Case I. Joint m of a truss connects four of its members which, as shown in Fig. 3.3a, lie on two intersecting straight lines. Joint m is not subjected to external forces. Referring to Fig. 3.3b from the equilibrium of the joint, we have

$$\Sigma \overline{F}_1 = 0 \qquad F_1^i \; = F_1^{i+2}$$
$$\Sigma \overline{F}_2 = 0 \qquad F_1^{i+1} = F_1^{i+3} \tag{3.7}$$

That is, the forces in the colinear members are equal.

Case II. Joint m of a truss connects three of its members, two of which are colinear. In Fig. 3.4a joint m is subjected to an external force acting in the direction of the third member. From the equilibrium of this joint it is apparent that the internal forces in the two colinear members must be equal. Moreover the force in the other member (the $i + 1$) must be equal to the external force acting on the joint. If as shown in Fig. 3.4b the joint is not subjected to an external force, member $i + 1$ is not subjected to an internal force.

Case III. Joint m of a truss connects two of its members and as shown in Fig. 3.5 is not subjected to an external force. By considering the equilibrium of the joint, we can readily see that if the two members are colinear, the internal forces acting in them are equal, while if the two members are not colinear, the internal forces in them must be zero.

For example, referring to Fig. 3.6, on the basis of case III, the forces in members 1 and 2 are zero. Moreover, on the basis of case II, the force in member 7 is zero, the forces in members 6 and 10 are equal, the force in member

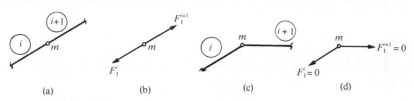

Figure 3.5   Joint connecting two members of a truss.

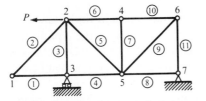

Figure 3.6   Geometry and loading of a truss.

3 is equal to the reaction of support 3, and the forces in members 1 and 4 are equal.

On the basis of the foregoing, the following procedure may be employed for analyzing a planar truss using the method of joints.

Step 1.   The reactions of the truss are computed. The reactions of a simple truss are established by considering its equilibrium as a whole. However, in order to compute the reactions of a compound truss, we may have to consider the equilibrium of portions of the truss in addition to the equilibrium of the truss as a whole (see Examples 4 and 5 at the end of this section).

Step 2.   The internal forces in the members of the truss are computed. For simple trusses this is accomplished by choosing a joint which is subjected to at most two unknown internal forces. The free-body diagram of this joint is drawn, and the equations of equilibrium for the forces acting on it are written. The solution of these equations gives the one or two unknown forces acting on the joint. The process is repeated for other joints until the internal forces in all the members of the truss are established. However, the internal forces in some members of certain compound trusses cannot be established by considering consecutively the equilibrium of one joint at a time (see Example 3 at the end of this section). The equations of equilibrium for a number of joints must be written and solved simultaneously. This can be avoided if the method of sections is used to establish the internal forces in certain members of the compound truss (see Example 3 at the end of this section). When the forces in these members are established, the forces in the remaining members of the compound truss can be computed using the method of joints.

In the sequel the analysis of planar trusses is illustrated by the following five examples.

Example 1. The internal forces in the members of a simple truss are established using the method of joints. Moreover, the method of sections is illustrated by recomputing the internal forces in two members of the truss.

Example 2. The reactions of a simple truss having an inclined upper chord are computed. Moreover the internal forces in two members of the truss are computed using the method of sections.

Example 3. The reactions of a compound truss and the internal forces in its members are computed. The reactions of this truss are established by considering the equilibrium of the whole truss. However, the internal forces in some members of the truss cannot be computed using the method of joints unless the internal force in one member of the truss is computed using the method of sections.

Example 4. The reactions of a compound truss and the internal forces in some of its members are calculated. This truss has four reactions which are computed by considering the equilibrium of the simple trusses from which the compound truss is made. Once the reactions of the truss are established, the internal forces in its members can be computed using the method of joints. However, in this example we are interested in computing the internal forces in a few members of the truss only. For this reason we use the method of sections.

Example 5. The reactions of a compound truss are computed. This truss has four reactions which are computed by considering the equilibrium of the two simple trusses from which the compound truss is made.

Example 6. The internal forces in the members of a simple truss are calculated for various angles of inclination of a supporting link. It is shown that for certain values of the angle of inclination of the supporting link of the truss, the internal forces in the members of the truss become very large; that is, the truss approaches a kinematically unstable geometry.

Example 1. Compute the force in each member of the simple truss shown in Fig. *a*.

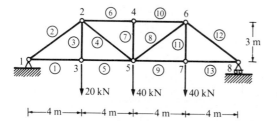

Figure a Geometry and loading of the truss.

solution First, we compute the reactions of the truss by referring to the free-body diagram of the whole truss shown in Fig. *b* and writing the equations of equilibrium.

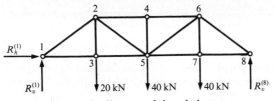

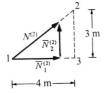

Figure b   Free-body diagram of the whole truss.

Thus,

$$\Sigma \bar{F}_h = 0 = R_h^{(1)} \tag{a}$$

$$\Sigma \bar{F}_v = 0 = R_v^{(1)} + R_v^{(8)} - 100 \tag{b}$$

$$\Sigma \bar{M}^{(8)} = 0 = 16R_v^{(1)} - 20(12) - 40(8) - 40(4) \tag{c}$$

or

$$R_v^{(1)} = 45 \text{ kN}$$

Substituting the value of $R_v^{(1)}$ into relation (b) we get

$$R_v^{(8)} = 55 \text{ kN}$$

The reaction $R_v^{(8)}$ may also be computed by setting the sum of moments about joint 1 equal to zero. Thus,

$$\Sigma \bar{M}^{(1)} = 16R_v^{(8)} - 40(12) - 40(8) - 20(4) = 0$$

or

$$R_v^{(8)} = 55 \text{ kN} \quad \textit{check}$$

The coincidence of the above result with that obtained by solving Eqs. (b) and (c) is a check on the calculations of $R_v^{(8)}$. It is advisable to check the values of the reactions prior to proceeding to the calculations of the internal forces in the members of the structure.

The forces in the members of the truss may be obtained by drawing the free-body diagrams of its joints and considering their equilibrium. Notice that in drawing free-body diagrams we assume, for convenience, that all unknown internal forces are tensile forces.

Figure c   Global components of force $N^{(2)}$.

Figure d   Free-body diagram of joint 1.

Equilibrium of Joint 1

Since the reactions of the truss have been established as shown in Fig. d, there are only two unknown internal forces ($N^{(2)}$, $N^{(1)}$) acting on joint 1; consequently, these forces may be established by considering the equilibrium of this joint. The global components of the force $N^{(2)}$ may be obtained from the similarity of the force triangle shown in Fig. c and the triangle of the truss 1,2,3 (see Fig. a). Thus,

$$\bar{N}^{(2)}_1 = \tfrac{4}{5}N^{(2)}$$

$$\bar{N}^{(2)}_2 = \tfrac{3}{5}N^{(2)}$$

Consequently, referring to Fig. d and setting equal to zero the sum of the vertical and horizontal components of the forces acting on joint 1, we obtain

$$\Sigma \bar{F}_h = 0 = \tfrac{4}{5}N^{(2)} + N^{(1)}$$

$$\Sigma \bar{F}_v = 0 = \tfrac{3}{5}N^{(2)} + 45$$

or

$$N^{(2)} = -(\tfrac{5}{3})45 = -75 \text{ kN} \qquad \text{or 75 kN compression}$$

$$N^{(1)} = 60 \text{ kN tension}$$

Equilibrium of Joint 3

Since the internal force $N^{(1)}$ has been established, as shown in Fig. e, there are only two unknown internal forces ($N^{(3)}$, $N^{(5)}$) acting on joint 3. Moreover, since the force $N^{(5)}$ and the 60-kN force are colinear and normal to the other forces acting at joint 3, it is apparent that

$$N^{(5)} = 60 \text{ kN tension}$$

$$N^{(3)} = 20 \text{ kN tension}$$

Figure e Free-body diagram of joint 3.

Figure f Free-body diagram of joint 2.

Equilibrium of Joint 2

As shown in Fig. f, since the internal force $N^{(2)}$ and $N^{(3)}$ are known, there are only two unknown forces ($N^{(6)}$, $N^{(4)}$) acting on joint 2. Consequently these forces may be established by considering the equilibrium of the joint. Thus

$$\Sigma \bar{F}_h = 0 = (\tfrac{4}{5})75 + N^{(6)} + \tfrac{4}{5}N^{(4)}$$

$$\Sigma \bar{F}_v = 0 = (\tfrac{3}{5})75 - 20 - \tfrac{3}{5}N^{(4)}$$

or

$$N^{(4)} = 41.7 \text{ kN tension}$$

and

$$N^{(6)} = -93.3 \text{ kN} \qquad \text{or 93.3 kN compression}$$

Equilibrium of Joint 4

As shown in Fig. g, three forces act on joint 4, two of which are colinear. This indicates that the third force must vanish. Thus

$$N^{(7)} = 0$$

$$N^{(10)} = -93.3 \text{ kN} \quad \text{or } 93.3 \text{ kN compression}$$

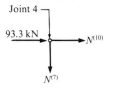

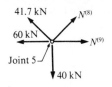

Figure g Free-body diagram of joint 4.   **Figure h** Free-body diagram of joint 5.

Equilibrium of Joint 5

Referring to Fig. h we have

$$\Sigma \bar{F}_h = 0 = (\tfrac{4}{5})41.7 + 60 - \tfrac{4}{5}N^{(8)} - N^{(9)}$$

$$\Sigma \bar{F}_v = 0 = (\tfrac{3}{5})41.7 + \tfrac{3}{5}N^{(8)} - 40$$

or

$$N^{(8)} = 25 \text{ kN tension}$$

$$N^{(9)} = 73.3 \text{ kN tension}$$

Equilibrium of Joint 7

Referring to Fig. b, we can see that

$$N^{(11)} = 40 \text{ kN tension}$$

$$N^{(13)} = N^{(9)} = 73.3 \text{ kN tension}$$

Equilibrium of Joint 8

Referring to Fig. i, we get

$$\Sigma \bar{F}_v = 0 = \tfrac{3}{5}N^{(12)} + 55$$

or

$$N^{(12)} = -91.7 \text{ kN} \quad \text{or } 91.7 \text{ kN compression}$$

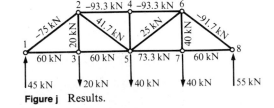

Figure i Free-body diagram of joint 8.   **Figure j** Results.

The same results are obtained from the second of the equilibrium equations of this joint. That is,

$$\Sigma \bar{F}_h = 0 = \tfrac{4}{5}N^{(12)} + 73.3$$

or
$$N^{(12)} = -91.7 \text{ kN}$$

This is a partial check of the calculations.

The results of the analysis of the truss are shown in Fig. j.

Notice that we could have obtained the force acting in a member of the truss by considering the equilibrium of an appropriate part of the truss. For instance, if we want to establish the force in member 4, we consider the part of the truss whose free-body diagram is shown in Fig. k. From the equilibrium of this part, we have

$$\Sigma \bar{F}_v = 0 = 45 - 20 - \tfrac{3}{5}N^{(4)}$$

hence
$$N^{(4)} = 41.7 \text{ kN tension}$$

If we want to compute the force in member 6, we can set equal to zero the sum of the moments about joint 5 of all the forces acting on the part shown in Fig. k. Thus,

$$\Sigma \bar{M}^{(5)} = 45(8) - 2(4) + 3N^{(6)}$$

hence
$$N^{(6)} = -93.3 \text{ kN} \qquad \text{or } 93.3 \text{ kN compression}$$

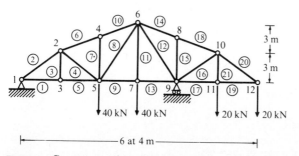

Figure k Free-body diagram of a part of the truss.

Example 2. Compute the forces in members 8 and 6 of the truss shown in Fig. a.

Figure a Geometry and loading of the truss.

solution Referring to Fig. a, we see that two of the internal forces acting at joints 3, 4, and 8 are colinear. Thus, the third force must vanish. Hence, the forces in members 3, 7, and 15 vanish. Moreover, referring to Fig. a, by inspection we conclude that

$$N^{(1)} = N^{(5)} \qquad N^{(6)} = N^{(10)} \qquad N^{(9)} = N^{(13)} \qquad N^{(14)} = N^{(18)} \qquad N^{(17)} = N^{(19)}$$

We first compute the reactions of the truss by considering the equilibrium of the whole truss. Thus, referring to Fig. b, we have

$$\Sigma \overline{M}^{(9)} = 16R_v^{(1)} - 40(8) - 40(4) + 20(4) + 20(8) \quad = 0$$

$$\Sigma \overline{M}^{(1)} = 20(24) + 20(20) - 16R_v^{(9)} + 40(12) + 40(8) = 0$$

consequently, $\qquad R_v^{(1)} = 15$ kN $\qquad R_v^{(9)} = 105$ kN

moreover,

$$\Sigma \overline{F}_v = 0 = 15 + 105 - 40 - 40 - 20 - 20 = 0 \qquad \textit{partial check}$$

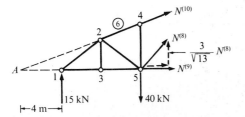

Figure b Free-body diagram of the whole truss.

To find the force in member 8, we consider the equilibrium of the part of the truss shown in Fig. c. In this figure, point A is the intersection of member 6 with the extension of the bottom chord of the truss. Setting the sum of the moments about point A equal to zero, we have

$$\Sigma \overline{M}^A = 0 = 15(4) + 12(3/\sqrt{13}) \, N^{(8)} - 40(12)$$

or $\qquad N^{(8)} = 42.06$ kN

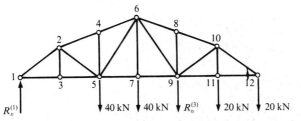

Figure c Free-body diagram of a part of the truss.

To obtain the force in member 6, we consider the equilibrium of the part of the truss shown in Fig. c. Setting the sum of the moments about joint 5 equal to zero, we get

$$\Sigma \overline{M}^{(5)} = 0 = 15(8) + 4.5 \left(\frac{4}{\sqrt{(1.5)^2 + 4^2}} \right) N^{(10)}$$

thus, $N^{(6)} = N^{(10)} = -28.48$ kN or 28.43 kN compression

Example 3. Compute the internal forces acting in the members of the roof truss shown in Fig. a.

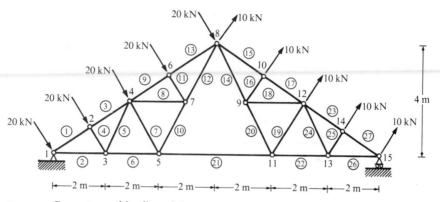

Figure a Geometry and loading of the truss.

solution This is a compound truss. However, its reactions may be established by considering the equilibrium of the free-body diagram of the whole truss shown in Fig. b. Thus,

$$\Sigma \overline{M}^{(1)} = 0 = 20(1.803)(4 + 3 + 2 + 1)$$
$$- 10[5(2.774) + 1.803(1 + 2 + 3 + 4)] - 12R_v^{(15)}$$

$$\Sigma \overline{M}^{(15)} = 0 = 10(1.803)(4 + 3 + 2 + 1)$$
$$- 20[5(2.774) + 1.803(1 + 2 + 3 + 4)] + 12R_v^{(1)}$$

$$\Sigma \overline{F}_h = 0 = \frac{4}{7.211} (100 + 50) + R_h^{(1)}$$

consequently,

$$R_v^{(15)} = 3.46 \text{ kN} \qquad R_v^{(1)} = 38.13 \text{ kN} \qquad R_h^{(1)} = -83.20 \text{ kN}$$

After the reactions of the truss are established, the forces in members 1 to 6, 22, 23, 25, 26, and 27 can be computed using the method of joints. For example the forces

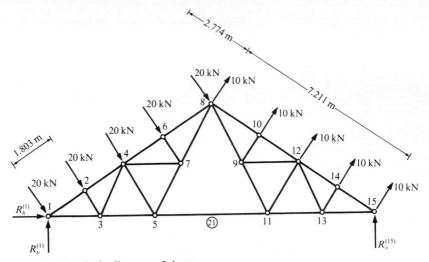

Figure b Free-body diagram of the truss.

in members 1 and 2 can be computed by considering the equilibrium of joint 1; the forces in members 3 and 4 can be computed by considering the equilibrium of joint 2; the forces in members 5 and 6 can be computed by considering the equilibrium of joint 3. However, after the forces in these members are established, there is no other joint which is subjected to two or less unknown internal forces. To overcome this difficulty, we compute the force in member 21 by considering the free-body diagram of the portion of the truss shown in Fig. c and setting the sum of the moments about joint 8 equal to zero. Thus,

$$\Sigma \overline{M}^{(8)} = 0 = 10(1.802)(4 + 3 + 2 + 1) + 3.46(6) - 4N^{(21)}$$

or $N^{(21)} = 50.26$ kN tension

The forces in the other members of the truss may be established by considering the equilibrium of its joints in the following sequence: 4, 7, 6, 8, etc. The results are shown in Fig. d.

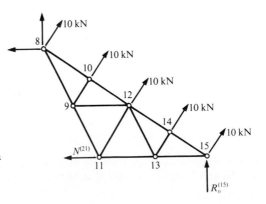

Figure c Free-body diagram of a portion of the truss.

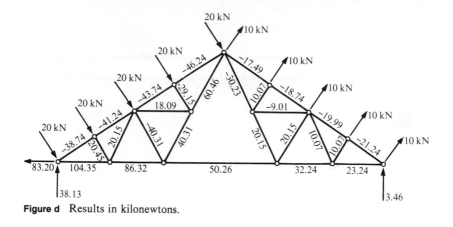

Figure d Results in kilonewtons.

Example 4. The truss of Fig. a is part of a bridge. The load is applied to the joints of the bottom chord of the truss by a system of stringers and floor beams (see Sec. 1.4.2). Compute the forces in members 7, 9, 10, 8, 20, 19, and 16 of the truss.

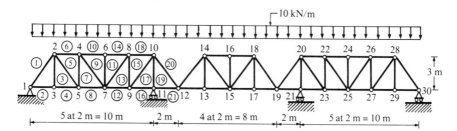

Figure a Geometry and loading of the truss.

solution This is a compound structure. To compute its reactions, we consider the free-body diagrams of the three simple trusses which comprise the compound truss shown in Fig. b. From the equilibrium of the middle simple truss, we have

$$R_v^{(12)} = R_v^{(9)} = \tfrac{10}{2}(8) = 40 \text{ kN}$$

From the equilibrium of the left-hand simple truss, we obtain

$$\Sigma \overline{M}^{(11)} = 0 = 10R_v^{(1)} + 40(2) - 12(10)(4)$$

$$\Sigma \overline{M}^{(1)} = 0 = 10R_v^{(11)} - 40(12) - 120(6)$$

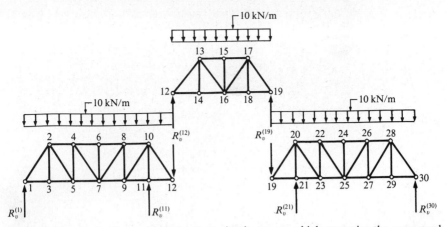

Figure b Free-body diagrams of the three simple trusses which comprise the compound truss.

consequently,

$$R_v^{(1)} = 40 \text{ kN} \qquad R_v^{(11)} = 120 \text{ kN}$$

Once the reactions of the truss have been established the internal forces in the members of the truss can be computed using the method of joints. However, since we are interested in computing the internal forces in a few members only, it is easier to use the method of sections. In the free-body diagram of the left-hand simple truss shown in Fig. c, the distributed load has been converted to concentrated forces acting on the joints of the truss. In order to find the forces in members 7 and 8, we consider the

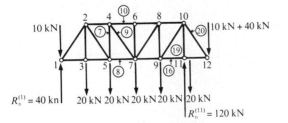

Figure c Free-body diagram of the left-hand simple truss.

equilibrium of the portion of the left-hand simple truss whose free-body diagram is shown in Fig. d. Thus,

$$\Sigma \overline{F}_v = 0 = 40 - 10 - 20 - 20 + N^{(7)}$$

$$\Sigma \overline{F}_h = 0 - N^{(6)} + N^{(8)}$$

$$\Sigma \overline{M}^{(2)} = 0 = 30(2) + 20(2) - 10(2) - 3N^{(8)}$$

or

$$N^{(7)} = 10 \text{ kN tension}$$

$$N^{(8)} = 26.7 \text{ kN tension}$$

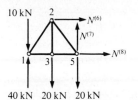

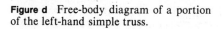

40 kN 20 kN 20 kN

Figure d Free-body diagram of a portion of the left-hand simple truss.

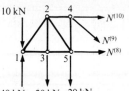

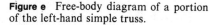

40 kN   20 kN   20 kN

Figure e Free-body diagram of a portion of the left-hand simple truss.

In order to find the forces in members 9 and 10, we consider the equilibrium of the portion on the left-hand, simple truss whose free-body diagram is shown in Fig. e. Thus,

$$\Sigma \bar{F}_h = 0 = N^{(10)} + N^{(8)} - \frac{4}{\sqrt{13}}(12.01)$$

$$\Sigma \bar{F}_v = 0 = 40 - 10 - 20 - 20 - \frac{3}{\sqrt{13}} N^{(9)}$$

or

$$N^{(10)} = -20 \text{ kN} \quad \text{or 20 kN compression}$$

$$N^{(9)} = -12.01 \text{ kN} \quad \text{or 12.02 kN compression}$$

In order to compute the forces in members 16, 19, and 20, we first consider the equilibrium of joint 12. Thus, referring to Fig. f, we have

$$\Sigma \bar{F}_h = 0 = 60.09 \left(\frac{2}{\sqrt{13}} \right) N^{(21)}$$

$$\Sigma \bar{F}_v = 0 = \frac{3}{\sqrt{3^2 + 2^2}} N^{(20)} - 50 \text{ kN}$$

or

$$N^{(21)} = -33.4 \text{ kN} \quad \text{or 33.4 kN compression}$$

$$N^{(20)} = 60.09 \text{ kN tension}$$

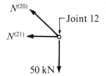

Figure f Free-body diagram of joint 12.

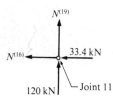

Figure g Free-body diagram of joint 11.

Using the above results and referring to Fig. g, from the equilibrium of joint 11 we can deduce that

$$N^{(16)} = N^{(21)} = -33.4 \text{ kN} \quad \text{or 33.4 kN compression}$$

$$N^{(19)} = -100 \text{ kN} \quad \text{or 100 kN compression}$$

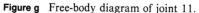

Example 5.   Compute the reactions of the truss loaded as shown in Fig. a.

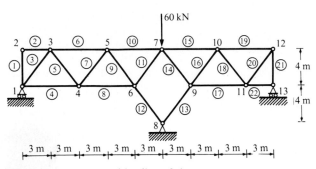

Figure a   Geometry and loading of the truss.

solution   This is a compound truss consisting of two simple trusses. As shown in Fig. b, there are four unknown reactions acting on the truss; therefore, they cannot be established by considering only the equilibrium of the truss as a whole. Referring to Fig. b, we have

$$\Sigma \bar{F}_h = 0 \qquad R_h^{(8)} = 0 \tag{a}$$

$$\Sigma \bar{F}_v = 0 \qquad R_v^{(1)} + R_v^{(8)} + R^{(13)} - 60 = 0 \tag{b}$$

$$\Sigma \bar{M}^{(13)} = 0 \qquad (12)(60) - 27R_v^{(1)} - 12R_v^{(8)} = 0 \tag{c}$$

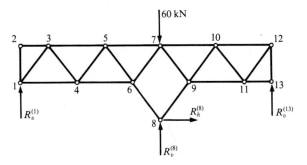

Figure b   Free-body diagram of the truss.

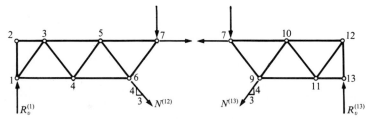

Figure c   Free-body diagrams of the two simple trusses.

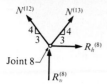

Figure d Free-body diagram of joint 8.

Referring to Fig. d and taking into account relation (a) from the equilibrium of joint 8, we obtain

$$\Sigma \bar{F}_h = 0 \qquad N^{(12)} = N^{(13)} \tag{d}$$

$$\Sigma \bar{F}_v = 0 \qquad R_v^{(8)} = \tfrac{6}{5} N^{(12)} \tag{e}$$

Referring to Fig. c, from the equilibrium of the left simple truss, we have

$$\Sigma M^{(7)} = 0 \qquad 15 R_v^{(1)} - 3(\tfrac{4}{5}) N^{(12)} - 4(\tfrac{3}{5}) N^{(12)} = 0$$

or

$$R_v^{(1)} = \tfrac{8}{25} N^{(12)} \tag{f}$$

Substituting relations (e) and (f) into (c), we obtain

$$N^{(12)} = 31.25 \text{ kN}$$

Substituting the above result into relations (f) and (e), we get

$$R_v^{(1)} = \tfrac{8}{25}(31.25) = 10 \text{ kN}$$

$$R_v^{(8)} = \tfrac{6}{5}(31.25) = 37.5 \text{ kN}$$

Substituting the above results into relation (b), we obtain

$$R_v^{(13)} = 12.5 \text{ kN}$$

Example 6. Compute the internal forces in the members of the truss shown in Fig. a for various values of the slope $1/a$ of member 14.

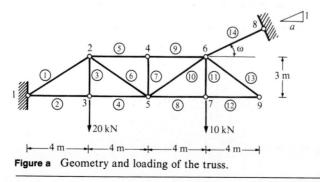

Figure a Geometry and loading of the truss.

solution   Referring to Fig. a, we have

$$\sin \omega = \frac{1}{\sqrt{1 + a^2}} \qquad \tan \omega = \frac{1}{a} \qquad \cos \omega = \frac{1}{\sqrt{1 + a^2}} \qquad \text{(a)}$$

Referring to Fig. b, we obtain

$$\Sigma \bar{F}_h = 0 = -R_h^{(1)} - R^{(8)} \cos \omega$$

$$\Sigma \bar{F}_v = 0 = R_v^{(1)} + R^{(8)} \sin \omega - 30 \qquad \text{(b)}$$

$$\Sigma \bar{M}^{(1)} = 0 = 20(4) + 10(12) + 3R^{(8)} \cos \omega - 12R^{(8)} \sin \omega$$

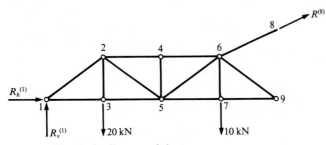

Figure b   Free-body diagram of the truss.

Using relations (a), we obtain from relations (b)

$$R^{(8)} = \frac{200}{12 \sin \omega - 3 \cos \omega} = \frac{200 \sqrt{1 + a^2}}{12 - 3a}$$

$$R_h^{(1)} = -\frac{200 \sqrt{1 + a^2}}{12 - 3a} \cos \omega = -\frac{200a}{12 - 3a}$$

$$R_v^{(1)} = 30 - \frac{200 \sqrt{1 + a^2}}{12 - 3a} \left(\frac{1}{\sqrt{1 + a^2}} \right) = \frac{160 - 90a}{12 - 3a}$$

We now proceed to compute the internal forces in the members of the truss.

Equilibrium of Joint 1

Referring to Fig. c, we get

$$\Sigma \bar{F}_v = 0 = \frac{3}{5} N^{(1)} + \frac{160 - 90a}{12 - 3a}$$

$$\Sigma \bar{F}_h = 0 = -\frac{200a}{12 - 3a} + \frac{4}{5} N^{(1)} + N^{(2)}$$

or

$$N^{(1)} = -\frac{5}{3}\left(\frac{160 - 90a}{12 - 3a}\right)$$

$$N^{(2)} = \frac{240a + 640}{3(12 - 3a)}$$

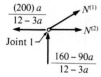

Figure c   Free-body diagram of joint 1.

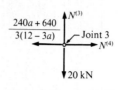

Figure d   Free-body diagram of joint 3.

Equilibrium of Joint 3
Referring to Fig. d, we obtain

$$N^{(4)} = N^{(2)} = \frac{240a + 640}{3(12 - 3a)}$$

$$N^{(3)} = 20 \text{ kN}$$

Equilibrium of Joint 2
Referring to Fig. e, we have

$$\Sigma \overline{F}_h = 0 = \frac{4}{3}\left(\frac{160 - 90a}{12 - 3a}\right) + \frac{4}{5}N^{(6)} + 3N^{(5)}$$

$$\Sigma \overline{F}_v = 0 = \frac{160 - 90a}{12 - 3a} - 20 - \frac{3}{5}N^{(6)}$$

or

$$N^{(6)} = -\frac{5}{3}\left(\frac{80 + 30a}{12 - 3a}\right)$$

$$N^{(5)} = \frac{480a - 320}{3(12 - 3a)}$$

Equilibrium of Joint 4
Referring to Fig. a, we can see that

$$N^{(9)} = N^{(5)} \qquad N^{(7)} = 0$$

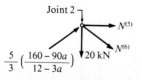

Figure e   Free-body diagram of joint 2.

TABLE a Forces (kN) in the Members of the Truss

	a:	-3	-2	-1	0	1	2	3	3.5	4	4.5
$N^{(1)} = -\dfrac{5}{3}\left(\dfrac{160-90a}{12-3a}\right)$		-34.1	-31.5	-27.8	-22.2	-13.0	5.6	61.1	172.2	∞	-272.2
$N^{(2)} = \dfrac{240a+640}{3(12-3a)}$		-1.26	2.96	8.89	17.78	32.59	62.23	151.12	328.89	∞	-382.23
$N^{(3)}$		20	20	20	20	20	20	20	20	∞	20
$N^{(4)} = N^{(2)}$		-1.26	2.96	8.89	17.78	32.59	62.23	151.12	328.89	∞	-382.23
$N^{(5)} = \dfrac{480a-320}{3(12-3a)}$		-27.9	-23.7	-17.8	-8.9	5.9	35.6	124.4	302.2	∞	-408.9
$N^{(6)} = -\dfrac{5}{3}\left(\dfrac{80+30a}{12-3a}\right)$		0.8	-1.8	-5.8	-11.1	-20.4	-38.9	-94.4	-205.5	∞	238.9
$N^{(7)} = 0$		0	0	0	0	0	0	0	0	0	0
$N^{(8)}$		0	0	0	0	0	0	0	0	0	0
$N^{(9)} = N^{(5)}$		-27.9	-23.7	-17.8	-8.9	5.9	35.6	124.4	302.2	∞	-408.9
$N^{(10)} = -N^{(6)}$		-0.8	1.8	5.8	11.1	20.4	38.9	94.4	205.5	∞	-238.9
$N^{(11)}$		10	10	10	10	10	10	10	10	10	10
$N^{(12)} = N^{(8)}$		0	0	0	0	0	0	0	0	0	0
$N^{(13)}$		0	0	0	0	0	0	0	0	0	0
$N^{(14)} = \dfrac{200\sqrt{1+a^2}}{12-3a}$		30.11	24.84	18.85	16.67	31.42	74.53	210.81	485.34	∞	-614.68

Equilibrium of Joint 5

Referring to Fig. f, we obtain

$$\Sigma \bar{F}_v = 0 - \frac{3}{5} N^{(10)} - \frac{80 + 30a}{12 - 3a}$$

$$\Sigma \bar{F}_h = 2\left(\frac{4}{3}\right)\frac{80 + 30a}{12 - 3a} - \frac{240a + 640}{3(12 - 3a)} + N^{(8)}$$

or

$$N^{(10)} = \frac{5}{3}\left(\frac{80 + 30a}{12 - 3a}\right)$$

$$N^{(8)} = 0$$

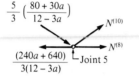

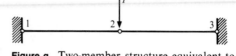

Figure f Free-body diagram of joint 5.

Figure g Two-member structure equivalent to the truss of Fig. a with $a = 4$.

Equilibrium of Joints 7 and 9

Referring to Fig. a for the equilibrium of joint 7, we get

$$N^{(11)} = 10 \text{ kN} \qquad N^{(2)} = N^{(8)}$$

Finally from the equilibrium of joint 9 we have

$$N^{(12)} = N^{(13)} = 0$$

The values of the interval forces in the members of the truss for various values of the angle of inclination of member 14 are given in Table a. For $a = 4$ the truss becomes kinematically unstable. Notice that for $a = 4$, points 1, 6, and 8 lie on a straight line. That is, the truss becomes equivalent to the two-member structure shown in Fig. g. As discussed in Sec. 1.10.3, this structure is kinematically unstable. Furthermore, referring to Table a, notice that the internal forces in some members of the truss increase as the angle of inclination of member 14 approaches that corresponding to kinematic instability.

3.3 Analysis of Statically Determinate Planar Beams and Frames

In order to compute the internal actions of statically determinate beams or frames, we first compute their reactions. For simple structures this is accomplished by considering the equilibrium of the whole structure. For compound structures, in addition to the equilibrium of the whole structure, it is necessary to consider the equilibrium of appropriate parts of the structure so as to take into account the conditions imposed by the internal hinges (see Examples 3 and 4 in Sec. 3.3.2).

Once the reactions of a beam or a frame have been computed, the internal actions in its members can be expressed as functions of the axial coordinate x_1. For instance, for the ith member of a planar structure we have

$$N^{(i)} = N^{(i)}(x_1) \qquad Q^{(i)} = Q^{(i)}(x_1) \qquad M^{(i)} = M^{(i)}(x_1) \qquad (3.8)$$

Generally, these functions are not continuous. For example, consider the beam shown in Fig. 3.7a; from the equilibrium of an infinitesimal portion of this beam which contains the concentrated force P^A, it is apparent (see Fig. 3.7b) that at the point of application of the concentrated force there is a discontinuity (jump) in the function $Q(x_1)$. Similarly, referring to Fig. 3.7c, we can see from the equilibrium of an infinitesimal portion of the beam which contains the concentrated moment M^B that there is a discontinuity (jump) in the function $M(x_1)$ at the point of application of the concentrated moment. The graphical representation of the functions $N^{(i)}(x_1)$, $Q^{(i)}(x_1)$, and $M^{(i)}(x_1)$ are referred to as *the axial force diagram, the shear diagram,* and *the moment diagram,* respectively.

The axial $\tau_{11}^{(i)}$ (in case $N^{(i)} = 0$) and shearing $\tau_{13}^{(i)}$ components of stress at a point of a cross section of the ith member are proportional to the moment $M^{(i)}$ and the shearing force $Q^{(i)}$, respectively, acting on this cross section [see relations (A.116)]. Consequently, the moment and shear diagrams constitute a pictorial representation of the variation of the axial $\tau_{11}^{(i)}$ and the shearing $\tau_{13}^{(i)}$ components of stress, respectively, along the length of the member.

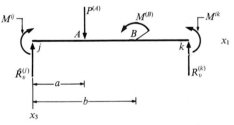

(a) Member subjected to a concentrated force and a moment

(b) Equilibrium of an infinitesimal portion of the member containing point A.

(c) Equilibrium of an infinitesimal portion of the member containing point B.

Figure 3.7 Discontinuity in the internal shearing force and moment.

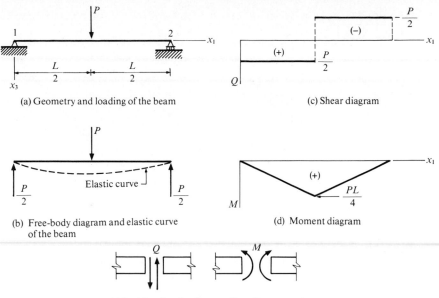

(a) Geometry and loading of the beam

(c) Shear diagram

(b) Free-body diagram and elastic curve
of the beam

(d) Moment diagram

(e) Positive shearing force and bending moment

Figure 3.8   Shear and moment diagrams for a beam.

We plot the moment diagram for a member using the sign convention of Fig. 1.33a. That is, the shearing force is considered positive if it acts as shown in Fig. 3.8e, while the moment is considered positive if it induces a tensile stress on the bottom fibers of the beam (see Fig. 3.8e). The shear and moment diagrams for the beam of Fig. 3.8a are plotted in Figs. 3.8c and d, respectively.

A distributed load may not be normal to the direction of the axis of the member on which it acts. Moreover, it may be given as a force per unit length in a direction other than that of the axis of the member on which it acts. In order to establish the shearing and axial forces in the member, it is desirable to transform the load first to a force per unit of length in the direction of the axis of the member and then into an equivalent set of two distributed loads, one normal and the other parallel to the axis of the member (see Example 2, Section 3.3.2). For example, consider the member shown in Fig. 3.9a whose x_1 axis is inclined by an angle ϕ with the horizontal axis. The member is subjected to a vertical distributed force of q kilonewtons per unit of horizontal length. As shown in Fig. 3.9b, the load q is first transformed into an equivalent vertical distributed load $q_v^{(e)}$ given in kilonewtons per unit of length along the axis of the member. Thus

$$q_v^{(e)} = \frac{qa}{A2} = q \cos \phi \qquad (3.9)$$

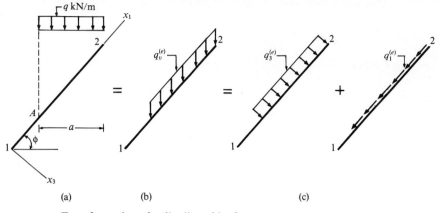

Figure 3.9 Transformation of a distributed load.

As shown in Fig. 3.9c the distributed load $q_v^{(e)}$ is transformed into a set of equivalent distributed loads $q_1^{(e)}$ and $q_3^{(e)}$ parallel and normal, respectively, to the axis of the member. That is

$$q_3^{(e)} = q_v^{(e)} \cos \phi = q \cos^2 \phi$$
$$q_1^{(e)} = q_v^{(e)} \sin \phi = q \sin \phi \cos \phi \tag{3.10}$$

3.3.1 Shear and moment diagrams for beams and frames by the method of sections

The shear and moment diagrams of a member may be constructed by considering the equilibrium of segments of the member in which the shearing force and bending moments are continuous functions of x_1. This method is referred to as the method of sections and is illustrated by the following example.

Example 1. Consider the beam subjected to the external actions shown in Fig. a. Plot its shear and moment diagrams and sketch its elastic curve.

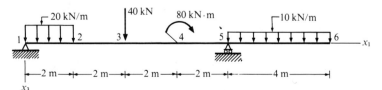

Figure a Geometry and loading of the beam.

solution First, the supports of the beam are examined in order to establish the reactions which they can transmit to the beam. The reaction at support 1 may have a

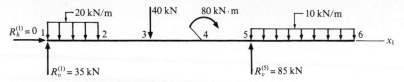

Figure b Free-body diagram of the beam.

vertical and a horizontal component, while the reaction at support 5 may have only a vertical component.

The free-body diagram of the beam is shown in Fig. b. The reactions are established by referring to Fig. b and considering the equilibrium of the beam as a whole. Setting the sum of the forces in the x_1 direction equal to zero, we find that the horizontal component of the reaction at support 1 vanishes. Setting the sum of the moments about any point equal to zero (for convenience we choose point 1), we find

$$\Sigma M^{(1)} = 0 = 20(2)(1) + 40(4) + 80 + (40)10 - 8R_v^{(5)}$$

therefore
$$R_v^{(5)} = 85 \text{ kN}$$

Setting the sum of the forces in the vertical direction equal to zero, we get

$$\Sigma F_v = 0 = R_v^{(1)} - 20(2) - 40 + 85 - 40$$

therefore
$$R_v^{(1)} = 35 \text{ kN}$$

The plus sign denotes that the reactions $R_v^{(1)}$ and $R_v^{(5)}$ are in the direction shown in Fig. b. To check the algebra involved in calculating the reactions, we may consider the sum of the moments about another point, say point 5. Thus

$$\Sigma M^{(5)} = 0 = 8R_v^{(1)} - 20(2)(7) - 40(4) + 80 + 2(40) \qquad check$$

We now proceed to express the shearing force and the bending moment in the beam as functions of x_1 by considering the equilibrium of segments of the beam in which the shearing force and bending moment are continuous functions of x_1. The free-body diagrams of such segments of the beam are shown in Fig. c. To maintain these segments in equilibrium, an internal shearing force Q and an internal bending moment M must act on the cross section where the beam is cut. In the free-body diagrams of Fig. c, the internal actions are shown as assumed positive.

Considering the equilibrium of the beam segment of length x_1 $(0 \leq x_1 \leq 2)$ shown in Fig. ca, we obtain

$$\Sigma F_v = 0 = 35 - 20x_1 - Q \qquad \text{or} \qquad Q = 35 - 20x_1 \text{ kN} \qquad (a)$$

$$\Sigma M^{(0)} = 0 = 35x_1 - 20x_1\left(\frac{x_1}{2}\right) - M \qquad \text{or} \qquad M = 35x_1 - 10x_1^2 \text{ kN·m} \qquad (b)$$

Considering the equilibrium of the beam segment of length x_1 $(2 \leq x_1 < 4)$ shown in Fig. cb, we obtain

$$\Sigma F_v = 0 = 35 - 20(2) - Q \qquad \text{or} \qquad Q = -50 \text{ kN} \qquad (c)$$

$$\Sigma M^{(0)} = 0 = 35x_1 - 20(2)(x_1 - 1) - M \qquad \text{or} \qquad M = -5x_1 + 40 \text{ kN·m} \qquad (d)$$

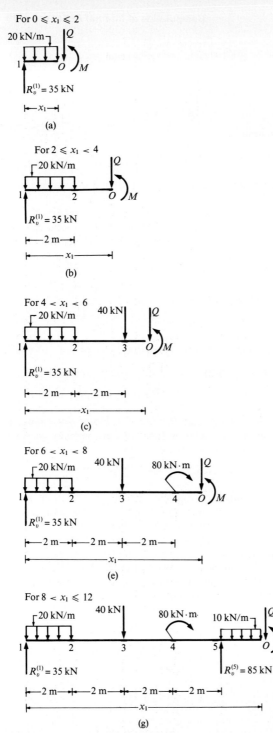

For $0 \leqslant x_1 \leqslant 2$

20 kN/m

$R_v^{(1)} = 35$ kN

x_1

(a)

For $2 \leqslant x_1 < 4$

20 kN/m

$R_v^{(1)} = 35$ kN

2 m

x_1

(b)

For $4 < x_1 < 6$

20 kN/m   40 kN

$R_v^{(1)} = 35$ kN

2 m   2 m

x_1

(c)

For $6 < x_1 < 8$

20 kN/m   40 kN   80 kN·m

$R_v^{(1)} = 35$ kN

2 m   2 m   2 m

x_1

(e)

For $8 < x_1 \leqslant 12$

20 kN/m   40 kN   80 kN·m   10 kN/m

$R_v^{(1)} = 35$ kN $R_v^{(5)} = 85$ kN

2 m   2 m   2 m   2 m

x_1

(g)

40 kN

5 kN   45 kN

3

(d) Free-body diagram of an infinitesimal segment containing point 3.

80 kN·m

70 kN·m   10 kN·m

4

(f) Free-body diagram of an infinitesimal segment containing point 4.

Figure c Free-body diagram of segments of the beam.

Considering the equilibrium of the beam segment of length x_1 $(4 < x_1 < 6)$ shown in Fig. cc, we have

$$\Sigma F_v = 0 = 35 - 20(2) - 40 - Q \quad \text{or} \quad Q = -45 \text{ kN} \qquad (e)$$

$$\Sigma M^{(0)} = 0 = 35x_1 - 20(2)(x_1 - 1)$$
$$- 40(x_1 - 4) - M \quad \text{or} \quad M = -45x_1 + 200 \quad \text{kN} \cdot \text{m} \qquad (f)$$

Considering the equilibrium of the beam segment of length x_1 $(6 < x_1 \leq 12)$ shown in Fig. cd, we obtain

$$\Sigma F_v = 0 = 35 - 20(2) - 40 - Q \quad \text{or} \quad Q = -45 \text{ kN} \qquad (g)$$

$$\Sigma M^{(0)} = 0 = 35x_1 - 20(2)(x_1 - 1)$$
$$- 40(x_1 - 4) + 80 - M \quad \text{or} \quad M = -45x_1 + 280 \quad \text{kN} \cdot \text{m} \qquad (h)$$

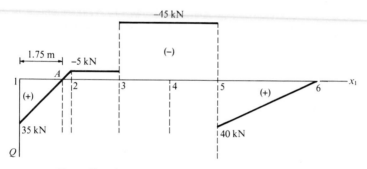

Figure d Shear diagram.

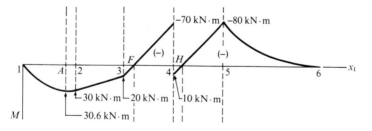

Figure e Moment diagram.

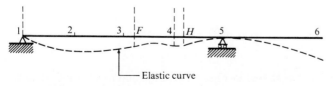

Figure f Sketch of the elastic curve of the beam.

Considering the equilibrium of the beam segment of length x_1 ($8 < x_1 \leq 12$) shown in Fig. ce, we obtain

$$\Sigma F_v = 0 = 35 - 20(2) - 40 + 85 - 10(x_1 - 8) - Q$$
$$\text{or} \quad Q = 120 - 10x_1 \quad \text{kN} \tag{i}$$

$$\Sigma M^{(0)} = 0 = 35x_1 - 20(2)(x_1 - 1) - 40(x_1 - 4) + 80$$
$$+ 85(x_1 - 8) - \frac{10(x_1 - 8)^2}{2} - M$$

$$\text{or} \qquad M = \frac{10x_1^2}{2} + 120x_1 - 720 \quad \text{kN·m} \tag{j}$$

The shear and moment diagrams are plotted in Figs. d and e, respectively. The shear diagram is a graphical representation of relations (a), (c), (e), (g), and (i), while the moment diagram is a graphical representation of relations (b), (d), (f), (h), and (j).

The sketch of the elastic curve of the beam is drawn in Fig. f by referring to its moment diagram (Fig. e). From point 1 to F and from point 4 to H, the moment is positive. That is, between these points the bottom fibers of the beam elongate, while its upper fibers shorten. From point F to 4 and from point H to 6, the moment is negative. That is, between these points, the bottom fibers of the beam shorten while the upper fibers elongate. Thus, the centers of curvature of the elastic curve from point 1 to F and from point 4 to H are located above the beam, while those of the elastic curve from points F to 4 and from point H to 6 are located below the beam. Hence, points F, 4, and H are inflection points of the elastic curve. That is, at these points the sign of the curvature of the elastic curve changes.

3.3.2   Shear and moment diagrams for beams and frames by the summation method

The method of plotting shear and moment diagrams presented in the previous section is time-consuming. For this reason, a more convenient method is discussed in this section. In this method, the relation between the external force and the shearing force (A.56), and the relation between the shearing force and the bending moment (A.59) derived in Sec. A.6 are employed. These relations have been obtained by considering the equilibrium of an infinitesimal segment of a member. For a member of a planar structure in the x_1x_3 plane they are

$$q = -\frac{dQ}{dx_1} \tag{3.11}$$

$$Q = \frac{dM}{dx_1} \tag{3.12}$$

Consider the planar beam shown in Fig. 3.10, subjected to external actions. If we integrate Eq. (3.11) over an arbitrary interval of the length of the beam from a point $x_1 = a$ to a point $x_1 = b$, we obtain

$$\int_a^b dQ = -\int_a^b q\, dx_1 \tag{3.13}$$

Referring to Fig. 3.10a, we can easily see that the quantity $q\, dx_1$ represents the cross-hatched portion of the area under the diagram of external forces. Thus, the right-hand integral of Eq. (3.13) is equal to the area under the diagram of external forces from point $x_1 = a$ to point $x_1 = b$. If the shearing force Q is a continuous function of x_1 between points $x_1 = a$ and $x_1 = b$, Eq. (3.13) gives

$$Q^{(b)} - Q^{(a)} = -(\text{the area under the diagram of external forces from}$$

$$x_1 = a \text{ to } x_1 = b) \tag{3.14}$$

The shearing force is discontinuous only at the points of a member where a concentrated external force acts (see Fig. 3.7b). Thus, if we know the shearing force $Q^{(a)}$ at a point $x_1 = a$ of a member and if a concentrated force does not act on the member in the interval from $x_1 = a$ to $x_1 = b$, then from Eq. (3.14), we can find directly the shearing force $Q^{(b)}$ at point $x_1 = b$ of the member. Integrating Eq. (3.12) from $x_1 = a$ to $x_1 = b$, we get

$$\int_a^b dM = \int_a^b Q\, dx_1 \tag{3.15}$$

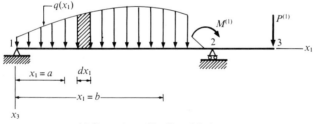

(a) Geometry and loading of the beam

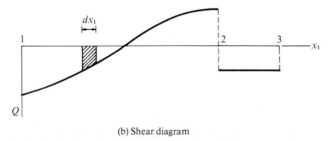

(b) Shear diagram

Figure 3.10 Loaded beam and its shear diagram.

Referring to Fig. 3.10b, we can easily see that the quantity $Q\, dx_1$ represents the cross-hatched portion of the area under the shear diagram. Thus, the integral at the right-hand side of Eq. (3.15) is equal to the area under the shear diagram from point $x_1 = a$ to $x_1 = b$. If the moment M is a continuous function of x_1 between points $x_1 = a$ and $x_1 = b$, relation (3.15) gives

$$M^{(b)} - M^{(a)} = \text{(area under the shear diagram from}$$
$$x_1 = a \text{ to } x_1 = b) \tag{3.16}$$

The moment is discontinuous only at the points of a member where a concentrated external moment acts (see Fig. 3.7c). Thus, if we know the moment $M^{(a)}$ at any point $x_1 = a$ of a member, and if a concentrated moment does not act on the member in the interval from $x_1 = a$ to $x_1 = b$, we can find directly from Eq. (3.16), the moment $M^{(b)}$ at point $x_1 = b$ of the member.

On the basis of the foregoing, it is apparent that the shear and moment diagrams for a member can be plotted starting from the left end of the member where the shear and moment are known and proceeding to the right using relations (3.14) and (3.16) respectively. This method of plotting the shear and moment diagrams is referred to as the *summation method*.

In the sequel we illustrate the analysis of statically determinate beams and frames and the plotting of their shear and moment diagrams by the summation method using the following examples.

Example 1.   The reactions of a simply supported beam with an overhang are established, and its shear and moment diagrams are plotted. The loading of the beam includes a concentrated force and a concentrated moment, as well as uniformly distributed forces.

Example 2.   The reactions of a simply supported, inclined beam are established, and its axial force, its shear, and its moment diagrams are plotted. The loading of the beam consists of a uniformly distributed force of 10 kN per unit length along the horizontal direction.

Example 3.   The reactions of a three-hinge frame are established, and its shear and moment diagrams are plotted. The elastic curve of the frame is sketched.

Example 4.   The reactions of a compound beam are established and its shear and moment diagrams are plotted. The elastic curve of the beam is also sketched.

Example 1.   Using the summation method [Eqs. (3.14) and (3.16)], plot the shear and moment diagrams for the beam shown in Fig. a.

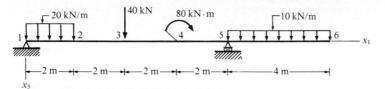

Figure a Geometry and loading of the beam.

solution First, we compute the reactions of the beam. This was done in the example of Sec. 3.3.1, and the results are shown in Fig. b.

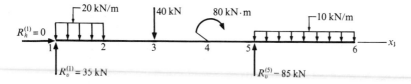

Figure b Free-body diagram of the beam.

Referring to Fig. b, we know that at point 1 the shearing force is 35 kN. Using Eq. (3.14), the shearing force at point 2 is smaller than that at point 1 by 40 kN, which is the area under the diagram of external forces from point 1 to point 2; that is, the shearing force at point 2 is equal to $35 - 40 = -5$ kN. The shear diagram from point 1 to point 2 is a straight line inasmuch as the external force is constant ($dQ/dx_1 = -q = -20$ kN/m). From point 2 to point 3, the shearing force remains constant since the area under the diagram of external forces is zero. At point 3, there is a discontinuity in the shear diagram equal to the external force of 40 kN. The shearing force changes from -5 kN to the left of point 3 to -45 kN to the right of point 3. This may be understood by considering the free-body diagram of a segment of the beam which includes point 3, shown in Fig. c. On the left face of this segment, a shearing force of 5 kN acts downward. It is apparent that an upward force of 45 kN must act on its right face in order for the segment to be in equilibrium. From point 3 to point 5, the shearing force remains constant and equal to -45 kN. At point 5, another discontinuity of the shear diagram equal to the reaction of 85 kN occurs. The shearing force changes from -45 kN at the left of point 5 to 40 kN at the right of point 5. The shearing force varies linearly from point 5 to point 6. It decreases by an amount equal to the area under the diagram of external forces, that is, by $(10)(4) = 40$ kN. Thus, the value of the shearing force at point 6 is equal to zero. This, in fact, is a check of the algebra, since by inspection it can be seen that the shearing force at

Figure c Free-body diagram of an infinitesimal segment of the beam containing point 3.

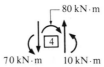

Figure d Free-body diagram of an infinitesimal segment of the beam containing point 4.

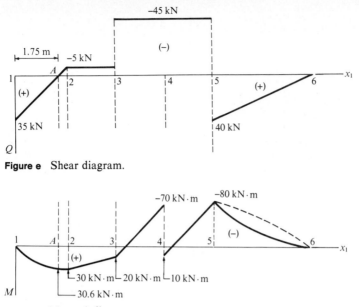

Figure e Shear diagram.

Figure f Moment diagram.

the free end of the beam (point 6) must vanish. The shear diagram is plotted in Fig. e. Notice that the shearing force Q vanishes at point A, located between points 1 and 2. Referring to Eq. (3.12), we can infer that at this point $dM/dx_1 = 0$. Consequently, at this point, $M(x_1)$ assumes an extremum value which is maximum in this case. The position of point A may be established from the similarity of the two triangles under the shear diagram between points 1 and A, and points A and 2. That is, the distance $\hat{x}_1$ from point 1 to point A is equal to 1.75 m $[\hat{x}_1/35 = (2 - \hat{x}_1)/5]$. Thus the area under the shear diagram from point 1 to A is equal to $(35)(1.75/2) = 30.6$ kN·m, whereas the area under the shear diagram from point A to point 2 is equal to $5(0.25/2) = 0.6$ kN·m.

Referring to Fig. b, we see that the moment at point 1 is zero. Therefore, using Eq. (3.16) at point A, the moment is equal to 30.6 kN·m, which is the area under the shear diagram from point 1 to point A. At point 2, the moment is equal to the moment at point A plus the area under the shear diagram from point A to point 2, which is equal to -0.6 kN·m. Therefore, the moment at point 2 is $30.6 - 0.6 = 30$ kN·m. From point 1 to point 2, the moment diagram is a second-degree curve because the shear diagram in this interval is a linear function of $x_1(dM/dx_1 = Q)$. Inasmuch as Q becomes less positive as x_1 increases, the slope of the moment diagram must decrease with x_1. The moment at point 3 differs from the moment at point 2 by -10 kN·m, which is the area under the shear diagram from point 2 to point 3. That is, the moment at point 3 is 20 kN·m. Since $Q = dM/dx_1$ is constant from point 2 to point 3, the moment diagram between these two points is a straight line. Similarly, the moment just to the left of point 4 differs from the moment at point 3 by -90 kN·m, which is the area under the shear diagram from point 3 to point 4. Therefore, the moment just to the left of point 4 is -70 kN·m. At point 4, there is a discontinuity in the moment diagram of 80 kN·m. The moment changes from -70 kN·m just to the left of point 4 to 10 kN·m just to the right of point 4. This becomes clear by considering the equilibrium of the beam segment which includes point 4 (see Fig. d). A counterclockwise moment of 70 kN·m acts on the left face of the segment. For the equilibrium of the segment, a counterclockwise moment of 10 kN·m must act on its right face. From point 4 to point 5, the shearing force is constant. Consequently, the moment diagram is a straight line. The moment at point 5 is equal to the moment just to the right of point 4, plus the area under the shear diagram from point 4 to point 5. This area is equal to −90 kN·m; that is, the moment at point 5 is equal to −80 kN·m (10 − 90 = −80 kN·m). From point 5 to point 6, the moment diagram is a second-degree curve since its slope $(dM/dx_1 = Q)$ is a linear function of x_1. The value of the moment at point 6 differs from that at point 5 by 80 kN·m, which is the area under the shear diagram from point 5 to point 6. That is, the moment at point 6 is zero. This concurs with the actual moment at the free end of the beam and is a partial check of the algebra performed in plotting the moment diagram.

The moment diagram is plotted in Fig. f. Notice that, inasmuch as Q becomes less positive as x_1 increases, the slope of the moment diagram $(dM/dx_1 = Q)$ from point 5 to point 6 must decrease with x_1. Thus, the full line curve in Fig. f is part of the moment diagram inasmuch as its slope decreases with x_1, whereas the slope of the broken line curve increases with x_1.

Example 2. Using the summation method [Eqs. (3.14) and (3.16)], plot the shear and moment diagrams for the beam shown in Fig. a.

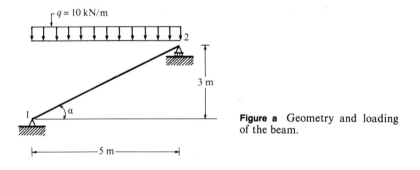

Figure a Geometry and loading of the beam.

solution First the supports are examined in order to establish the reactions which they can transmit to the beam. At support 1, the reaction can act in any direction required for the equilibrium of the beam. Thus, the reaction at this support can have two independent components, an axial and a transverse. At support 2, the reaction must be vertical, and thus, as shown in Fig. b, its axial and transverse components are not independent.

The load is given in kilonewtons per unit length in the horizontal direction. We convert this load to kilonewtons per meter in the direction along the length of the beam. Thus, we have

$$q_v^{(e)}(5.83) = q(5)$$

or

$$q_v^{(e)} = \frac{10(5)}{5.83} = 8.576 \text{ kN/m} \tag{a}$$

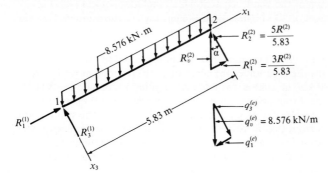

Figure b   Free-body diagram of the beam.

Moreover, we compute the axial and transverse components of the load. Thus,

$$q_1^{(e)} = -8.576 \left(\frac{3}{5.83}\right) = 4.413 \text{ kN/m} \qquad (b)$$

$$q_3^{(e)} = 8.576 \left(\frac{5}{5.83}\right) = 7.355 \text{ kN/m} \qquad (c)$$

In order to compute the reactions of the beam we refer to its free-body diagram, shown in Fig. b. Considering the equilibrium of the beam, we have

$$\Sigma M^{(2)} = 0 = 5.83 R_3^{(1)} - \frac{7.355(5.83)^2}{2}$$

$$\Sigma M^{(1)} = 0 = 5.83 R_3^{(2)} - \frac{7.355(5.83)^2}{2}$$

Consequently $R_3^{(1)} = 21.4 \text{ kN} \qquad R_3^{(2)} = 21.4 \text{ kN}$

Moreover $R_1^{(2)} = \frac{3}{5.83} R_v^{(2)} = \frac{3}{5} R_3^{(2)} = \frac{3}{5}(21.4) = 12.9 \text{ kN}$

and $\Sigma F_1 = 0 = R_1^{(1)} - 4.413(5.83) + R_1^{(2)} = 0$

or $R_1^{(1)} = 12.9 \text{ kN}$

The above results could be obtained in a simpler manner by noting that, if we disregard the effect of the change of length of the beam due to the loading, the distribution of the internal actions in the beam under consideration is identical to that in the same beam pinned to rigid supports at both ends. This beam has a center of symmetry (the midpoint of its span) (see Fig. cb) and, moreover, the loading is antisymmetric with respect to this center. Thus, as discussed in Chap. 2, the reactions of the beam are, as shown in Fig. cb, antisymmetric. Hence,

$$R_v^{(1)} = R_v^{(2)} = q_v^{(e)} \frac{5.83}{2} = 8.576 \left(\frac{5.83}{2}\right) = 25 \text{ kN}$$

Consequently $$R_3^{(1)} = R_3^{(2)} = 25 \left(\frac{5}{5.83} \right) = 21.4 \text{ kN}$$

and $$R_1^{(1)} = R_1^{(2)} = 25 \left(\frac{3}{5.83} \right) = 12.9 \text{ kN}$$

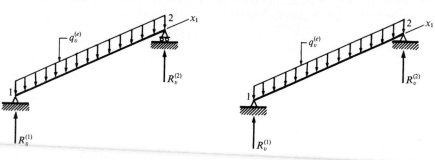

(a) Unsymmetric beam   (b) Symmetric beam

Figure c Equivalent symmetric beam.

We proceed to plot the shear and moment diagrams. At $x_1 = 0$ the shearing force is equal to

$$Q^{(1)} = R_3^{(1)} = 21.4 \text{ kN}$$

From Eq. (3.12), we have

$$\frac{dQ}{dx_1} = -q_3^{(e)} = -7.355 \text{ kN}$$

Thus, the shear diagram is a straight line. At any point x_1, the shearing force Q is equal to the shearing force at point 1, minus the area under the load diagram, (7.355 x_1). Thus, at point 2 the shearing force is equal to

$$Q^{(2)} = 21.40 - 7.355(5.83) = -21.4 \text{ kN} = -R_3^{(2)}$$

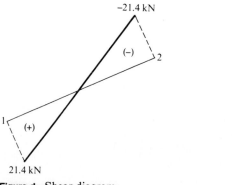

−21.4 kN

(−)   2

1

(+)

21.4 kN

Figure d Shear diagram.

2

1   (+)

31.34 kN·m

Figure e Moment diagram.

The shear diagram is plotted in Fig. d.

Since the shear diagram is a straight line, the moment diagram is a second-degree curve $(dM/dx_1 = Q)$. The moment at point 1 is equal to zero. As x_1 increases, the slope of the moment diagram must decrease because the shearing force decreases. At the midspan, the moment assumes its maximum value because the shearing force vanishes $(dM/dx_1 = Q = 0)$. At this point, the moment is equal to the moment at point 1 plus the area under the shear diagram from point 1 to the midspan. Thus,

$$M_{\text{at midspan}} = 0 + 21.4 \left(\frac{5.83}{4} \right) = 31.3 \text{ kN·m}$$

At point 2, the moment is equal to the moment at the midspan, minus the area under the shear diagram from the midspan to point 2. Thus,

$$M^{(2)} = 31.3 - 21.4 \left(\frac{5.83}{2} \right) = 0$$

This result was anticipated inasmuch as the moment at a pinned support vanishes. The moment diagram is plotted in Fig. e.

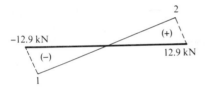

Figure f Axial force diagram.

The axial force at any point x_1 is equal to

$$N = -R_1^{(1)} + q_1^{(e)}x_1 = -12.9 + 4.413x_1$$

It is plotted in Fig. f.

Example 3.   Using the summation method, plot the moment and shear diagram of the frame of Fig. a and sketch its elastic curve.

solution   First we compute the reactions of the frame by referring to its free-body diagram shown in Fig. b and considering its equilibrium. Thus,

$$\Sigma M^{(1)} = 8R_v^{(5)} - 20(8)(4) - 40(4)$$

$$\Sigma M^{(5)} = 0 = 8R_v^{(1)} + 40(4) - 20(8)(4)$$

Consequently,

$$R_v^{(5)} = 100 \text{ kN} \qquad R_v^{(1)} = 60 \text{ kN}$$

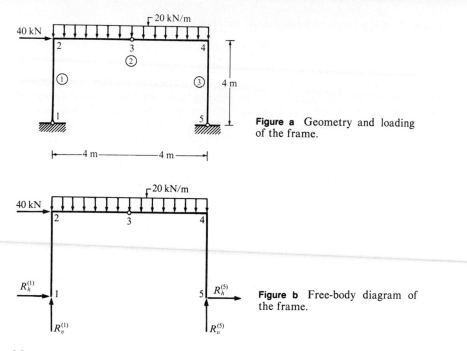

Figure a Geometry and loading of the frame.

Figure b Free-body diagram of the frame.

Moreover,

$$\Sigma \bar{F}_v = 0 = R_v^{(1)} + R_v^{(5)} - 20(8) = 100 + 60 - 160 = 0 \ \ check$$

$$\Sigma \bar{F}_h = 0 = 40 + R_h^{(1)} - R_h^{(5)} \tag{a}$$

To compute the horizontal component of the reactions at the supports, we use the condition imposed by the internal action release mechanism (hinge) at point 3. Thus, we consider the equilibrium of the portion 1,2,3 of the frame, whose free-body diagram is shown in Fig. c. Hence,

$$\Sigma M^{(3)} = 0 = 60(4) - 4R_h^{(1)} - 20(4)(2)$$

or $$R_h^{(1)} = 20 \ kN$$

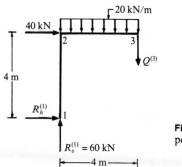

Figure c Free-body diagram of a portion of the frame.

Substituting this result into relation (a) we obtain

$$R_h^{(5)} = -60 \text{ kN}$$

The results are shown in Fig. d. The shear diagram is plotted in Fig. e. Referring to Fig. d, we note that at support 1, the shearing force is equal to -20 kN. Since member 1 is not subjected to external forces, the shearing force remains constant throughout its length $(dQ/dx_1 = 0)$. The shearing force acting at the left end of member 2 is equal to 60 kN. Since the external force acting on this member is constant, the shear diagram is a straight line $(dQ/dx_1 = -20)$. The shearing force decreases by an amount equal to the area under the diagram of external forces, that is by $20(8) = 160$ kN. Thus, the shearing force at the right end of member 2 is equal to -100 kN. From geometric considerations, referring to Fig. e, we can establish that the shearing force vanishes at point A, located 3 m $[\hat{x}_1/60 = (8 - \hat{x}_1)/100]$ from joint 2. The shearing force at the upper end of member 3 is equal to 60 kN. Since this member is not subjected to external forces, the shearing force is constant throughout its length.

The moment diagram is plotted in Fig. f. The moment at joint 1 is zero. The moment diagram for member 1 is a straight line because $dM/dx_1 = Q = -20$. The moment at joint 2 is equal to the area under the shear diagram from joint 1 to joint 2. Thus, referring to Fig. e, the moment at joint 2 is equal to -80 kN·m $[-20(4) = -80]$. Since the shearing force varies linearly along the length of member 2, the moment diagram is a second-degree curve. This curve has a maximum at point A (3 m from joint 2) where the shearing force vanishes $(dM/dx_1 = 0)$. The value of the moment at point A is equal to the moment at joint 2 $(-80$ kN·m$)$ plus the area under the shear diagram from joint 2 to point A $[60(3)/2 = 90]$. Thus, the moment at point A is equal to 10 kN·m. From point A to point 4, the moment decreases by an amount equal to the area under the shear diagram from point A to joint 4 $[(100(5)/2)| = 250]$. Thus, the moment at joint 4 is equal to -240 kN·m. The moment in member 3 increases from -240 kN·m at joint 4 to zero at joint 5. This increase is equal to the area under the shear diagram from joint 4 to joint 5.

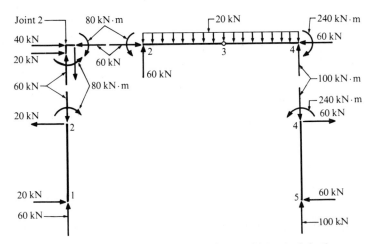

Figure d Free-body diagrams of the members and joint 2 of the frame.

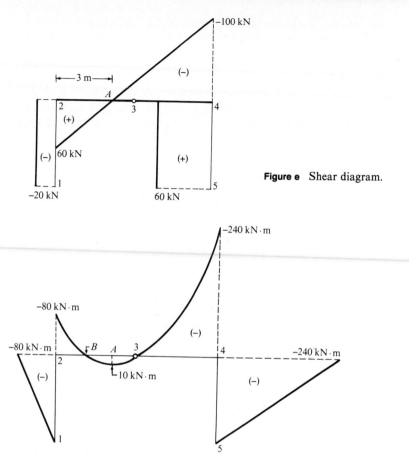

Figure e   Shear diagram.

Figure f   Moment diagram.

The sketch of the elastic curve of the frame is drawn in Fig. g by referring to its moment diagram (Fig. f). From point 1 to point B and from point 3 to point 5 the moment is negative. That is between these points, the interior fibers of the members of the frame have shortened while their exterior fibers have elongated. From point B to point 3 the moment is positive. That is between these points the interior fibers of member 2 have elongated while its exterior fibers have shortened. Thus, point B is an

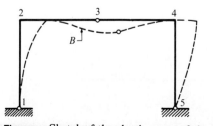

Figure g   Sketch of the elastic curve of the frame.

inflection point, while at point 3 the curvature of the elastic curve changes sign. Notice that because of the presence of the hinge at point 3, the slope of the elastic curve at this point is not continuous.

Example 4. Determine the reactions at the supports of the three-span beam with suspended center span shown in Fig. a. Plot its shear and moment diagrams, and sketch its elastic curve.

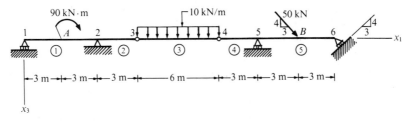

Figure a Geometry and loading of the beam.

solution This is a statically determinate compound structure. Its free-body diagram is shown in Fig. b. There are five unknown reactions. They are computed referring to the free-body diagrams of the parts of the beam shown in Fig. c. In drawing these diagrams, we took into account that the internal moments at the two hinges vanish.

Notice, that by Newton's law of action and reaction, once the directions of the internal actions at an end, say j of a member of a continuous beam are assumed, the internal actions at the end k of the adjacent member of the beam must be equal and opposite.

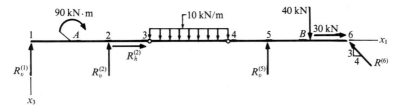

Figure b Free-body diagram of the beam.

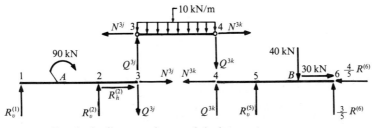

Figure c Free-body diagram of parts of the beam.

Referring to Fig. c, from the equilibrium of member 3 we obtain

$$\Sigma \overline{F}_h = 0 \qquad\qquad N^{3j} = N^{3k}$$

$$\Sigma M^{(4)} = 0 \qquad 6Q^{3k} + 10(6)(3) = 0 \qquad\qquad\text{(a)}$$

$$\Sigma M^{(3)} = 0 \qquad 6Q^{3j} - 10(6)(3) = 0$$

Thus,
$$Q^{3k} = -30 \text{ kN} \qquad\qquad\qquad\text{(b)}$$

$$Q^{3j} = 30 \text{ kN} \qquad\qquad\qquad\text{(c)}$$

To check the algebra involved in calculating the shearing forces Q^{3j} and Q^{3k}, we consider the sum of the forces in the vertical direction. Thus,

$$\Sigma \overline{F}_v = -Q^{3j} + 60 + Q^{3k} = -30 + 60 - 30 = 0 \qquad check$$

Referring to Fig. c, from the equilibrium of the portion of the beam from joint 4 to joint 6, we get

$$\Sigma M^{(6)} = (-30)(9) + 6R_v^{(5)} - 40(3) = 0$$

$$\Sigma M^{(5)} = -30(3) + 40(3) - (6)\tfrac{3}{5} R^{(6)}$$

$$\Sigma \overline{F}_h = N^{3k} - 30 + \tfrac{4}{5}R^{(6)} = N^{3k} - 30 + \tfrac{20}{3}$$

Consequently

$$R_v^{(5)} = 65 \text{ kN} \qquad R^{(6)} = \tfrac{25}{3} = 8.33 \text{ kN} \qquad N^{3k} = \tfrac{70}{3} = 23.33 \text{ kN}$$

To check the algebra in calculating the reactions $R_v^{(5)}$ and $R^{(6)}$, we consider the sum of the forces in the vertical direction. Thus,

$$\Sigma \overline{F}_v = -30 + 65 - 40 + 5 = 0 \qquad check$$

Referring to Fig. c, from the equilibrium of the portion of the beam from joint 1 to joint 3, and using result (a), we have

$$\Sigma \overline{F}_h = R_h^{(2)} + N^{3j} = R_h^{(2)} + 23.33 = 0$$

$$\Sigma M^{(1)} = 9Q^{3j} - 6R_v^2 + 90 = 0 \qquad\qquad\text{(d)}$$

or
$$\Sigma M^{(2)} = 3Q^{3j} + 90 + 6R_v^{(1)} = 0$$

Thus,

$$R_h^{(2)} = -23.33 \text{ kN} \qquad R_v^{(2)} = 60 \text{ kN} \qquad R_v^{(1)} = -30 \text{ kN}$$

moreover
$$\Sigma \overline{F}_v = 0 = -30 + 60 - 30 = 0 \qquad check$$

The minus sign in the result for $R_h^{(2)}$ indicates that this reaction acts in a direction opposite to the assumed one. The results are shown in Fig. d.

The shear diagram is plotted in Fig. e. Referring to Fig. d, we see that the shearing force at point 1 is equal to -30 kN and remains constant throughout the length of member 1. At support 2 there is a jump in the value of the shearing force equal to the

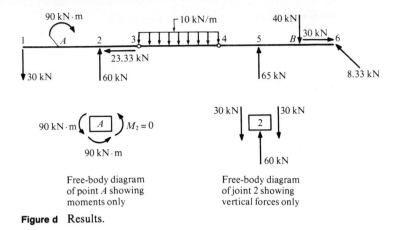

Free-body diagram
of point A showing
moments only

Free-body diagram
of joint 2 showing
vertical forces only

Figure d Results.

reaction $R_v^{(2)} = 60$ kN. Thus, to the left of support 2, the shearing force is equal to -30 kN, while to the right it is equal to 30 kN. This becomes apparent by considering the free-body diagram of joint 2 shown in Fig. d. The shearing force in member 2 is constant. At point 3 of member 3, the value of the shearing force is 30 kN. From point 3 to point 4, the shearing force decreases by an amount equal to the area under the diagram of external forces [10(6) = 60]. Thus, at point 4 the shearing force is equal to -30 kN. The shearing force in member 4 is constant and equal to -30 kN. At support 5, the shearing force changes by an amount equal to the reaction $R_v^{(5)} = 65$ kN. That is, to the left of support 5, the shearing force is equal to -30 kN, while to the right it is equal to 35 kN. The shearing force remains constant from support 5 to point B, where it decreases by an amount equal to the vertical component of the external force at point B. Thus, to the left of point B it is equal to 35 kN, while to the right of point B it is equal to -5 kN. From point B to support 6, the shearing force remains constant and it is equal to -5 kN. As anticipated, this value is equal to the vertical component of the reaction at support 6 [(25/3)(3/5) = 5]. If it were not equal, it would have been an indication of an error in the arithmetic involved in plotting the shear diagram.

The moment diagram is plotted in Fig. f. Referring to Fig. d, we see that the moment at support 1 is equal to zero. From support 1 to point A, the moment decreases by an amount equal to the area under the shear diagram [30(3) = 90 kN $\cdot$ m]. Thus, the moment to the left of point A is equal to -90 kN $\cdot$ m. Referring to the free-body diagram of the infinitesimal segment of the beam at point A shown in Fig. d, we see that there is a jump of 90 kN $\cdot$ m in the internal moment at point A. Thus, the moment to the right of point A is equal to zero. The moment diagram from point A to joint 2 is a straight line because the shearing force is constant. The value of the moment at joint 2 is equal to the area under the shear diagram from point A to joint 2 [$-30(3) = -90$ kN $\cdot$ m]. The value of the moment at joint 3 is equal to the moment at joint 2 plus the area under the shear diagram from joint 2 to joint 3 [30(3) = 90 kN $\cdot$ m]. Thus, the moment at joint 3 is zero as it should be at a hinge. Since the shear diagram in member 3 varies linearly with the axial coordinate, the moment diagram must be a second-degree curve. Its maximum value occurs at the midspan where the shearing force vanishes, and it is equal to 30(3)/2 = 45 kN $\cdot$ m.

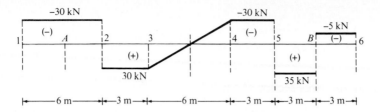

Figure e Shear diagram.

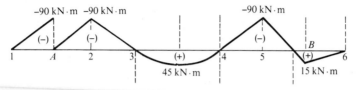

Figure f Moment diagram.

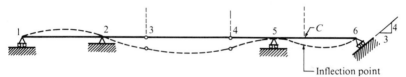

Figure g Sketch of the elastic curve of the beam.

The sketch of the elastic curve of the beam is drawn in Fig. g by referring to the moment diagram of Fig. f.

3.4 Analysis of Statically Determinate Arches

When a vertical load acts on a horizontal beam, it produces only vertical reactions at its supports. However, when a vertical load acts on an arch, it produces both horizontal and vertical reactions at its supports. The horizontal reactions produce a moment which tends to decrease the moment due to the vertical reactions; consequently, the resulting moment at any cross section of an arch is less than that in the corresponding cross section of a straight beam spanning the same length as the arch and subjected to the same loads (see Fig. 3.11a and b). Referring to the free-body diagrams of the arch and the beam shown in Fig. 3.11c and 3.11d, respectively, and setting the sum of the moments about points 1 and 3 equal to zero, we can readily see that the vertical components of the reactions of the arch are equal to those of the beam. The horizontal component of the reaction of the arch can be established by considering the equilibrium of its left half whose free-body diagram is shown in Fig. 3.11e.

That is

$$\Sigma M^{(2)} = 0 \qquad H = \frac{R_v^{(1)}L}{2h} = \frac{Pb}{2h} \tag{3.17}$$

The moment at any cross section of the beam is equal to

$$M^B(x_1) = R_v^{(1)}\bar{x}_1 \qquad 0 < x_1 < L - b \tag{3.18}$$

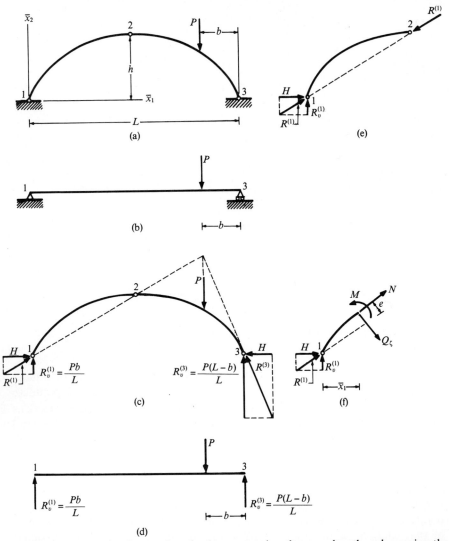

Figure 3.11 Analysis of an arch and a beam spanning the same length and carrying the same load.

Referring to Fig. 3-11f, notice that the moment at any cross section of the arch is equal to

$$M^A(\bar{x}_1) = R_v^{(1)}\bar{x}_1 - H\bar{x}_2 = R_e^{(1)} \qquad 0 < \bar{x}_1 < L - b \qquad (3.19)$$

Comparing Eqs. (3.18) and (3.19), we see that the moment in the arch of Fig. 3.11a is less than that of the beam of Fig. 3.11b by $H\bar{x}_2$. That is the horizontal reaction reduces the moment in the arch. Notice that since the three forces $R^{(1)}$, P, and $R^{(3)}$ acting on the arch are in equilibrium, as shown in Fig. 3.11c, their lines of action meet at one point. Moreover, since portion 1,2 of the arch is in equilibrium, the moment of the reaction $R^{(1)}$ about point 2 vanishes. Consequently, as shown in Fig. 3.11c and e, the line of action of R must pass through point 2. Thus, referring to Fig. 3.11f, notice that the moment of any point $\bar{x}_1$ of the arch is equal to the reaction $R^{(1)}$ times the distance e of the axis of the arch from the chord connecting the hinge at point 2 and support 1.

The analysis of statically determinate arches is illustrated by the following examples.

Example 1.   Find the internal actions in the semicircular, three-hinged arch shown in Fig. a, subjected to a concentrated force at its crown.

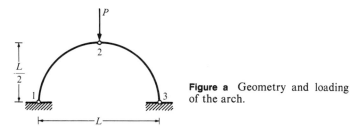

Figure a Geometry and loading of the arch.

---

solution   The free-body diagram of the arch is shown in Fig. b. Because of the symmetry of the structure and the loading, the vertical and horizontal components of the reactions $R^{(1)}$ and $R^{(3)}$ must form symmetric pairs. Thus, the vertical components of each reaction is equal to $P/2$.

Notice that in order for the three forces ($R^{(1)}$, P, and $R^{(3)}$) acting on the arch to be in equilibrium, they must meet at one point. Moreover, in order for portion 1,2 of the arch to be in equilibrium, the moment of the reaction $R^{(1)}$ about point 2 must vanish. Consequently, as shown in Fig. b, the line of action of $R^{(1)}$ must pass through point 2. Thus,

$$H = P/2$$

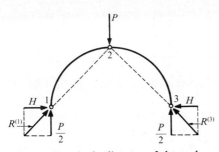

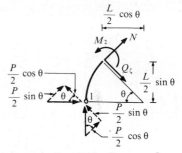

Figure b Free-body diagram of the arch.

Figure c Free-body diagram of a portion of the arch.

From the equilibrium of the portion of the arch shown in Fig. c, we have

$$\Sigma F_t = 0 = N + \frac{P}{2} \cos \theta + \frac{P}{2} \sin \theta$$

$$\Sigma F_\varsigma = 0 = Q_\varsigma - \frac{P}{2} \sin \theta + \frac{P}{2} \cos \theta$$

$$\Sigma M_2^{(\theta)} = 0 = M_2 + \frac{P}{2} \left(\frac{L}{2} \right) \sin \theta - \frac{P}{2} \left(\frac{L}{2} \right) (1 - \cos \theta)$$

Thus, for $0 < \theta < \pi/2$

$$N = -\frac{P}{2} (\cos \theta + \sin \theta)$$

$$Q_\varsigma = \frac{P}{2} (\sin \theta - \cos \theta)$$

$$M_2 = \frac{PL}{4} [1 - \sin \theta - \cos \theta]$$

Example 2. It is desired to design a parabolic, three-hinged arch to carry a highway over a river. It is required that the arch have an opening L. Moreover, its supports have a difference in elevation a. Furthermore, it is required that the height of the crown of the arch is h above its lower support. Find the internal actions in the arch when subjected to a uniform load, as shown in Fig. a.

solution The equation of the axis of the arch has the following form:

$$\bar{x}_2 = A\bar{x}_1^2 + B\bar{x}_1 \qquad (a)$$

The constants A and B can be obtained in terms of the geometric parameters of the arch by applying the following geometric conditions of the arch:

at $\qquad\qquad \bar{x}_1 = L \qquad \bar{x}_2 = a$

Hence $\qquad\qquad a = AL^2 + BL \qquad\qquad (b)$

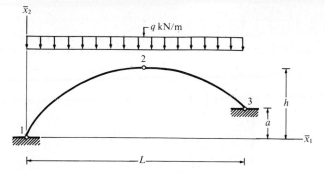

Figure a　Geometry and loading of the arch.

At the crown we have

$$\frac{d\bar{x}_2}{dx_1} = 0$$

Thus, denoting the abscissa of the crown by $\bar{x}_1 = L_1$, we obtain

$$L_1 = -\frac{B}{2A} \tag{c}$$

Moreover at the crown $x_2 = h$, hence

$$h = AL_1^2 + BL_1 \tag{d}$$

From relations (b) to (d) we get

$$B = \frac{2h}{L}\left(1 + \sqrt{1 - \frac{a}{h}}\right) \tag{e}$$

$$A = -\frac{h}{L^2}\left(1 + \sqrt{1 - \frac{a}{h}}\right)^2 \tag{f}$$

$$L_1 = \frac{L}{1 + \sqrt{1 - a/h}} \tag{g}$$

Referring to Fig. b, from the equilibrium of the arch we have

$$\Sigma M^{(3)} = 0 \qquad LR_v^{(1)} - \tfrac{1}{2}qL^2 - Ha = 0$$

$$\Sigma M^{(2)} = 0 \qquad L_1 R_v^{(1)} - \tfrac{1}{2}qL_1^2 - Hh = 0$$

or

$$R_v^{(1)} = \frac{qL}{2}\left(\frac{1 - L_1^2 a/L^2 h}{1 - L_1 a/Lh}\right) \tag{h}$$

$$H = \frac{qL_1 L}{2h}\left(\frac{1 - L_1/L}{1 - L_1 a/Lh}\right) \tag{i}$$

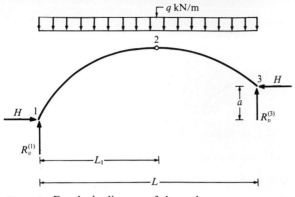

q kN/m

Figure b Free-body diagram of the arch.

In order to establish the internal actions at a cross section of the arch, we consider the equilibrium of its portion, shown in Fig. c. Thus,

$$\Sigma F_t = 0 \qquad N + H \cos \phi + R_v^{(1)} \sin \phi - q\bar{x}_1 \sin \phi = 0$$

$$\Sigma F_{\bar{s}} = 0 \qquad Q_{\bar{s}} + H \sin \phi - R_v^{(1)} \cos \phi + q\bar{x}_1 \cos \phi = 0 \qquad \text{(j)}$$

$$\Sigma M^{(0)} = 0 \qquad M + H\bar{x}_2 - R^{(1)}x_1 + \tfrac{1}{2}q\bar{x}_1^2 = 0$$

Substituting relations (h) and (i) into (j), we obtain

$$N = -\frac{qL}{2} \left\{ \frac{\dfrac{L_1}{h}\left(1 - \dfrac{L_1}{L}\right)}{\left(1 - \dfrac{L_1 a}{Lh}\right)} \cos \phi + \left[\frac{\left(1 - \dfrac{L_1^2 a}{L^2 h}\right)}{\left(1 - \dfrac{L_1 a}{Lh}\right)} - \frac{2\bar{x}_1}{L} \right] \sin \phi \right\}$$

$$Q_{\bar{s}} = \frac{qL}{2} \left\{ -\frac{\dfrac{L_1}{h}\left(1 - \dfrac{L_1}{L}\right)}{\left(1 - \dfrac{L_1 a}{Lh}\right)} \sin \phi + \left[\frac{1 - \dfrac{L_1^2 a}{L^2 h}}{\left(1 - \dfrac{L_1 a}{Lh}\right)} - \frac{2\bar{x}_1}{L} \right] \cos \phi \right\} \qquad \text{(k)}$$

$$M = \frac{qL^2}{2} \left\{ -\frac{\dfrac{L_1}{h}\left(1 - \dfrac{L_1}{L}\right)\bar{x}_2}{\left(1 - \dfrac{L_1 a}{Lh}\right)} + \left[\frac{\left(1 - \dfrac{L_1^2 a}{L^2 h}\right)}{\left(1 - \dfrac{L_1 a}{Lh}\right)} - \frac{\bar{x}_1}{L} \right] \bar{x}_1 \right\}$$

The angle ϕ is defined in Fig. c and is equal to

$$\tan \phi = \frac{d\bar{x}_2}{d\bar{x}_1} = 2A\bar{x}_1 + B$$

$$= -\frac{2h}{L}\left(1 + \sqrt{1 - \frac{a}{h}}\right)^2 \frac{\bar{x}_1}{L} + \frac{2h}{L}\left(1 + \sqrt{1 - \frac{a}{h}}\right) \qquad \text{(l)}$$

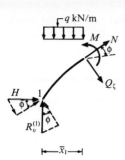

Figure c Free-body diagram of a portion of the arch.

For a symmetric arch ($a = 0$, $L_1 = \frac{1}{2}L$) Eqs. (a), (l), and (k) reduce to

$$\bar{x}_2 = 4h\left(\frac{\bar{x}_1}{L} - \frac{\bar{x}_1^2}{L^2}\right)$$

$$\tan \phi = 4h\left(1 - \frac{2\bar{x}_1}{L}\right)$$

$$N = -\frac{qL}{2}\left[\frac{L}{4h}\cos \phi + \left(1 - \frac{2\bar{x}_1}{L}\right)\sin \phi\right]$$

$$Q_\zeta = 0 \qquad M = 0$$

Thus, the shearing force and the moment at any point of a symmetric parabolic arch subjected to a uniform vertical load vanish. In general part of the load (the dead load) acting on an arch is uniform. Consequently this part of the load produces only an axial compressive force on the cross sections of the arch. The remaining load produces both an axial compressive force and a bending moment. The latter induces a distribution of compressive stresses on a portion of the cross section and a distribution of tensile stresses on the remaining cross section. If the maximum tensile stress induced by the bending moment on a cross section is equal to or less than the uniform compressive stress induced by the axial force, this cross section is subjected only to compressive stresses. Consequently, it is possible to design an arch whose particles under all loading conditions are subjected only to compressive stresses. This is desirable, particularly in concrete arches, since concrete can resist only a small amount of tensile stress before it cracks.

3.5 Analysis of Cables

As discussed in Sec. 1.4.1, cables are members of many engineering structures such as suspension bridges, cable-stayed bridges, cable roofs, radio antennas, etc. When a cable supports a force distributed uniformly along its length, as, for example, its own weight, it assumes the form of a catenary. However, the form of a cable subjected to its own weight may be approximated by a second-degree parabola when its sag is small compared to its length.

Consider the cable shown in Fig. 3.12 suspended between points 1 and 2. If

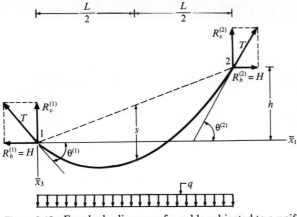

Figure 3.12 Free-body diagram of a cable subjected to a uniform load.

the length of this cable is large compared to its sag, its weight per unit horizontal length may be assumed uniform. Referring to Fig. 3.12, we have, from the equilibrium of the cable,

$$\Sigma \bar{F}_h = 0 \qquad R_h^{(1)} = R_h^{(2)} = H$$

$$\Sigma \bar{M}^{(2)} = 0 = Hh + LR_v^{(1)} - q\frac{L^2}{2}$$

or
$$R_v^{(1)} = \frac{qL}{2} - \frac{Hh}{L} \qquad (3.20)$$

From the free-body diagram of the portion of the cable shown in Fig. 3.13, we get

$$\Sigma \bar{M}^{(A)} = 0 = R_v^{(1)}\bar{x}_1 - H\bar{x}_3 - \frac{q\bar{x}_1^2}{2} \qquad (3.21)$$

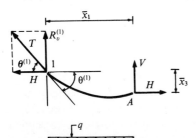

Figure 3.13 Free-body diagram of a portion of the cable.

Substituting relation (3.20) into (3.21), we obtain

$$H\left(\bar{x}_3 + \frac{\bar{x}_1 h}{L}\right) = \frac{q\bar{x}_1}{2}(L - \bar{x}_1) \tag{3.22}$$

Referring to Fig. 3.12, we can readily see that for

$$\bar{x}_1 = \tfrac{1}{2}L \qquad \bar{x}_3 = s - \tfrac{1}{2}h \tag{3.23}$$

Substituting relations (3.23) into (3.22), we obtain

$$H = \frac{qL^2}{8s} \tag{3.24}$$

Substituting relation (3.24) into (3.22), we get

$$\bar{x}_3 = \frac{4s\bar{x}_1(L - \bar{x}_1)}{L^2} - \frac{\bar{x}_1 h}{L} \tag{3.25}$$

This is the equation of a second-degree parabola. The slope of the cable is equal to

$$\frac{d\bar{x}_3}{d\bar{x}_1} = \frac{4s}{L}(L - 2\bar{x}_1) - \frac{h}{L} \tag{3.26}$$

The slope of the cable at its supports is

$$\tan \theta^{(1)} = \frac{d\bar{x}_3}{d\bar{x}_1}\bigg|_{\bar{x}_1 = 0} = \frac{4s}{L} - \frac{h}{L}$$
$$\tan \theta^{(2)} = \frac{d\bar{x}_3}{d\bar{x}_1}\bigg|_{\bar{x}_1 = L} = -\frac{4s}{L} - \frac{h}{L} \tag{3.27}$$

The same result may be obtained by referring to Fig. 3.12 and using relations (3.20) and (3.24). Thus

$$\tan \theta^{(1)} = \frac{(qL/2) - (Hh/L)}{H} = \frac{qL}{2H} - \frac{h}{L} = \frac{4s}{L} - \frac{h}{L} \tag{3.28}$$

The length of the cable L_c is equal to

$$L_c = \int dL_c = \int \sqrt{d\bar{x}_1{}^2 + d\bar{x}_3{}^2}$$

$$= 2\int_0^{L/2}\left[\sqrt{1 + \left(\frac{d\bar{x}_3}{d\bar{x}_1}\right)^2}\right]d\bar{x}_1 \tag{3.29}$$

Using the binomial expansion and retaining the first three terms, we see that relation (3.29) reduces to

$$L_c = 2 \int_0^{L/2} \left[1 + \frac{1}{2} \left(\frac{d\bar{x}_3}{d\bar{x}_1} \right)^2 - \frac{1}{8} \left(\frac{d\bar{x}_3}{d\bar{x}_1} \right)^4 \right] d\bar{x}_1 \qquad (3.30)$$

For cables subjected to a uniformly distributed force, relation (3.26) may be substituted into (3.30) to yield

$$
\begin{aligned}
L_c &= 2 \int_0^{L/2} \left\{ 1 + \frac{1}{2} \left[\frac{4s}{L^2} (L - 2\bar{x}_1 - \frac{h}{L} \right]^2 \right. \\
&\quad \left. - \frac{1}{8} \left[\frac{4s}{L^2} (L - 2\bar{x}_1) - \frac{h}{L} \right]^4 \right\} d\bar{x}_1 \\
&= L \left[1 - \frac{L}{24s} \left(\frac{h}{L} \right)^3 + \frac{L}{160s} \left(\frac{h}{L} \right)^5 \right. \\
&\quad \left. + \frac{L}{24s} \left(\frac{4s}{L} - \frac{h}{L} \right)^3 - \frac{L}{160s} \left(\frac{4s}{L} - \frac{h}{L} \right)^5 \right]
\end{aligned}
\qquad (3.31)
$$

If the ends of the cable are at the same elevation (see Fig. 3.14), relations (3.25), (3.26), and (3.31) reduce to

$$\bar{x}_3 = \frac{4s\bar{x}_1(L - \bar{x}_1)}{L^2} \qquad (3.32)$$

$$\frac{d\bar{x}_3}{d\bar{x}_1} = \frac{4s}{L^2} (L - 2\bar{x}_1) \qquad (3.33)$$

$$L_c = L \left[1 + \frac{8}{3} \left(\frac{s}{L} \right)^2 - \frac{32}{5} \left(\frac{s}{L} \right)^5 \right] \qquad (3.34)$$

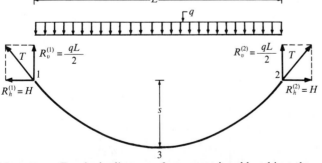

Figure 3.14   Free-body diagram of a symmetric cable subjected to a uniform load.

Example 1. A cable supporting a uniform load of 12 kN per horizontal meter is suspended between two points 80 m apart horizontally, and 20 m apart vertically. The tension in the cable is adjusted until the sag at the midspan is 6 m. Compute the maximum tension in the cable.

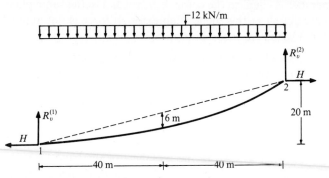

Figure a Geometry and loading of the cable.

solution Referring to relation (3.24), we have

$$H = \frac{qL^2}{8s} = \frac{12(80)^2}{8(6)} = 200 \text{ kN}$$

Considering the free-body diagram of the cable shown in Fig. a, we obtain

$$\Sigma M^{(1)} = 0 \qquad 80R_v^{(2)} - 20H - 12(80)(40) = 0$$

$$\Sigma M^{(2)} = 0 \qquad 80R_v^{(1)} + 20H - 12(80)(40) = 0$$

or $\qquad R_v^{(2)} = 530 \text{ kN} \qquad R_v^{(1)} = 430 \text{ kN}$

Thus $\qquad T_{max} = \sqrt{(530)^2 + (200)^2} = 566.48 \text{ kN}$

3.6 Analysis of Statically Determinate Space Trusses

Some space trusses can be analyzed by subdividing them into several planar trusses. However, as indicated in Sec. 1.4.2, there are trusses which must be analyzed as space trusses. In this section, we present an introduction to the analysis of such trusses. In general, the analysis of space trusses is considerably more difficult than the analysis of planar trusses, and the aid of a computer is desirable. For this reason, we limit our presentation to the analysis of very simple space trusses.

We can arrive at the following useful conclusions by considering the equilibrium of a joint of a space truss:

1. When only one (say the ith) of all members of a truss framing into a joint does not lie in the same plane, the component normal to that plane of the force in the ith member is equal to the component normal to the same plane of the external force acting at this joint. If an external force is not applied to the joint, the force in the ith member is zero.

2. When only two noncolinear members of a truss are framing into a joint and no external force acts on this joint, the forces in these members vanish.

In order to establish the reactions of a statically determinate space truss and the internal actions in its members, we do the following:

1. We choose a convenient set of global axes.

2. We compute the direction cosines of the members of the truss from the coordinates of its joints. That is, referring to Fig. 3.15, the direction cosines of a member of a space truss are equal to:

$$\cos \phi_{11} = \frac{\bar{x}_1^k - \bar{x}_1^j}{L} \qquad \cos \phi_{12} = \frac{\bar{x}_2^k - \bar{x}_2^j}{L} \qquad \cos \phi_{13} = \frac{\bar{x}_3^k - \bar{x}_3^j}{L} \qquad (3.35)$$

3. We compute the global components of the axial force acting in each member of the truss. That is, referring to Fig. 3.15, we have

$$\bar{F}_1^k = N \cos \phi_{11} \qquad \bar{F}_2^k = N \cos \phi_{12} \qquad \bar{F}_3^k = N \cos \phi_{13} \qquad (3.36)$$

4. We write the equations of equilibrium for each joint of the truss. That is,

$$\Sigma \bar{F}_1 = 0 \qquad \Sigma \bar{F}_2 = 0 \qquad \Sigma \bar{F}_3 = 0 \qquad (3.37)$$

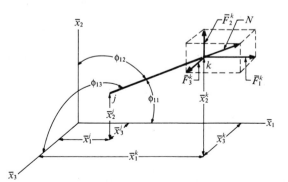

Figure 3.15 Global components of the axial force in a member of a space truss.

Thus, for a truss with n joints we write $3n$ equations of equilibrium. For a statically determinate truss, the number of these equations is equal to the sum of the unknown reactions and the unknown internal forces in its members.

5. We solve the system of linear, algebraic equations established in item 4. When there is a digital computer available, the solution of this system of equations can be readily obtained. When we perform the analysis of the truss by hand or with a desk calculator, we solve three equations of equilibrium of a joint at a time. Each time we consider the equations of equilibrium of a joint which involves only three unknown member forces. To accomplish this it may be necessary to compute first the reactions of the truss by considering its equilibrium as a whole. This, however, is not always necessary or possible.

The analysis of statically determinate space trusses is illustrated by the following two examples.

Example 1. The internal forces in the members of a tripod are established when it is subjected to an external force acting at its unsupported joint.

Example 2. The internal forces in the members of a space truss are established. The sides of the truss have variable slopes.

Example 1. Compute the internal forces in the members of the tripod shown in Fig. a. The force P is equal to

$$[P] = \begin{bmatrix} -20 \\ -40 \\ 10 \end{bmatrix} \text{ kN}$$

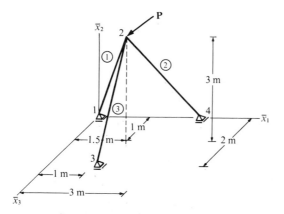

Figure a Geometry and loading of the tripod.

---

solution The internal forces in the members of the tripod can be computed by considering the equilibrium of joint 2. In order to write the equations of equilibrium for this joint, we adhere to the following steps.

1. We compute the length of the members of the tripod. Thus,

$$L_1 = L_2 = \sqrt{(1.5)^2 + (1)^2 + (3)^2} = 3.5 \text{ m}$$

$$L_3 = \sqrt{(1)^2 + (1)^2 + (3)^2} = 3.3166 \text{ m}$$

2. We compute the direction cosines of the members of the tripod. Hence,

$$\cos \phi_{11} = \frac{-1.5}{3.5} = -0.42857 \qquad \cos \phi_{12} = \frac{-3}{3.5} = -0.85714$$

$$\cos \phi_{13} = \frac{-1}{3.5} = -0.28571$$

$$\cos \phi_{21} = \frac{1.5}{3.5} = 0.42857 \qquad \cos \phi_{22} = \frac{-3}{3.5} = -0.85714$$

$$\cos \phi_{23} = \frac{-1}{3.5} = -0.28571$$

$$\cos \phi_{31} = \frac{-0.5}{3.3166} = -0.15076 \qquad \cos \phi_{32} = \frac{-3}{3.3166} = -0.90454$$

$$\cos \phi_{33} = \frac{1}{3.3166} = 0.30151$$

3. We find the global components of the axial forces in the members of the tripod. Thus

$$\overline{F}_1^{(1)} = -0.42857 N^{(1)} \qquad \overline{F}_1^{(2)} = 0.42857 N^{(2)} \qquad \overline{F}_1^{(3)} = -0.15076 N^{(3)}$$

$$\overline{F}_2^{(1)} = -0.85714 N^{(1)} \qquad \overline{F}_2^{(2)} = -0.85714 N^{(2)} \qquad \overline{F}_2^{(3)} = -0.9045 N^{(3)}$$

$$\overline{F}_3^{(1)} = 0.28571 N^{(1)} \qquad \overline{F}_3^{(2)} = -0.28571 N^{(2)} \qquad \overline{F}_3^{(3)} = 0.30151 N^{(3)}$$

4. We write the equations of equilibrium for joint 2. Hence,

$$\Sigma \overline{F}_1 = -0.42857 N^{(1)} + 0.42857 N^{(2)} - 0.15076 N^{(3)} - 20 = 0$$

$$\Sigma \overline{F}_2 = -0.8571 N^{(1)} - 0.85714 N^{(2)} - 0.90454 N^{(3)} - 40 = 0$$

$$\Sigma \overline{F}_3 = 0.28571 N^{(1)} - 0.28571 N^{(2)} + 0.30151 N^{(3)} + 10 = 0$$

5. We solve the above system of three equations. Thus,

$$N^{(1)} = -19.44 \text{ kN} \qquad \text{or } 19.44 \text{ kN compression}$$

$$N^{(2)} = 13.61 \text{ kN tension} \quad .$$

$$N^{(3)} = -38.69 \text{ kN} \qquad \text{or } 38.69 \text{ kN compression}$$

Example 2. Compute the internal forces in the members of the space truss shown in Fig. a.

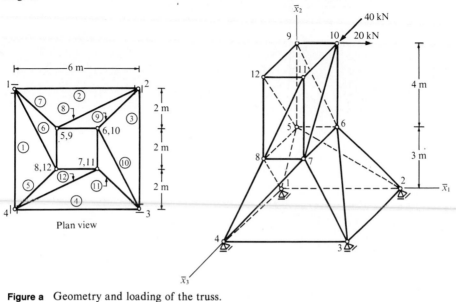

Figure a Geometry and loading of the truss.

solution This is a kinematically stable truss, having 12 joints and 28 members, and the reactions which can be exerted by its supports can have eight unknown components. Thus, we can write $3(12) = 36$ independent equations by considering the equilibrium of the 12 joints of the truss, involving the $28 + 8 = 36$ unknown reactions and internal actions.

Notice that since all the members which frame into joint 9 lie in the same plane, except member 9,12 and since an external force does not act on this joint, the force in member 9,12 vanishes. Moreover, since all members framing into joint 12 lie in the same plane, except members 12,5 and 9,12, and since no external force acts on this joint, and moreover since member 9,12 does not carry an internal force, the force in member 12,5 vanishes. Finally, the external forces in members 12,8 and 12,11 vanish because these two members are not colinear. Similarly, from the equilibrium of joint 11, we may conclude that the forces in members 8,11; 11,7; and 11,10 vanish. Furthermore, from the equilibrium of joint 8, we may conclude that the force in member 7,8 vanishes.

In Fig. b, the members of the truss which do not carry force have been removed. In order to establish the internal forces in the other members of the truss, we adhere to the following steps:

1. We compute the lengths of each member of the truss.
2. We compute the direction cosines of each member of the truss.
3. We find the global components of the axial forces $\mathbf{F}$ acting at end k of each member. If we denote the axial force acting on the end k of the ith member by $\mathbf{F}^{(i)}$, the axial force acting on the end j of the same member is $-\mathbf{F}^{(i)}$.

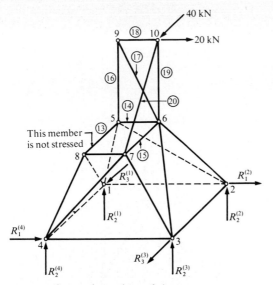

Figure b Stressed members of the truss.

4. We write the equations of equilibrium for each joint.
5. We solve the resulting system of equations.

Steps 1, 2, and 3 are performed in Table a. Referring to the table and denoting the value of the axial force in member i by $N^{(i)}$ and the unit vectors in the $\bar{x}_k(k = 1, 2, 3)$ directions by $i_k(k = 1, 2, 3)$, we can write the equations of equilibrium for each joint.

Joint 1

$$\Sigma \mathbf{F} = \mathbf{F}^{(1)} + \mathbf{F}^{(6)} + \mathbf{F}^{(7)} + R_2^{(1)}i_2 + R_3^{(1)}i_3 = 0$$

or

$$\Sigma \bar{F}_1 = N^{(2)} + 0.3714N^{(6)} + 0.4851N^{(7)} = 0 \tag{a}$$

$$\Sigma \bar{F}_2 = 0.5571N^{(6)} + 0.7276N^{(7)} + R_2^{(1)} = 0 \tag{b}$$

$$\Sigma \bar{F}_3 = N^{(1)} + 0.7428N^{(6)} + 0.4851N^{(7)} + R_3^{(1)} = 0 \tag{c}$$

Joint 2

$$\Sigma \mathbf{F} = -\mathbf{F}^{(2)} + \mathbf{F}^{(3)} + \mathbf{F}^{(8)} + \mathbf{F}^{(9)} + R_2^{(2)}i_2 + R_1^{(2)}i_1 = 0$$

or

$$\Sigma \bar{F}_1 = -N^{(2)} - 0.7428N^{(8)} - 0.4851N^{(9)} + R_1^{(2)} = 0 \tag{d}$$

$$\Sigma \bar{F}_2 = 0.5571N^{(8)} + 0.7276N^{(9)} + R_2^{(2)} = 0 \tag{e}$$

$$\Sigma \bar{F}_3 = N^{(3)} + 0.3714N^{(8)} + 0.4851N^{(9)} = 0 \tag{f}$$

TABLE a Computation of the Global Components of the Forces in the Members

Member	Length, m	$\lambda_{11} = \dfrac{\bar{x}_1^k - \bar{x}_1^j}{L_i}$	$\lambda_{12} = \dfrac{\bar{x}_2^k - \bar{x}_2^j}{L_i}$	$\lambda_{13} = \dfrac{\bar{x}_3^k - \bar{x}_3^j}{L_i}$	$\bar{F}_1^{(i)}$	$\bar{F}_2^{(i)}$	$\bar{F}_3^{(i)}$
1	6	0	0	1	0	0	$N^{(1)}$
2	6	1	0	0	$N^{(2)}$	0	0
3	6	0	0	1	0	0	$N^{(3)}$
4	6	-1	0	0	$-N^{(4)}$	0	0
5	4.1231*	0.4851	0.7276	-0.4851	$0.4851N^{(5)}$	$0.7276N^{(5)}$	$-0.4851N^{(5)}$
6	5.3852†	0.3714	0.5571	0.7428	$0.3714N^{(6)}$	$0.5571N^{(6)}$	$0.7428N^{(6)}$
7	4.1231	0.4851	0.7276	0.4851	$0.4851N^{(7)}$	$0.7276N^{(7)}$	$0.4851N^{(7)}$
8	5.3852	-0.7428	0.5571	0.3714	$-0.7428N^{(8)}$	$0.5571N^{(8)}$	$0.3714N^{(8)}$
9	4.1231	-0.4851	0.7276	0.4851	$-0.4851N^{(9)}$	$0.7276N^{(9)}$	$0.4851N^{(9)}$
10	5.3852	-0.3714	0.5571	-0.7428	$-0.3714N^{(10)}$	$0.5571N^{(10)}$	$-0.7428N^{(10)}$
11	4.1231	-0.4851	0.7276	-0.4851	$-0.4851N^{(11)}$	$0.7276N^{(11)}$	$-0.4851N^{(11)}$
12	5.3852	0.7428	0.5571	-0.3714	$0.7428N^{(12)}$	$0.5571N^{(12)}$	$-0.3714N^{(12)}$
13	2	0	0	1	0	0	$N^{(13)}$
14	2	1	0	0	$N^{(14)}$	0	0
15	2	0	0	1	0	0	$N^{(15)}$
16	4		1	0	0	$N^{(16)}$	0
17	4.4721‡	-0.4472	0.8944	0	$-0.4472N^{(17)}$	$0.8944N^{(17)}$	0
18	2	1	0	0	$N^{(18)}$	0	0
19	4	0	1	0	0	$N^{(19)}$	0
20	4.4721	0	0.8944	-0.4472	0	$0.8944N^{(20)}$	$-0.4472N^{(20)}$

* $\sqrt{2^2 + 3^2 + 2^2} = 4.1231.$

† $\sqrt{2^2 + 3^2 + 4^2} = 5.3852.$

‡ $\sqrt{4^2 + 2^2} = 4.4721.$

Joint 3

$$\Sigma \mathbf{F} = \mathbf{F}^{(3)} + \mathbf{F}^{(10)} + \mathbf{F}^{(11)} + \mathbf{F}^{(4)} + R_3^{(3)}\mathbf{i}_2 + R_3^{(3)}\mathbf{i}_3 = 0$$

or

$$\Sigma \overline{F}_1 = -0.3714N^{(10)} - 0.4851N^{(11)} - N^{(4)} = 0 \tag{g}$$

$$\Sigma \overline{F}_2 = 0.5571N^{(10)} + 0.7276N^{(11)} + R_2^{(3)} = 0 \tag{h}$$

$$\Sigma \overline{F}_3 = -N^{(3)} - 0.7428N^{(10)} - 0.4851N^{(11)} + R_3^{(3)} = 0 \tag{i}$$

Joint 4

$$\Sigma \mathbf{F} = -\mathbf{F}^{(1)} + \mathbf{F}^{(5)} + \mathbf{F}^{(12)} - \mathbf{F}^{(4)} + R_2^{(4)}\mathbf{i}_2 + R_1^{(4)}\mathbf{i}_1 = 0$$

or

$$\Sigma \overline{F}_1 = 0.4851N^{(5)} + 0.7428N^{(12)} + N^{(4)} + R_1^{(4)} = 0 \tag{j}$$

$$\Sigma \overline{F}_2 = 0.7276N^{(5)} + 0.5571N^{(12)} + R_2^{(4)} = 0 \tag{k}$$

$$\Sigma \overline{F}_3 = N^{(1)} - 0.4851N^{(5)} - 0.3714N^{(12)} = 0 \tag{l}$$

Joint 5

$$\Sigma \mathbf{F} = \mathbf{F}^{(13)} + \mathbf{F}^{(14)} - \mathbf{F}^{(7)} - \mathbf{F}^{(8)} + \mathbf{F}^{(16)} = 0$$

or

$$\Sigma \overline{F}_1 = N^{(14)} - 0.4851N^{(7)} + 0.7428N^{(8)} = 0 \tag{m}$$

$$\Sigma \overline{F}_2 = -0.7276N^{(7)} - 0.5571N^{(8)} + N^{(16)} = 0 \tag{n}$$

$$\Sigma \overline{F}_3 = N^{(13)} - 0.4851N^{(7)} - 0.3714N^{(8)} = 0 \tag{o}$$

Joint 6

$$\Sigma \mathbf{F} = -\mathbf{F}^{(14)} + \mathbf{F}^{(19)} + \mathbf{F}^{(15)} - \mathbf{F}^{(10)} - \mathbf{F}^{(9)} + \mathbf{F}^{(17)} = 0$$

or

$$\Sigma \overline{F}_1 = -N^{(14)} + 0.3714N^{(10)} + 0.4851N^{(9)} - 0.4472N^{(17)} = 0 \tag{p}$$

$$\Sigma \overline{F}_2 = N^{(19)} - 0.5571N^{(10)} - 0.7276N^{(9)} + 0.8944N^{(17)} = 0 \tag{q}$$

$$\Sigma \overline{F}_3 = N^{(15)} + 0.7428N^{(10)} - 0.4851N^{(9)} = 0 \tag{r}$$

Joint 7

$$\Sigma \mathbf{F} = -\mathbf{F}^{(11)} - \mathbf{F}^{(12)} - \mathbf{F}^{(15)} + \mathbf{F}^{(20)} = 0$$

$$\Sigma \overline{F}_1 = 0.4851N^{(11)} - 0.7428N^{(12)} = 0 \tag{s}$$

$$\Sigma \overline{F}_2 = -0.7276N^{(11)} - 0.5571N^{(12)} + 0.8944N^{(20)} = 0 \tag{t}$$

$$\Sigma \overline{F}_3 = 0.4851N^{(11)} + 0.3714N^{(12)} - N^{(15)} - 0.4472N^{(20)} = 0 \tag{u}$$

Joint 8

$$\Sigma \mathbf{F} = -\mathbf{F}^{(5)} - \mathbf{F}^{(6)} - \mathbf{F}^{(13)} = 0$$

$$\Sigma \overline{F}_1 = -0.4851N^{(5)} - 0.3714N^{(6)} = 0 \tag{v}$$

$$\Sigma \overline{F}_2 = -0.7276N^{(5)} - 0.5571N^{(6)} = 0 \tag{w}$$

$$\Sigma \overline{F}_3 = 0.4851N^{(5)} - 0.7428N^{(6)} - N^{(13)} = 0 \tag{x}$$

Joint 9

$$\Sigma \overline{F}_1 = N^{(18)} + 0.4472N^{(17)} = 0 \tag{y}$$

$$\Sigma \overline{F}_2 = -N^{(16)} - 0.8944N^{(17)} = 0 \tag{z}$$

Joint 10

$$\Sigma \overline{F}_1 = -N^{(18)} + 20 = 0 \tag{aa}$$

$$\Sigma \overline{F}_2 = -0.8944N^{(20)} - N^{(19)} \tag{ab}$$

$$\Sigma \overline{F}_3 = 0.4472N^{(20)} + 40 = 0 \tag{ac}$$

From Eqs. (ac), (ab), and (aa), we get

$$N^{(20)} = -\frac{40}{0.4472} = -89.45 \text{ kN} \qquad \text{or } 89.45 \text{ kN compression}$$

$$N^{(19)} = 80 \text{ kN tension}$$

$$N^{(18)} = 20 \text{ kN tension}$$

Using the above results and Eqs. (y) and (z), we obtain

$$N^{(17)} = -\frac{N^{(18)}}{0.4472} = \frac{-20}{0.4472} = -44.72 \text{ kN} \qquad \text{or } 44.72 \text{ kN compression}$$

$$N^{(16)} = 40 \text{ kN tension}$$

From Eqs. (s) and (t), we get

$$N^{(12)} = -47.87 \text{ kN} \qquad \text{or } 47.87 \text{ kN compression}$$

$$N^{(11)} = -73.30 \text{ kN} \qquad \text{or } 73.30 \text{ kN compression}$$

Using the above results in Eq. (u), we obtain

$$N^{(15)} = -13.33 \text{ kN} \qquad \text{or } 13.33 \text{ kN compression}$$

Equations (q) and (r) yield

$$-0.5571N^{(10)} - 0.7276N^{(9)} + 40 = 0$$

$$0.7428N^{(10)} - 0.4851N^{(9)} - 13.33 = 0$$

$$\text{or } N^{(9)} = 27.5 \text{ kN tension}$$

$$N^{(10)} = 35.9 \text{ kN tension}$$

From Eq. (p), we get

$$N^{(14)} = 46.66 \text{ kN tension}$$

From Eqs. (m) and (n), we obtain

$$N^{(7)} = 68.71 \text{ kN tension}$$

$$N^{(8)} = -17.94 \text{ kN} \qquad \text{or } 17.94 \text{ kN compression}$$

Using the above results in Eq. (o), we get

$$N^{(13)} = 26.67 \text{ kN tension}$$

Using the above result in Eqs. (w) and (x), we obtain

$$N^{(5)} = 18.32 \text{ kN tension}$$

$$N^{(6)} = -23.93 \text{ kN} \qquad \text{or } 23.93 \text{ kN compression}$$

Equations (a) and (b) yield

$$N^{(2)} = -24.44 \text{ kN} \qquad \text{or } 24.44 \text{ kN compression}$$

$$R_2^{(1)} = -36.66 \text{ kN}$$

From Eq. (c), we get

$$N^{(1)} + R_3^{(1)} + 15.55 = 0 \qquad\qquad (ad)$$

Equations (e), (d), and (f) yield

$$R_2^{(2)} = -10.01 \text{ kN} \qquad R_1^{(2)} = -24.42 \text{ kN}$$

From Eqs. (g), (h), and (i), we obtain

$$N^{(4)} = 22.22 \text{ kN tension}$$

$$R_2^{(3)} = 33.33 \text{ kN}$$

$$R_3^{(3)} = -15.56 \text{ kN}$$

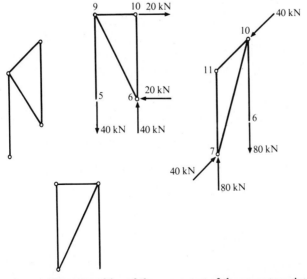

Figure c Decomposition of the upper part of the space truss into planar trusses.

Equations (j) and (k) yield

$$R_1^{(4)} = 4.45 \text{ kN}$$

$$R_2^{(4)} = 13.34 \text{ kN}$$

$$N^{(1)} = 8.89 \text{ kN tension}$$

From Eq. (ae) we get

$$R_3^{(1)} = -24.44 \text{ kN}$$

Notice that the upper part of the space truss may be analyzed by considering each of its sides as a planar truss. The external forces acting on each of these planar trusses may be established by inspection and are shown in Fig. c. The reactions of the planar trusses are transmitted to the bottom part of the space truss. This part of the space truss could also be analyzed by considering its sides as planar trusses. However in this case, the sides are not parallel planes and thus, the external forces acting on them cannot be established by inspection.

3.7 Analysis of Statically Determinate Space Frames

Generally, the analysis of space frames is considerably more difficult than the analysis of planar frames and the aid of a computer is highly desirable. For this reason, we limit our presentation to the analysis of very simple space frames.

When analyzing a space frame, we choose a convenient set of global axes for the frame and a set of local axes for each of its members. Moreover, we establish the direction cosines of the local axes of each member with respect to the global axes of the frame. Referring to Fig. 3.16, the direction cosines of the local system of axes x_i relative to the global system of axes $\bar{x}_j$ are defined as

$$\lambda_{ij} = \cos \phi_{ij} \qquad i, j = 1, 2, 3 \tag{3.35}$$

We use these direction cosines to transform the components of the end actions of the members of the frame from local to global and vice versa.

Denoting by $\mathbf{i}_i$ and $\bar{\mathbf{i}}_j$ the unit vectors acting along the x_i and $\bar{x}_j$ axes, respectively, and referring to Fig. 3.16, we have

$$\mathbf{i}_1 = \lambda_{11}\bar{\mathbf{i}}_1 + \lambda_{12}\bar{\mathbf{i}}_2 + \lambda_{13}\bar{\mathbf{i}}_3 \tag{3.36}$$

$$\mathbf{i}_2 = \lambda_{21}\bar{\mathbf{i}}_1 + \lambda_{22}\bar{\mathbf{i}}_2 + \lambda_{23}\bar{\mathbf{i}}_3$$

$$\mathbf{i}_3 = \lambda_{31}\bar{\mathbf{i}}_1 + \lambda_{32}\bar{\mathbf{i}}_2 + \lambda_{32}\bar{\mathbf{i}}_3$$

or $\qquad \mathbf{i}_i = \sum_{j=1}^{3} \lambda_{ij}\bar{\mathbf{i}}_j \tag{3.37}$

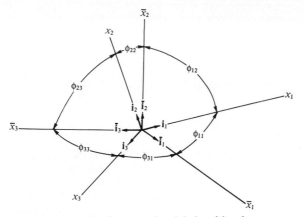

Figure 3.16 Angles between the global and local axes.

similarly, we obtain

$$\bar{\mathbf{i}}_j = \sum_{i=1}^{3} \lambda_{ij}\mathbf{i}_i \tag{3.38}$$

Let us consider the vector $\mathbf{F}$ whose components relative to the systems of axes x_i and $\bar{x}_j$ are denoted by F_i and $\bar{F}_j$, respectively; thus,

$$\mathbf{F} = \sum_{i=1}^{3} F_i\mathbf{i}_i = \sum_{j=1}^{3} \bar{F}_j\bar{\mathbf{i}}_j \tag{3.39}$$

Substituting relation (3.38) into (3.39), we obtain

$$\sum_{i=1}^{3} F_i\mathbf{i}_i = \sum_{j=1}^{3} \sum_{i=1}^{3} \bar{F}_j\lambda_{ij}\mathbf{i}_i \tag{3.40}$$

Consequently,
$$F_i = \sum_{j=1}^{3} \lambda_{ij}\bar{F}_j \tag{3.41}$$

Similarly, substituting relation (3.37) into (3.39), we get

$$\bar{F}_i = \sum_{j=1}^{3} \lambda_{ji}F_j \tag{3.42}$$

Relations (3.41) and (3.42) can be written in matrix form as

$$\begin{Bmatrix} F_1 \\ F_2 \\ F_3 \end{Bmatrix} = \begin{bmatrix} \lambda_{11} & \lambda_{12} & \lambda_{13} \\ \lambda_{21} & \lambda_{22} & \lambda_{23} \\ \lambda_{31} & \lambda_{32} & \lambda_{33} \end{bmatrix} \begin{Bmatrix} \bar{F}_1 \\ \bar{F}_2 \\ \bar{F}_3 \end{Bmatrix} \tag{3.43}$$

and
$$\begin{Bmatrix} \bar{F}_1 \\ \bar{F}_2 \\ \bar{F}_3 \end{Bmatrix} = \begin{bmatrix} \lambda_{11} & \lambda_{21} & \lambda_{31} \\ \lambda_{12} & \lambda_{22} & \lambda_{32} \\ \lambda_{13} & \lambda_{23} & \lambda_{33} \end{bmatrix} \begin{Bmatrix} F_1 \\ F_2 \\ F_3 \end{Bmatrix} \tag{3.44}$$

or
$$\{F\} = [\Lambda_S]\{\bar{F}\} \tag{3.45}$$

and
$$\{\bar{F}\} = [\Lambda_S]^T\{F\} \tag{3.46}$$

where
$$[\Lambda_S] = \begin{bmatrix} \lambda_{11} & \lambda_{12} & \lambda_{13} \\ \lambda_{21} & \lambda_{22} & \lambda_{23} \\ \lambda_{31} & \lambda_{32} & \lambda_{33} \end{bmatrix} \tag{3.47}$$

Relations (3.45) and (3.46) represent the transformation relations of the components of a vector referred to two orthogonal systems of axes (x_i and $\bar{x}_j$). The (3 × 3) matrix $[\Lambda_S]$ is referred to as the *transformation matrix* of the rectangular system of axes x_i relative to the rectangular system of axes $\bar{x}_j$. From relations (3.45) and (3.46) it is apparent that

$$[\Lambda_S]^T = [\Lambda_S]^{-1}$$

Relations (3.45) or (3.46) are employed to transform to local or global components, the global or local components, respectively, of the internal forces and moments acting at the ends of the members of space framed structures.

In the free-body diagram of a member of a space frame, we show the local components of internal actions, while in the free-body diagram of a joint of a space frame, we show the global components of the actions acting on it. These actions are equal and opposite to the end actions of the members of the frame which are connected to the joint. On the free-body diagrams of the members and joints of a space frame, it is more convenient to represent moments by their vector instead of by their curl.

In the sequel, we illustrate the analysis of a space frame by an example.

Example 1.   Compute the end actions in the members of the cantilever frame loaded as shown in Fig. a.

---

solution   The actions at the end j of member 3 are computed by considering its equilibrium. That is, referring to the local axes shown in Fig. b, we have

$$\Sigma F_2 = 0 \qquad Q_2^{3j} = (2)(15) = 30 \text{ kN}$$

$$\Sigma F_3 = 0 \qquad Q_3^{3j} = 20 \text{ kN}$$

$$\Sigma M_2^{(3)} = 0 \qquad M_2^{3j} = -20(2) = -40 \text{ kN·m} \tag{a}$$

$$\Sigma M_3^{(3)} = 0 \qquad M_3^{3j} = (15)(2)(1) = 30 \text{ kN·m}$$

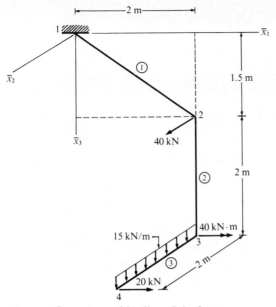

Figure a Geometry and loading of the frame.

In Fig. b, the actions exerted on joint 3 by member 3 are equal and opposite to the actions at the end j of member 3 given by relations (a). The actions exerted on joint 3 by member 2 have been established by considering the equilibrium of this joint. These actions are equal and opposite to those acting at the end k of member 2. Thus, referring to Fig. b, we have

$$Q_3^{2k} = -20 \text{ kN} \qquad N^{2k} = 30 \text{ kN}$$

$$M_3^{2k} = -70 \text{ kN·m} \quad M_1^{2k} = 40 \text{ kN·m}$$

(b)

Notice that the bending moment M_3^{3j} of member 3 is transformed to a torsional moment M_1^{2k} at the end k of member 2.

The actions at the end j of member 2 are obtained by considering its equilibrium. That is, referring to the local axes shown in Fig. b and using results (b), we have

$$\Sigma F_1 = 0 \qquad N^{2j} = N^{2k} = 30 \text{ kN}$$

$$\Sigma F_2 = 0 \qquad Q_3^{2j} = Q_3^{2k} = -20 \text{ kN}$$

$$\Sigma M_1^{(2)} = 0 \qquad M_1^{2j} = M_1^{2k} = -40 \text{ kN·m}$$

(c)

$$\Sigma M_2^{(2)} = 0 \qquad M_2^{2j} = M_2^{2k} = -70 \text{ kN·m}$$

$$\Sigma M_3^{(2)} = 0 \qquad M_3^{2j} = 2Q_3^{2k} = -40 \text{ kN·m}$$

In Fig. b, the actions exerted on joint 2 by member 2 are equal and opposite to the actions at the end j of member 2 given by relations (c). The global components of the actions exerted on joint 2 by member 1 have been established by considering the equi-

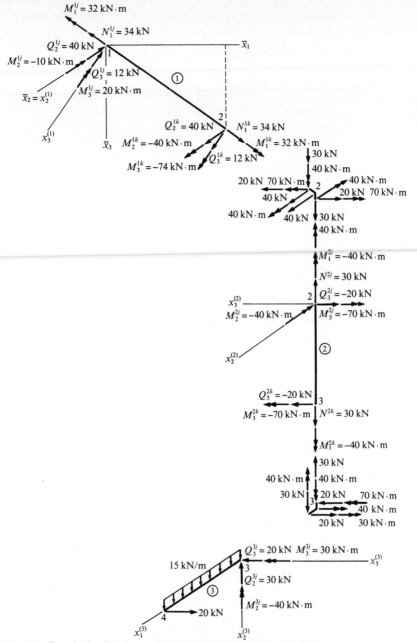

Figure b Free-body diagrams of the members and joints of the frame.

librium of joint 2. These actions are equal and opposite to the global components of the actions at the end k of member 1. Thus

$$\overline{N}^{1k} = 20 \text{ kN} \qquad \overline{Q}_2^{1k} = 40 \text{ kN} \qquad \overline{Q}_3^{1k} = 30 \text{ kN}$$

$$\overline{M}_1^{1k} = 70 \text{ kN·m} \qquad \overline{M}_2^{1k} = -40 \text{ kN·m} \qquad \overline{M}_3^{1k} = -40 \text{ kN·m} \tag{d}$$

The global components of the forces and moments acting at end k of member 1 are transformed to local components. Referring to Fig. a, we see that transformation matrix for member 1 is

$$[\Lambda_S] = \begin{bmatrix} 0.8 & 0 & 0.6 \\ 0 & 1 & 0 \\ -0.6 & 0 & 0.8 \end{bmatrix} \tag{e}$$

Using relations (d) and (e) into (3.35), the local components of the actions at end k of member 1 are

$$\begin{Bmatrix} N^{1k} \\ Q_2^{1k} \\ Q_3^{1k} \end{Bmatrix} = [\Lambda_S] \begin{Bmatrix} \overline{N}^{1k} \\ \overline{Q}_2^{1k} \\ \overline{Q}_3^{1k} \end{Bmatrix} = \begin{bmatrix} 0.8 & 0 & 0.6 \\ 0 & 1 & 0 \\ -0.6 & 0 & 0.8 \end{bmatrix} \begin{Bmatrix} 20 \\ 40 \\ 30 \end{Bmatrix} = \begin{Bmatrix} 34 \\ 40 \\ 12 \end{Bmatrix} \tag{f}$$

moreover,

$$\begin{Bmatrix} M_1^{1k} \\ M_2^{1k} \\ M_3^{1k} \end{Bmatrix} = \begin{bmatrix} 0.8 & 0 & 0.6 \\ 0 & 1 & 0 \\ -0.6 & 0 & 0.8 \end{bmatrix} \begin{Bmatrix} 70 \\ -40 \\ -40 \end{Bmatrix} = \begin{Bmatrix} 32 \\ -40 \\ -74 \end{Bmatrix} \tag{g}$$

The actions at the end of end j of member 1 are obtained by considering its equilibrium. That is, referring to the local axes shown in Fig. b and using results (f) and (g), we have

$$\Sigma F_1 = 0 \qquad N^{1j} = N^{1j} = 34 \text{ kN}$$

$$\Sigma F_2 = 0 \qquad Q_2^{1j} = Q_2^{1k} = 40 \text{ kN}$$

$$\Sigma F_3 = 0 \qquad Q_3^{1j} = Q_3^{1k} = 12 \text{ kN}$$

$$\Sigma M_1 = 0 \qquad M_1^{1j} = M_1^{1k} = 32 \text{ kN} \tag{h}$$

$$\Sigma M_2 = 0 \qquad M_2^{1j} = M_2^{1k} + 2.5Q_3^{1k} = -40 + 30 = -10 \text{ kN·m}$$

$$\Sigma M_3 = 0 \qquad M_3^{1j} = M_3^{1k} - 2.5Q_2^{1k} = -74 + 100 = 26 \text{ kN·m}$$

The global components of the reactions at support 1 are equal to minus the global components of the end actions at end j of member 1. Thus,

$$\begin{Bmatrix} R_1^{(1)} \\ R_2^{(1)} \\ R_3^{(1)} \end{Bmatrix} = -[\Lambda_S]^T \begin{Bmatrix} N_1^{1j} \\ Q_2^{1j} \\ Q_3^{1j} \end{Bmatrix} = - \begin{bmatrix} 0.8 & 0 & -0.6 \\ 0 & 1 & 0 \\ 0.6 & 0 & 0.8 \end{bmatrix} \begin{Bmatrix} 134 \\ 40 \\ 12 \end{Bmatrix} = - \begin{Bmatrix} 20 \\ 40 \\ 30 \end{Bmatrix}$$

$$\begin{Bmatrix} M_1^{(1)} \\ M_2^{(1)} \\ M_3^{(1)} \end{Bmatrix} = -[\Lambda_S]^T \begin{Bmatrix} M_1^{1j} \\ M_2^{1j} \\ M_3^{1j} \end{Bmatrix} = - \begin{bmatrix} 0.8 & 0 & -0.6 \\ 0 & 1 & 0 \\ 0.6 & 0 & 0.8 \end{bmatrix} \begin{Bmatrix} 32 \\ -10 \\ 26 \end{Bmatrix} = - \begin{Bmatrix} 10 \\ -10 \\ 40 \end{Bmatrix}$$

3.8 Problems

1 to 13. Determine the reactions of the supports and the internal forces in the members of the trusses resulting from the loading indicated in Fig. 3.P1. Repeat for Figs. 3.P2 to 3.P13.

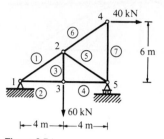

Figure 3.P1

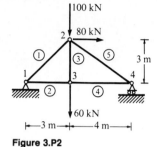

Figure 3.P2

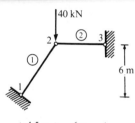

Figure 3.P3

Figure 3.P4

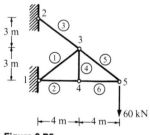

Figure 3.P5

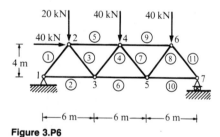

Figure 3.P6

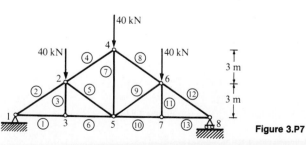

Figure 3.P7

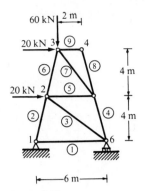

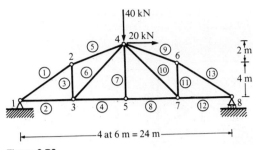

Figure 3.P8

Figure 3.P9

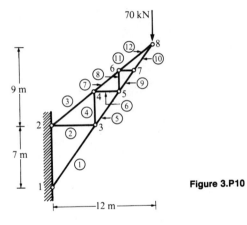

Figure 3.P10

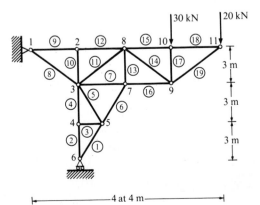

Figure 3.P11

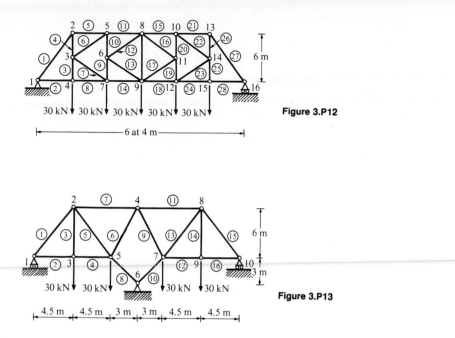

Figure 3.P12

30 kN 30 kN 30 kN 30 kN 30 kN

6 at 4 m

Figure 3.P13

4.5 m   4.5 m   3 m   3 m   4.5 m   4.5 m

14. Compute the internal force in the numbered members of the truss resulting from the loading indicated in Fig. 3.P14.

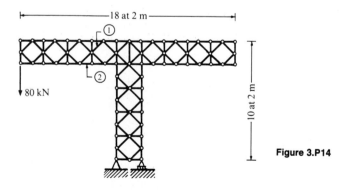

18 at 2 m

80 kN

10 at 2 m

Figure 3.P14

15 to 19. Use the method of sections to determine the internal force in the numbered members of the truss resulting from the loading indicated in Fig. 3.P15. Repeat for Fig. 3.P16 to Fig. 3.P19.

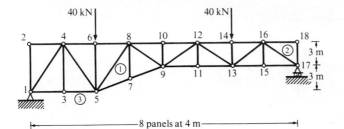

Figure 3.P15

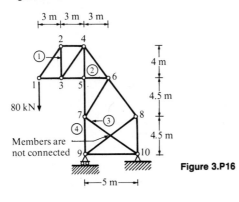

Figure 3.P16

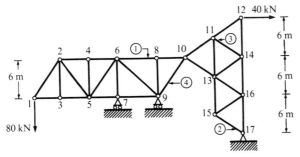

Figure 3.P17

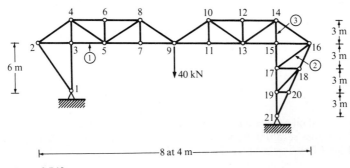

Figure 3.P18

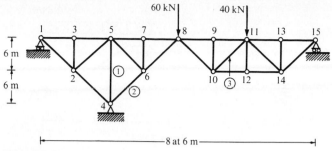

Figure 3.P19

20 to 36. For the structure shown in Fig. 3.P20, plot the shear and moment diagrams and sketch the elastic curve which results from the loading indicated. Repeat for Figs. 3.P21 to 3.P36.

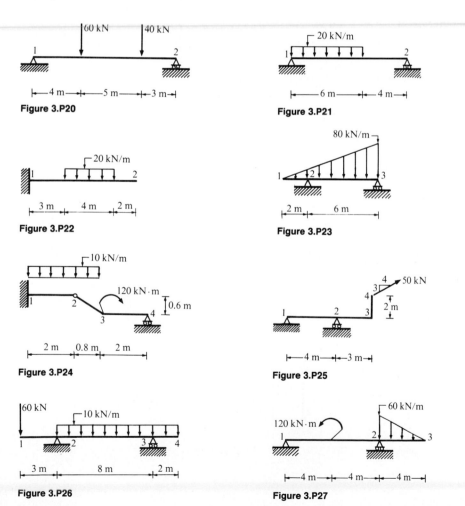

Figure 3.P20

Figure 3.P21

Figure 3.P22

Figure 3.P23

Figure 3.P24

Figure 3.P25

Figure 3.P26

Figure 3.P27

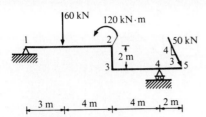

Figure 3.P28

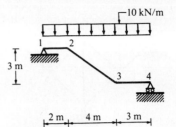

Figure 3.P29

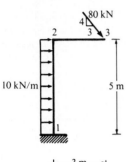

Figure 3.P30

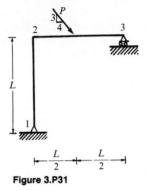

Figure 3.P31

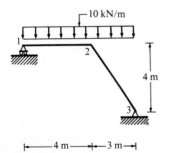

Figure 3.P32

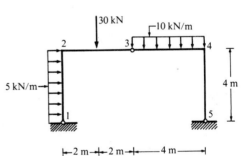

Figure 3.P33

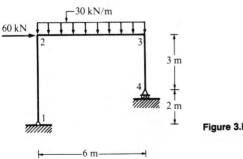

Figure 3.P34

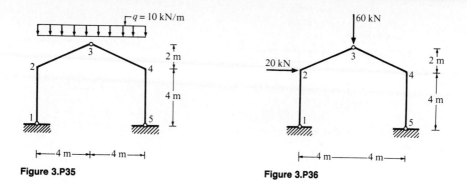

Figure 3.P35

Figure 3.P36

37 to 43.   Find the reactions of the structure resulting from the loading indicated in Fig. 3.P37. Plot the shear and moment diagrams for the structure. Repeat with Figs. 3.P38 to 3.P43.

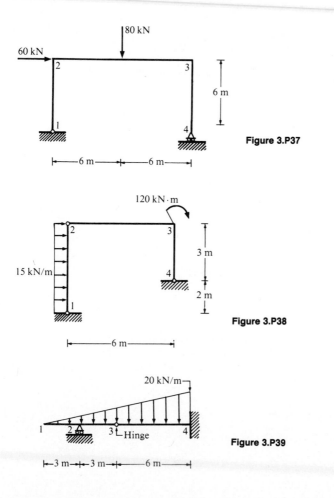

Figure 3.P37

Figure 3.P38

Figure 3.P39

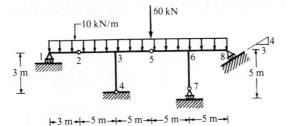

Figure 3.P40

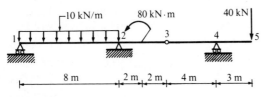

Figure 3.P41

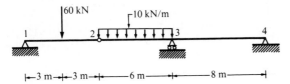

Figure 3.P42

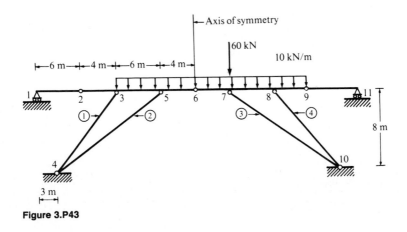

Figure 3.P43

44.   A floor rests on stringers supported by floor beams which in turn are supported on two girders placed 4 m apart. A uniform load of 40 kN/m² acts on the 3 × 4 portion of the floor and a concentrated force of 160 kN acts at the midpoint between the two girders, as shown in Fig. 3.P44. Plot the shear and moment diagrams for the one girder.

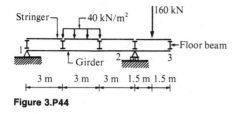

Figure 3.P44

45 and 46.   Compute the reactions and sketch the moment diagram for the arch loaded as shown in Fig. 3.P45. Repeat with Fig. 3.P46.

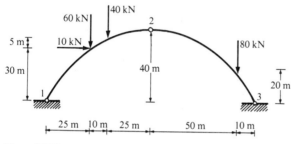

Figure 3.P45

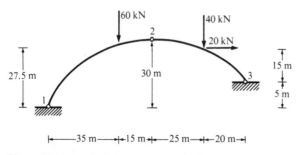

Figure 3.P46

47.   Determine the reaction and the tension in each segment of the cable of Fig. 3.P47. The length of the cable is 70 m.

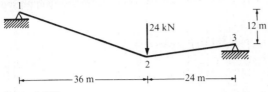

Figure 3.P47

48.   A cable supporting a uniform load of 10 kN per horizontal meter is supported between two points, 60 m apart horizontally and 16 m apart vertically. Find the maximum tension in the cable when the sag at the middle of the cable is 10 m.

49 to 53.   Find the internal forces in the members of the space truss resulting from the forces indicated in Fig. 3.P49. For the schematic representation of the supports see Fig. 1.7. Repeat using Figs. 3.P50 to 3.P53.

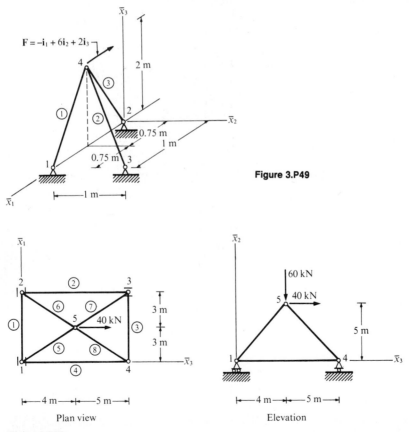

Figure 3.P49

Plan view

Elevation

Figure 3.P50

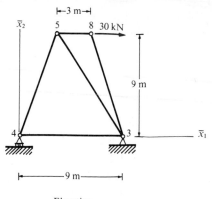

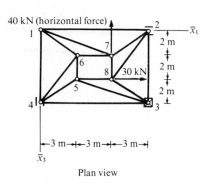

Plan view

Figure 3.P51

Elevation

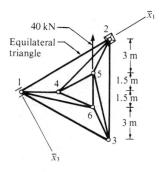

Elevation

Figure 3.P52

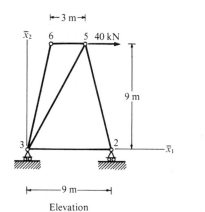

Plan view

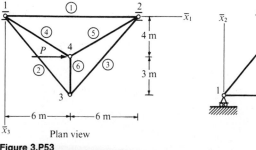

Plan view

Figure 3.P53

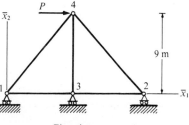

Elevation

54 and 55.   Find the internal forces in the members of the space frame shown in Fig. 3.P54. Repeat with Fig. 3.P55.

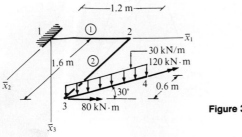

Figure 3.P54

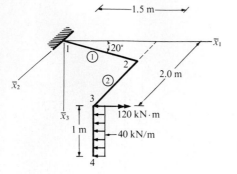

Figure 3.P55

Photographs and Brief Descriptions of Steel Bridges

The Henry Hudson Bridge (Fig. 3.17) connects Manhattan island with the Bronx. The loads from its two decks are transferred by steel columns to the fixed-at-both-ends twin steel arches. The support-to-support length of the arches is 735 ft (224 m). The bracing of the arches can be seen in the second photograph. The lower deck was completed in 1936. The upper deck was constructed later.

(a)

(b)

Figure 3.17 The Henry Hudson Bridge of New York City. *(Courtesy Steinman Boyton Gronquist and Birdstill, Consulting Engineers New York, N.Y.)*

The Fremond Bridge (Fig. 3.18) is one of the longest tied arch
bridges in the world. Its main span is 1255 ft (382.52 m). It has two
decks: an orthotropic steel upper deck and a reinforced concrete
lower deck carried on steel hangers and girders. Each deck carries
four lanes of traffic. It opened to traffic in 1973. In the second
photograph the 902 ft (274.93 m) long center arch of the main span
weighing 6000 tons is lifted into position. This lift was accomplished
by placing a battery of hydraulic jacks at each corner.

(a)

(b)

Figure 3.18 The Fremond Bridge in Portland, Oregon. *(Courtesy Oregon State Highway Dept. and Parsons Brickerhoff, Consulting Engineers, New York, N.Y.)*

The substructure of the Stanislaus River Bridge (Fig. 3.19) consists of five hollow, rectangular, reinforced concrete piers that are tapered in both directions. The tallest pier is 400 ft (122 m). The superstructure consists of a lightweight concrete deck supported on two continuous, parallel welded steel box girders. The girders were constructed by welding together prefabricated elements up to 40 ft

(a)

(b)

Figure 3.19 The Stanislaus River Bridge on Route 49, California. *(Courtesy Dept. of Transportation of the State of California.)*

(12 m) long. Each element of the three shortest spans was raised in place and welded to a previously connected element utilizing falsework piers. The elements of the three longest (152, 168, and 152 m) spans were welded together on falsework near the ground to form larger units of superstructure up to 340 ft (112.6 m) long (see photograph). These units were subsequently jacked vertically and were welded into their proper position.

The Bayonne Bridge (Fig. 3.20) connects the city of Bayonne, N.J. with Port Richmond in Staten Island, N.Y. The length of the steel truss arch is 1675 ft (510.54 m). The height of the arch above the water at the crown is 325 ft (99 m). It was completed in 1931.

(a)

(b)

Figure 3.20   The Bayonne Bridge. *(Courtesy the Port Authority of New York and New Jersey.)*

The George P. Colemen Memorial Bridge (Fig. 3.21) is the world's largest double swing, span highway bridge. It has a total length from abutment to abutment of 3750 ft (1143 m) and a width of 26 ft (7.92 m), providing two opposing traffic lanes. The center spans are each 500 ft (152.4 m) in length and when open provide a clear channel of 450 ft (137.17 m).

(a)

(b)

Figure 3.21   The George P. Colemen Memorial Bridge at Yorktown, Va. *(Courtesy Parsons Brickerhoff, Consulting Engineers, New York, N.Y.)*

The Carquinez Narrows Bridge (Fig. 3.22) consists of a four-span continued truss of total length of 3350 ft (1021 m). Each of the two main spans has a length of 1100 ft (335.3 m). It was completed in 1927.

(a)

(b)

Figure 3.22 The Carquinez Narrows Bridge in California. *(Courtesy Steinman Boynton Gronquist and Birdsill Consulting Engineers, New York, N.Y.)*

Influence Lines

4.1 Introduction

In the previous chapter, we were concerned with the determination of the internal actions acting on the cross sections of members of statically determinate structures subjected to nonmoving loads. In this chapter we deal with the determination of the maximum internal actions at cross sections of members of statically determinate structures subjected to moving and movable loads. In this endeavor, certain diagrams referred to as influence lines are of considerable assistance.

In general, the value of a component of a reaction of a structure or of a component of the internal actions acting at a cross section of a member of a structure depends on the location of the live loads acting on the structure. Thus, in order to establish the maximum value of a component of a reaction of a structure or of a component of the internal actions acting at a cross section of a member of a structure, it is necessary to establish the location of the live loads which renders this quantity a maximum, as well as the value of this maximum.

Let us consider a planar structure and let us denote by X a specific quantity, such as a component of a reaction or a component of the internal actions acting at a specific cross section of one of its members. For instance, referring to the structure of Fig. 4.1, the quantity X could be the moment or the shearing force at joint 3. Suppose that a vertical unit force moves slowly on the parts on this structure on which moving loads act. For instance, if a crane operates on the member 4,7 of the frame of Fig. 4.1, the unit force moves on member 4,7. The unit force is taken in the direction of the moving loads which usually is the vertical. Let us specify the position of the unit force on the structure by the coordinate x_1 (see Fig. 4.1a). It is apparent that, generally, as a result of the aforementioned loading, the magnitude $\hat{X}$ of the quantity X will be a function of x_1. The graphical representation of $\hat{X}(x_1)$ is referred to as the *influence*

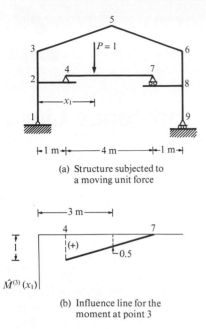

(a) Structure subjected to
a moving unit force

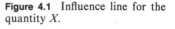

(b) Influence line for the
moment at point 3

Figure 4.1 Influence line for the
quantity X.

line for the quantity X. Thus, an influence line for a quantity may be constructed by plotting the values of the function $\hat{X}(x_1)$ for the successive positions of the unit force.

On the basis of the foregoing, it is apparent that the coordinate $\hat{X}(x_1^{(1)})$ of the influence line for a quantity X in a member of a structure represents the value of X when the structure is loaded only by a unit force at $x_1 = x_1^{(1)}$. For instance, the influence line for the moment at joint 3 of the structure of Fig. 4.1a is shown in Fig. 4.1b. The coordinate of this influence line for $x_1 = 3$ m is 0.5. This indicates that when the unit force is at the midpoint ($x_1 = 3$ m) of beam 4,7, the moment at joint 3 is 0.5 kN·m.

It becomes apparent in the following section that the influence line for a quantity X of a statically determinate structure is made from straight-line segments.

4.2 Influence Lines for Planar Beams, Frames, and Arches

In this section, the construction of influence lines for statically determinate planar beams, frames, and arches is illustrated by the following five examples.

Example 1. The influence lines for certain quantities of a simply supported beam are constructed by considering the equilibrium of segments of the beam when the beam is subjected to a unit force.

Example 2. The influence lines for certain quantities of a simply supported beam with an overhang are constructed.

Example 3. The influence lines for certain quantities of a compound beam are constructed.

Example 4. The influence lines for the axial force, shearing force, and the moment at two cross sections of a frame are constructed.

Example 5. The influence lines for certain quantities of a three-hinged arch are constructed.

Example 1. Consider the simply supported beam of span L shown in Fig. a, and construct the influence lines for the reaction at support 2 and the shearing force and bending moment at point A.

Figure a Geometry of the beam.

solution Consider a unit force moving on this beam and let us specify its position by the coordinate x_1, as shown in Fig. b. The ordinate $\hat{R}^{(1)}$ of the influence line for the reaction $R^{(1)}$ at any point x_1 is equal to the reaction of the beam at its support 1 when the unit force is placed at point x_1. From the equilibrium of the beam, referring to Fig. b we have

$$\Sigma M^{(2)} = 0 \qquad \hat{R}^{(1)} = \frac{L - x_1}{L}$$

$$\Sigma M^{(1)} = 0 \qquad \hat{R}^{(2)} = \frac{x_1}{L} \tag{a}$$

The influence line for the reaction $R^{(2)}$ is plotted in Fig. c.

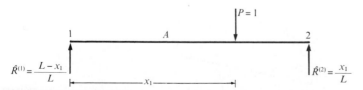

Figure b Free-body diagram of the beam, subjected to a unit force.

Figure c   Influence line for the reaction at support 2.

The equations of the influence lines for the shearing force and bending moment at point A are obtained by considering the equilibrium of the portions of the beam whose free-body diagrams are shown in Figs. d and e. They are

$$
\hat{Q}^A =
\begin{cases}
-\dfrac{x_1}{L} & \text{for } x_1 < a \\[2ex]
\dfrac{L - x_1}{L} & \text{for } x_1 > a
\end{cases}
$$

$$
\hat{M}^{(A)} =
\begin{cases}
\dfrac{x_1 b}{L} & \text{for } x_1 \le a \\[2ex]
\dfrac{(L - x_1)a}{L} & \text{for } x_1 \ge a
\end{cases}
$$

(b)

The influence lines for the shearing force and the bending moment at point A are plotted in Figs. f and g, respectively.

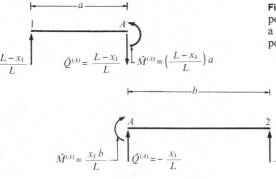

Figure d   Free-body diagram of a portion of the beam subjected to a unit force located to the right of point A as shown in Fig. b.

$$\hat{Q}^{(A)} = \frac{L - x_1}{L} \qquad \hat{M}^{(A)} = \left(\frac{L - x_1}{L}\right)a$$

$$\hat{M}^{(A)} = \frac{x_1 b}{L} \qquad \hat{Q}^{(A)} = -\frac{x_1}{L}$$

Figure e   Free-body diagram of a portion of the beam subjected to a unit force located to the left of point A.

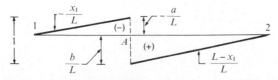

Figure f   Influence line for the shearing force at point A.

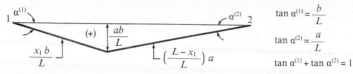

$$\tan \alpha^{(1)} = \frac{b}{L}$$

$$\tan \alpha^{(2)} = \frac{a}{L}$$

$$\tan \alpha^{(1)} + \tan \alpha^{(2)} = 1$$

Figure g Influence line for the bending moment at point A.

Example 2. Consider the beam shown in Fig. a, and construct the influence lines for the reaction at point 1, the shearing force to the right of point 2, the bending moment at point 2, and the shearing force and bending moment at point A.

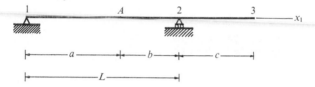

Figure a Geometry of the beam.

solution As shown in Fig. b, the position of the unit load is specified by the coordinate x_1 measured from point 2. Referring to the free-body diagram of the beam shown in Fig. b and taking moments about point 2, the reaction at point 1 corresponding to any position of the unit force is given as

$$\hat{R}^{(1)} = -\frac{x_1}{L}$$

The influence line for the reaction $R^{(1)}$ is plotted in Fig. c.

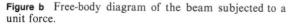

Figure b Free-body diagram of the beam subjected to a unit force.

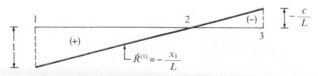

Figure c Influence line for the reaction at support 1.

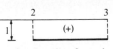

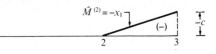

Figure d   Influence line for the shearing force just to the right of point 2.

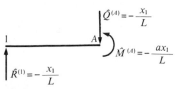

Figure e   Influence line for the moment at point 2.

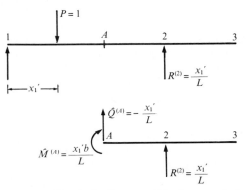

Figure f   Free-body diagram for a portion of the beam subjected to a unit force located to the right of point A as shown in Fig. b.

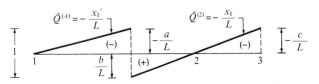

Figure g   Free-body diagram of the beam and of a portion of the beam subjected to a unit force located to the left of point A.

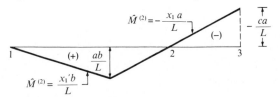

Figure h   Influence line for the shearing force at point A.

Figure i   Influence line for the moment at point A.

The influence lines for the shearing force just to the right of point 2 and the moment at point 2 are plotted in Figs. d and e, respectively. The ordinates of these influence lines are given as

$$\hat{Q} = \begin{cases} 1 & \text{for } x_1 > 0 \\ 0 & \text{for } x_1 < 0 \end{cases}$$

and

$$\hat{M} = \begin{cases} -x_1 & \text{for } x_1 \geq 0 \\ 0 & \text{for } x_1 \leq 0 \end{cases}$$

The free-body diagram of the portion of beam from point 1 to point A, when the unit force is located to the right of point A, is shown in Fig. f. The free-body diagram of the portion of the beam from point A to point 3, when the unit force is located to the left of point A, is shown in Fig. g. From the equilibrium of these portions of the beam, the ordinates of the influence lines for the shearing force and bending moment at point A may be established as functions of the position of the unit force. The influence lines for the shearing force and the bending moment at point A are plotted in Figs. h and i, respectively.

Example 3. Consider the beam shown in Fig. a, and plot the influence lines for the reaction at support 2, for the shearing force and the bending moment at point A, the bending moment at support 2, and the shearing force and the bending moment at point B.

Figure a Geometry of the beam.

solution This is a compound structure. Consider a vertical unit force moving on this beam. When the unit force moves from point 1 to point 3, its position is specified by the coordinate x_1, shown in Fig. b. When the unit force moves from point 3 to point 4, its position is specified by the coordinate x_1', shown in Fig. c. From the equilibrium of the beam, referring to Figs. b and c, we may obtain the equations for the influence line for the reaction $R^{(2)}$.

$$\hat{R}^{(2)} = \begin{cases} \dfrac{x_1}{L} & 0 < x_1 < \dfrac{3L}{2} \\[2ex] \dfrac{3}{2}\dfrac{x_1'}{L} & 0 < x_1' < L \end{cases} \tag{a}$$

The influence line for $\hat{R}^{(2)}$ is plotted in Fig. d.

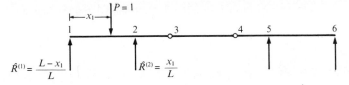

$$\hat{R}^{(1)} = \frac{L - x_1}{L} \qquad \hat{R}^{(2)} = \frac{x_1}{L}$$

Figure b   Free-body diagram of the beam subjected to a unit force between points 1 and 3.

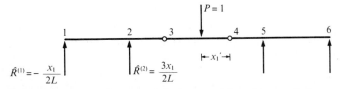

$$\hat{R}^{(1)} = -\frac{x_1}{2L} \qquad \hat{R}^{(2)} = \frac{3x_1}{2L}$$

Figure c   Free-body diagram of the beam subjected to a unit force between points 3 and 4.

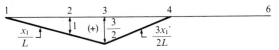

Figure d   Influence line for the reaction at point 2.

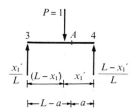

Figure e   Free-body diagram of beam 3,4 subjected to a unit force.

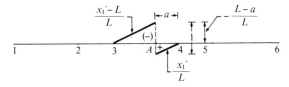

Figure f   Influence line for the shearing force at point A.

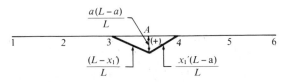

Figure g   Influence line for the moment at point A.

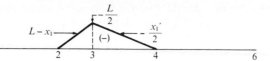

Figure h Influence line for the moment at point 2.

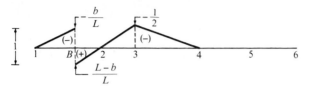

Figure i Influence line for the shearing force at point B.

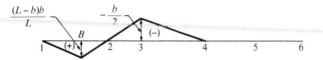

Figure j Influence line for the moment at point B.

The equations of the influence lines for the shearing force and the moment at point A are obtained by considering the equilibrium of beam 3,4, whose free-body diagram is shown in Fig. e. They are

$$\hat{Q}^A = \begin{cases} \dfrac{x_1'}{L} & 0 \le x_1' \le a \\[2mm] \dfrac{x_1' - L}{L} & a \le x_1' \le L \end{cases} \tag{b}$$

$$\hat{M}^A = \begin{cases} \dfrac{x_1'(L - a)}{L} & 0 \le x_1' < a \\[2mm] \dfrac{(L - x_1')a}{L} & a < x_1' < L \end{cases} \tag{c}$$

The influence lines for the shearing force and moment at point A are plotted in Figs. f and g, respectively.

The equations of the influence line for the moment at point 2 are obtained by considering the equilibrium of the beam subjected to a unit force at the positions shown in Figs. b and c. Thus,

$$\hat{M}^{(2)} = \begin{cases} 0 & 0 \le x_1 \le L \\[2mm] L - x_1 & L \le x_1 \le \dfrac{3}{2} L \\[2mm] -\dfrac{x_1'}{2} & 0 \le x_1' \le L \end{cases} \tag{d}$$

The influence line for the moment at point 2 is shown in Fig. h.

The equations of the influence lines for the shearing force and moment at point B are obtained from the influence line for $R^{(1)}$. Thus, referring to Figs. b and c, we have

$$
\hat{Q}^{(B)} = \begin{cases} \dfrac{L - x_1}{L} - 1 = -\dfrac{x_1}{L} & 0 \le x_1 \le b \\[2ex] \dfrac{L - x_1}{L} & b \le x_1 \le \dfrac{3L}{2} \\[2ex] -\dfrac{x_1'}{2L} & 0 \le x_1' \le L \end{cases} \tag{e}
$$

$$
\hat{M}^{(B)} = \begin{cases} \dfrac{(L - x_1)b}{L} - (b - x_1) = x_1 \dfrac{(L - b)}{L} & 0 \le x_1 \le b \\[2ex] \dfrac{(L - x_1)b}{L} & b \le x_1 \le \dfrac{3L}{2} \\[2ex] \dfrac{x'b}{2L} & 0 \le x_1' \le L \end{cases} \tag{f}
$$

The influence lines for the shearing force and bending moment at point B are plotted in Figs. i and j, respectively.

Example 4.   A crane moves on beam 3,10 of the frame of Fig. a. Plot the influence lines for the axial force, the shearing force, and the moment at points A and B of the frame. Point B is located just below joint 2.

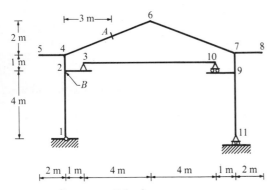

Figure a   Geometry of the frame.

---

solution   The unit force moves on the simply supported beam 3,10. As shown in Fig. b, its position is specified by the coordinate x_1. The ordinates of the influence line for

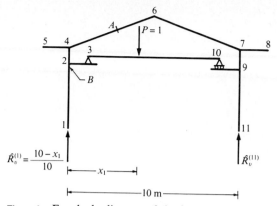

Figure b Free-body diagram of the frame.

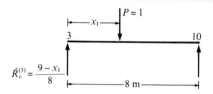

Figure c Free-body diagram of the simply supported beam 3,10.

Figure d Influence line for the axial force at point B.

the reactions of the simply supported beam 3,10 may be obtained by considering its equilibrium. Thus, referring to Fig. c, we get

$$\hat{R}_v^{(3)} = \frac{9 - x_1}{8} \qquad 1 \le x_1 \le 9 \tag{a}$$

The coordinates of the influence line for the reaction of the frame $R_v^{(1)}$ may be obtained by considering the equilibrium of the frame. Thus, referring to Fig. b, we obtain

$$\hat{R}_v^{(1)} = \frac{10 - x_1}{10} \qquad 1 \le x_1 \le 9 \tag{b}$$

Referring to Fig. b, we can easily see that the shearing force and bending moment at point B vanish, while the axial force is equal to $- \hat{R}_v^{(1)}$. The influence line for the axial force at point B is shown in Fig. d. The coordinates of the influence line for the axial force, the shearing force, and the bending moment at point A may be obtained by

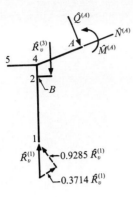

Figure e Free-body diagram of a portion of the frame.

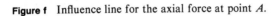

Figure f Influence line for the axial force at point A.

Figure g Influence line for the shearing force at point A.

Figure h Influence line for the moment at point A.

considering the equilibrium of the portion of the frame whose free-body diagram is shown in Fig. e. Thus,

$$\Sigma F_N = 0 \qquad \hat{N}^{(A)} + 0.3714[\hat{R}_v^{(1)} - \hat{R}_v^{(3)}] = 0$$

$$\Sigma F_Q = 0 \qquad \hat{Q}^{(A)} - 0.9285[\hat{R}_v^{(1)} - \hat{R}_v^{(3)}] = 0$$

$$\Sigma M^{(A)} = 0 \qquad 3\hat{R}_v^{(1)} - 2\hat{R}_v^{(3)} - \hat{M}^{(A)} = 0$$

or

$$\hat{N}^{(A)} = 0.3714[\hat{R}_v^{(3)} - \hat{R}_v^{(1)}] = 0.3714\left[\frac{9 - x_1}{8} - \frac{10 - x_1}{10}\right] = 0.009285(5 - x_1)$$

$$\hat{Q}^{(A)} = 0.9285[\hat{R}_v^{(3)} - \hat{R}_v^{(1)}] = 0.0232(x_1 - 5)$$

$$\hat{M}^{(A)} = 3\hat{R}_v^{(1)} - 2\hat{R}_v^{(3)} = \frac{15 - x_1}{20}$$

The influence lines for the axial force, the shearing force, and the bending moment at point A are shown in Figs. f to h.

Example 5. As mentioned in Sec. 1.4.4, arches are often used as structural components of bridges. In this case, the moving loads (vehicles) are transferred to the arch by a system of columns if the pavement is above the arch or by cables if the pavement is suspended from the arch. Consider the three-hinged arch shown in Fig. a. Plot the influence lines for the reactions at its support 1 and the internal actions at point A.

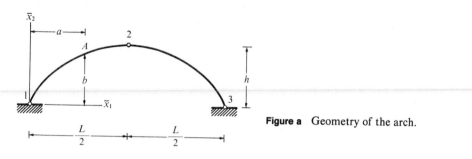

Figure a Geometry of the arch.

solution The coordinates of the influence lines for the vertical component of the reactions of the arch may be computed by considering the equilibrium of the arch whose free-body diagram is shown in Fig. b. Referring to this figure, we obtain

$$\Sigma M^{(3)} = 0 \qquad \hat{R}_v^{(1)} = \frac{L - \bar{x}_1}{L} \tag{a}$$

$$\Sigma M^{(1)} = 0 \qquad \hat{R}_v^{(3)} = \frac{\bar{x}_1}{L} \tag{b}$$

It is apparent that the influence line for the vertical components of the reactions of the arch are the same as those of an auxiliary, simply supported beam of length equal to the opening of the arch. The influence lines for the reaction $R_v^{(1)}$ are shown in Fig. e.

Referring to Fig. b, we have, from the equilibrium of the half arch,

$$\Sigma M^{(2)} = 0 \qquad \hat{R}_h^{(3)} = \frac{L\bar{x}_1}{2h} \qquad \hat{R}_v^{(3)} = \frac{\bar{x}_1}{2h} \qquad 0 \le \bar{x}_1 \le \frac{L}{2}$$

$$\Sigma M^{(2)} = 0 \qquad \hat{R}_h^{(1)} = \frac{L\bar{x}_1}{2h} \qquad \hat{R}_v^{(1)} = \frac{L - \bar{x}_1}{2h} \qquad \frac{L}{2} \le \bar{x}_1 \le L \tag{c}$$

The influence line for the reaction $\hat{R}_h^{(1)}$ is shown in Fig. f.

Referring to the free-body diagram of the portion of the arch shown in Fig. d, we have

$$\Sigma M^{(A)} = 0 \qquad \hat{M}^{(A)} = a\hat{R}_v^{(1)} - b\hat{R}_h^{(1)} = \hat{M}^{(A)} - b\hat{R}_h^{(1)} \tag{d}$$

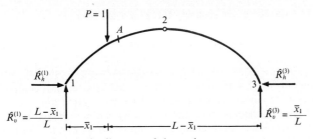

Figure b   Free-body diagram of the arch.

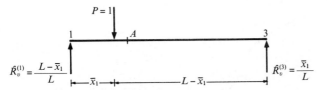

Figure c   Free-body diagram of the auxiliary beam.

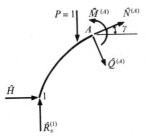

Figure d   Free-body diagram of a portion of the arch.

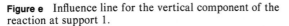

Figure e   Influence line for the vertical component of the reaction at support 1.

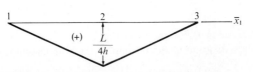

Figure f   Influence line for the horizontal component of the reaction at support 1.

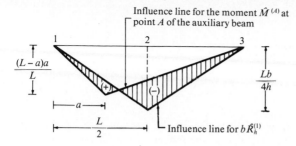

Influence line for the moment $\hat{M}^{(A)}$ at point A of the auxiliary beam

$\dfrac{(L-a)a}{L}$

$\dfrac{Lb}{4h}$

(+)

(−)

$\dfrac{L}{2}$

a

Influence line for $b\hat{R}_h^{(1)}$

Figure g Computation of the influence line for $M^{(A)}$ [see relation (d)].

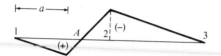

a

1

A

2

(−)

3

(+)

Figure h Influence line for $M^{(A)}$.

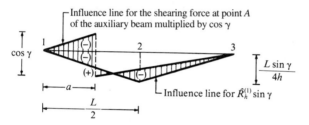

Influence line for the shearing force at point A of the auxiliary beam multiplied by $\cos\gamma$

$\cos\gamma$

1

(−)

(−)

2

3

(+)

(−)

$\dfrac{L\sin\gamma}{4h}$

Influence line for $\hat{R}_h^{(1)}\sin\gamma$

a

$\dfrac{L}{2}$

Figure i Computation of the influence line for $Q^{(A)}$.

(−)

1

2

3

a

Figure j Influence line for $Q^{(A)}$.

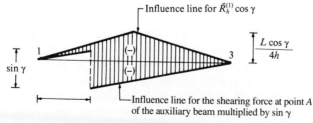

Influence line for $\hat{R}_h^{(1)}\cos\gamma$

1

(−)

(−)

3

$\dfrac{L\cos\gamma}{4h}$

$\sin\gamma$

Influence line for the shearing force at point A of the auxiliary beam multiplied by $\sin\gamma$

Figure k Computation of the influence line for $N^{(A)}$.

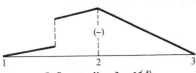

Figure I Influence line for $N^{(A)}$.

where $\hat{\hat{M}}^{(A)}$ is the coordinate of the influence line for the moment at point A of the auxiliary beam of Fig. c. Referring to Fig. g, the influence line for the moment at point A of the arch is sketched in Fig. h. Similarly, from the equilibrium of the portion of the arch shown in Fig. d, we have

$$\Sigma F_N = 0 \qquad \hat{R}_h^{(1)} \cos \gamma + \hat{N}^{(A)} + \hat{R}_v^{(1)} \sin \gamma - \sin \gamma = 0$$

$$\Sigma F_Q = 0 \qquad \hat{R}_h^{(1)} \sin \gamma + \hat{Q}^{(A)} - \hat{R}_v^{(1)} \cos \gamma + \cos \gamma = 0$$

or

$$\hat{N}^{(A)} = \sin \gamma \hat{\hat{Q}}^{(A)} - \hat{R}_h^{(1)} \cos \gamma \tag{e}$$

$$\hat{Q}^{(A)} = \cos \gamma \hat{\hat{Q}}^{(A)} - \hat{R}_h^{(1)} \cos \gamma \tag{f}$$

where $\hat{\hat{Q}}^{(A)}$ is the coordinate of the influence line for the shearing force at point A of the auxiliary beam of Fig. c. Referring to Figs. i and k the influence lines for the axial and the shearing force at point A of the arch are sketched in Figs. j and l, respectively.

4.3 Influence Lines for Girders with a System of Stringers and Floor Beams

As discussed in Sec. 1.4.3, usually the moving loads do not act directly on the girders of bridges. They act on stringers that are supported by floor beams which, in turn, are supported on the girders (see Fig. 1.26).

In this section, we prove that when the unit force moves on the stringers, the influence lines for a quantity X (shearing force or bending moment) of a girder (statically determinate or indeterminate) is a polygon inscribed in the influence line for the quantity X, obtained by assuming that the unit load moves directly on the girder.

Consider a quantity X (shearing force or bending moment) acting on a cross section of a girder located between the m and the $m + 1$ floor beams. The influence line for this quantity obtained when the unit force moves directly on the girder is shown in Fig. 4.2c. Consider a unit force located on the stringers between the m and $m + 1$ floor beams, and denote by e_m and $L_s - e_m$ the distance of the unit force from the m and $m + 1$ floor beams, respectively. Referring to Fig. 4.2b, we see that the reaction on the mth floor beam is equal to $(L_s - e_m)/L_s$, while the reaction on the $(m + 1)$th floor beam is equal to

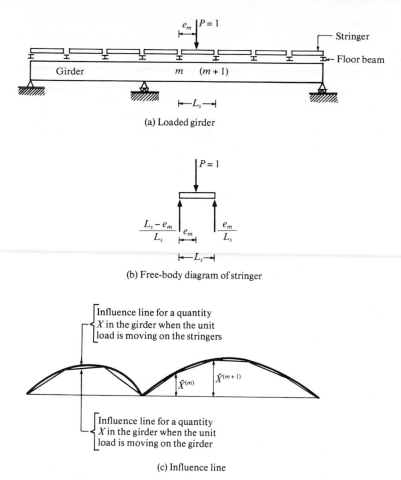

(a) Loaded girder

(b) Free-body diagram of stringer

(c) Influence line

Figure 4.2 Influence line for the quantity X in a girder with a system of stringers and floor beams.

e_m/L_s. Thus, for this position of the unit force, the coordinate of the influence line of the quantity X is

$$\hat{X} = \frac{L_s - e_m}{L_s} \hat{X}^{(m)} + \frac{e_m}{L_s} \hat{X}^{(m+1)} = \hat{X}^m + \frac{e_m}{L_s}(\hat{X}^{(m+1)} - \hat{X}^m) \quad (4.1)$$

where $\hat{X}^m$ and $\hat{X}^{(m+1)}$ are the coordinates at the m and $m + 1$ floor beams, respectively, of the influence line of the quantity X when the unit force moves directly on the girder.

Notice that the influence line for any quantity X in a statically determinate structure is made of straight line segments. However, the influence line for a quantity in a statically indeterminate structure could be a curve. From relation

(4.1) it is apparent that, generally, when the unit force moves on the stringers the influence line for any quantity X in a girder (statically determinate or indeterminate) is a straight line between any two adjacent floor beams. At the floor beams, this influence line assumes values equal to those of the influence line for the quantity X when the unit force moves directly on the girder. Thus, the influence line for a quantity X (shearing force or bending moment) acting at a cross section of a girder (statically determinate or indeterminate), obtained when the unit force is moving on the stringers, is a polygon inscribed in the influence line curve for the quantity X, obtained when the unit force moves directly on the girder.

Consider the influence line for any quantity X acting on a cross section between the m and the $m + 1$ floor beams of a statically determinate girder, obtained with the unit force traveling on the stringers. Moreover, consider the influence line for the same quantity obtained with the unit force traveling directly on the girder. Since these influence lines are composed of straight line segments, their portions to the left of the mth or to the right of the $(m + 1)$th floor beams are identical (see Fig. e, example 1).

Example 1. Construct the influence lines for the shearing force in panel CD and for the moment at points D and m of the girder shown in Fig. a.

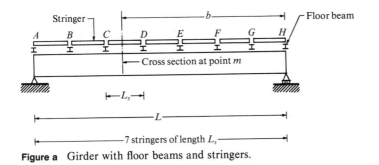

Figure a Girder with floor beams and stringers.

---

solution Suppose that the girder is cut in two parts at cross section m. The free-body diagrams of these two parts are shown in Fig. b when the girder is loaded with a unit force located to the left of panel CD. Considering the equilibrium of the right part of the girder shown in Fig. b, the shearing force and the bending moment acting on cross section m of the girder are:

$$\hat{Q} = -\frac{x_1}{L} \qquad 0 \le x_1 \le 2L_s$$

$$\hat{M} = \frac{x_1 b}{L}$$

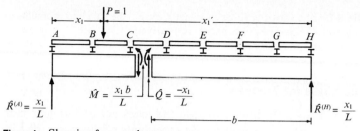

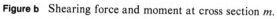

Figure b Shearing force and moment at cross section m.

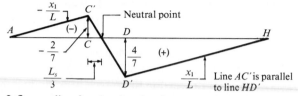

Figure c Influence line for the shearing force at any cross section in panel CD.

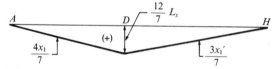

Figure d Influence line for the moment at point D.

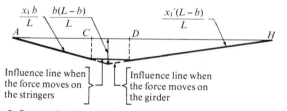

Figure e Influence line for the moment at point m.

Moreover, when the unit force moves on stringer CD, the force exerted on the girder by the floor beam at D is equal to

$$P^{(D)} = \frac{x_1}{L_s} - 2$$

Thus referring to Fig. b, we see that the shearing force and the bending moment acting on cross section m of the girder are:

$$\hat{Q} = -\hat{R}^{(H)} + P^{(D)} = -\frac{x_1}{L} + \frac{x_1}{L_s} - 2$$

$$\hat{M} = \frac{x_1 b}{L} - \left(\frac{x_1}{L_s} - 2\right)(b - 4L_s)$$

The influence lines for the shearing force in panel CD and the bending moments at points D and m are plotted in Figs. c, d, and e, respectively.

4.4   Influence Lines for Trusses

As discussed in Sec. 1.4.2, in bridge trusses, the loads act on stringers supported on floor beams which are connected to the joints of the truss (see Fig. 1.24). Thus the procedure for drawing the influence lines for the internal forces in the members of trusses is analogous to that for drawing influence lines for the internal actions at points of girders with a system of floor beams and stringers.

In this section, the procedure for drawing influence lines for the internal forces in the members of a truss is illustrated by two examples. In the first example, the influence lines are drawn for the forces in some members of a truss with horizontal upper and lower chord members. In the second example, the influence lines are drawn for the forces in some members of a truss whose lower chord members are not horizontal.

Example 1. Construct the influence lines for the forces in members 13, 12, 14, 11, and 3 of the Pratt truss shown in Fig. a.

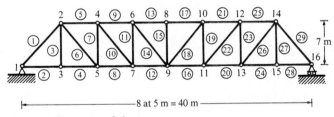

Figure a Geometry of the truss.

solution In order to establish the ordinates of the influence line for the force in member 13, we consider the free-body diagrams of the two parts of the truss, shown in Fig. b. Referring to this figure, we can see that for any position of the unit force along the span of the truss, the value of the force in member 13 is equal to minus the moment about joint 9 of the external forces (unit force and reaction) acting to the left or to the right of joint 9 divided by the height of the truss. The ordinates of the influence line for the force in member 13 in the interval for $x_1 = 0$ m to $x_1 = 20$ m are obtained by considering the equilibrium of the right-hand part of the truss shown in Fig. b. Thus,

$$\Sigma M^{(9)} = 0 \qquad \hat{N}^{(13)} = -\frac{x_1}{40}\left(\frac{20}{7}\right) = -\frac{x_1}{14} \qquad 0 \le x_1 \le 20$$

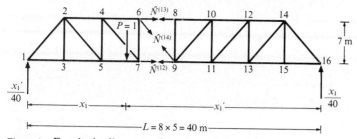

Figure b Free-body diagrams of portions of the truss.

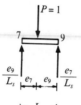

Figure c Free-body diagram of the stringer between joints 7 and 9.

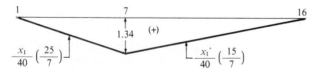

Figure d Influence line for the force in member 13.

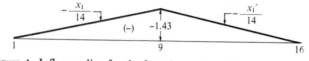

Figure e Influence line for the force in member 12.

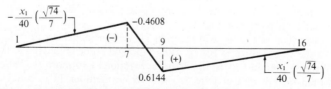

Figure f Influence line for the force in member 14.

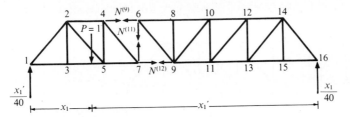

Figure g　Free-body diagram of portions of the truss.

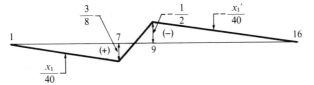

Figure h　Influence line for the force in member 11.

Figure i　Influence line for the force in member 3.

The ordinates of the influence line for the force in member 13 in the interval $x_1 = 20$ m to $x_1 = 40$ m are obtained by considering the equilibrium of the left-hand part of the truss shown in Fig. b. Thus,

$$\Sigma M^{(9)} = 0 \qquad \hat{N}^{(13)} = \frac{x_1'}{40}\left(\frac{20}{7}\right) = -\frac{x_1'}{14} \qquad 0 \le x_1' \le 20$$

The influence line for the force in member 13 is plotted in Fig. d.

Referring to Fig. b, the value of the force in member 12 is equal to the moment of the external forces acting to the left or to the right of joint 7 about joint 6 divided by the height of the truss. Thus, the influence line for the force in member 12 is as shown in Fig. e.

When the unit force is located to the left of joint 7, the vertical component of the force in member 14 is equal to minus the value of the reaction at support 16. When the unit force is located to the right of joint 9, the vertical component of the force in member 14 is equal to the value of the reaction at support 1. When the unit force is located on the stringer between joints 7 and 9, the vertical component of the force in member 14 is equal to minus the value of the reaction at support 16 of the truss, plus the force e_7/L_s transferred at joint 9 by the floor beam (see Fig. c). The influence line for the force in member 14 is plotted in Fig. f.

To obtain the influence line for the force in member 11, we consider the equilibrium of the two portions of the truss shown in Fig. g. When the unit force is located to the left of joint 7, the force in member 11 is equal to the reaction at support 16. When the unit force is located to the right of joint 9, the force in member 11 is equal to minus the reaction at support 1. When the unit force is located on the stringer between joints 7 and 9, the force in member 11 is equal to the value of the reaction at support 16 minus the force e_7/L_s (see Fig. c). The influence line for the force in member 11 is plotted in Fig. h.

To obtain the influence line for the force in member 3, we consider the equilibrium of joint 3. From inspection, it can be seen that the force in member 3 vanishes when an external force does not act on joint 3. Thus, the force in member 3 vanishes when the unit force is located to the right of joint 5, and it is equal to unity when the unit force is located at joint 3. The influence line for the force in member 3 is plotted in Fig. i.

Example 2. Consider the truss shown in Fig. a, whose lower chord members are not horizontal. The loads travel on stringers supported on floor beams, supported at the joints of the upper chord of the truss. Plot the influence lines for the forces in members 13, 14, 15, and 17.

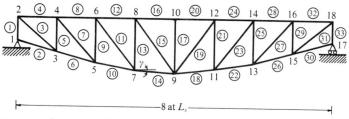

Figure a   Geometry of the truss.

solution   In order to establish the internal forces in members 13 and 14, consider the free-body diagrams of the two portions of the truss shown in Fig. b, subjected to a unit load. Referring to Fig. b, we see from the equilibrium of the left-hand portion of the truss that the horizontal component of the force in the lower chord member, 14, is equal to the moment about joint 8 of the external forces (reaction and unit force) acting to the left or to the right of joint 8 divided by the distance h_8 from joint 7 to joint 8. Thus, the influence line for the force in member 14 has the same shape as the influence line for the moment at point A of a simply supported beam whose length is equal to that of the truss (see Fig. c). The distance of point A from support 1 of the beam is equal to the horizontal distance of joint 7 of the truss from its support 1. The value of the ordinates of the influence line for the force in member 14 is

$$\hat{N}^{(14)} = \frac{5x_1}{8h_8 \cos \gamma} \qquad 0 \le x_1 \le -3L_s$$

$$\hat{N}^{(14)} = \frac{3x_1'}{8h_8 \cos \gamma} \qquad 0 \le x_1' \le 5L_s$$

The influence line for the force in member 14 is shown in Fig. d. Referring to Fig. b, we can see that when the unit force is located to the right of joint 8, the force in member 13 is equal to minus the moment of the reaction $R^{(1)}$ about point B divided

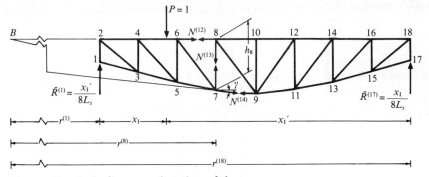

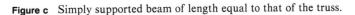

Figure b   Free-body diagrams of portions of the truss.

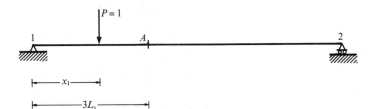

Figure c   Simply supported beam of length equal to that of the truss.

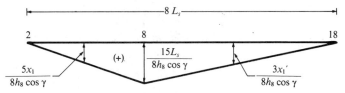

Figure d   Influence line for the force in member 14.

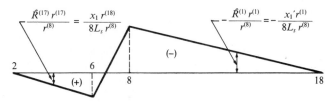

Figure e   Influence line for the force in member 13.

by the distance $r^{(8)}$ from point B to joint 8. Point B is the point of intersection of the extension of member 14 with the extension of the upper chord. When the unit force is located to the left of point 6, the force in member 13 is equal to the moment of the reaction $R^{(17)}$ divided by the distance $r^{(8)}$ between joint 8 and point B. The influence line for member 13 is shown in Fig. e.

The influence line for the force in member 15 may be obtained by considering the equilibrium of the two portions of the truss whose free-body diagrams are shown in

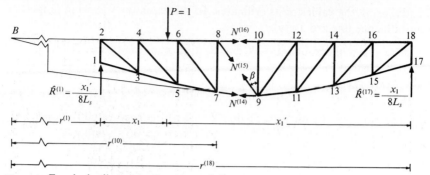

Figure f Free-body diagrams of portions of the truss.

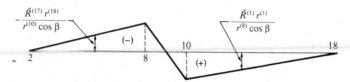

Figure g Influence line for the force in member 15.

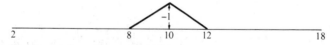

Figure h Influence line for the force in member 17.

Fig. f. Referring to this figure, we can see that when the unit force is located to the right of joint 10, the vertical component of the force in member 15 is equal to the moment of the reaction $R^{(1)}$ about point B divided by the distance $r^{(8)}$ from point B to joint 8. When the unit force is located to the left of joint 8, the vertical component of the force in member 15 is equal to minus the moment of the reaction $R^{(17)}$ about point B divided by the distance $r^{(10)}$ from point B to joint 10. The influence line for the force in member 15 is shown in Fig. g.

The influence line for the force in member 17 may be obtained by considering the equilibrium of joint 10. It is plotted in Fig. h.

4.5 Properties of Influence Lines

The influence line $\hat{X}(x_1)$ may be used to determine qualitatively the position of the live loads which will result in the maximum value of the quantity X (axial force, shearing force, moment). The maximum value of X is then computed by placing the loads in this position and analyzing the structure. In this case, only the shape of the influence line is required and, therefore, it is not necessary to compute its ordinates. Moreover, for any given loading condition the influence line $\hat{X}(x_1)$ may be used to establish the value of the quantity X

directly, using the values of its ordinates. In this case, the following properties of an influence line are useful:

1. The value of the quantity X due to a concentrated force P acting at a point $x_1^{(i)}$ is equal to

$$X = P\hat{X}(x_1^{(i)}) \tag{4.2}$$

2. The value of the quantity X due to a set of concentrated forces P_1, P_2, ..., P_n, acting at points $x_1^{(1)}$, $x_1^{(2)}$, ..., $x_1^{(n)}$, respectively, is equal to

$$X = P_1\hat{X}(x_1^{(1)}) + P_2\hat{X}(x_1^{(2)}) + \cdots + P_n\hat{X}(x_1^{(n)}) \tag{4.3}$$

3. The value of the quantity X due to the distributed force $q(x_1)$ is equal to

$$X = \int_{x_1^L}^{x_1^R} q\hat{X}(x_1) \, dx_1 \tag{4.4}$$

This relation can be deduced by considering the member shown in Fig. 4.3, subjected to a distributed load $q(x_1)$ extending from $x_1 = x_1^L$ to $x_1 = x_1^R$. The portion of this load from point x_1 to $x_1 + dx_1$ may be treated as a concentrated force of magnitude $q \, dx_1$ and, consequently, from relation (4.2), the value of the quantity X due to this force is equal to

$$dX = (q \, dx_1)\hat{X} \tag{4.5}$$

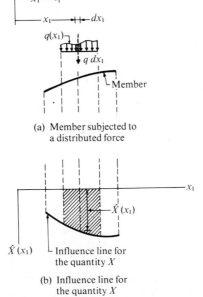

(a) Member subjected to
    a distributed force

(b) Influence line for
    the quantity X

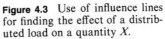

Figure 4.3 Use of influence lines for finding the effect of a distributed load on a quantity X.

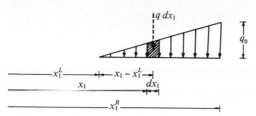

Figure 4.4 Triangular loading.

Thus,

$$X = \int_{x_1^L}^{x_1^R} q\hat{X}\, dx_1$$

If q is a uniform load, the above relation yields

$$X = q \int_{x_1^L}^{x_1^R} \hat{X}\, dx_1 \tag{4.6}$$

Referring to Fig. 4.3, we can easily see that the value of the integral is equal to the area under the influence line, from $x_1 = x_1^L$ to $x_1 = x_1^R$.

If $q(x_1)$ is a triangular load, referring to Fig. 4.4 we have

$$q = \frac{q_0(x_1 - x_1^L)}{x_1^R - x_1^L} \tag{4.7}$$

Substituting relation (4.7) into (4.4) we find that the value of the quantity X due a triangular load is equal to

$$X = \frac{q_0}{x_1^R - x_1^L} \int_{x_1^L}^{x_1^R} (x_1 - x_1^L)\hat{X}\, dx_1 \tag{4.8}$$

The value of the integral is equal to the moment of the portion of the area under the influence line from $x_1 = x_1^L$ to $x_1 = x_1^R$ about point $x_1 = x_1^L$.

4.6   Use of Influence Lines for Obtaining the Maximum Value of a Quantity Due to Moving Loads

The loadings which must be considered in the design of a structure are usually given by design specifications or building codes. For instance, the loading shown in Fig. 4.5 is taken from the Standard Specifications for Highway Bridges (12th ed.) of the American Association of State Highway and Transportation Officials (AASHTO). According to these specifications, the maxi-

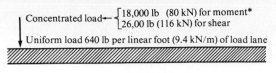

Concentrated load $\leftarrow$ $\begin{cases} 18{,}000 \text{ lb} & (80 \text{ kN}) \text{ for moment*} \\ 26{,}00 \text{ lb} & (116 \text{ kN}) \text{ for shear} \end{cases}$

Uniform load 640 lb per linear foot (9.4 kN/m) of load lane

HS 20-44 (MS 18) loading

* For the loading of continuous spans involving lane loading refer
to Article 1.2.8(C) which provides for additional concentrated load.

(a) Lane loading

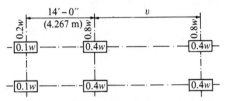

HS 20-44 (MS 18)   8,000 lb (36 kN)   32,000 lb+ (144 kN)   32,000 lb+(144 kN)

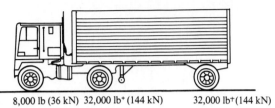

w = combined weight on the first two axes which is the same
as for the corresponding H(M) truck

v = variable spacing- 14 ft to 30 ft (4.267 to 9.144 m) inclusive.
Spacing to be used is that which produces maximum stresses.

+ In the design of timber floors and orthotropic steel decks
(excluding tranverse beams) for H20 (M 18) loading, one axle
load of 24,000 lb (108 kN) or two axle loads of 16,000 lb
(72 kN) each, spaced 4 ft (1.219 m) apart may be used,
whichever produces the greater stress, instead of the 32,000 lb
(144 kN) axle shown.

(b) Truck loading

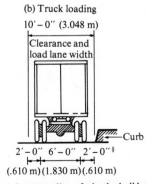

10′ – 0″ (3.048 m)

Clearance and
load lane width

Curb

2′ – 0″ 6′ – 0″ 2′ – 0″ §

(.610 m)(1.830 m)(.610 m)

§ For slab design, the center line of wheels shall be assumed to be
1 ft (.305 m) from face of curb. [See Art. 1.3.2(B).]

(c) Lane arrangement

Figure 4.5 Highway bridge loading taken from *Standard Specifications for
Highway Bridges*, 1977, 12th ed. (*Courtesy of the American Association of
State Highway and Transportation Officials, Washington, D.C.*)

mum value of a quantity at a cross section of a stringer, floor beam, or girder of a bridge can result from one of the following two loading conditions:

1. Lane loading, which consists of a uniform and a concentrated load, as shown in Fig. 4.5a.
2. Truck loading as shown in Fig. 4.5b

The loading shown in Fig. 4.5 is for a single lane of traffic which has a width of 10 ft (3.048 m). Thus, for a bridge of n lanes, n of these lane loadings must be arranged on its deck in order to establish the loading on each stringer, floor beam, or girder.

In this section, we determine the maximum value of a quantity (shearing force or bending moment) at a cross section of a structure due to certain simple types of moving loads, using the influence line for the quantity under consideration.

4.6.1 Sufficiently long uniform line load and one concentrated live load

Often, the live loads acting on a structure involve a uniform load q (kilonewtons per meter) extending over a length longer than the span on which the load travels as well as a concentrated load P (see Fig. 4.5a). In this case, the position of these loads on the structure corresponding to the maximum value of a quantity X (shearing force, axial force, or bending moment) acting on a cross section of a member of this structure may be obtained by inspection of the influence line $\hat{X}(x_1)$.

Example 1. Consider the girder of Fig. a. Find the maximum value of the shearing force in panel CD and the bending moment at point D due to a long, uniform load q and a concentrated force P.

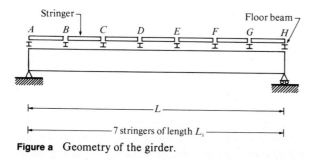

Figure a Geometry of the girder.

---

solution The influence line for the shearing force in panel CD and the bending moment at point D are shown in Figs. b and c, respectively. From inspection of Fig.

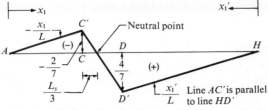

Figure b Influence line for the shearing force in panel CD.

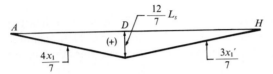

Figure c Influence line for the moment at point D.

b, it can be seen that the maximum value of the shearing force in panel CD of the girder may be obtained by loading the portion of the span of the girder from the neutral point to the end H with the uniform load, and placing the concentrated load at point D, where the ordinate of the influence line is maximum. Thus,

$$Q_{\text{max at } CD} = \begin{pmatrix} \text{area under the influence} \\ \text{line from the neutral} \\ \text{point to point } H \end{pmatrix} q + \frac{4P}{7} = \frac{4L_s q}{3} + \frac{4P}{7} \qquad \text{(a)}$$

Similarly, referring to Fig. c, it is apparent that the maximum value of the bending moment in the girder under the floor beam D may be obtained by loading the entire span of the structure with the uniform load and placing the concentrated load at point D. Thus,

$$M_{\text{max at point } D} = \frac{1}{2}(7L_s)\left(\frac{12L_s}{7}\right)q + \frac{12L_s P}{7} \qquad \text{(b)}$$

$$= 6L_s^2 q + \frac{12L_s P}{7}$$

4.6.2   Short uniform live load

Consider a quantity X at a cross section of a structure whose influence line has a vertex (see Fig. 4.6). Suppose that a uniform load of length c travels on the structure. Assume that c is sufficiently small and that the vertex of the influence line is located at such a point along the path of travel of the load that when the load is placed in the neighborhood of the vertex, all of it is on the

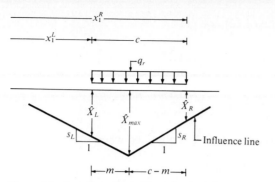

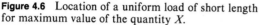

Figure 4.6 Location of a uniform load of short length for maximum value of the quantity X.

structure. Referring to Fig. 4.6, the value of the quantity X for the position of the live load specified by the distance m is equal to

$$X = \left(\begin{array}{c}\text{area under the influence}\\\text{line from } x_1 = x_1^L \text{ to } x_1 = x_1^R\end{array}\right) q$$

$$= \left[\left(\frac{\hat{X}_L + \hat{X}_{\max}}{2}\right) m + \left(\frac{\hat{X}_R + \hat{X}_{\max}}{2}\right)(c - m)\right] q$$

$$= \left[\left(\frac{\hat{X}_L - \hat{X}_R}{2}\right) m + \left(\frac{\hat{X}_R + \hat{X}_{\max}}{2}\right) c\right] q \qquad (4.9)$$

Denoting the slope of the influence line to the right and the left of its vertex by S_R and S_L, respectively, and referring to Fig. 4.6, we get

$$\hat{X}_R = \hat{X}_{\max} - S_R(c - m)$$
$$\hat{X}_L = \hat{X}_{\max} - S_L m \qquad (4.10)$$

Substituting relations (4.10) into (4.9), we obtain

$$X = \left[\frac{S_L m^2 + S_R(c - m)^2}{2}\right] q \qquad (4.11)$$

The maximum value of this quantity is obtained for the value of m which satisfies the following relation:

$$\frac{dX}{dm} = 0 = 2mS_L - 2S_R(c - m)$$

or
$$m = \frac{cS_R}{S_R + S_L} \qquad (4.12)$$

Substituting relation (4.12) into (4.10), we get

$$\hat{X}_L = \hat{X}_{max} - \frac{cS_L S_R}{S_L + S_R}$$

$$\hat{X}_R = \hat{X}_{max} - \frac{cS_L S_R}{S_L + S_R} \qquad (4.13)$$

Thus, the maximum value of the quantity X is obtained when

$$\hat{X}_L = \hat{X}_R \qquad (4.14)$$

Substituting relation (4.14) into (4.9), we obtain

$$X_{max} = \left(\frac{\hat{X}_R + \hat{X}_{max}}{2}\right) cq \qquad (4.15)$$

Example 2. For the beam shown in Fig. a, compute the maximum moment at point A due to a moving uniform load of magnitude 20 kN/m and length 1 m.

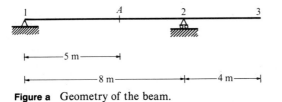

Figure a Geometry of the beam.

solution The influence line for the moment at point A is shown in Fig. b. The maximum value of the moment at A occurs when the load is placed on the beam so that

$$\hat{X}_L = \hat{X}_R \qquad (a)$$

Referring to Fig. b, from geometric considerations, we have

$$\frac{\hat{X}_L}{1.875} = \frac{5 - a}{5} \qquad (b)$$

$$\frac{\hat{X}_R}{1.875} = \frac{3 - (1 - a)}{3} = \frac{2 + a}{3} \qquad (c)$$

Using relation (a), we get from the above relations

$$a = \tfrac{5}{8} \qquad (d)$$

Substituting relation (d) into (b) or (c), we obtain

$$\hat{X}_L = \hat{X}_R = \tfrac{7}{8}(1.875) = 1.64 \qquad (e)$$

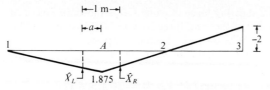

Figure b Influence line for the moment at point A.

Substituting relation (e) into (4.15), we obtain

$$M^{(A)}_{max} = \left(\frac{1.64 + 1.875}{2}\right) 20 = 35.15 \text{ kN}\cdot\text{m}$$

4.6.3 Series of concentrated live loads

There are systems of live loads which consist of a series of concentrated loads such as those transmitted by the wheels of trains or trucks. When there are many concentrated live loads located relatively close together, as in the case of railroad trains, the load is often simulated by an equivalent uniformly distributed load plus a concentrated load. In this section, we illustrate the use of influence lines in computing the maximum value of an internal action at a cross section of a member of a structure, subjected to a series of concentrated live loads, by the following two examples:

Example 1.   Consider a truck traveling from right to left on a bridge consisting of two girders with floor beams and stringers as shown in Fig. a. The spacing and magnitude of the wheel loads are shown in Fig. b. Compute the maximum value of the shearing force in panel CD and of the moment at point D.

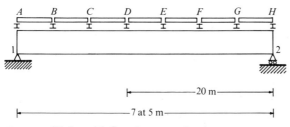

Figure a Girder with floor beams and stringers.

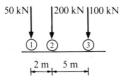

Figure b Wheel load on one girder.

solution The influence line for the shearing force in panel *CD* is shown in Fig. c. As can be seen from this influence line, the shearing force in panel *CD* increases as the first wheel approaches panel point *D*. When wheel 1 is on panel point *D*, the value of the shearing force in panel *CD* may be computed by referring to Fig. c. Thus,

$$Q_{CD} = 50\left(\frac{20}{35}\right) + 200\left(\frac{18}{35}\right) + 100\left(\frac{13}{35}\right) = \frac{1180}{7} = 168.6 \text{ kN}$$

As the truck proceeds to the left, the value of the ordinate of the influence line at the position of wheel 1 will decrease, while the values of the ordinates of the influence line at the position of wheels 2 and 3 will increase. Referring to Fig. c, notice that the shearing force in panel *CD* when wheel 2 is on panel point *D* is equal to

$$Q_{CD} = 50\left(\frac{8}{35}\right) + 200\left(\frac{20}{35}\right) + 100\left(\frac{15}{35}\right) = \frac{1180}{7} = 168.6 \text{ kN}$$

It is apparent that in this example, the increase in shearing force due to the movement of wheels 2 and 3 balances the decrease in shearing force due to the movement of wheel 1. As the truck continues to move to the left, it can be seen by inspection of the influence line of Fig. c that the value of the shearing force in panel *CD* will decrease. Consequently the maximum value of the shearing force in panel *CD* is 168.6 kN.

Notice that the maximum shearing force in a panel usually occurs when the first or the second wheel is placed at one of its panel points.

The influence line for the bending moment at panel point *D* is shown in Fig. d. It can be seen from this influence line that the moment at panel point *D* increases as the first wheel approaches panel point *D*. When wheel 1 is at panel point *D*, the moment is equal to

$$M^{(D)} = 50\left(\frac{60}{7}\right) + 200\left(\frac{54}{7}\right) + 100\left(\frac{39}{7}\right) = 2530 \text{ kN}\cdot\text{m}$$

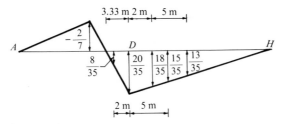

Figure c Influence line for the shearing force at any cross section in panel *CD*.

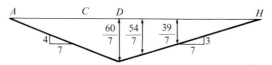

Figure d Influence line for the moment at point *D*.

The increase or decrease of the moment as the truck proceeds to the left may be obtained from the slopes of the influence line. Referring to Fig. d, these slopes are $-4/7$ for part AD and $3/7$ for part DH. Thus, as the truck proceeds to the left, the effect of wheel 1 will decrease at a rate of $50(4/7) = 200/7$ (kN·m/m), while the combined effect of wheels 2 and 3 will increase at a rate of $(200 + 100)(3/7) = 900/7$ kN·m/m. That is, the total moment at panel point D will increase by $700/7$ kN·m when wheel 1 moves 1 m to the left of panel point D. This moment will continue to increase at the same rate until wheel 2 reaches panel point D. Then, the total moment at panel point D will be equal to

$$M^{(D)} = 2530 + 2(100) = 2730 \text{ kN·m}$$

Referring to Fig. d, as the truck continues to move to the left, we can see that the moment at panel point D will decrease. Consequently, the maximum value of the moment at point D is 2730 kN·m.

From this example it is apparent that, generally, by inspection of the influence lines, it is possible to start with a load position close to that which yields a maximum value of the shearing force or the bending moment at a cross section and then establish the maximum value by computing the change of the shearing force or the moment as the loads move. This procedure may be employed even for long series of moving loads, as in the case of railroad trains.

Example 2.   Compute the maximum and minimum values of the internal forces in members 7, 8, 9, 10, 16, 19, and 20 of the truss of Fig. a due to a dead load of 10 kN/m and the series of concentrated moving loads shown in Fig. b.

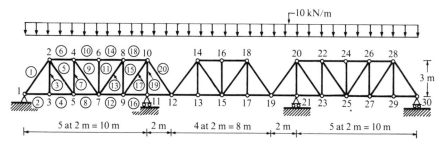

Figure a   Geometry and dead load of the structure.

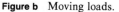

Figure b   Moving loads.

solution The internal forces in members 7, 8, 9, 10, 16, 19, and 20 of the truss due to the dead load of 10 kN/m have been established in Example 4 of Sec. 3.2 and are given in the table below.

Member no.	Force due to the dead load, kN
7	10.0
8	26.7
9	−12.0
10	−20.0
16	−33.4
19	−100.0
20	60.9

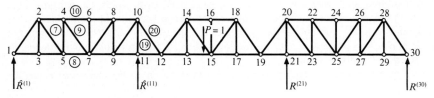

Figure c Free-body diagram of the truss.

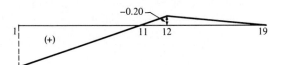

Figure d Influence line for the reaction $R^{(1)}$.

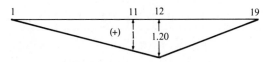

Figure e Influence line for the reaction $R^{(11)}$.

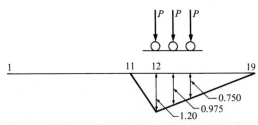

Figure f Influence line for the force in member 20.

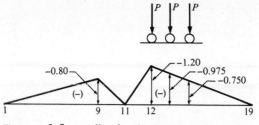

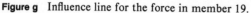

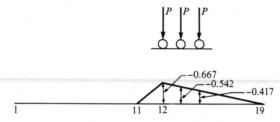

Figure g Influence line for the force in member 19.

Figure h Influence line for the force in member 16.

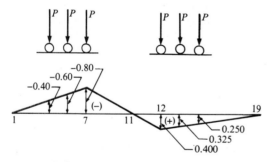

Figure i Influence line for the force in member 10.

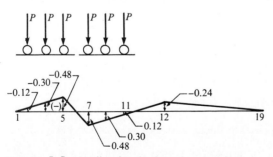

Figure j Influence line for the force in member 9.

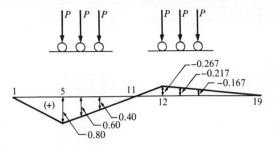

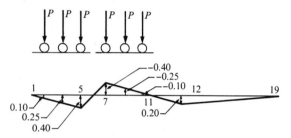

Figure k Influence line for the force in member 8.

Figure l Influence line for the force in member 7.

In order to establish the position of the set of concentrated moving loads for maximum (tension) and minimum (compression) values of the internal forces in members 7, 8, 9, 10, 16, 19, 20 of the structure, the influence lines for the forces in these members are plotted in Figs. f to l. Moreover, in these figures the location of the loads is indicated for maximum or minimum value of the internal force in the aforementioned members. Referring to Figs. f to l and using the values of the forces due to the dead load from the table above, we get

$$\text{max } N^{(7)} = 10.0 + 50(0.10 + 0.25 + 0.40) = 47.50 \text{ kN}$$

$$\text{min } N^{(7)} = 10.0 - 50(0.10 + 0.25 + 0.40) = -27.50 \text{ kN}$$

$$\text{max } N^{(8)} = 26.7 + 50(0.80 + 0.60 + 0.40) = 116.70 \text{ kN}$$

$$\text{min } N^{(8)} = 26.7 - 50(0.267 + 0.257 + 0.167) = -5.85 \text{ kN}$$

$$\text{max } N^{(9)} = -12.0 + 50(0.48 + 0.3 + 0.12) = 33.00 \text{ kN}$$

$$\text{min } N^{(9)} = -12.0 - 50(0.48 + 0.30 + 0.12) = -57.00 \text{ kN}$$

$$\text{max } N^{(10)} = -20.0 + 50(0.40 + 0.325 + 0.250) = 28.75 \text{ kN}$$

$$\text{min } N^{(10)} = -20.0 - 50(0.80 + 0.60 + 0.40) = -110.00 \text{ kN}$$

$$\text{min } N^{(16)} = -33.4 - 50(0.667 + 0.542 + 0.417) = -114.70 \text{ kN}$$

$$\text{max } N^{(19)} = -100 - 50(1.20 + 0.975 + 0.750) = -246.25 \text{ kN}$$

$$\text{max } N^{(20)} = 60.9 + 50(1.20 + 0.975 + 0.750) = 207.15 \text{ kN}$$

4.7   Absolute Maximum for Shearing Force and Bending Moment in a Beam

In the previous section, we presented methods for computing the maximum shearing force and moment at a specific cross section of a beam or the maximum axial force at a specific member of a truss due to moving loads. In this section, we present methods for determining the cross section of a beam where the shearing force or the moment assumes a value which is larger than that at any other cross section of the beam when the beam is subjected to a set of moving loads. This value of the shearing force or moment is referred to as its *absolute maximum*. The absolute maximum value of the shear force or moment occurs at a certain cross section when the moving loads are at a certain position. Each member of a structure must be designed so that the dimensions of any cross section along its length are such that the maximum values of the components of stress obtained by combining the maximum effect of the axial force, the shearing force, and the bending moment which can occur at this cross section do not exceed their allowable values given by specifications or design codes. If a member has a constant cross section, the maximum shearing or the maximum normal component of stress occurs at the cross sections where the shearing force or moment, respectively, assumes its absolute, maximum value. However, if a member has a variable cross section, the maximum shearing or the maximum normal component of stress does not always occur at the cross sections of the member where the shearing force or the moment, respectively, assumes its absolute, maximum value. Thus, in this case, in addition to the absolute maximum values of the shearing force and moment, their maximum values may be required at a number of cross sections along the length of the member. In fact, it may be necessary to plot the envelope of maximum values for each internal action of the member. The *envelope of maximum values* of a component of internal action of a member is a curve whose ordinates represent the maximum values of this component of internal action resulting from all possible combinations of the loads which are anticipated to act on the member (see Example 2 at the end of this section).

Frequently, the cross section where the absolute maximum shearing force or moment in a member may occur is identified by inspection of the corresponding influence line. However, if this cross section cannot be identified by inspection of the influence line, it is necessary to compute and compare the maximum shearing forces or moments at various cross sections where the absolute maximum shearing force or moment is likely to occur.

The absolute maximum shearing force in a beam due to a given set of moving loads will occur at a cross section located on one side of one of its supports. Thus, we must compute the maximum shearing force at the cross sections located to the left and to the right of each support of the beam. The greatest of these values is the absolute maximum shearing force due to the given set of moving loads.

The absolute maximum moment in a simply supported beam or a simply supported girder subjected to a uniform live load, or to a single concentrated live load, occurs at the midspan. The absolute maximum moment in a beam or a girder with overhangs is likely to occur at a section adjacent to one support. The absolute maximum moment due to a series of concentrated loads moving on a simply supported beam occurs under one of the loads.

In the sequel, we present a procedure for finding the location of a set of concentrated loads moving on a simply supported beam which will result in maximum moment under one of these loads. The absolute maximum moment in the beam can be established by finding the maximum moment under several loads and comparing the results.

Consider the simply supported beam, shown in Fig. 4.7, subjected to a set of concentrated moving loads $P^{(1)}$, $P^{(2)}$, ..., $P^{(n)}$. We will find the position of this set of loads on the beam so that the maximum moment occurs under load $P^{(i)}$. Let the position of this set of loads on the beam be specified by the coordinate x_1' measured from the center of the span to load $P^{(i)}$. Denote the position of the resultant R of the set of concentrated loads from load $P^{(i)}$ by d. The value of the reaction $R^{(1)}$ of the beam may be established by taking moments about point 2. Thus,

$$R^{(1)} = \frac{R(L/2 + x_1' - d)}{L} \tag{4.16}$$

The moment $M^{(i)}$ under the load $P^{(i)}$ is equal to

$$M^{(i)} = R^{(1)}\left(\frac{L}{2} - x_1'\right) - [P^{(1)}a_1 + P^{(2)}a_2 + \cdots + P^{(i-1)}a_{(i-1)}] \tag{4.17}$$

The value of x_1' for maximum value $M^{(i)}$ may be obtained by setting

$$\frac{dM^{(i)}}{dx_1'} = \frac{dR^{(1)}}{dx_1'}\left(\frac{L}{2} - x_1'\right) - R^{(1)} = 0 \tag{4.18}$$

Substituting relation (4.16) into the above, we get

$$x_1' = \tfrac{1}{2}d \tag{4.19}$$

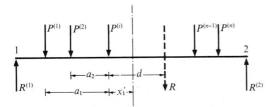

Figure 4.7 Simply supported beam subjected to a series of concentrated moving loads.

Thus, the maximum moment directly beneath one of a series of concentrated loads moving on a simply supported beam occurs when the center of the span is halfway between this load and the resultant of all the loads on the span.

Example 1. For the simply supported beam shown in Fig. a, find the absolute maximum bending moment due to a truck, which is represented by the series of concentrated loads shown in Fig. b and which moves from point 2 to point 1.

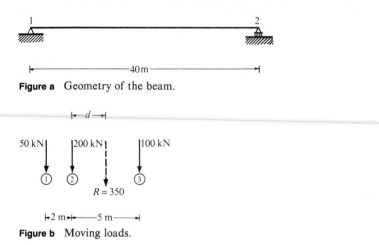

Figure a Geometry of the beam.

Figure b Moving loads.

---

solution We assume that the maximum moment occurs under wheel 2. We first compute the distance d between wheel 2 and the resultant force of all the moving loads. Thus,

$$d = \frac{-50(2) + 100(5)}{350} = \frac{8}{7} \text{ m}$$

For the maximum moment at wheel 2, the loads are placed as shown in Fig. c. The reaction $R^{(1)}$ for this position of the loads is equal to

$$R^{(1)} = \frac{(350)(136/7)}{40} = \frac{70}{8}\left(\frac{136}{7}\right) = 170 \text{ kN}$$

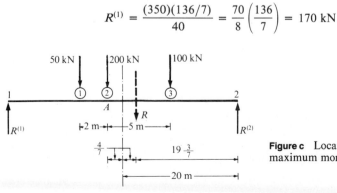

Figure c Location of the loads for maximum moment under load 2.

The maximum moment under wheel 2 is

$$M_{\max} = R^{(1)}(136/7) - 50(2) = 3202.9 \text{ kN} \cdot \text{m}$$

If we check the maximum moments under the other two wheels, we will find that they are smaller than the one obtained above. Thus, the absolute maximum moment in the beam of Fig. a due to the moving loads of Fig. b occurs at point A (see Fig. c) when wheel 2 is at this point. The value of the absolute maximum moment in the beam is 3202.9 kN·m.

Example 2. For the one girder of the bridge of Fig. a, plot the envelope of the maximum values of the moment due to the live loads given by loading HS 20-4A(MS18) of the Standard Specifications for Highway Bridges of the American Association of State Highway and Transportation officials (AASHTO) given in Fig. 4.5. Take the width of each lane as 3.3 m. The weight of the bridge is 110 kN/m. Give the ordinates of the envelope every 2.5 m.

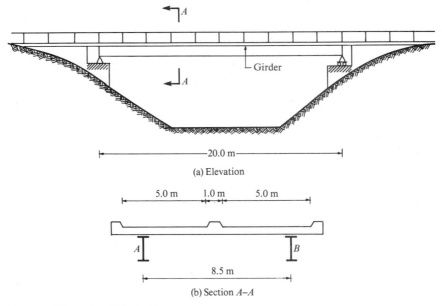

(a) Elevation

(b) Section A–A

Figure a Geometry of the bridge.

solution In order to produce maximum effect on girder B, the lane loading of Fig. 4.5a must be placed on the deck of the bridge in the position shown in Fig. b. Referring to this figure for maximum moment for the lane loading of Fig. 4.5a, we have

Concentrated Force (80 kN per lane)

$$\Sigma M^A = 0 \qquad 80(2.10 + 8.10) - 8.5R^B = 0 \qquad \text{or} \qquad R^B = 96 \text{ kN}$$

Distributed Force (9.4 kN/m per lane)

$$\Sigma M^A = 0 \qquad 9.4(2.10 + 8.10) - 8.5R^B = 0 \qquad \text{or} \qquad R^B = 11.28 \text{ kN/m}$$

Moreover in order to produce maximum effect on girder B, the loading of the one truck of Fig. 4.5b must be placed on the deck of the bridge in the position shown in Fig. c. Referring to this figure for maximum moment in girder B for the one truck of Fig. 4.5b we have

$$\Sigma M^A = 0 \qquad P(7.31 + 9.14) - 8.5R^B = 0 \qquad \text{or} \qquad R^B = 1.9354P$$

where P is the wheel loading.

The loading for maximum moment on girder B due to one truck driving on the lane which is located close to this girder is shown in Fig. ea. However, the bridge is carrying a two-way road and consequently it is possible to have a second truck traveling in the direction opposite to that of the first one. This truck must be placed on the deck of the bridge as shown in Fig. d. Referring to this figure for maximum moment in girder B we have

$$\Sigma M^A = 0 \qquad P(1.31 + 3.14) - 8.5R^B = 0 \qquad \text{or} \qquad R^B = 0.652354P$$

Thus for maximum moment the girder must be subjected to the following loading.

1. To a dead load of $q = 110/2 = 55$ kN/m
2. To a live load consisting of one of the following:
   a. A combination of a concentrated force of 96 kN and a uniform load of 11.28 kN/m.
   b. The two series of wheel loads traveling in opposite directions as shown in Fig. e.

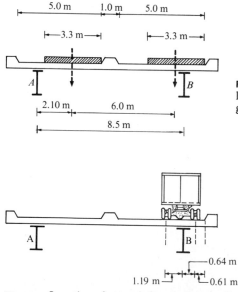

Figure b Location of the lane loading for maximum effect on girder B.

Figure c Location of a truck for maximum effect on girder B.

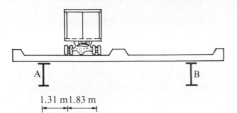

Figure d Location of a truck for maximum effect on girder *B*.

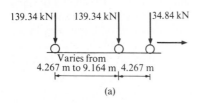

(a)

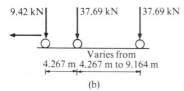

(b)

Figure e Wheel loads for maximum moment on girder *B*.

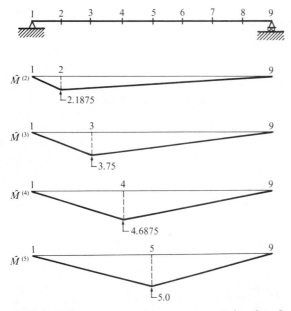

Figure f Influence lines for the moment at points 2 to 5 of the beam.

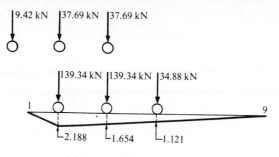

Figure g Position of trucks for maximum moment of point 2.

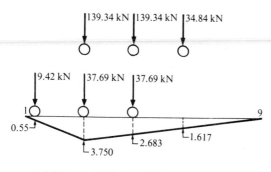

Figure h Position of trucks for maximum moment of point 3.

The girder is subdivided into eight equal parts by points 2 to 8. The influence lines for the moment at points of the girder are shown in Fig. f. Referring to this figure, we see that the moments at the points of the girder due to the dead load are equal to

$$M^{(2)} = \frac{2.1875(55)(20)}{2} = 1204.13 \text{ kN} \cdot \text{m}$$

$$M^{(3)} = \frac{3.75(55)(20)}{2} = 2062.50 \text{ kN} \cdot \text{m}$$

$$M^{(4)} = \frac{4.6875(55)(20)}{2} = 2578.13 \text{ kN} \cdot \text{m}$$

$$M^{(5)} = \frac{5(55)(20)}{2} = 2750.00 \text{ kN} \cdot \text{m}$$

Moreover, referring to Fig. f, we see that at points 2 to 5 of the girder the maximum moment due to the combination of a concentrated force of 96 kN and a uniform load of 11.28 kN/m is equal to

$$\max M^{(2)} = 96(2.19) + 11.28(\tfrac{1}{2})2.19(20) = 457.27 \text{ kN·m}$$

$$\max M^{(3)} = 96(3.75) + 11.28(\tfrac{1}{2})3.75(20) = 783.00 \text{ kN·m}$$

$$\max M^{(4)} = 96(4.69) + 11.28(\tfrac{1}{2})4.69(20) = 979.27 \text{ kN·m}$$

$$\max M^{(5)} = 96(5) + 11.28(\tfrac{1}{2})5(20) = 1044.00 \text{ kN·m}$$

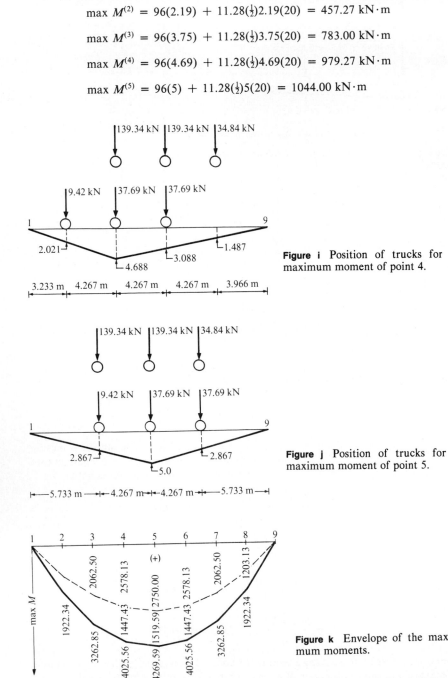

Figure i Position of trucks for maximum moment of point 4.

Figure j Position of trucks for maximum moment of point 5.

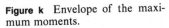

Figure k Envelope of the maximum moments.

To obtain at points 2 to 5 of the girder the maximum moment due to the track loading, we must place two trucks traveling in opposite directions in the positions shown in Figs. g to j. Thus referring to these figures, we have

$$\max M^{(2)} = (139.34 + 37.69)(2.188 + 1.654) + 34.84(1.121) = 719.21 \text{ kN} \cdot \text{m}$$

$$\max M^{(3)} = 9.42(0.55) + (139.34 + 37.69)(3.75 + 2.683) + 34.84(1.617)$$

$$= 1200.35 \text{ kN} \cdot \text{m}$$

$$\max M^{(4)} = 9.42(2.021) + (139.34 + 37.69)(4.688 + 3.088) + 34.84(1.487)$$

$$= 1447.43 \text{ kN} \cdot \text{m}$$

$$\max M^{(5)} = (139.34 + 9.42)2.867 + (139.34 + 37.69)5 + (34.84 + 37.69)2867$$

$$= 1519.59 \text{ kN} \cdot \text{m}$$

It is apparent that the maximum moments at all points of the girder are produced by the truck loading. Thus the maximum moments due to the truck loading are combined with those due to the dead load to get the maximum moments at the points of the girder. The envelope of maximum moments is plotted in Fig. k.

4.8 Problems

1 and 2.   Draw the influence lines for the reaction at support 2 as well as the shearing force and moment at point A and the moment at point 2 of the beam shown in Fig. 4.P1. Repeat with Fig. 4.P2.

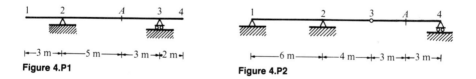

Figure 4.P1

Figure 4.P2

3 and 4.   Draw the influence lines for the reactions at support 1 and the shearing force and moment at point A of the beam shown in Fig. 4.P3. Repeat with Fig. 4.P4.

Figure 4.P3

Figure 4.P4

5.   For the beam shown in Fig. 4.P5, draw the influence lines for the reactions at points 1 and 3 and the shearing force and moment at point A when the unit load moves from point 4 to 5 on the upper deck.

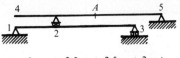

Figure 4.P5

6 and 7.   For the girder of Fig. 4.P6 draw the influence lines for the shearing force in panel *CD*, the moment at panel point *C*, and the moment in the middle point of panel *CD* when the load moves on the stringers. Repeat with Fig. 4.P7.

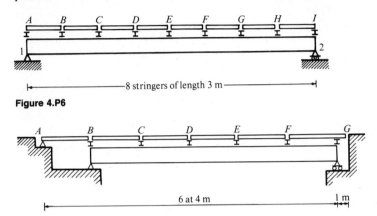

Figure 4.P6

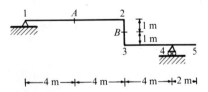

Figure 4.P7

8.   Draw the influence lines for the reaction at point 4 as well as the shearing force and the moment at points *A* and *B* of the structure shown in Fig. 4.P8. The loads travel on the horizontal members.

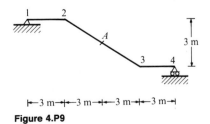

Figure 4.P8

9.   Draw the influence lines for the reaction at point 1 as well as the shearing force and moment at point *A* of the structure shown in Fig. 4.P9.

Figure 4.P9

10. Draw the influence lines for the axial force in member 2, the shearing force and bending moments at point A, and the bending moment at point 8 of the structure shown in Fig. 4.P10.

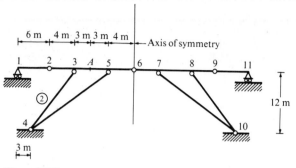

Figure 4.P10

11. Draw the influence lines for the shearing force and the moment at points A, B, and C of the structure shown in Fig. 4.P11. Points B and C are just below joints 3 and 6, respectively.

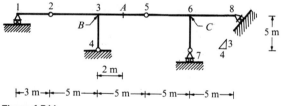

Figure 4.P11

12. Draw the influence lines for the axial force, the shearing force, and the moment at points A, B, and C of the structure shown in Fig. 4.P12. The load is moving on the beam 4,10. Points B and C are just below joints 2 and 3, respectively.

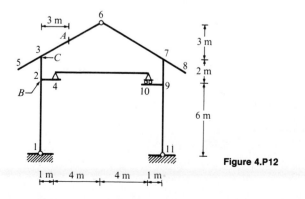

Figure 4.P12

13 to 17.   Draw the influence lines for the forces in the numbered members of the truss shown in Fig. 4.P13. The load is moving on a floor system resting on the bottom chord of the truss. Repeat with Figs. 4.P14 to 4.P17.

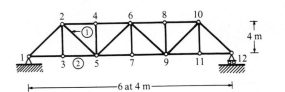

Figure 4.P13

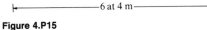

Figure 4.P14

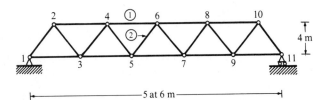

Figure 4.P15

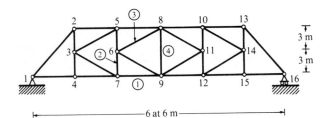

Figure 4.P16

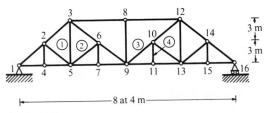

Figure 4.P17

18. For the trussed frame of Fig. 4.P18 draw the influence lines for the forces in members 1, 2, and 3 when the load is moving on a floor system which rests at the top of the trussed frame.

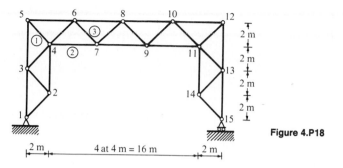

Figure 4.P18

19 to 24. For point A of the structure of Fig. 4.P1 compute the maximum value of the shearing force and the moment due to a long uniform moving load of 120 kN/m and a concentrated moving force of 60 kN. Repeat with the structures of Figs. 4.P2, 4.P3, 4.P4, 4.P8, and 4.P11.

25 and 26. For the girder of Fig. 4.P7 compute the maximum value of the shearing force in panel CD and the maximum moment at panel point C due to a long uniform moving load of 30 kN/m and a concentrated moving force of 40 kN. Repeat for the girder of Fig. 4.P6.

27. For the structure of Fig. 4.P10 compute the maximum values of the axial force in member 2 and the shearing force and moment at point A due to a long uniform moving load of 20 kN/m and a concentrated moving force at 50 kN.

28. For points A and B of the structure of Fig. 4.P11 compute the maximum values of the shearing force and bending moments due to a long uniform moving load of 30 kN/m and a concentrated moving force of 40 kN/m.

29 to 33. For members 1, 2, and 3 of the truss of Fig. 4.P13 compute the maximum value of the axial force due to a long uniform moving load of 10 kN/m and a concentrated moving force of 60 kN. Repeat for members 1, 2 and 3 of the trusses of Figs. 4.P14 to 4.P17.

34 and 35. For points 2 and A of the beam of Fig. 4.P2 compute the maximum values of the moment due to a 1-m-long uniform moving load of 20 kN/m. Repeat for points 2 and A of the beam of Fig. 4.P1.

36. For point 1 and A of the beam of Fig. 4.P3, compute the maximum values of the moment due to a 1-m-long moving uniform load of 20 kN/m.

37 to 39. For the numbered members of the truss of Fig. 4.P13, compute the maximum values of the axial force due to a 1-m-long moving uniform load of 20 kN/m. Repeat for the numbered members of the trusses of Figs. 4.P14 and 4.P16.

40 and 41. For the beam of Fig. 4.P1 compute the maximum values of the moment at point 2 and the shearing force and moment at point A due to a truck traveling from right to left. The spacing and magnitude of the wheel loads are shown in Fig. 4.P40. Repeat for the beam of Fig. 4.P2.

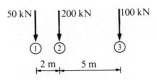

Figure 4.P40

42 to 44. Compute the maximum values of the axial forces in the numbered members of the truss of Fig. 4.P13 due to a truck traveling from right to left. The spacing and magnitude of the wheel loads are shown in Fig. 4.P42. Repeat for the numbered members of the trusses of Figs. 4.P14 and 4.P16.

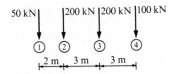

Figure 4.P42

45. Determine the absolute maximum moment in the beam of Fig. 4.P45 due to a truck traveling from right to left. The spacing and magnitude of the wheel loads are shown in Fig. 4.P45.

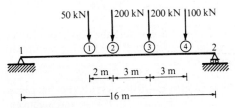

Figure 4.P45

46. Plot the envelope of the maximum absolute values of the shearing force and the maximum positive and negative values of the moment in the beam subjected to the dead load shown in Fig. 4.P46 and to a truck moving from points 1 to 3. The wheel loads are those shown in Fig. 4.P42.

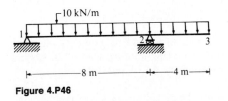

Figure 4.P46

Photographs and Brief Description of a Concrete Box Girder Bridge

The Columbia River flows from Canada through the state of Washington. For its last 200 mi before it reaches the Pacific Ocean, it marks the border between the states of Washington and Oregon.

The twin bridges of the Columbia River consist of two parallel, similar, but separate structures (Fig. 4.8). The bridges opened to traffic in December 1983. The superstructure of each bridge carries four lanes of traffic and consists of a closed, trapezoidal, box-section girder with internal struts or webs (see Fig. 4.13). The longest span

Figure 4.8 The twin bridges of the Columbia River near Portland, Oregon. *(Courtesy Mr. Allan C. Harwood of the Highway Division of the Department of Transportation of the State of Oregon.)*

is 600 ft (183 m). Each girder is supported on a series of reinforced concrete piers. The land piers are supported on spread footings, bearing directly on a cemented, gravel layer. The river piers are supported on steel H piles. After the piles were driven and cut off to grade underwater, a welded steel form with the reinforcing cage inside it was lowered into position and a 9-ft-deep concrete seal was poured. After the concrete seal hardened, the form was dewatered (see Fig. 4.10) and a reinforcing steel mat was placed at the top of the piles before concrete was poured to fill the form.

Each girder was constructed as a series of double cantilevers (see

Figure 4.9 View of the twin bridges during construction. The Columbia River flows from Canada through the state of Washington. *(Courtesy Mr. Allan C. Harwood of the Highway Division of the Department of Transportation of the State of Oregon.)*

Figure 4.10 View inside the welded steel form after the concrete seal had hardened and the form was dewatered. *(Courtesy Mr. Allan C. Harwood of the Highway Division of the Department of Transportation of the State of Oregon.)*

Figs. 4.9 and 4.12). The construction of each double cantilever
started from the pier on which it is supported and proceeded
simultaneously on both sides toward the cantilevers which were
constructed starting from the neighboring piers. The cantilevers stop
approximately 1 m from the middle of the span between the
adjacent piers and the gap is filled with concrete. The reinforced
concrete cantilever box girders of the smaller spans were constructed
from prefabricated concrete sections 3.65 m long and up to 5.18 m
deep (see Fig. 4.11). The reinforced concrete cantilever box girders
of the larger spans were cast in place one section at a time using
special metal forms that move hydraulically. Their depth varied
from 5.18 to 9.79 m (see Fig. 4.13).

Figure 4.11 A prefabricated concrete block is lifted into position from a
barge by two hydraulic jacks. In order to ensure a water tight joint, an
expoxy adhesive was applied to the contact surface between adjacent
blocks. *(Courtesy Mr. Allan C. Harwood of the Highway Division of the
Department of Transportation of the State of Oregon.)*

Figure 4.12 Construction of the double cantilevers. *(Courtesy Mr. Allan C. Harwood of the Highway Division of the Department of Transportation of the State of Oregon.)*

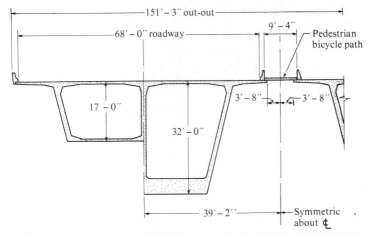

Figure 4.13 Cross section of the box girder for the 600-ft main span. The depth of the girder at the pier is 32 ft and 17 ft at midspan. *(Courtesy Mr. Allan C. Harwood of the Highway Division of the Department of Transportation of the State of Oregon.)*

Chapter 5

Energy Methods for Computing Components of Displacement of Points of a Structure

5.1 Introduction

In the computation of the internal actions of a statically determinate structure, it is not necessary to specify the material from which its members are made. However, as can be deduced from physical intuition, in order to compute the components of displacement of the cross sections of a structure, it is necessary to specify the material from which its members are made. Thus, we limit our attention to structures whose members are made of *isotropic, linearly elastic materials* (see Sec. A.5).

Because the dimensions of the cross sections of line members are small compared to their length, the normal component of stress τ_{22} and τ_{33} acting on the planes normal to the x_2 and x_3 axes, respectively, and the shearing component of stress† $\tau_{23} = \tau_{32}$ are negligible compared to the other components of stress (see Fig. 5.1). That is

$$\tau_{22} = \tau_{33} = \tau_{23} = 0 \tag{5.1}$$

Taking relation (5.1) into account, we see that the relations among the components of stress and strain for a particle of a framed structure whose members

---

† τ_{23} is the shearing component of stress acting on the plane normal to the x_2 axis in the direction of the x_3 axis.

are made of isotropic linearly elastic materials [see relations (A.49)] are

$$e_{11} = \frac{\tau_{11}}{E} + \alpha(T - T_0) + e_{11}^I$$

$$e_{12} = \frac{\tau_{12}}{2G} \qquad e_{13} = \frac{\tau_{13}}{2G} \qquad (5.2)$$

where E, G, and α are the modulus of elasticity, the shear modulus, and the coefficient of linear, thermal expansion, respectively, of the material from which the structure is made. T is the temperature of the particle at the present state; T_0 is the temperature of the particle at the stress free state; e_{11}^I is the normal component of the initial strain which was present in the member in the stress-free state.

In structural analysis, the movement due to deformation of a cross section of a member of a space framed structure is specified by the three components of the displacement vector of its centroid, referred to a set of retangular axes, and by the three components of its rotation about the same axes. Moreover, the movement due to the deformation of a cross section of a member of a planar framed structure is specified by the two components (in the plane of the structure) of the displacement vector of its centroid and by the component of rotation about the axis normal to the plane of the structure (see Fig. 5.2). The components of the displacement vector of the centroid of a cross section of a member of a structure are referred to as the *components of translation of this cross section*. Inasmuch as a cross section of a structure is represented by a point on the line diagram of the structure, we refer to the components of translation and rotation of the cross section represented by point A on the line diagram of the structure as the components of translation and rotation of point A of the structure.

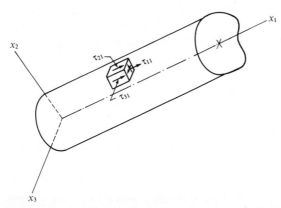

Figure 5.1 Components of stress acting on an element of a member of a framed structure.

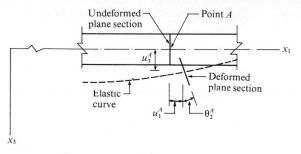

Figure 5.2 Components of displacement of a cross section of a member of a planar structure.

When a straight, line member is subjected to external centroidal axial forces, its axis elongates or shrinks but does not bend. Moreover, plane sections normal to the axis of the member prior to deformation remain plane subsequent to deformation and normal to its deformed axis. That is, in this case, the transverse components of translation u_2 and u_3 and the components of rotation θ_1, θ_2, θ_3 of any cross section of the member vanish. Moreover, referring to relation (A.74a), we see that the component of translation u_1 of a cross section of a member made of an isotropic linearly elastic material is related to the axial component of the internal force N acting on this cross section by the following relation:

$$\frac{du_1}{dx_1} = \frac{N}{EA} \tag{5.3}$$

where A is the area of the cross section.

When a prismatic, straight, line member is subjected only to internal torsional moments, its axis does not elongate or bend. Thus, the components of translation u_1, u_2, and u_3 vanish. However, a centroidal straight line normal to the axis of the member prior to deformation does not in general remain straight subsequent to deformation. It becomes a curve whose projection onto a plane normal to the axis of the member is a centroidal straight line which rotated by an angle θ_1 about the axis of the member subsequent to deformation. Every centroidal line of a plane normal to the axis of a member prior to deformation rotates by the same angle θ_1 about this axis subsequent to deformation. Referring to relation (A.100), we see that the angle θ_1 of a cross section of a member made of an isotropic linearly elastic material is related to the internal torsional moment M_1 acting on this cross section by the following relation:

$$\frac{d\theta_1}{dx_1} = \frac{M_1}{KG} \tag{5.4}$$

where K is referred to as the *torsional constant* of the member and is dependent upon the geometry of its cross section. The torsional constant of cylindrical (prismatic) members having a circular (hollow or full) cross section is equal to the polar moment of inertia of their cross section about its centroid. Formulas for the torsional constants of prismatic members whose cross sections have geometries of practical interest are given in Sec. A.11.

On the basis of the foregoing discussion, plane sections normal to the axis of a prismatic straight line member subjected to torsional moments generally do not remain plane after deformation; they rotate about the axis of the member, and they warp. Only plane sections normal to the axis of prismatic members of circular (full or hollow) cross section remain plane subsequent to deformation.

In structural analysis the warping of the cross sections of the members of a structure subjected to torsional moments is disregarded when their deformed configuration is specified. However, the effect of the warping of the cross section of the members of a structure is taken into account in the computation of their torsional constants. (See Sec. A.11.)

When line members are subjected to bending without twisting[†] their axis does not elongate and their cross sections do not twist. Moreover, because the dimensions of the cross sections of line members are small compared to their length, the following assumptions as to the geometry of their deformed configuration can be made.

1. Plane sections normal to the axis of a line member prior to deformation can be considered plane subsequent to deformation; that is, the warping of the cross sections of a line member is assumed negligible.

2. Plane sections normal to the axis of a line member before deformation can be considered normal to its deformed axis subsequent to deformation. This implies that the effect of the shearing components of strain[‡] e_{12} and e_{13} on the transverse components of translation of the points of the member is negligible.

Actually, when a line member is subjected to end bending moments, plane sections normal to its axis remain plane subsequent to deformation and normal to its deformed axis. However, when a line member is subjected to transverse external forces, its cross sections warp and do not remain normal to its deformed axis. Nevertheless, when the length of a line member is considerably larger than its other dimensions, the warping of its cross sections and the change of the angle between its axis and its cross sections due to its deforma-

---

† In order that a line member be subjected to bending without twisting, the external moments acting on it should not have a torsional component. Moreover, the plane of the external forces acting on it must pass through the shear center of the cross section of the member (see Sec. A.10).

‡ The shearing component of strain e_{12} at a particle is equal to half the change of the angle, due to deformation, between two infinitesimal materials line segments which are parallel to x_1 and x_2 axes prior to deformation.

tion do not affect appreciably its components of translation and its normal components of strain; thus they are disregarded.

The theory based on the first assumption only is referred to as the *Timoshenko theory of beams,* while the theory based on both assumptions is referred to as the *classical theory of beams.* Thus, in the classical theory of beams, the effect of the shearing deformation of a line member on its transverse components of displacement is disregarded, while in the Timoshenko theory of beams, this effect is taken into account.

Referring to relations (A.68b and (A.68c), we see that the components of rotation θ_2 and θ_3 of a cross section of a member made of an isotropic linearly elastic material are related to the components of the moments M_2 and M_3 acting on this cross section as follows:

$$\frac{d\theta_2}{dx_1} = \frac{M_2}{EI_2} \tag{5.5a}$$

$$\frac{d\theta_3}{dx_1} = \frac{M_3}{EI_3} \tag{5.5b}$$

where I_2 or I_3 are the moments of inertia of the cross section of the member about the x_2 or x_3 axis, respectively.

When the effect of shear deformation is taken into account (Timoshenko theory of beams), referring to relations (A.63b) and (A.63c) and (A.68d) and (A.68e) we may express the components of rotation of a cross section as

$$\theta_2 = -\frac{du_3}{dx_1} + 2e_{13} = -\frac{du_3}{dx_1} + \frac{\lambda_3 Q_3}{GA} \tag{5.6a}$$

$$\theta_3 = \frac{du_2}{dx_1} - 2e_{12} = \frac{du_2}{dx_1} - \frac{\lambda_2 Q_2}{GA} \tag{5.6b}$$

where Q_2/A or Q_3/A represents the average shearing component of stress τ_{12} or τ_{13}, respectively, acting on the cross section; λ_2 and λ_3 are correction factors by which the average value of the shearing component of stress must be multiplied in order to obtain its maximum value at the centroid. The coefficients λ_2 and λ_3 depend primarily on the geometry of the cross section of the member (see Sec. A.7).

When the effect of shear deformation is disregarded (classical theory of beams), referring to relations (5.6), we see that the components of rotation θ_2, θ_3 of a cross section are equal to the slope of the elastic curve. That is

$$\theta_2 = -\frac{du_3}{dx_1} \tag{5.7a}$$

$$\theta_3 = \frac{du_2}{dx_1} \tag{5.7b}$$

Substituting relations (5.7) into (5.5), we obtain

$$\frac{du_2^2}{dx_1^2} = \frac{M_3}{EI} \tag{5.8a}$$

$$\frac{du_3^2}{dx_1^2} = -\frac{M_2}{EI} \tag{5.8b}$$

Thus, on the basis of the classical theory of beams, the components of rotation θ_2 and θ_3 can be established if the components of translation u_2 and u_3 are known functions of x_1. This implies that on the basis of the classical theory of beams, the deformed configuration of a member of a space framed structure is completely specified if the components of translation u_1, u_2, and u_3 of its axis and the component of rotation θ_1 about its axis are known functions of the axial coordinate x_1. Moreover, choosing the x_3 axis in the plane of the structure, the deformed configuration of a member of a planar structure is completely specified if the components of translation u_1 and u_3 of its axis are known functions of x_1.

As discussed in Secs. A.8 and A.11, the components of displacement of the points of a member of a space structure can be established uniquely if the loading of the member is known and if the value of one of each of the following pairs of variables is known at each end of the member:

$$u_1 \text{ or } N \qquad \theta_1 \text{ or } M_1$$
$$u_2 \text{ or } Q_2 \qquad \theta_2 \text{ or } M_2 \tag{5.9}$$
$$u_3 \text{ or } Q_3 \qquad \theta_3 \text{ or } M_3$$

Moreover, the components of displacement of the points of a member of a planar structure can be established uniquely if the loading of the member is known and the value of one of each of the following pairs of variables is known at each end of the member.

$$u_1 \text{ or } N \qquad u_3 \text{ or } Q_3 \qquad \theta_2 \text{ or } M_2 \tag{5.10}$$

The computation of components of displacement (translation and rotation) of points of framed structures is part of structural analysis and design. For instance, as will become apparent in Chap. 7, in order to analyze a statically indeterminate structure by the force method, it is necessary to compute components of displacements of a number of its points. Moreover, components of displacements of points of parts of a structure are computed to ensure their smooth connection during construction. Furthermore, components of displacements of points of a structure are computed to ensure that they are not excessive. Excessive displacements of members of a structure could result in cracking of attached components (floors, ceilings, walls, pavements, etc) made of brittle materials, such as concrete, brick, or plaster. Moreover, excessive dis-

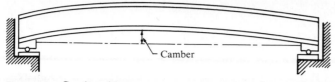

Figure 5.3. Cambered beam.

placements make a structure appear unsafe. For this reason, in order to reduce their apparent deflections, beams and trusses are often *cambered*. That is, they are fabricated with an initial deflection equal and opposite to their anticipated deflection when subjected to the dead load and part of the live load (see Fig. 5.3). Thus, under this loading, they appear as if they have not deflected at all.

As illustrated in the examples of Sec. A.8, the components of displacements of the points of a member of a statically determinate structure may be established by integrating the differential equations relating the components of displacements to the components of internal actions [i.e., Eqs. (5.3) to (5.6) or Eqs. (5.3), (5.4), and (5.8)]. The constants of integration are evaluated from the given displacement conditions at the supports of the structure and from the requirement that the components of displacement at adjacent points be equal (see examples of Sec. A.8). However, this approach can become very cumbersome when the loading on the members of the structure is complex or when the structure consists of more than one or two members.

In this and the subsequent chapters, we present methods for computing components of displacement of points of a structure. In Secs. 5.4 to 5.7 we derive the principle of "virtual" work for framed structures disregarding the effect of shear deformation. In Sec. 5.8 we present the method of virtual work (method of "virtual" forces). In Secs. 5.9 to 5.12 we apply this method to compute the components of displacements of points of structures. In this chapter, we employ the method of virtual work in computing the components of displacement of points of statically determinate structures. In Sec. 7.11, we employ the method of virtual work in computing the components of displacements of points of statically indeterminate structures. In Sec. 5.13 we derive the principle of virtual work for framed structures, including the effect of shear deformation. Moreover, in Sec. 5.15 from the principle of virtual work, we derive Castigliano's second theorem and we employ it to compute components of translation and rotation of points of statically determinate framed structures. In Sec. 5.16 we derive the Betti-Maxwell reciprocal theorem.

5.2 Work Performed by Concentrated Actions

The increment of work dW of a concentrated force $\mathbf{P} = P_1\mathbf{i}_1 + P_2\mathbf{i}_2 + P_3\mathbf{i}_3$ acting on a point of a body when that point is displaced by an infinitesimal amount $d\mathbf{r} = dx_1\mathbf{i}_1 + dx_2\mathbf{i}_2 + dx_3\mathbf{i}_3$ is equal to

$$dW = \mathbf{P} \cdot d\mathbf{r} = P_1\, dx_1 + P_2\, dx_2 + P_3\, dx_3 = P\, dr\cos\phi \quad (5.11)$$

where $\mathbf{P} \cdot d\mathbf{r}$ is the dot product of the vectors $\mathbf{P}$ and $d\mathbf{r}$ and ϕ is the angle between the vectors $\mathbf{P}$ and $d\mathbf{r}$ (see Fig. 5.4). From Eq. (5.11) it is apparent that the work dW is positive if the angle ϕ is less than $\pi/2$, negative if ϕ is greater than $\pi/2$, and vanishes if $\phi = \pi/2$.

The work of a group of forces is equal to the sum of the work of each of the forces of the group. Moreover, the work of a force acting on a point of a body when that point is displaced along a curve AB (see Fig. 5.4) is equal to the sum of the increments of work $\mathbf{P} \cdot d\mathbf{r}$ performed by the force as its point of application moves through each successive infinitesimal distance $d\mathbf{r}$ making up this curve. Thus,

$$W = \int_A^B \mathbf{P} \cdot d\mathbf{r} \quad (5.12)$$

In general, the work W of a force depends on the path of motion of the force. However, the work of certain forces depends only on their original and final positions and not on the path which they have followed in moving from the original to the final position. Such forces are referred to as *conservative. In this text we consider only conservative forces.* The work performed by a conservative force $\mathbf{P} = P_1\mathbf{i}_1 + P_2\mathbf{i}_2 + P_3\mathbf{i}_3$ acting on a point of a rigid body which translates by $\mathbf{u} = u_1\mathbf{i}_1 + u_2\mathbf{i}_2 + u_3\mathbf{i}_3$ is equal to

$$W = P_1u_1 + P_2u_2 + P_3u_3 = \mathbf{P} \cdot \mathbf{u} \quad (5.13)$$

As discussed in Sec. 1.3.1, a concentrated moment $\mathbf{M}$ is a convenient representation of forces of very high intensity, which are distributed over a very small area and change abruptly in direction. The resultant of these forces vanishes, while their moment about any point is equal to $\mathbf{M}$. These forces may be replaced by an equivalent system of two equal and opposite concentrated forces of very high intensity acting in the plane normal to the vector of the moment $\mathbf{M}$ and located at a very small distance ΔL apart. For example, referring to

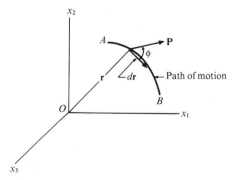

Figure 5.4 Force whose point of application moves from point A to point B.

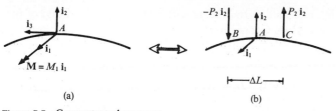

(a)

(b)

Figure 5.5 Concentrated moment.

Fig. 5.5*a*, we see that the concentrated moment $\mathbf{M} = M_1\mathbf{i}_1$ which is applied at point A of a rigid body may be replaced by an equivalent system of two equal and opposite forces $P_2\mathbf{i}_2$ and $-P_2\mathbf{i}_2$ located as shown in Fig. 5.5*b* in the plane normal to the unit vector $\mathbf{i}_1$. The magnitude of these forces is such that their moment is equal to M_1. When the rigid body moves, the work of the concentrated moment must be equal to the sum of the work of these two forces. Thus, referring to Fig. 5.5, we may deduce that only when the rigid body rotates about an axis parallel to the vector of the moment $\mathbf{M}$ the two forces and thus the moment $\mathbf{M}$ perform work. For instance when the rigid body rotates by a sufficiently small rotation $\theta_1\mathbf{i}_1$ the work of the moment $\mathbf{M} = M_1\mathbf{i}_1$ is equal to

$$W = P_2 u_2^B - P_2 u_2^C = P_2\,\Delta L\,\theta_1 = M_1\theta_1 \tag{5.14}$$

where u_2^B and u_2^C are the components of displacement of point B and C, respectively, in the direction of the two forces.

In general the work performed by a moment $\mathbf{M} = M_1\mathbf{i}_1 + M_2\mathbf{i}_2 + M_3\mathbf{i}_3$ acting on a rigid body which undergoes a sufficiently small rotation† $\theta = \theta_1\mathbf{i}_1 + \theta_2\mathbf{i}_2 + \theta_3\mathbf{i}_3$ is equal to

$$W = M_1\theta_1 + M_2\theta_2 + M_3\theta_3 = \mathbf{M} \cdot \theta \tag{5.15}$$

5.2.1 Work of a concentrated action as it is applied to a framed structure

Consider a simply supported beam subjected to a concentrated force as shown in Fig. 5.6 whose magnitude increases from zero. The beam is made of a lin-

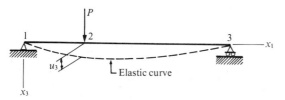

Figure 5.6 Beam subjected to a concentrated force.

---

† Small rotations are vector quantities; large rotations are not.

early elastic material, and its deformation is within the range of validity of the theory of small deformation. Consequently, the relation between the force P and the resulting translation (deflection) u_3 of the point of application of the force is linear. That is

$$P = Ku_3 \qquad (5.16)$$

We denote by P' the value of the force at some time during the process of loading and by u_3' the corresponding value of the translation of the point of application of the force. We assume that as a result of an increment dP' of the force P' the value of the translation u_3' increases to $u_3' + du_3'$. The additional work performed by the force P' due to the additional translation du_3' is given as

$$dW = P' \, du_3' \qquad (5.17)$$

The total work performed by the external force during the process of loading is equal to

$$W = \int_0^{u_3} dW = \int_0^{u_3} P' \, du_3' = \int_0^{u_3} Ku_3' \, du_3' = \tfrac{1}{2}Ku_3^2 = \tfrac{1}{2}Pu_3 \quad (5.18)$$

In general the work performed by a force $\mathbf{P} = P_1\mathbf{i}_1 + P_2\mathbf{i}_2 + P_3\mathbf{i}_3$ as it is applied to a point of a framed structure made of a linearly elastic material when the deformation of the structure is within the range of validity of the theory of small deformation is equal to

$$W = \tfrac{1}{2}(P_1u_1 + P_2u_2 + P_3u_3) = \tfrac{1}{2}\mathbf{P} \cdot \mathbf{u} \qquad (5.19)$$

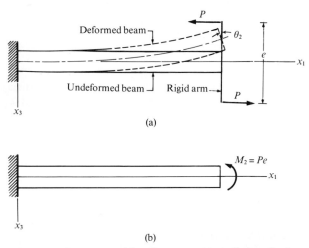

Figure 5.7   Concentrated bending moment applied to the free end of a cantilever beam.

where $\mathbf{u} = u_1\mathbf{i}_1 + u_2\mathbf{i}_2 + u_3\mathbf{i}_3$ is the translation vector of the point of application of the force $\mathbf{P}$.

Consider a cantilevered beam subjected to an end moment M_2 whose magnitude increases from zero. As shown in Fig. 5.7, this moment is equivalent to two equal and opposite forces acting on a rigid arm attached to the free end of the beam and having a moment equal to M_2 ($Pe = M_2$). The work performed by the moment M_2, as it is applied to the beam, is equal to the work performed by the two forces. Referring to Fig. 5.7a, we see that each force moves in its direction by $\theta_2 e/2$. Thus, using relation (5.19) we have

$$W = \tfrac{1}{2}P(e\theta_2) = \tfrac{1}{2}M_2\theta_2 \qquad (5.20a)$$

Consider a cantilevered beam subjected to a torsional moment M_1 at its free end. As shown in Fig. 5.8 this moment is equivalent to two equal and opposite tangential forces acting at the end face of the beam and having a moment equal to M_1 ($Pe = M_1$). The work performed by the moment M_1 as it is applied to the beam is equal to the work performed by the two forces. Hence referring to Fig. 5.8a and relation (5.19), we have

$$W = \tfrac{1}{2}P(e\theta_1) = \tfrac{1}{2}M_1\theta_1 \qquad (5.20b)$$

Thus in general the work performed by a moment $\mathbf{M} = M_1\mathbf{i}_1 + M_2\mathbf{i}_2 + M_3\mathbf{i}_3$ as it is applied to a point of a framed structure made of a linearly elastic mate-

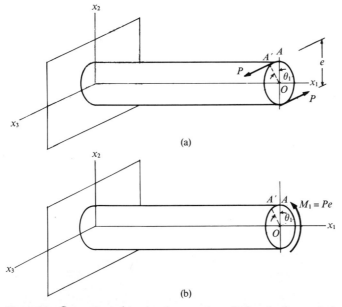

(a)

(b)

Figure 5.8 Concentrated torsional moment applied to the free end of a cantilever beam.

rial when the deformation of the structure is within the range of validity of the theory of small deformation is equal to

$$W = \tfrac{1}{2}(M_1\theta_1 + M_2\theta_2 + M_3\theta_3) = \tfrac{1}{2}\mathbf{M} \cdot \boldsymbol{\theta} \tag{5.21}$$

where $\boldsymbol{\theta} = \theta_1\mathbf{i}_1 + \theta_2\mathbf{i}_2 + \theta_3\mathbf{i}_3$ is the rotation vector at the point of application of the moment $\mathbf{M}$.

5.3 Flexibility and Stiffness Form of the Action-Displacement Relations

Consider a simply supported beam subjected to a force $P^{(1)}$ and a moment $P^{(2)}$ as shown in Fig. 5.9a. The beam is made of a linearly elastic material, and its deformation is within the range of validity of the theory of small deformation. Consequently, the relations between the actions $P^{(1)}$ and $P^{(2)}$ and the corresponding resulting displacements $\Delta^{(1)}$ and $\Delta^{(2)}$ are linear. Thus they can be written as follows:

In flexibility form

$$\Delta^{(1)} = F_{11}P^{(1)} + F_{12}P^{(2)}$$

$$\Delta^{(2)} = F_{21}P^{(1)} + F_{22}P^{(2)}$$

or
$$\begin{Bmatrix} \Delta^{(1)} \\ \Delta^{(2)} \end{Bmatrix} = \begin{bmatrix} F_{11} & F_{12} \\ F_{21} & F_{22} \end{bmatrix} \begin{Bmatrix} P^{(1)} \\ P^{(2)} \end{Bmatrix} \tag{5.22}$$

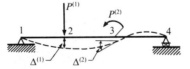

(a) Beam subjected to a force and a moment.

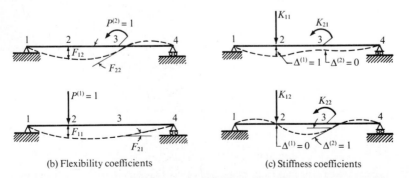

(b) Flexibility coefficients   (c) Stiffness coefficients

Figure 5.9 Physical significance of the flexibility and stiffness coefficients of a beam.

In stiffness form

$$P^{(1)} = K_{11}\Delta^{(1)} + K_{12}\Delta^{(2)}$$

$$P^{(2)} = K_{21}\Delta^{(1)} + K_{22}\Delta^{(2)}$$

or
$$\left\{ \begin{array}{c} P^{(1)} \\ P^{(2)} \end{array} \right\} = \left[\begin{array}{cc} K_{11} & K_{12} \\ K_{21} & K_{22} \end{array} \right] \left\{ \begin{array}{c} \Delta^{(1)} \\ \Delta^{(2)} \end{array} \right\} \qquad (5.23)$$

The coefficients F_{ij} (i, j = 1, 2) are called *the flexibility coefficients of the structure corresponding to the actions $P^{(1)}$ and $P^{(2)}$.* The coefficients K_{ij} (i, j = 1, 2) are called the *stiffness coefficients of the structure corresponding to the displacements $\Delta^{(1)}$ and $\Delta^{(2)}$.* The physical significance of the flexibility coefficients of the beam of Fig. 5.9a subjected to the actions $P^{(1)}$ and $P^{(2)}$ is shown in Fig. 5.9b. For instance, the flexibility coefficients F_{12} and F_{22} are equal to the translation (deflection) of point 2 of the beam and the rotation of point 3, respectively, when the beam is subjected only to the moment $P^{(2)} = 1$. The flexibility coefficient F_{21} is the rotation of point 3 of the beam when it is subjected only to the force $P^{(1)} = 1$. The physical significance of the stiffness coefficients of the beam subjected to the actions $P^{(1)}$ and $P^{(2)}$ is shown in Fig. 5.9c. For instance, the stiffness coefficients K_{11} and K_{21} are equal to the actions corresponding to $P^{(1)}$ and $P^{(2)}$, respectively, which must be applied to points 2 and 3 of the beam in order to deform it in a way that the translation of point 2 is equal to unity while the rotation of point 3 vanishes.

In general when a structure is made of a linearly elastic material and its deformation is within the range of validity of the theory of small deformation, the relation between a set of actions $P^{(1)}$, $P^{(2)}$, ..., $P^{(n)}$ acting on the structure and the corresponding resulting displacements $\Delta^{(1)}$, $\Delta^{(2)}$, ..., $\Delta^{(n)}$ can be written in flexibility form as

$$\{\Delta\} = [F]\{P\} \qquad (5.24)$$

$\Delta^{(i)}$ (the term in the ith row of the matrix $\{\Delta\}$) is the component of displacement of the point of application of the component of action $P^{(i)}$ (the term in the ith row of the matrix $\{P\}$), in the direction of $P^{(i)}$. The matrix $[F]$ is called *the flexibility matrix of the structure corresponding to the actions $P^{(1)}$, $P^{(2)}$, ..., $P^{(n)}$.* The relations between the actions $P^{(1)}$, $P^{(2)}$, ..., $P^{(n)}$ acting on the structure and the corresponding resulting displacements $\Delta^{(1)}$, $\Delta^{(2)}$, ..., $\Delta^{(n)}$ may also be written in stiffness form as

$$\{P\} = [K]\{\Delta\} \qquad (5.25)$$

The matrix $[K]$ is called *the stiffness matrix of the structure corresponding to the displacements $\Delta^{(1)}$, $\Delta^{(2)}$, ..., $\Delta^{(n)}$.* Comparing relations (5.24) and (5.25), we obtain

$$[K] = [F]^{-1} \qquad (5.26)$$

Consider the beam of Fig. 5.9a. The work performed by the actions $P^{(1)}$ and $P^{(2)}$ as they are applied on the beam is independent of the order in which the actions are applied. Referring to Fig. 5.9b, we readily see that if the force $P^{(1)}$ is applied on the beam first, the corresponding translation of point 2 is equal to $P^{(1)}F_{11}$ while the rotation of point 3 is equal to $P^{(1)}F_{21}$. Thus, the work performed by the force $P^{(1)}$ as it is applied to the beam is equal to

$$W_{11} = \tfrac{1}{2}P^{(1)}P^{(1)}F_{11} \qquad (5.27a)$$

When the moment $P^{(2)}$ is subsequently applied to the beam, the rotation of point 3 increases by $P^{(2)}F_{22}$ while the translation of point 2 increases by $P^{(2)}F_{12}$. Thus, as the moment $P^{(2)}$ is applied to the beam, the work of the actions $P^{(1)}$ and $P^{(2)}$ is equal to

$$W_{12} = P^{(1)}P^{(2)}F_{12} \qquad (5.27b)$$

$$W_{22} = \tfrac{1}{2}P^{(2)}P^{(2)}F_{22} \qquad (5.27c)$$

Consequently, the total work of the two actions is equal to

$$W = W_{11} + W_{21} + W_{22}$$
$$= \tfrac{1}{2}P^{(1)}P^{(1)}F_{11} + P^{(1)}P^{(2)}F_{12} + \tfrac{1}{2}P^{(2)}P^{(2)}F_{22} \qquad (5.27d)$$

If the moment $P^{(2)}$ is applied to the beam first, the corresponding rotation of point 3 is equal to $P^{(2)}F_{22}$ while the translation of point 2 is equal to $P^{(2)}F_{12}$. Thus, the work performed by the moment $P^{(2)}$ as it is applied to the beam is equal to

$$W_{22} = \tfrac{1}{2}P^{(2)}P^{(2)}F_{22} \qquad (5.28a)$$

When the force $P^{(1)}$ is subsequently applied to the beam, the translation of point 2 increases by $P^{(1)}F_{11}$, while the rotation of point 3 increases by $P^{(1)}F_{21}$. Thus, as force $P^{(1)}$ is applied to the beam, the work of the actions $P^{(1)}$ and $P^{(2)}$ is equal to

$$W_{21} = P^{(1)}P^{(2)}F_{21} \qquad (5.28b)$$

$$W_{11} = \tfrac{1}{2}P^{(1)}P^{(1)}F_{11} \qquad (5.28c)$$

Consequently the total work of the two actions is equal to

$$W = W_{22} + W_{21} + W_{11}$$
$$= \tfrac{1}{2}P^{(2)}P^{(2)}F_{22} + P^{(1)}P^{(2)}F_{21} + \tfrac{1}{2}P^{(1)}P^{(1)}F_{11} \qquad (5.28d)$$

Since the total work of the actions $P^{(1)}$ and $P^{(2)}$ must be independent of the order in which they are applied, referring to relations (5.27*d*) and (5.28*d*), we have

$$F_{21} = F_{12} \tag{5.29}$$

That is, the flexibility matrix of the beam corresponding to the actions $P^{(1)}$ and $P^{(2)}$ is symmetric. Referring to relation (5.26), we can easily see that the stiffness matrix of the beam is also symmetric. In general the flexibility and the stiffness matrices of a structure corresponding to the forces $P^{(1)}, P^{(2)}, \ldots, P^{(n)}$ are square symmetric matrices. That is

$$[F] = [F]^T \qquad [K] = [K]^T \tag{5.30}$$

where the matrices $[F]^T$ and $[K]^T$ represent the transpose of the matrices $[F]$ and $[K]$, respectively.

5.4 Bernoulli's Principle of Virtual Displacements for Rigid Bodies

Consider a rigid body in equilibrium (see Fig. 5.10) under the influence of a system of external actions and reactions consisting of:

1. A distribution of body forces $\tilde{\mathbf{B}}(x_k)$ exerted on the particles of the body because of its presence in a force field, such as the gravitational field. The distribution of the body forces ($\tilde{\mathbf{B}}$) is usually given in units of force per unit volume.

2. A distribution of surface actions. Generally, these may include N concentrated forces $\tilde{\mathbf{P}}^{(i)}$, M concentrated moments $\tilde{\mathbf{M}}^{(i)}$, and distributed forces $\tilde{\mathbf{q}}$ given in units of force per unit surface.

3. Appropriate reactions (forces and moments) exerted on the body by its supports or other constraints.

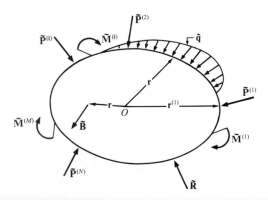

Figure 5.10 Rigid body in equilibrium.

Denoting by $\tilde{\mathbf{R}}$ the resultant of all reacting forces and by $\tilde{\mathbf{M}}^R$ the resultant of the moments of all reactions about an origin O, and referring to Fig. 5.10, we express the equilibrium of the actions acting on the rigid body under consideration by the following relations:

$$\sum_{i=1}^{N} \tilde{\mathbf{P}}^{(i)} + \iiint_V \tilde{\mathbf{B}} \, dV + \iint_S \tilde{\mathbf{q}} \, dS + \tilde{\mathbf{R}} = 0 \qquad (5.31)$$

and

$$\sum_{i=1}^{N} \mathbf{r}^{(i)} \times \tilde{\mathbf{P}}^{(i)} + \sum_{i=1}^{M} \tilde{\mathbf{M}}^{(i)} + \iiint_V \mathbf{r} \times \tilde{\mathbf{B}} \, dV$$

$$+ \iint_S \mathbf{r} \times \tilde{\mathbf{q}} \, dS + \tilde{\mathbf{M}}^R = 0 \qquad (5.32)$$

Equation (5.31) indicates that the sum of all forces acting on the body vanishes, while Eq. (5.32) indicates that the sum of the moments of all actions about an origin O vanishes. $\mathbf{r}$ is the position vector of a point from the origin O. Each of the two vector equations, (5.31) and (5.32), is equivalent to three scalar equations. For instance, denoting the position vector as $\mathbf{r} = x_1\mathbf{i}_1 + x_2\mathbf{i}_2 + x_3\mathbf{i}_3$, Eq. (5.32) is equivalent to the following three equations:

$$\sum_{i=1}^{N} (\tilde{P}_3^{(i)}x_2^{(i)} - \tilde{P}_2^{(i)}x_3^{(i)}) + \sum_{i=1}^{M} \tilde{M}_1^{(i)} + \iint_S (\tilde{q}_3 x_2 - \tilde{q}_2 x_3) \, dS$$

$$+ \iiint_V (\tilde{B}_3 x_2 - \tilde{B}_2 x_3) \, dV + \tilde{M}_1^R = 0 \qquad (5.33a)$$

$$\sum_{i=1}^{N} (\tilde{P}_1^{(i)}x_3^{(i)} - \tilde{P}_3^{(i)}x_1^{(i)}) + \sum_{i=1}^{M} \tilde{M}_2^{(i)} + \iint_S (\tilde{q}_1 x_3 - \tilde{q}_3 x_1) \, dS$$

$$+ \iiint_V (\tilde{B}_1 x_3 - \tilde{B}_3 x_1) \, dV + \tilde{M}_2^R = 0 \qquad (5.33b)$$

$$\sum_{i=1}^{N} (\tilde{P}_2^{(i)}x_1^{(i)} - \tilde{P}_1^{(i)}x_2^{(i)}) + \sum_{i=1}^{M} \tilde{M}_3^{(i)} + \iint_S (\tilde{q}_2 x_1 - \tilde{q}_1 x_2) \, dS$$

$$+ \iiint_V (\tilde{B}_2 x_1 - \tilde{B}_1 x_2) \, dV + \tilde{M}_3^R = 0 \qquad (5.33c)$$

Relations (5.33) indicate that the x_1, x_2, x_3 components of the sum of the moments about an origin O of all the actions acting on the body vanish.

We assume that the rigid body under consideration is subjected to a translation without rotation, specified by the vector $\tilde{\mathbf{u}}$. The applied actions are maintained constant during this translation which is called "virtual," with the connotation of "possible" as opposed to the term "actual." The work W performed by the system of actions acting on the body as a result of this virtual translation is referred to as *virtual work* and is given by

$$W = \sum_{i=1}^{N} \tilde{\mathbf{P}}^{(i)} \cdot \tilde{\mathbf{u}} + \iiint_{V} \tilde{\mathbf{B}} \cdot \tilde{\mathbf{u}} \, dV + \iint_{S} \tilde{\mathbf{q}} \cdot \tilde{\mathbf{u}} \, dS + \tilde{\mathbf{R}} \cdot \tilde{\mathbf{u}}$$

$$= \left(\sum_{i=1}^{N} \tilde{\mathbf{P}}^{(i)} + \iiint_{V} \tilde{\mathbf{B}} \, dV + \iint_{S} \tilde{\mathbf{q}} \, dS + \tilde{\mathbf{R}} \right) \cdot \tilde{\mathbf{u}} = 0$$

(5.34)

The above result has been obtained using Eq. (5.31). If the virtual translation satisfies the conditions imposed by the supports and other constraints of the body, we have the advantage that the reactions associated with nonmoving constraints (workless constraints) do not contribute to relation (5.34).

Relation (5.34) indicates that when a rigid body is in equilibrium under the influence of a system of actions consisting of body forces, surface actions, and reactions, the total work W performed by this system of actions during an arbitrary virtual translation of the rigid body must vanish. The converse is also true. That is, the sum of the body forces of the surface forces and of the reacting forces acting on a rigid body must be equal to zero, if the total work W performed by this system of forces during an arbitrary virtual translation $\tilde{\mathbf{u}}$ of the rigid body vanishes. The validity of this statement becomes apparent by noting that since the virtual translation $\tilde{\mathbf{u}}$ is completely arbitrary, the term in parentheses must vanish in order for relation (5.34) to be satisfied for any value of $\tilde{\mathbf{u}}$.

Let us suppose that the rigid body under consideration is subjected to a sufficiently small virtual rotation about the x_3 axis, specified by the angle $\Delta\tilde{\theta}_3$. Consider a point A on the surface of the rigid body shown in Fig. 5.11, and let

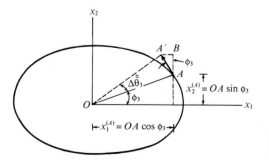

Figure 5.11 Rigid body subjected to a rotation about the x_3 axis.

OA be the line perpendicular to the x_3 axis from point A. As a result of the rotation $\Delta\tilde{\theta}_3$, point A moves to A'. If this rotation is sufficiently small, line AA' may be considered perpendicular to OA. Thus, referring to Fig. 5.11, we have:

$$A'B = AA' \sin\phi_3 \tag{5.35}$$

$$AB = AA' \cos\phi_3 \tag{5.36}$$

moreover

$$AA' = OA\,\Delta\tilde{\theta}_3 \tag{5.37}$$

Hence, the components of displacement $\tilde{u}_1^A$ and $\tilde{u}_2^A$ of point A are equal to

$$-\tilde{u}_1^A = BA' = \Delta\tilde{\theta}_3\,OA\sin\phi_3 = \Delta\tilde{\theta}_3\,x_2^A$$
$$\tilde{u}_2^A = AB = \Delta\tilde{\theta}_3\,OA\cos\phi_3 = \Delta\tilde{\theta}_3\,x_1^A \tag{5.38}$$

Thus, the components of displacement of any point of the body due to a virtual rotation $\Delta\tilde{\theta}_3$ are equal to

$$\tilde{u}_1 = -\Delta\tilde{\theta}_3\,x_2 \qquad \tilde{u}_2 = \Delta\tilde{\theta}_3\,x_1 \tag{5.39}$$

On the basis of the foregoing discussion, the work performed by the previously described system of actions applied to the rigid body under consideration as it rotates about the x_3 axis by the virtual rotation $\Delta\tilde{\theta}_3$ is equal to

$$\Delta W = \sum_{i=1}^{N}\left(-\tilde{P}_1^{(i)}\right)\Delta\tilde{\theta}_3\,x_2^{(i)} + \tilde{P}_2^{(i)}\,\Delta\tilde{\theta}_3\,x_1^{(i)}) + \sum_{i=1}^{M}\Delta\tilde{\theta}_3\,\tilde{M}_3^{(i)}$$

$$-\iint_S \tilde{q}_1\,\Delta\tilde{\theta}_3\,x_2\,dS + \iint_S \tilde{q}_2\,\Delta\tilde{\theta}_3\,x_1\,dS$$

$$-\iiint_V \tilde{B}_1\,\Delta\tilde{\theta}_3\,x_2\,dV + \iiint_V \tilde{B}_2\,\Delta\tilde{\theta}_3\,x_1\,dV + \Delta\tilde{\theta}_3\,\tilde{M}_3^R \tag{5.40}$$

or

$$\Delta W = \Delta\tilde{\theta}_3\left[\sum_{i=1}^{N}\left(-\tilde{P}_1^{(i)}x_2^{(i)} + \tilde{P}_2^{(i)}x_1^{(i)}\right) + \sum_{i=1}^{M}\tilde{M}_3^{(i)} - \iint_S \tilde{q}_1 x_2\,dS\right.$$

$$\left. + \iint_S \tilde{q}_2 x_1\,dS - \iiint_V \tilde{B}_1 x_2\,dV + \iiint_V \tilde{B}_2 x_1\,dV + \tilde{M}_3^R\right] = 0$$

$$\tag{5.41}$$

The above result was obtained using Eq. (5.33c). On the basis of the foregoing discussion, we may conclude that when a rigid body is in equilibrium, under a system of body forces, surface actions, and reactions, the work ΔW of this system of actions during an arbitrary, sufficiently small virtual rotation must vanish. The converse is also true. That is, the sum of the moments about an origin O of a system of body forces, surface actions, and reactions acting on a rigid body must equal zero if the total work performed by this system of actions during any arbitrary, sufficiently small, virtual rotation of this body vanishes. The validity of this statement is apparent by noting that since the virtual rotation $\Delta \tilde{\theta}_3$ is arbitrary, the term in the brackets of relation (5.41) must vanish in order that this relation be satisfied for any value of $\Delta \tilde{\theta}_3$.

Inasmuch as any displacement of a rigid body can be decomposed into the sum of a translation and a rotation about one or more axes, we can make the following statements.

Consider a rigid body in equilibrium under the influence of a system of body forces, surface actions, and reactions. The work of this system of actions vanishes when this rigid body is subjected to a sufficiently small virtual displacement. The converse is also valid. That is, a rigid body is in equilibrium under the influence of a system of body forces, surface actions, and reactions if the total work performed by this system of actions during every sufficiently small virtual displacement vanishes. These statements constitute the principle of virtual displacements for rigid bodies. Notice, that if the virtual displacement satisfies the conditions imposed by the supports or other constraints of the body, the reactions associated with nonmoving constraints (workless constraints) do not contribute to the virtual work equation.

It can be readily shown that the principle of virtual displacements is also valid for a system of interconnected rigid bodies. The principle of virtual displacements can be employed in establishing the required equations for finding unknown forces acting on a rigid body by subjecting this body to known virtual displacements. The smallest number of displacement components which must be given in order that the displaced configuration of a body be specified represents the number of degrees of freedom which the body possesses. This is always equal to the number of independent equations of equilibrium which can be written for the body. Therefore, it is apparent that the number of unknown actions which can be established employing the principle of virtual displacements is equal to the number of unknown actions which can be established by considering the equilibrium of the body and its parts.

The application of the principle of virtual displacements is illustrated by the following two examples.

Example 1. The reactions of a beam are computed.

Example 2. One of the reactions of a two member truss is computed directly.

Example 1. Compute the reactions of the beam shown in Fig. a using the principle of virtual displacements.

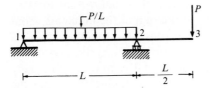

Figure a Geometry and loading of the beam.

solution The free-body diagram of the beam is shown in Fig. b. Moreover a virtual displacement of the beam is shown in Fig. c. It is specified by the components of translation $\tilde{u}^{(1)}$ and $\tilde{u}^{(2)}$ of points 1 and 2, respectively. The component of translation of any other point of the beam is equal to

$$\tilde{u} = \tilde{u}^{(1)} + \frac{(\tilde{u}^{(2)} - \tilde{u}^{(1)})x_1}{L} \tag{a}$$

Applying the principle of virtual displacements, we obtain:

$$R^{(1)}\tilde{u}^{(1)} + R^{(2)}\tilde{u}^{(2)} - \int_0^L \frac{P}{L}\left(\tilde{u}^{(1)} + \frac{\tilde{u}^{(2)} - \tilde{u}^{(1)}}{L}x_1\right)dx_1 - \frac{P}{2}(3\tilde{u}^{(2)} - \tilde{u}^{(1)}) = 0$$

Integrating and simplifying, we get

$$R^{(1)}\tilde{u}^{(1)} + (R^{(2)} - 2P)\tilde{u}^{(2)} = 0 \tag{b}$$

Since the virtual displacements $\tilde{u}^{(1)}$ and $\tilde{u}^{(2)}$ are arbitrary, their multipliers must each equal zero. Thus from relation (b) we obtain

$$R^{(1)} = 0 \tag{c}$$

$$R^{(2)} = 2P \tag{d}$$

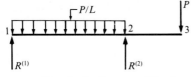

Figure b Free-body diagram of the beam.

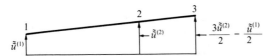

Figure c Virtual displacement of the beam.

These are the two equations of equilibrium for the beam. Equation (c) can also be obtained by setting the sum of the moments about point 2 equal to zero. Equation (d) can also be obtained by setting the sum of the moments about point 1 equal to zero.

Example 2. Consider the truss shown in Fig. a consisting of two members pinned together at joint 2 and to the ground at points 1 and 3. Compute the horizontal reaction at point 3 using the principle of virtual displacements.

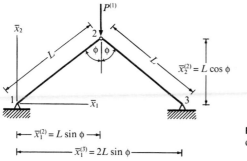
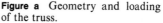

Figure a Geometry and loading of the truss.

solution The free-body diagram of the truss is shown in Fig. b. Referring to this figure, it can be seen that the configuration of the truss is specified if the two components of translation of joints 1 and 3 are specified. That is, the degree of freedom of the truss is four. Consequently, using the principle of virtual displacements, we can

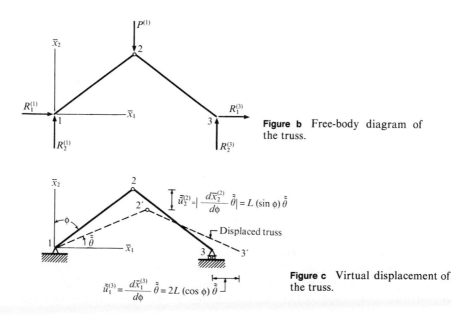

Figure b Free-body diagram of the truss.

Figure c Virtual displacement of the truss.

establish four independent equations involving the four unknown reactions. However, in this example we are interested in establishing only the horizontal reaction at point 3. An equation involving only this reaction can be established using the principle of virtual displacements by choosing the virtual displacement of the truss shown in Fig. c. This virtual displacement is completely specified by the infinitesimal change $\Delta\tilde{\tilde{\theta}}$ of the angle ϕ. Notice, that the reactions at joint 1, and the vertical reaction at joint 3 do not perform work during this virtual displacement, and, consequently, the application of the principle of virtual displacements gives a relation between the external force $P^{(1)}$ and the reaction $R_1^{(3)}$ only. That is

$$P^{(1)}\tilde{u}_2^{(2)} + R_1^{(3)}\tilde{u}_1^{(3)} = 0$$

Referring to Fig. c, we can rewrite the above relation as

$$L(\sin\phi)\tilde{\tilde{\theta}}P^{(1)} + 2L(\cos\phi)\tilde{\tilde{\theta}}R_1^{(3)} = 0$$

or

$$R_1^{(3)} = -\tfrac{1}{2}(\tan\phi)P^{(1)}$$

The minus sign indicates that $R_1^{(3)}$ acts in the direction opposite to that assumed in Fig. b.

5.5   Statically Admissible Stress Distribution and Geometrically Admissible Displacement Distribution

Consider a deformable body in equilibrium under the influence of a given system of external actions and corresponding support reactions. In general, we can find an infinite number of distributions of the components of stress which satisfy the requirements for the equilibrium of any part of this body. These distributions of the components of stress are called *statically admissible*. The actual distribution of the components of stress is the statically admissible distribution which yields such components of strain that relations (1.13) can be integrated to give a continuous distribution of components of displacement which, moreover, assume the specified values on the points of the surface of the body where components of displacement are specified.

In the analysis of framed structures, we are interested in establishing the distribution of the components of internal actions acting on the cross sections of their members, from which we can compute the components of stress [see relations (A.116)]. In statically determinate framed structures only the actual internal actions are in equilibrium with the given external actions, while in statically indeterminate framed structures, we can find an infinite number of distributions of internal actions which satisfy the equations of equilibrium for any portion of the structure. These distributions of internal actions are referred to as *statically admissible*. In addition to being statically admissible, the actual distribution of internal actions must yield components of displacement which are continuous functions of the space coordinates and satisfy the boundary conditions of the structure.

When a framed structure is externally statically indeterminate an infinite number of sets of reactions can be found which are in equilibrium with the given external actions acting on the structure. These sets of reactions are referred to as *statically admissible*. The one set of statically admissible reactions which yields components of displacement compatible with the constraints of the structure is the actual set of reactions. For example, consider the beam shown in Fig. 5.12a and the two statically admissible distributions of moments shown in Fig. 5.12c and e. They correspond to the applied force and to the two statically admissible sets of reactions shown in Fig. 5.12b and d.

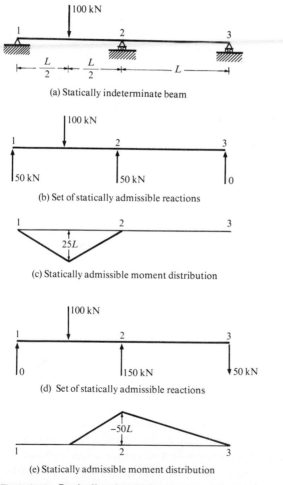

(a) Statically indeterminate beam

(b) Set of statically admissible reactions

(c) Statically admissible moment distribution

(d) Set of statically admissible reactions

(e) Statically admissible moment distribution

Figure 5.12   Statically admissible sets of reactions and corresponding moment distributions.

An admissible displacement field $\hat{\mathbf{u}}(x_k)$ for a body is defined as having the following properties.

1. The components of displacement could be any functions of the space coordinates which have continuous first derivatives, at every point of the body.

2. The magnitude of the components of displacement is within the range of validity of the theory of small deformation. In this range of deformation, the magnitude of the rotation of a particle (element) of the body may be considered of the order required for the validity of the principle of virtual displacements. When a body in equilibrium under the influence of external actions is subjected to an admissible displacement field, its particles translate and rotate as rigid bodies and deform. Inasmuch as the principle of virtual displacements is valid, the work performed by the forces acting on any particle of the body due to the rigid body displacement of the particle must vanish.

A geometrically admissible displacement field $\hat{\mathbf{u}}(x_k)$ for a body is defined as having the following properties.

1. It is admissible.

2. The components of displacement assume the specified values at the points of the body where the components of displacement are known. That is, at a fixed support, the components of displacement of a geometrically admissible displacement field must vanish, while at a support at which a settlement in the x_1 direction is specified, the component of displacement $\hat{u}_1$ must be equal to the specified settlement.

The components of strain obtained from a geometrically admissible displacement field on the basis of relations (1.13) are referred to as *geometrically admissible*.

A geometrically admissible displacement field is not necessarily the actual displacement field. Moreover, a statically admissible stress field and a geometrically admissible displacement field may be independently prescribed. However, if the stress field in a particular body, obtained on the basis of the appropriate relations between the components of stress and strain from a geometrically admissible strain field, is statically admissible to the applied actions, then this geometrically admissible strain field is the actual strain field. Similarly, if the strain field obtained from a stress field which is statically admissible to the actions acting on a body yields a geometrically admissible displacement field, then this stress field is the actual stress field corresponding to the actions acting on the body.

5.6 Principle of Virtual Work for Deformable Bodies

Let us consider a body subjected to external actions and a set of statically admissible reactions, and denote by $\tilde{\tau}_{ij}$ the corresponding statically admissible components of stress (see Fig. 5.13). Let us assume that the aforementioned body is subjected to a sufficiently small additional deformation characterized by the geometrically admissible components of displacement $\tilde{u}_i$ and strain $\tilde{e}_{ij}$. Generally, this deformation could be unrelated to the system of actions originally acting on the body. As a result of this sufficiently small deformation ($\tilde{u}_i$, $\tilde{e}_{ij}$), any infinitesimal element of the body translates and rotates as a rigid body and deforms. Thus the total work dW_T of the components of stress $\tilde{\tau}_{ij}$ and of the body force acting on an infinitesimal element of the body, due to the previously described deformation, consists of two parts; the work dW_d due to the deformation of the element and the work $(dW_T - dW_d)$ due to its displacement (rotation and translation) as a rigid body. Since the element is in equilibrium, on the basis of the principle of virtual displacements discussed in Sec. 5.4, the work performed by all the forces acting on the element due to its sufficiently small displacement as a rigid body must vanish. Consequently,

$$dW_T = dW_d \qquad (5.42)$$

The total work W_T performed by the statically admissible components of stress acting on all the infinitesimal elements comprising the body, due to the geometrically admissible deformation, is equal to

$$W_T = \iiint_V dW_d \qquad (5.43)$$

However, two adjacent elements have a common boundary and the components of stress acting on this boundary of the one element are equal and opposite to

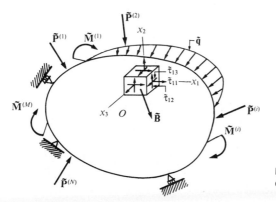

Figure 5.13 Deformable body in equilibrium.

those acting on the boundary of the other element. Thus, the sum of the work performed by these equal and opposite components of stress vanishes. Consequently, W_T consists only of the work of the components of stress acting on the faces of elements which are part of the surface of the body and of the work of the body forces. Thus, W_T is equal to the work of the known external actions (body forces and surface actions) acting on the body and of the reactions of its supports. If the supports of the body do not move, the work of the reactions vanishes. On the basis of the previous discussion, Eq. (5.43) becomes

$$W_{\text{ext.act.}} + W_{\text{reac.}} = \int\int\int_V dW_d \qquad (5.44)$$

Relation (5.44) is referred to as *the principle of virtual work for deformable bodies.*

5.7 Principle of Virtual Work for Framed Structures Disregarding the Effect of Shear Deformation

Consider a framed structure composed of NM members (straight or curved with small h/R ratio) subjected to a general loading, consisting of:

1. Distributed external forces $\tilde{\mathbf{q}}(x_1)$ (including its weight) given per unit axial length of the member
2. Distributed external moments $\tilde{\mathbf{m}}(x_1)$ given per unit axial length of the member
3. Concentrated external forces $\tilde{\mathbf{P}}^{(i)}$ ($i = 1, 2, \ldots, N$)
4. Concentrated external moments $\tilde{\mathbf{M}}^{(i)}$ ($i = 1, 2, \ldots, M$)
5. Change of temperature
6. Translation and rotation of the supports $\Delta^{(s)}$ ($s = 1, 2, \ldots, S$)
7. Initial stress (see Sec. 1.3.4)

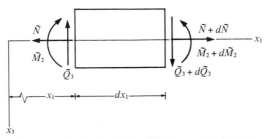

Figure 5.14 Statically admissible internal actions acting on an element of a member of a planar structure.

For this loading, we choose a set of statically admissible reactions of the supports of the structure and establish the corresponding statically admissible internal actions.

Consider an element of length dx_1 of a member of the framed structure under consideration. The statically admissible internal actions acting on its face at x_1 are denoted by $\tilde{N}$, $\tilde{Q}_2$, $\tilde{Q}_3$, $\tilde{M}_1$, $\tilde{M}_2$, and $\tilde{M}_3$, while those at its face at $x_1 + dx_2$ are denoted by $\tilde{N} + d\tilde{N}$, $\tilde{Q}_2 + d\tilde{Q}_2$, $\tilde{Q}_3 + d\tilde{Q}_3$, $\tilde{M}_1 + d\tilde{M}_1$, $\tilde{M}_2 + d\tilde{M}_2$, and $\tilde{M}_3 + d\tilde{M}_3$ (see Fig. 5.14).

Suppose that the structure under consideration is subjected to a geometrically admissible displacement field not necessarily related to the previously described loading. That is, the displacement field could be the result of another loading. As discussed in Sec. 5.2 to our order of approximation, the deformed configuration of a member of a framed structure is such that plane sections normal to their axis before deformation can be considered plane after deformation. Thus the deformation of the element under consideration is completely specified by the components of translation and rotation of its two faces. The components of displacement of the face at x_1 of this element corresponding to the geometrically admissible displacement field are denoted by $\tilde{\tilde{u}}_1$, $\tilde{\tilde{u}}_2$, $\tilde{\tilde{u}}_3$, $\tilde{\tilde{\theta}}_1$, $\tilde{\tilde{\theta}}_2$, and $\tilde{\tilde{\theta}}_3$, while those of its face at $x_1 + dx_1$ are denoted by $\tilde{\tilde{u}}_1 + d\tilde{\tilde{u}}_1$, $\tilde{\tilde{u}}_2 + d\tilde{\tilde{u}}_2$, $\tilde{\tilde{u}}_3 + d\tilde{\tilde{u}}_3$, $\tilde{\tilde{\theta}}_1 + d\tilde{\tilde{\theta}}_1$, $\tilde{\tilde{\theta}}_2 + d\tilde{\tilde{\theta}}_2$, and $\tilde{\tilde{\theta}}_3 + d\tilde{\tilde{\theta}}_3$. The deformed configuration of the element corresponding to the geometrically admissible displacement field may be obtained by subjecting it to a rigid-body translation specified by the components of translation $\tilde{\tilde{u}}_1$, $\tilde{\tilde{u}}_2$, $\tilde{\tilde{u}}_3$; to a rigid-body rotation about the centroid of its face at x_1, specified by the components of rotation $\tilde{\tilde{\theta}}_1$, $\tilde{\tilde{\theta}}_2$, $\tilde{\tilde{\theta}}_3$; and to a deformation. The latter includes the twist of the element, the change of the length of the longitudinal fibers of the element, as well as its transverse shear deformation. The twist of the element is specified by the relative rotation $d\theta_1$ of its faces (see Fig. 5.15). The change of length of the longitudinal fibers of the element is specified by the change of the length $d\tilde{\tilde{u}}_1$ of the axis of the element and by the relative rotations $d\tilde{\tilde{\theta}}_2$, $d\tilde{\tilde{\theta}}_3$ of its faces (see Fig. 5.16).

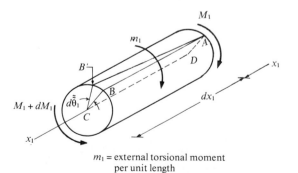

m_1 = external torsional moment per unit length

Figure 5.15 Deformation of an element of a member of a structure subjected to torsional moments.

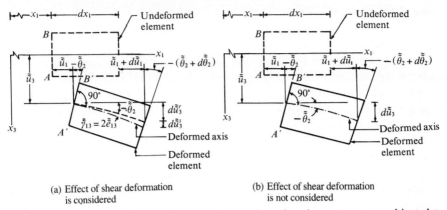

(a) Effect of shear deformation is considered

(b) Effect of shear deformation is not considered

Figure 5.16 Deformation of an element of a member of a planar structure subjected to bending without twisting.

In Fig. 5.16a an element of length dx_1 of a member of a structure subjected to bending without twisting is shown in the undeformed and the deformed configuration. In this figure the effect of shear deformation of the element is taken into account and consequently its cross sections do not remain normal to its deformed axis after deformation. The angle between a cross section and the axis of the element after deformation is equal to $90° + 2\tilde{e}_{13}$. Referring to Fig. 5.16a, it can be seen that the increment of the component of translation $d\tilde{u}_3$ generally consists of two parts. The one part $d\tilde{u}_3^r$ is the result of the rotation† $-\tilde{\theta}_2$ of the element as a rigid body, while the other part $d\tilde{u}_3^s$ is the result of the shear deformation of the element. That is,

$$d\tilde{u}_3 = d\tilde{u}_3^r + d\tilde{u}_3^s \tag{5.45}$$

Similarly
$$d\tilde{u}_2 = d\tilde{u}_2^r + d\tilde{u}_2^s \tag{5.46}$$

Referring to relation (5.6a) and Fig. 5.16a, we have

$$d\tilde{u}_3^r = -\tilde{\theta}_2 \, dx_1 \tag{5.47}$$

$$d\tilde{u}_3^s = 2\tilde{e}_{13} \, dx_1 \tag{5.48}$$

Moreover referring to relation (5.6b), we obtain

$$d\tilde{u}_2^s = \tilde{\theta}_3 \, dx_1 \tag{5.49}$$

$$d\tilde{u}_2^s = 2\tilde{e}_{12} \, dx_1 \tag{5.50}$$

On the basis of the foregoing, when the structure under consideration is subjected to a geometrically admissible displacement field, the work of deforma-

---

† Counterclockwise rotations are considered positive.

tion of the statically admissible internal actions acting on an element of the structure is equal to

$$dW_d = \tilde{N} \, d\tilde{\tilde{u}}_1 + \tilde{M}_1 \, d\tilde{\tilde{\theta}}_1 + \tilde{M}_2 \, d\tilde{\tilde{\theta}}_2 + \tilde{M}_3 \, d\tilde{\tilde{\theta}}_3 + \tilde{Q}_2 \, d\tilde{\tilde{u}}_2^s + \tilde{Q}_3 \, d\tilde{\tilde{u}}_3^s$$

$$(5.51)$$

The classical theory of beams is based on the assumption that when a member is subjected to bending without twisting, plane sections normal to its axis prior to deformation can be considered plane and normal to the deformed axis of the member subsequent to deformation. That is, it is assumed that the change due to the deformation of the right angle between a cross section and the axis of a member is negligible. This implies that to the order of approximation of the classical theory of beams, when the structure under consideration is subjected to a geometrically admissible displacement field, the components of translation du_3^s and du_2^s are very small. Consequently the work of the shearing forces is negligible, and relation (5.51) reduces to

$$dW_d = \tilde{N} \, d\tilde{\tilde{u}}_1 + \tilde{M}_1 \, d\tilde{\tilde{\theta}}_1 + \tilde{M}_2 \, d\tilde{\tilde{\theta}}_2 + \tilde{M}_3 \, d\tilde{\tilde{\theta}}_3$$

Hence the total work of deformation of a structure consisting of NM members may be approximsted as

$$W_d = \sum_{k=1}^{NM} \left[\int_0^L \left(\tilde{N} \frac{d\tilde{\tilde{u}}_1}{dx_1} + \tilde{M}_1 \frac{d\tilde{\tilde{\theta}}_1}{dx_1} + \tilde{M}_2 \frac{d\tilde{\tilde{\theta}}_2}{dx_1} + \tilde{M}_3 \frac{d\tilde{\tilde{\theta}}_3}{dx_1} \right) dx_1 \right]^{(k)} \quad (5.52)$$

The superscript (k) indicates that the quantities of the terms inside the bracket pertain to the kth member.

We denote by $\tilde{\tilde{\mathbf{u}}}^{(i)}$ the displacement vector corresponding to the geometrically admissible displacement field $\tilde{\tilde{\mathbf{u}}}$ at the point of application of the concentrated force $\tilde{\mathbf{P}}^{(i)}$, and by $\tilde{\tilde{\theta}}^{(i)}$ the rotation vector corresponding to the displacement field $\tilde{\tilde{\mathbf{u}}}$ at the point of application of the concentrated moment $\tilde{\mathbf{M}}^{(i)}$. Thus the work of the force $\tilde{\mathbf{P}}^{(i)}$ and the moment $\tilde{\mathbf{M}}^{(i)}$ when the structure is subjected to the geometrically admissible displacement field $\tilde{\tilde{\mathbf{u}}}$ is equal to $\tilde{\mathbf{P}}^{(i)} \cdot \tilde{\tilde{\mathbf{u}}}^{(i)}$ and $\tilde{\mathbf{M}}^{(i)} \cdot \tilde{\tilde{\theta}}^{(i)}$, respectively. For the previously described loading, statically admissible internal actions, and geometrically admissible displacement field, by substituting relation (5.52) into the principle of virtual work (5.44), we obtain

$$\sum_{i=1}^{N} \tilde{\mathbf{P}}^{(i)} \cdot \tilde{\tilde{\mathbf{u}}}^{(i)} + \sum_{i=1}^{M} \tilde{\mathbf{M}}^{(i)} \cdot \tilde{\tilde{\theta}}^{(i)} + \sum_{k=1}^{NM} \left[\int_0^L (\tilde{\mathbf{q}} \cdot \tilde{\tilde{\mathbf{u}}} + \tilde{\mathbf{m}} \cdot \tilde{\tilde{\theta}}) \, dx_1 \right]^{(k)}$$

$$+ \sum_{s=1}^{S} \tilde{R}^{(s)} \tilde{\tilde{\Delta}}^{(s)} = \sum_{k=1}^{NM} \left[\int_0^L \left(\tilde{N} \frac{d\tilde{\tilde{u}}_1}{dx_1} + \tilde{M} \frac{d\tilde{\tilde{\theta}}_1}{dx_1} + \tilde{M}_2 \frac{d\tilde{\tilde{\theta}}_2}{dx_1} \right. \right.$$

$$\left. \left. + \tilde{M}_3 \frac{d\tilde{\tilde{\theta}}_3}{dx_1} \right) dx_1 \right]^{(k)} \quad (5.53)$$

The superscript (k) indicates that the quantities of the terms inside the bracket pertain to the kth member. $\tilde{R}^{(s)}$ $(s = 1, 2, \ldots, S)$ are the statically admissible reactions (forces or moments) of the supports of the structure corresponding to the given loading. Moreover, $\tilde{\tilde{\Delta}}^{(s)}$ $(s = 1, 2, \ldots, S)$ are the specified components of displacements (translations and rotations) of the supports of the structure. The reaction $\tilde{R}^{(s)}$ is considered positive in the direction of the component of displacement $\tilde{\tilde{\Delta}}^{(s)}$.

Relation (5.53) represents the principle of virtual work for a structure composed of NM straight, line members made of any material when the effect of shear deformation is negligible. It relates a set of external actions and the corresponding statically admissible internal actions to a geometrically admissible displacement field.

5.8 Method of Virtual Work

In this section, we present the *method of virtual forces,* also known as the *method of virtual work,* or as the *dummy load method,* as employed in establishing:

1. The component of translation u_m in the direction of the unit vector **m** at any point A of a framed structure whose members are made of isotropic, linearly elastic materials.

2. The component of rotation θ_m about an axis specified by the unit vector **m** at any point A of a framed structure whose members are made of isotropic, linearly elastic materials.

The component of displacement u_m and the component of rotation θ_m of a point of a structure may be due to external actions, to a change of temperature, to a distribution of initial stress, or to specified movement of the supports of the structure. The structure may be statically determinate or statically indeterminate.

In order to establish the component of displacement u_m, or the component of rotation θ_m at any point A of a structure, we use the following distributions of displacements and internal actions in Eq. (5.53).

1. As the geometrically admissible displacement field, we choose the actual (real) displacement field of the members of the structure subjected to the given loading.

2. As the statically admissible distribution of internal actions $\tilde{N}$, $\tilde{M}_1$, $\tilde{M}_2$, and $\tilde{M}_3$, we choose a distribution of internal actions in the members of the structure which is statically admissible to a virtual loading. If we want to establish the translation u_m at point A of the structure, the virtual loading consists of a unit force acting at point A in the direction of the unit vector **m.** If we want to establish the rotation θ_m at point A of the structure, the virtual loading consists of a unit moment acting at point A in the direction specified by the

unit vector **m**. For example, the virtual loading for computing the components of translation and rotation in the directions of the unit vector **m** at point 3 of the beam of Fig. 5.17a is shown in Fig. 5.17b and c, respectively. Moreover the virtual loading for computing the horizontal component of translation and the rotation of joint 3 of the planar frame of Fig. 5.18a is shown in Fig. 5.18b and c, respectively.

On the basis of the foregoing, Eq. (5.53) reduces to

$$d + \sum_{s=1}^{S} \tilde{R}^{(s)} \Delta^{(s)} = \sum_{k=1}^{NM} \left[\int_0^L \left(\tilde{N} \frac{du_1}{dx_1} + \tilde{M}_1 \frac{d\theta_1}{dx_1} \right. \right.$$
$$\left. \left. + \tilde{M}_2 \frac{d\theta_2}{dx_1} + \tilde{M}_3 \frac{d\theta_3}{dx_1} \right) dx_1 \right]^{(k)} \qquad (5.54)$$

where d = either u_m or θ_m
$\Delta^{(s)}$ = given components of displacements of supports of structure
$\tilde{R}^{(s)}$ = corresponding statically admissible reactions of supports of structure subjected to virtual loading

and the superscript (k) indicates that the quantities of the terms inside the bracket pertain to the kth member. Notice that the units of the first term of

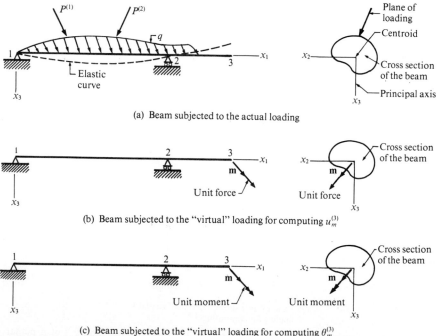

(a) Beam subjected to the actual loading

(b) Beam subjected to the "virtual" loading for computing $u_m^{(3)}$

(c) Beam subjected to the "virtual" loading for computing $\theta_m^{(3)}$

Figure 5.17   Computation of the components of displacement $u_m^{(3)}$ and $\theta_m^{(3)}$ of a beam.

the left-hand side of relation (5.54) are not compatible with those of the other terms. This is due to the fact that the first term is actually multiplied by the unit load.

When a structure is subjected only to external actions and to movement of its supports, the quantities which specify the deformation of an element (du_1, $d\theta_1$, $d\theta_2$, $d\theta_3$) can be expressed in terms of the internal actions N, M_1, M_2, and M_3 acting on this element using relations (5.3) to (5.5). Thus for this case Eq. (5.54) can be rewritten as

$$d + \sum_{s=1}^{S} \tilde{R}^{(s)}\Delta^{(s)} = \sum_{k=1}^{K} \left[\int_0^L \left(\frac{N\tilde{N}}{EA} + \frac{M_1\tilde{M}_1}{KG} + \frac{M_2\tilde{M}_2}{EI_2} + \frac{M_3\tilde{M}_3}{EI_3} \right) dx_1 \right.$$

$$\tag{5.55}$$

This relation can be employed in computing a component of translation or rotation of a point of a structure subjected to external actions and to movement of its supports by adhering to the following steps.

Step 1. We consider the structure subjected to the given external actions and to the given movement of its supports, and we compute the internal actions N, M_1, M_2, and M_3 in each member of the structure as functions of the axial coordinate of the member. In Chap. 3 we discuss how to compute the values

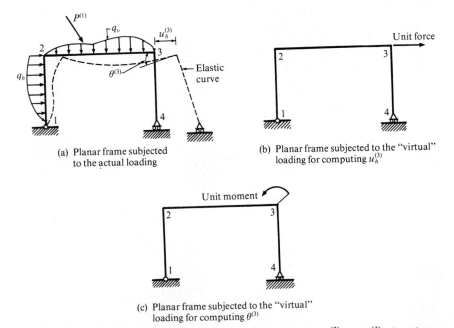

(a) Planar frame subjected
to the actual loading

(b) Planar frame subjected to the "virtual"
loading for computing $u_h^{(3)}$

(c) Planar frame subjected to the "virtual"
loading for computing $\theta^{(3)}$

Figure 5.18. Computation of the components of displacement $u_h^{(3)}$ and $\theta^{(3)}$ of a planar frame.

of N, M_1, M_2, and M_3 in statically determinate structures. In Chaps. 7 and 8 we present methods for computing these quantities in statically indeterminate structures.

Step 2.   We consider the structure subjected to the virtual loading described previously, and we compute a set of statically admissible internal actions $\tilde{N}$, $\tilde{M}_1$, $\tilde{M}_2$, and $\tilde{M}_3$ in each member of the structure as functions of the axial coordinate of the member. If some supports of the structure translate and rotate by a given amount, we compute in this step the reactions $\tilde{R}^{(s)}$ ($s = 1, 2, \ldots, S$) corresponding to the given translations or rotations of the supports of the structure. The reactions $\tilde{R}^{(s)}$ are considered positive in the direction of the corresponding component of displacement $\tilde{\Delta}^{(s)}$.

Step 3.   We substitute the internal actions established in steps 1 and 2, and the reactions $\tilde{R}^{(s)}$ established in step 2 in relation (5.55) and we evaluate the resulting integrals to obtain the desired component of translation or rotation.

5.8.1 Effect of temperature and initial displacements

In general the real displacement field **u** of the members of a structure could be the results of the following causes.

1. The external actions acting on the structure
2. Movement of the supports of the structure
3. The change of temperature $\Delta T(x_1, x_2, x_3)$ if any, in the members of the structure from its temperature at the unstressed state.
4. The initial displacement (initial strain) field of the members of the structure, that is, the components of displacement (components of strain) if any, before the structure was stressed (see Sec. 1.3.4).

We assume that the change of temperature is such that the component of rotation θ_1 is not affected by it. Moreover, we do not consider the effect of initial rotation θ_1^I. On this basis, the quantities which specify the deformation of an element of a member can be expressed as

$$du_1 = du_1^A + du_1^T + du_1^I$$

$$d\theta_2 = d\theta_2^A + d\theta_2^T + d\theta_2^I$$

$$d\theta_3 = d\theta_3^A + d\theta_3^T + d\theta_3^I$$

$$d\theta_1 = d\theta_1^A$$

(5.56)

The quantities du_1^A, $d\theta_1^A$, $d\theta_2^A$, and $d\theta_3^A$ specify the deformation of an element of a structure due to the internal actions acting on it. They can be expressed in terms of the internal actions using relations (5.3) to (5.5). The quantities

du_1^I, $d\theta_2^I$, and $d\theta_3^I$ specify the initial deformation of an element that is its deformation before the element was stressed (see Sec. 1.3.4). Referring to Fig. A.13, we can express them as

$$\frac{du_1^I}{dx_1} = e_{11}^I$$

$$\frac{d\theta_2^I}{dx_1} = \frac{1}{\rho_3^I} = k_3^I \tag{5.57}$$

$$\frac{d\theta_3^I}{dx_1} = \frac{1}{\rho_2} = k_2^I$$

where e_{11}^I = uniform axial component of initial strain of member
k_2^I and k_3^I = curvatures of projections of initially bent axis of member on x_1x_2 and x_3x_1 planes, respectively
ρ_2^I and ρ_3^I = radius of curvature of projections of initially bent axis of member on x_1x_2 and x_3x_1 planes, respectively

A member of a structure which is manufactured with a length which is either shorter or longer than that required to fit may be regarded as being subjected to a uniform axial component of initial strain. In this case we have

$$e_{11}^I = \frac{\text{manufactured length} - \text{required length}}{\text{required length}}$$

Moreover, a member of a structure which is manufactured with a curvature k_2^I when it is supposed to be straight may be regarded as being subjected to an initial curvature k_2^I. Furthermore a prestressed member of a structure, as for example a pretensioned steel bar of a concrete beam (see Sec. 1.3.4) may be regarded as being subjected to a uniform axial component of initial strain. In this case we have

$$e_{11}^I = -\frac{\Delta L}{L}$$

when ΔL is the increase of the length L of the member due to the pretensioning.

The quantities du_1^T, $d\theta_2^T$, and $d\theta_3^T$ specify the deformation of an element of a structure due to the change of temperature. In what follows we express these quantities in terms of the change of temperature which induced them. For this purpose we introduce the following notation.

1. T_0 is the uniform temperature at the stress-free state, that is, the temperature during construction of the structure.

2. $T_2^{(+)}$ and $T_2^{(-)}$ are the temperatures at the points of the cross section of a member where the positive and negative x_2 axis, respectively, intersect its perimeter (see Fig. 5.19a). We define the difference in temperature δT_2 as

$$\delta T_2 = T_2^{(+)} - T_2^{(-)} \tag{5.58}$$

3. $T_3^{(+)}$ and $T_3^{(-)}$ are the temperatures at the points of the cross section of a member where the positive and negative x_3 axis, respectively, intersect its perimeter (see Fig. 5.19a). We define the difference in temperature δT_3 as

$$\delta T_3 = T_3^{(+)} - T_3^{(-)} \tag{5.59}$$

4. T_c is the temperature at the centroid of the cross section of the member. Thus, the change of temperature at the centroid of the cross section of the member is equal to

$$\Delta T_c = T_c - T_0 \tag{5.60}$$

5. α is the coefficient of linear expansion of the material from which the member is made. Thus the change, due to an increase of temperature ΔT, of a fiber of a member of length dx_1 is equal to

$$\alpha \, \Delta T \, dx_1$$

We assume that the variation of the temperature is such that plane sections normal to the axis of a member, before the temperature changes, remain plane and normal to the deformed axis of the member after the temperature changes (see Fig. 5.20). This implies that the change of temperature is a linear function of the coordinates x_2 and x_3, and it does not induce shear deformation. On the

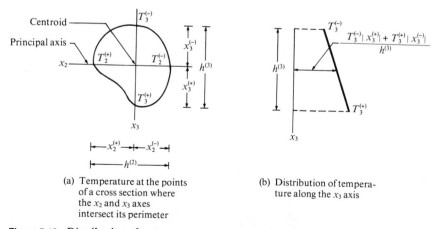

(a) Temperature at the points of a cross section where the x_2 and x_3 axes intersect its perimeter

(b) Distribution of temperature along the x_3 axis

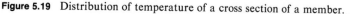

Figure 5.19 Distribution of temperature of a cross section of a member.

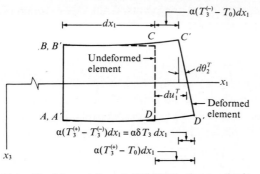

Note: The deformed element $A'B'C'D'$ is placed so that its
face $A'B'$ coincides with that of the undeformed element

Figure 5.20   Effect of a difference in temperature.

basis of this assumption, referring to Fig. 5.19b, the temperature T_c is equal to

$$T_c = \frac{T_3^{(-)}|x_3^{(+)}| + T_3^{(+)}|x_3^{(-)}|}{h^{(3)}} = \frac{T_2^{(-)}|x_2^{(+)}| + T_2^{(+)}|x_2^{(-)}|}{h^{(2)}} \quad (5.61)$$

Notice that, as expected, when three of the temperatures $T_2^{(+)}$, $T_2^{(-)}$, $T_3^{(+)}$, $T_3^{(-)}$ are given, the fourth may be established from relation (5.61).

Referring to Fig. 5.20, we have

$$du_1^T = \alpha \, \Delta T_c \, dx_1 \quad (5.62)$$

$$d\theta_2^T = \frac{\alpha \, \delta T_3}{h^{(3)}} \, dx_1 \quad (5.63)$$

similarly,

$$d\theta_3^T = -\frac{\alpha \, \delta T_2}{h^{(2)}} \, dx_1 \quad (5.64)$$

Subsituting relations (5.3) to (5.5), (5.57), and (5.62) to (5.64) into (5.56), we obtain

$$\frac{du_1}{dx_1} = \frac{N}{EA} + \alpha \, \Delta T_c + e_{11}^I$$

$$\frac{d\theta_2}{dx_1} = \frac{M_2}{EI_2} + \frac{\alpha \, \delta T_3}{h^{(3)}} + k_3^I$$

$$\frac{d\theta_3}{dx_1} = \frac{M_3}{EI_3} - \frac{\alpha \, \delta T_2}{h^{(2)}} + k_2^I \qquad (5.65)$$

$$\frac{d\theta_1}{dx_1} = \frac{M_1}{KG}$$

Substituting relations (5.65) into (5.54), we can write the principle of virtual work for a framed structure consisting of NM members made of isotropic linearly elastic materials as

$$
d + \sum_{s=1}^{S} \tilde{R}^{(s)} \Delta^{(s)} = \sum_{k=1}^{NM} \left[\int_0^L \left(\frac{N\tilde{N}}{EA} + \frac{M_1\tilde{M}_1}{KG} + \frac{M_2\tilde{M}_2}{EI_2} + \frac{M_3\tilde{M}_3}{EI_3} \right) dx_1 \right.
$$

$$
+ \int_0^L \left(\alpha \tilde{N} \, \Delta T_c + \alpha \tilde{M}_2 \frac{\delta T_3}{h^{(3)}} - \alpha \tilde{M}_3 \frac{\delta T_2}{h^{(2)}} \right) dx_1
$$

$$
\left. + \int_0^L \left(\tilde{N} e_{11}^I + \tilde{M}_2 k_2^I + \tilde{M}_3 k_2^I \right) dx_1 \right]^{(k)} \tag{5.66}
$$

In the above relation, the quantities without a tilde are the actual quantities, while those with a tilde are the virtual quantities; d is either u_m or θ_m; the superscript (k) indicates that the quantities of the terms inside the bracket pertain to the kth member; $\Delta^{(s)}$ are the given components of displacements of the supports of the structure; and $\tilde{R}^{(s)}$ are the corresponding statically admissible reactions of the supports of the structure subjected to the virtual loading.

Relation (5.66) is valid for structures composed of NM straight and curved (with small h/R ratio) members made of an isotropic linearly elastic material. However, for curved members the following substitutions must be made in relation (5.66): x_1 becomes s, x_3 becomes ζ; subscripts: 1 becomes t, 3 becomes ζ. The coordinate s is measured along the curved axis of the member; ζ is measured along the normal to the axis of the member (see Fig. 1.31b); L is the total length of the axis of the member.

5.9 Application of the Method of Virtual Work to Planar Trusses

In this section, the method of virtual work is employed in establishing the component of translation $u_m^{(i)}$ in the direction specified by the unit vector **m** of joint i of a truss (see Fig. 5.21) or the rotation of one of its members. The truss is

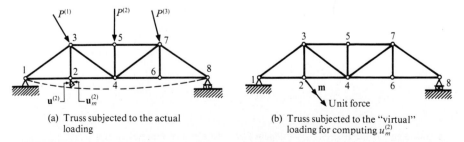

(a) Truss subjected to the actual loading

(b) Truss subjected to the "virtual" loading for computing $u_m^{(2)}$

Figure 5.21 Computation of the component of translation of joint 2 in the direction of **m**.

subjected to a general loading (given forces, change of temperature, settlement of its supports, a distribution of initial stress). The members of the truss are made of an isotropic linearly elastic material.

As mentioned previously it is assumed that the members of a truss are in a state of uniform uniaxial tensile or compressive stress. Thus, for this case, Eq. (5.66) reduces to

$$u_m^{(i)} + \sum_{s=1}^{S} \tilde{R}^{(s)}\Delta^{(s)} = \sum_{k=1}^{NM} \left[\tilde{N}L \left(\frac{N}{AE} + \alpha\,\Delta T_c + e_{11}^I \right) \right]^{(k)} \qquad (5.67)$$

$$e_{11}^I = \frac{\Delta L}{L} \qquad (5.68)$$

where ΔL is the difference of the actual (as manufactured) length of the kth member of the truss from its required length. This difference in length could be the result of manufacturing error. Moreover, the members of the top chord of a truss are often intentionally manufactured longer than required in order to produce an initial upward deflection (camber) of the bottom chord of the truss (see Example 2 at the end of this section). Thus when the truss is subjected to the dead load and part of the live load, it seems it has not deflected at all.

In order to determine the total translation of a joint of a truss, its horizontal and vertical components are computed using relation (5.67) and the results are added vectorially, as shown in Fig. 5.22.

Relation (5.67) may also be used to establish the rotation of a member of a truss. Consider member i of a truss and $\mathbf{m}$ the unit vector normal to it. In Fig. 5.23 member i is shown in its undeformed and deformed configurations. The displacement of its ends j and k are denoted by $\mathbf{u}^{ij} = \overline{jj''}$ and $\mathbf{u}^{ik} = \overline{kk'''}$, respectively, and are decomposed into a component along the axis of the member and a component normal to it. Referring to Fig. 5.23 we have

$$\mathbf{u}^{ij} = \overrightarrow{jj'} + u_m^{ij}\mathbf{m} \qquad \mathbf{u}^{ik} = \overrightarrow{kk'} + u_m^{ik}\mathbf{m} \qquad (5.69)$$

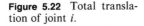

Figure 5.22 Total translation of joint i.

Figure 5.23 Rotation of a member of a truss.

Inasmuch as we limit the magnitude of deformation to the range of validity of the theory of small deformation, referring to Fig. 5.23 we see that the rotation $\Delta\theta$ of member i of the truss is equal to

$$\Delta\theta = \frac{k''k'''}{L_i} = \frac{u_m^{ik} - u_m^{ij}}{L_i} \tag{5.70}$$

In order to obtain the components of translation u_m^{ij} or u_m^{ik}, we must use as the virtual loading a unit force normal to the direction of member i, applied to the joint of the truss at the end j or k of the member, respectively. Thus, using Eq. (5.67), we get

$$u_m^{ij} + \sum_{s=1}^{S} (\tilde{R}^{(s)})^j \Delta^{(s)} = \sum_{p=1}^{NM} \left[(\tilde{N})^j \left(\frac{N}{AE} + \alpha\,\Delta T_c + e_{11}^I \right) L \right]^{(p)}$$

$$u_m^{ik} + \sum_{s=1}^{S} (\tilde{R}^{(s)})^k \Delta^{(s)} = \sum_{p=1}^{NM} \left[(\tilde{N})^k \left(\frac{N}{AE} + \alpha\,\Delta T_c + e_{11}^I \right) L \right]^{(p)} \tag{5.71}$$

$(\tilde{R}^{(s)})^j$ are the statically admissible reactions of the truss and $(\tilde{N})^j$ is the internal axial force in member i of the truss when a unit force normal to member i is applied to the joint of the truss to which the end j of member i is connected. Subtracting the first from the second of relations (5.71), dividing by the length of the member L_i, and using Eq. (5.70), we obtain

$$\Delta\theta + \sum_{s=1}^{S} \left[\frac{(\tilde{R}^{(s)})^k - (\tilde{R}^{(s)})^j}{L_i} \right] \Delta^{(s)}$$

$$= \sum_{p=1}^{NM} \left\{ \left[\frac{(\tilde{N})^k - (\tilde{N})^j}{L_i} \right] \left(\frac{N}{AE} + \alpha\,\Delta T_c + e_{11}^I \right) L \right\}^{(p)} \tag{5.72}$$

or $\quad \Delta\theta + \sum_{s=1}^{S} (\tilde{R}^{(s)})' \Delta^{(s)} = \sum_{p=1}^{NM} \left[(\tilde{N})' \left(\frac{N}{AE} + \alpha\,\Delta T_c + e_{11}^I \right) L \right]^{(p)} \tag{5.73}$

It is apparent that $(\tilde{R}^{(s)})' = [(\tilde{R}^{(s)})^k - (\tilde{R}^{(s)})^j]/L_i$ ($s = 1, 2, \ldots, S$) are statically admissible reactions at the supports of the truss and $(\tilde{N})' = [(\tilde{N})^k - (\tilde{N})^j]/L_i$ are statically admissible internal forces in the members of the truss, subjected to a virtual loading, which consists of two opposite forces of magnitude $1/L_i$ acting in the direction normal to member i. One of these forces acts on the joint at end j of member i, while the other acts on the joint at end k of member i. For instance, the virtual loading for finding the rotation of member 7,8 of a truss is shown in Fig. 5.24.

In an analogous fashion, it can be shown that if we want to find the relative movement u_{mn} of joints m and n of a truss, due to its deformation, we use as virtual loading two unit forces applied to joints m and n and acting in opposite

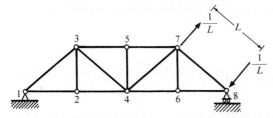

Figure 5.24 Virtual loading for finding the rotation of member 7, 8 of a truss.

directions along line *mn*. In Fig. 5.25 the virtual loading for finding the relative movement $u_{2,7}$ of joints 2 and 7 of the truss is shown.

In the sequel, we illustrate the application of the method of virtual work in computing the components of translation of joints of planar trusses by the following two examples.

Example 1. The total translation of a joint of a planar cantilever truss and the rotation of one of its members are computed. The truss is subjected to external forces, to an increase of the temperature of the members of its bottom chord, and to a movement of one of its supports.

Example 2. A simply supported truss is cambered by making the members of its upper chord longer by a specified amount. The vertical components of displacements of the joints of the bottom chord of the truss are computed.

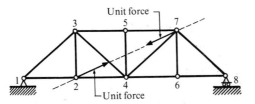

Figure 5.25 Virtual loading for finding the relative movement of two joints of a truss.

Example 1. Consider the truss shown in Fig. a. The cross-sectional areas of its top and bottom chords are equal to 2.4×10^3 mm², while the cross-sectional areas of its diagonal and vertical members are equal to 1.6×10^3 mm². The modulus of elasticity of the material from which the members of the truss are made is $E = 200{,}000$ MPa. The coefficient of linear expansion is $\alpha = 10^{-5}/°C$. Compute the total displacement of joint 7 and the rotation of member 8 due to:

1. The forces shown in Fig. a.
2. Increase of temperature of the members of the bottom chord by $\Delta T = 30°C$.
3. Movement of support 1 by 5 mm downward.

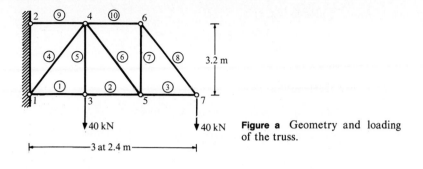

40 kN   40 kN   **Figure a** Geometry and loading
                        of the truss.

├──────3 at 2.4 m──────┤

solution This truss is statically determinate. Consequently, we can readily compute its reactions and the internal forces in its members resulting from the application of the given loading by considering the equilibrium of the truss as a whole and of its joints.

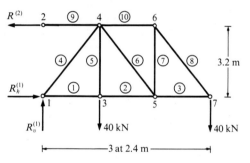

$R^{(1)}_v$ 40 kN   40 kN   **Figure b** Free-body diagram of
                                            the truss.

├──────3 at 2.4 m──────┤

Equilibrium of the Truss (refer to Fig. *b*)

$$\Sigma M^{(1)} = 0 \qquad -3.2R^{(2)} + (40)(2.4) + 40(7.2) = 0$$

or
$$R^{(2)} = -120 \text{ kN}$$

$$\Sigma \bar{F}_v = 0 \qquad R^{(1)}_v = 80 \text{ kN}$$

$$\Sigma \bar{F}_h = 0 \qquad R^{(1)}_h = 120 \text{ kN}$$

Equilibrium of Joint 1 (refer to Fig. *c*)

$$\Sigma \bar{F}_v = 0 \qquad N^{(4)} = -\tfrac{5}{4}(80) = -100 \text{ kN}$$

$$\Sigma \bar{F}_h = 0 \qquad N^{(1)} = \tfrac{3}{5}(100) - 120 = -60 \text{ kN}$$

Equilibrium of Joint 3 (refer to Fig. *d*)

$$\Sigma \bar{F}_v = 0 \qquad N^{(5)} = 40 \text{ kN}$$

$$\Sigma \bar{F}_h = 0 \qquad N^{(2)} = -60 \text{ kN}$$

Equilibrium of Joint 7 (refer to Fig. *e*)

$$\Sigma \bar{F}_v = 0 \qquad N^{(8)} = \tfrac{5}{4}(40) = 50 \text{ kN}$$

$$\Sigma \bar{F}_h = 0 \qquad N^{(3)} = -\tfrac{3}{5}(50) = -30 \text{ kN}$$

Equilibrium of Joint 6 (refer to Fig. *f*)

$$\Sigma \bar{F}_v = 0 \qquad N^{(7)} = -\tfrac{4}{5}(50) = -40 \text{ kN}$$

$$\Sigma \bar{F}_h = 0 \qquad N^{(10)} = \tfrac{3}{5}(50) = 30 \text{ kN}$$

Equilibrium of Joint 4 (refer to Fig. *g*)

$$\Sigma \bar{F}_v = 0 \qquad \tfrac{4}{5}N^{(6)} - \tfrac{4}{5}(100) + 40 = 0 \qquad \text{or} \qquad N^{(6)} = 50 \text{ kN}$$

$$\Sigma \bar{F}_h = 0 \qquad N^{(9)} = 30 + \tfrac{3}{5}(100) + \tfrac{3}{5}(50) = 120 \text{ kN}$$

The internal forces N in the members of the truss are tabulated in column I of Table a on page 280.

To find the total translation of joint 7, we must establish its horizontal and vertical components. This is accomplished by using the virtual loadings of Figs. h and i, respectively. For each of these virtual loadings, the internal force $\hat{N}$ in each member of the truss is computed and tabulated in either column VII or VIII of Table a.

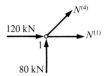

Figure c Free-body diagram of joint 1.

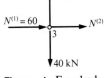

Figure d Free-body diagram of joint 3.

Figure e Free-body diagram of joint 7.

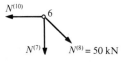

Figure f Free-body diagram of joint 6.

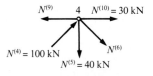

Figure g Free-body diagram of joint 4.

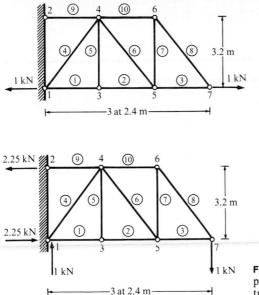

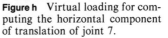

Figure h   Virtual loading for computing the horizontal component of translation of joint 7.

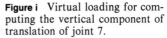

Figure i   Virtual loading for computing the vertical component of translation of joint 7.

To find the rotation of member 8, the virtual loading shown in Fig. j is employed. For this virtual loading, the internal forces $\tilde{N}$ in each member of the truss are computed and tabulated in column IX of Table a. Notice, that in obtaining the value of the virtual loading of Fig. j, the length of the member was converted into millimeters.

The numerical results from Table a are substituted in relation (5.67) to obtain the vertical and horizontal components of translation of joint 7 and in relation (5.73) to obtain the rotation of member 8. In relation (5.67), $\Delta^{(s)}$ is the given settlement of support 1 ($\Delta^{(s)} = 5$ mm). Hence, $\tilde{R}^{(1)}$ is the vertical reaction at support 1 when the truss is subjected to a unit force at joint 7, as shown in Figs. h and i. In relation (5.73), $(\tilde{R}^{(1)})'$ is the vertical reaction at support 1 when the truss is subjected to a unit couple, as shown in Fig. j. For the virtual loading of Figs. h and j, $\tilde{R}_v^{(1)}$ and $(\tilde{R}_v^{(1)})'$ vanish, while for the virtual loading of Fig. i, $\tilde{R}_v^{(1)} = -1$. Thus, the settlement of support 1 does not affect the horizontal component of displacement of joint 7 or the rotation of member 8. Substituting the results of Table a into relation (5.67), we get

$$u_h^{(7)} = \sum_{k=1}^{10} \left[\tilde{N}L \left(\frac{N}{AE} + \alpha\,\Delta T \right) \right]^{(k)} = 1.41 \text{ mm}$$

or

$$u_v^{(7)} - 1(5) = \sum_{k=1}^{10} \left[\tilde{N}L \left(\frac{N}{AE} + \alpha\,\Delta T \right) \right]^{(k)} = 3.3 \text{ mm}$$

$$u_v^{(7)} = 8.3 \text{ mm}$$

Thus, the total displacement of joint 7 is

$$u^{(7)} = \sqrt{(u_h^{(7)})^2 + (u_v^{(7)})^2} = \sqrt{(1.41)^2 + (8.3)^2} = 8.42 \text{ mm}$$

TABLE a

Member	N, kN (I)	A, mm² (II)	$N/EA \times 10^3$ (III)	$N/EA + \alpha\,\Delta T$, $\times 10^3$ (IV)	L, mm (V)	(IV) × (V), mm (VI)	$\tilde{N}$ loading Fig. h (VII)	$\tilde{N}$ loading Fig. i (VII)	$\tilde{N} \times 10^3$ loading Fig. j (IX)	(VI) × (VII), mm (X)	(VI) × (VIII), mm (XI)	(VII) × (IX) $\times 10^2$, (XII)
1	−60	2400	−0.125	0.175	2400	0.42	1.0	−1.5	−0.3125	0.42	−0.63	−0.013125
2	−60	2400	−0.125	0.175	2400	0.42	1.0	−1.5	−0.3125	0.42	−0.63	−0.013125
3	−30	2400	−0.0625	0.2375	2400	0.57	1.0	−0.75	−0.3125	0.57	−0.4275	−0.0178125
4	−100	1600	−0.3125	−0.3125	4000	1.252	0	−1.25	0	0	1.5625	0
5	40	1600	0.125	0.125	3200	0.40	0	0	0	0	0	0
6	50	1600	0.15625	0.15625	4000	0.624	0	1.25	0	0	0.78125	0
7	−40	1600	−0.125	−0.125	3200	0.40	0	−1.0	0	0	0.40	0
8	50	1600	0.15625	0.15625	4000	0.624	0	1.25	0.1875	0	0.78125	0.0117187
9	120	2400	0.250	0.250	2400	0.60	0	2.25	0.3125	0	1.350	0.01875
10	30	2400	0.0625	0.0625	2400	0.15	0	0.75	0.3125	0	0.1125	0.0046875

$$\sum_{k=1}^{NM} \left[\tilde{N} \left(\frac{N}{EA} + \alpha\,\Delta T \right) L \right]^{(k)}$$

	(X)	(XI)	(XII)
→	1.41	3.30	−0.0089062

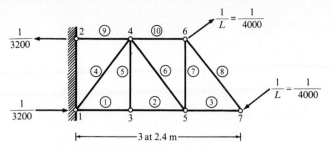

Figure j Virtual loading for finding the rotation of member 8.

Substituting the results of Table a into relation (5.73), we obtain the rotation of member 8 as

$$\Delta\theta = \sum_{k=1}^{10} \left[\tilde{N}L \left(\frac{N}{AE} + \alpha \, \Delta T \right) \right]^{(k)} = -0.89062 \times 10^{-4} \text{ rad}$$

The minus sign indicates that member 8 rotates in the direction opposite to that of the applied unit torque of Fig. j, that is, counterclockwise.

Thus, we have established that joint 7 moves 8.3 mm downward and 1.41 mm to the right, while member 8 rotates by 0.8925×10^{-4} rad counterclockwise.

Example 2.   The truss of Fig. a is to be cambered by making its members 4 and 8 longer by 9 mm and its members 2 and 10 longer by 12 mm. Compute the vertical components of displacement of joints 3 and 5 due to the change of length of the four members of the upper chord.

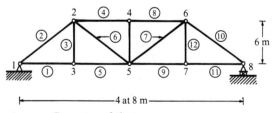

Figure a   Geometry of the truss.

solution   To find the vertical components of displacement of joints 3 and 5, we use the virtual loadings of Figs. b and c, respectively. For each of these loadings, the internal forces in members 2, 4, 8, and 10 are tabulated in Table a. We do not have to

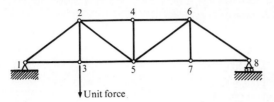

Figure b Truss subjected to the virtual loading for computing $u_3^{(3)}$.

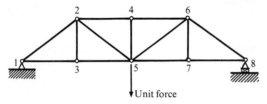

Figure c Truss subjected to the virtual loading for computing $u_3^{(5)}$.

TABLE a

| Member | ΔL | Unit force at joint 3 | | Unit force at joint 5 | |
		$\tilde{N}$	$\tilde{N} \Delta L$	$\tilde{N}$	$\tilde{N} \Delta L$
2	12	$-\frac{5}{4}$	-15	$-\frac{5}{6}$	-10
4	9	$-\frac{2}{3}$	-6	$-\frac{4}{3}$	-12
8	9	$-\frac{2}{3}$	-6	$-\frac{4}{3}$	-12
10	12	$-\frac{5}{12}$	-5	$-\frac{5}{6}$	-10
Total			-32		-44

include the internal forces in the other members of the truss since the length of these members did not change ($\Delta L = 0$). Referring to Table a from relation (5.67), we get

$$u_v^{(3)} = -32 \text{ mm} \quad \text{or} \quad 32 \text{ mm} \uparrow$$

$$u_v^{(5)} = -44 \text{ mm} \quad \text{or} \quad 44 \text{ mm} \uparrow$$

5.10   Application of the Method of Virtual Work to Beams and Frames

In this section, we employ the method of virtual work (5.66) to establish the components of displacement (translation and rotation) of points of beams and frames whose members are made of isotropic, linearly elastic materials by adhering to one of the following procedures.

TABLE 5.1 Evaluation of $\displaystyle\int_0^L F(x_1)\bar{F}(x_1)\,dx_1$

$\bar{F}(x_1)$ $\quad$ $F(x_1)$	$\tilde{a}$ $\quad$ (+) $\quad$ $\bar{b}$ $\qquad L$
a $\quad$ (+) $\quad$ b $\qquad L$	$\dfrac{L}{6}\,[\tilde{a}(2a+b)$ $+\,\bar{b}(a+2b)]$
c $\quad$ (+) d $\quad$ e $\qquad \dfrac{L}{2}\;\dfrac{L}{2}$ Second-degree curve	$\dfrac{L}{6}\,[\tilde{a}(c+2d)$ $+\,\bar{b}(2d+e)]$
$\lceil$ Tangent $\qquad$ (+) e $\qquad L$ Second-degree curve	$\dfrac{Le\,(\tilde{a}+3\bar{b})}{12}$

1. *Direct integration (see Examples 1 and 2).* Direct integration involves the following steps.

Step 1. Establish the axial forces and the moments in each member of the structure subjected to the actual loading as functions of the axial coordinate of the member.

Step 2. Establish statically admissible axial forces and moments in each member of the structure subjected to the virtual loading as functions of the axial coordinate of the member.

Step 3. Substitute the axial forces and the moments obtained in steps 1 and 2 into Eq. (5.66) and evaluate the integrals by direct integration.

2. *Use of Table† 5.1 (see Example 3).* The use of Table 5.1 involves the following steps.

Step 1. Plot the moment diagram for each member of the structure subjected to the actual loading.

---

† The same table is also available on the inside of the back cover.

Step 2. Plot the moment diagram for each member of the structure subjected to the virtual loading.

Step 3. Evaluate the integrals using Table 5.1 and the moment diagrams established in steps 1 and 2. Table 5.1 gives the values of the integral of the product of two functions $F(x_1)$ and $\tilde{F}(x_1)$. For example the functions $F(x_1)$ and $\tilde{F}(x_1)$ for the second term of the first integral on the right-hand side of relation (5.66) are

$$F(x_1) = \frac{M_2}{EI} \qquad \tilde{F}(x_1) = \tilde{M}_2$$

3. *Use of the fact that the internal actions in the members of the structure subjected to the virtual loading vary linearly with x_1* (see Example 4).   When a structure is subjected to virtual loading, the internal axial force $\tilde{N}$ and the internal torsional moment $\tilde{M}_1$ are constants while the internal bending moments $\tilde{M}_2$ and $\tilde{M}_3$ are either constant or vary linearly with x_1. This is so because the virtual loading is either a concentrated force or a concentrated moment. Thus denoting $\tilde{F}(x_1)$ as an internal action of a member of the structure subjected to the virtual loading, we have

$$\tilde{F}(x_1) = ax_1 + b \tag{5.74}$$

Substituting relation (5.74) into the integral of the product of the functions $F(x_1)$ and $\tilde{F}(x_1)$, we obtain

$$\int_{L_1}^{L_2} F(x_1)\tilde{F}(x_1)\, dx_1 = a \int_{L_1}^{L_2} x_1 F(x_1)\, dx_1 + b \int_{L_1}^{L_2} F(x_1)\, dx_1 \tag{5.75}$$

The first integral on the right-hand side of the above relation represents the moment of the area under the graph of the function $F(x_1)$ from $x_1 = L_1$ to $x_1 = L_2$ about $x_1 = 0$. The second integral on the right-hand side of the above relation represents the area under the graph of the function $F(x_1)$ from $x_1 = L_1$ to $x_1 = L_2$. Denoting this area by A_F and the x_1 coordinate of the centroid of this area by $\bar{x}_1$, we can write relation (5.75) as

$$\int_{L_1}^{L_2} F(x_1)\tilde{F}(x_1)\, dx_1 = (a\bar{x}_1 + b)A_F = \tilde{F}(\bar{x}_1)A_F \tag{5.76}$$

That is, the value of the integral of the product of the functions $F(x_1)$ and $\tilde{F}(x_1)$ from $x_1 = L_1$ to $x_1 = L_2$ is equal to the product of the area under the graph of the function $F(x_1)$ from $x_1 = L_1$ to $x_1 = L_2$ and the value at $x_1 = \bar{x}_1$ of the function $\tilde{F}(x_1)$.†

---

† The areas under the graphs of certain functions and the location of the centroids of these areas are tabulated in Table B on the inside of the back cover.

4. *Numerical integration* (see Sec. 5.11).   In this procedure we adhere to the same steps as in direct integration. However, in step 3 we integrate the integrals numerically using one of the known formulas such as trapezoidal rule or Simpson's rule.

The effect of the axial deformation of members of beams and frames on the components of translation and rotation of their cross sections is actually very small, and it is in general disregarded. The effect of axial deformation of the members of a frame is established in Example 3 at the end of this section.

The method of virtual work is illustrated by the following examples.

Example 1.   The deflection and rotation of the free end of a cantilever beam are computed when the beam is subjected to a difference in temperature between its upper and lower fibers. The integrals are evaluated by direct integration.

Example 2.   The deflection and the rotation of the free end of a simply supported beam with an overhang are computed when the beam is subjected to the following loading cases.

1. To a uniform load along a portion of its length

2. To a difference in temperature between its upper and lower fibers

3. To a settlement of one of its supports

The integrals are evaluated by direct integration.

Example 3.   The horizontal displacement of a joint of a three-hinged frame and the change of slope of its elastic curve at the intermediate hinge are computed. The frame is subjected to the following loading cases.

1. To external actions (For this loading the integrals are evaluated using Table 5.1 or Table A on the inside of the back cover.)

2. To a difference in temperature between the internal and external fibers of its members

3. To a vertical and horizontal movement of a support

The effect of the axial deformation of the members of the frame is considered.

Example 4.   The deflection of the free end of a simply supported beam with an overhang is computed when it is subjected to external forces. The integrals are evaluated using the linearity of the moment in the beam when subjected to the virtual loading.

Example 5.   The total displacement of the unsupported end of a cantilever space frame is computed. The frame is subjected to an external force. Some members of the frame are subjected to torsion and to bending. The integrals are evaluated using the formulas of Table 5.1 or Table A on the inside of the back cover.

Example 1. Consider a cantilever beam of length L having a constant cross section of a given geometry. Establish the deflection and the rotation of the free end of the beam when the difference of temperature at its points, where the positive and negative centroidal axes x_3 intersect the perimeter of its cross section, is $\delta T = T_b - T_t$ (see Fig. 5.19a).

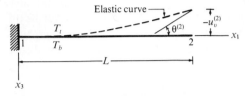

Figure a Geometry, loading, and elastic curve of the beam.

---

solution The deflection and the rotation, due to the given change of temperature, of point 2 of the beam may be established from relation (5.66) using the virtual loading of Figs. b and c, respectively. Thus,

$$u_v^{(2)} = -\int_0^L \frac{\alpha\,\delta T}{h} x_1'\,dx_1' = -\frac{\alpha\,\delta T}{2h} L^2 \tag{a}$$

$$\theta^{(2)} = \int_0^L \frac{\alpha\,\delta T}{h}\,dx_1' = \frac{\alpha\,\delta T}{h} L \tag{b}$$

The negative sign of the deflection $u_v^{(2)}$ indicates that its direction is opposite to that of the unit force of Fig. b; that is, point 2 moves upward.

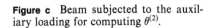

Figure b Beam subjected to the virtual loading for computing $u_v^{(2)}$.

Figure c Beam subjected to the auxiliary loading for computing $\theta^{(2)}$.

Example 2. For point 3 of the beam shown in Fig. a, compute the deflection and the rotation due to following loading cases.

1. The forces shown in Fig. a

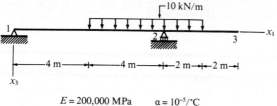

$$E = 200,000 \text{ MPa} \qquad \alpha = 10^{-5}/°\text{C}$$
$$I = 360 \times 10^6 \text{ mm}^4 \qquad h = 300 \text{ mm}$$

Figure a Geometry and loading of the beam.

2. A difference in temperature between the top and bottom fibers of the beam (T_t = 40°C, T_b = 10°C)
3. A 5 mm downward settlement of support 2

---

solution

1. *Computation of the deflection and rotation of point 3 of the beam due to the forces shown in Fig. a.* Referring to Fig. b, we may obtain the following distribution of moments due to the forces shown in Fig. a.

$$M = 7.5x_1 \qquad\qquad 0 \le x_1 \le 4$$

$$M = 7.5x_1 - \frac{10(x_1 - 4)^2}{2} \qquad 4 \le x_1 \le 8$$

$$M = -\frac{10(x_1')^2}{2} \qquad\qquad 0 \le x_1' \le 2$$

$$M = 0 \qquad\qquad 0 \le x_1'' \le 2$$

Notice that in order to reduce the required algebra, the moment in the beam segment 2,3 was expressed in terms of the coordinate x_1', shown in Fig. b.

To compute the deflection of point 3 of the beam, the virtual loading shown in Fig. c is employed. The distribution of moments corresponding to this loading is

$$\tilde{M} = -\frac{x_1}{2} \qquad\qquad 0 \le x_1 \le 8$$

$$\tilde{M} = -(x_1' + 2) \qquad 0 \le x_1' \le 2 \qquad\qquad (b)$$

$$\tilde{M} = -x_1'' \qquad\qquad 0 \le x_1'' \le 2$$

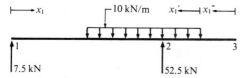

Figure b Free-body diagram of the beam loaded with the actual loading.

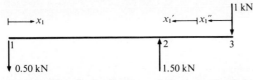

Figure c Free-body diagram of the beam loaded with the virtual loading for computing the deflection of point 3.

Applying the method of virtual work (5.66), we have

$$u_v^{(3)} = \frac{10^{12}}{EI} \left\{ - \int_0^4 7.5x_1 \left(\frac{x_1}{2} \right) dx_1 - \int_4^8 \left[7.5x_1 - \frac{10(x_1 - 4)^2}{2} \right] \frac{x_1}{2} \, dx_1 \right. $$

$$\left. + \int_0^2 \frac{10(x_1')^2}{2} (x_1' + 2) \, dx_1' \right\} \tag{c}$$

In the above relation, if E is in megapascals and I is in mm^4, the deflection $u_v^{(3)}$ is in millimeters. Integrating relation (c), we get

$$u_v^{(3)} = - \frac{710 \times 10^{12}}{3EI} = - \frac{710 \times 10^{12}}{3(200,000)(3.6 \times 10^8)} = -3.29 \text{ mm} \tag{d}$$

The minus sign indicates that point 5 moves in the direction opposite to the unit load shown in Fig. c. That is, it moves upward.

In order to compute the rotation of the elastic curve at point 3 of the beam, the virtual loading shown in Fig. d is employed. The distribution of moments corresponding to this loading is:

$$\tilde{M} = \frac{x_1}{8} \qquad 0 \le x_1 \le 8$$

$$\tilde{M} = 1 \qquad 0 \le x_1' \le 2 \tag{e}$$

Applying the method of virtual work (5.66), we have

$$\theta^{(3)} = \frac{10^9}{EI_2} \left\{ \int_0^4 7.5x_1 \frac{x_1}{8} \, dx_1 + \int_4^8 \left[7.5x_1 - \frac{10(x_1 - 4)^2}{2} \right] \frac{x_1}{8} \, dx_1 \right. $$

$$\left. + \int_0^2 \frac{-10(x_1')^2}{2} \, dx_1' \right\}$$

$$= \frac{53.33 \times 10^9}{EI_2} = \frac{53.33 \times 10^9}{(200,000)(3.6 \times 10^8)} = 7.41 \times 10^{-4} \text{ rad} \tag{f}$$

The plus sign indicates that the rotation $\theta^{(3)}$ of the beam is in the direction of the applied unit moment in Fig. d. The deformed configuration of the beam is shown in Fig. e.

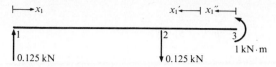

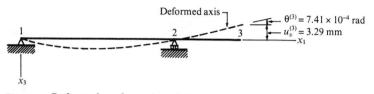

Figure d Free-body diagram of the beam loaded with the virtual loading for computing the rotation at point 3.

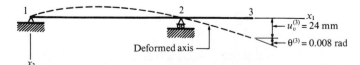

Figure e Deformed configuration of the beam subjected to the given external forces.

2. *Computation of the deflection and rotation of point 3 of the beam due to the difference in temperature between the top and bottom fibers.* The deflection and the rotation of point 3 of the beam due to the given change of temperature, may be established from relation (5.66) using the virtual loading of Figs. c and d, and, consequently, the distribution of moments given by relations (b) and (e), respectively. Thus,

$$u_v^{(3)} = \int_0^L \alpha \tilde{M} \frac{\delta T}{h} \, dx_1 = -\int_0^8 \frac{\alpha}{h} \frac{x_1}{2} (10 - 40) \, dx_1 - \int_0^4 \frac{\alpha}{h} (10 - 40) x_1'' \, dx_1''$$

$$= \frac{720\alpha}{h} = \frac{720 \times 10^{-5}}{0.30} = 24 \text{ mm} \downarrow \tag{g}$$

$$\theta^{(3)} = \int_0^L \alpha \tilde{M} \frac{\delta T}{h} \, dx_1 = \int_0^8 \frac{\alpha}{h} \frac{x_1}{8} (10 - 40) \, dx_1 - \int_0^4 \frac{\alpha}{h} 30 \, dx_1$$

$$= -\frac{240\alpha}{h} = -\frac{240 \times 10^{-5}}{0.30} = 0.008 \text{ rad clockwise} \tag{h}$$

The minus sign indicates that the rotation $\theta^{(3)}$ of the beam is as shown in Fig. f, in the direction opposite to the applied unit moment in Fig. d.

3. *Computation of the deflection and rotation of point 3 of the beam due to the settlement of support 2.* We use relation (5.66) to compute the deflection and the rotation of point 3 of the beam due to the 5 mm settlement of support 3. Thus, we

Figure f Deformation of the beam due to the change of temperature.

have

$$u_v^{(3)} + 5\tilde{R}^{(s)} = 0 \qquad \theta^{(3)} + 5[\tilde{R}^{(s)}]' = 0$$

where $\tilde{R}^{(s)}$ and $[\tilde{R}^{(s)}]'$ are the reactions at support 2 when the beam is subjected to the virtual loading shown in Figs. c and d, respectively. The reactions $\tilde{R}^{(s)}$ and $[\tilde{R}^{(s)}]'$ are positive in the direction of the settlement $\Delta^{(s)}$, i.e., when they act downward. Thus:

$$\tilde{R}^{(s)} = -1.5 \qquad [\tilde{R}^{(s)}]' = 0.5 \tag{j}$$

consequently,

$$u_v^{(3)} = 7.5 \text{ mm} \tag{k}$$

and

$$\theta^{(3)} = -\frac{0.005}{8} = -0.000625 \text{ rad}$$

The minus sign indicates that the rotation $\theta^{(3)}$ of the beam is opposite to the unit moment shown in Fig. d. That is the rotation $\theta^{(3)}$ is clockwise. The same results may be obtained by referring to Fig. g and noting that because of the settlement of the support, the beam moves as a rigid body.

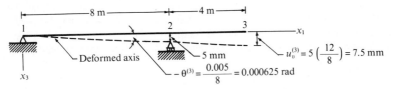

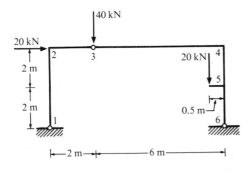

Figure g Deformation of the beam due to the settlement of support 2.

Example 3. Consider the three-hinged frame shown in Fig. a, and for the following loading cases, compute the horizontal component of translation of joint 4 and the change of the slope of its elastic curve at joint 3.

1. The forces shown in Fig. a

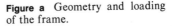

Figure a Geometry and loading of the frame.

2. Difference in temperature between the external and internal fibers of the members of the frame. It is given that the temperatures of the external and internal fibers of the members of the frame are $T_e = 10°C$ and $T_i = 40°C$, respectively. The temperature during construction was $T_o = 10°C$.

3. A 5 mm downward settlement of support 6 and a 10 mm horizontal movement to the right of the same support

The members of the frame are made of standard I-beams whose cross sections have the following properties:

$I = 576.8 \times 10^6$ mm⁴

$A = 19.8 \times 10^3$ mm²

Depth of cross section $h = 400$ mm

Coefficient of thermal expansion $\alpha = 10^{-5}/°C$

Modulus of elasticity $E = 210,000$ MPa

Consider and discuss the effect of the axial deformation of the members of the frame.

solution In this example, we use the formulas given in Table A on the inside of the back cover to evaluate the integrals in the expressions for the translation and rotation (5.56). For this reason, we plot the moment diagrams of the frame subjected to the actual and the virtual loadings.

1. *For the external forces shown in Fig. a, compute the horizontal component of translation of joint 4 of the frame and the change of slope of its elastic curve at point 3.* The reactions of the frame subjected to the given external forces may be computed by referring to Fig. b. Setting the sum of the forces acting in the vertical and horizontal direction equal to zero, we obtain

$$\Sigma \overline{F}_h = 0 \qquad R_h^{(6)} + R_h^{(1)} = -20$$
$$\Sigma \overline{F}_v = 0 \qquad R_v^{(1)} + R_v^{(6)} = 60 \tag{a}$$

In order to compute the four components of the reactions of the frame we must use the fact that the moment vanishes at the hinge at point 3. Thus considering the equi-

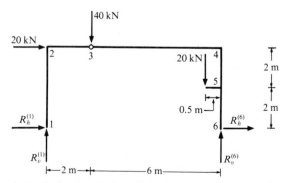

Figure b Free-body diagram of the frame loaded with the actual loading.

librium of the moments acting on parts 1,2,3 and 3,4,6 of the frame, we have

$$\text{Part 1,2,3} \qquad \Sigma M^{(3)} = 2R_v^{(1)} - 4R_h^{(1)} = 0$$

$$\text{Part 3,4,6} \qquad \Sigma M^{(3)} = 6R_v^{(6)} + 4R_h^{(6)} - 20(5.5) = 0 \qquad \text{(b)}$$

Solving Eqs. (a) and (b) simultaneously, we obtain

$$R_h^{(1)} = 10.6 \text{ kN} \qquad R_h^{(6)} = 30.6 \text{ kN} \qquad R_v^{(1)} = 21.2 \text{ kN} \qquad R_v^{(6)} = 38.8 \text{ kN} \qquad \text{(c)}$$

The free-body diagrams of the members of the frame loaded with the actual loading are shown in Fig. c. The corresponding moment diagram of the frame is plotted in Fig. d. Referring to Fig. c, notice that the moment in the column 4,6 is equal to

$$M = -30.6x_1' \qquad 0 \le x_1' < 2 \qquad \text{(d)}$$

$$M = -30.6x_1' + 10 \qquad 2 < x_1' \le 4 \qquad \text{(e)}$$

In order to simplify the evaluation of the integral in relation (5.66) using Table A on the inside of the back cover, we plot in Fig. c each term of Eq. (e) separately. The resulting moment diagram is called *the moment diagram by parts*.

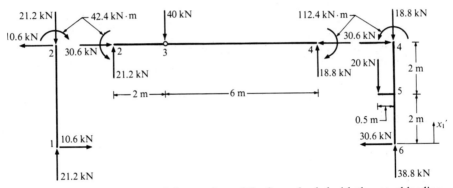

Figure c   Free-body diagrams of the members of the frame loaded with the actual loading.

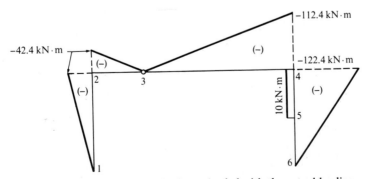

Figure d   Moment diagram of the frame loaded with the actual loading.

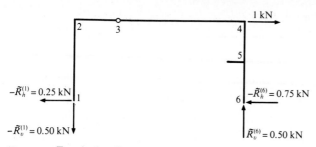

Figure e Free-body diagram of the frame loaded with the virtual loading for computing $u_h^{(4)}$.

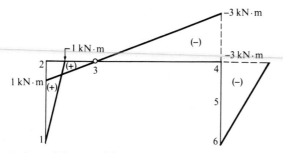

Figure f Moment diagram of the frame loaded with the virtual loading for computing $u_h^{(4)}$.

In order to compute the horizontal component of translation of joint 4 of the frame, the virtual loading shown in Fig. e is employed, while in order to compute the change of slope of the elastic curve of the frame at point 3, the virtual loading shown in Fig. g is employed. The moment diagrams of the frame due to these loadings are shown in Figs. f and h, respectively. Referring to relation (5.66) and using Table A on the inside of the back cover in conjunction with Figs. d and f, we get:

$$u_h^{(4)} = \frac{10^6}{AE} \left[-21.2(\tfrac{1}{2})(4) - 30.6(\tfrac{1}{4})(8) + 18.8(\tfrac{1}{2})(2) \right.$$

$$\left. + 38.8(\tfrac{1}{2})(2) \right] + \frac{10^{12}}{EI_2} \left\{ -\tfrac{4}{6}(42.4)(2) - \tfrac{2}{6}(1)(2)(42.4) \right.$$

$$+ \tfrac{6}{6}(112.4)(2)(3) + \tfrac{4}{6}(3)(2)(122.4) - \tfrac{2}{6}[3[2(10) + 10]$$

$$\left. + 1.5[2(10) + 10]] \right\}$$

$$= -\frac{46 \times 10^6}{EA} + \frac{1038.2 \times 10^{12}}{EI_2}$$

$$= -\frac{46 \times 10^6}{210,000(1.98 \times 10^4)} + \frac{1038.2 \times 10^{12}}{210,000(5.768 \times 10^8)}$$

$$= -0.01 + 8.57 = 8.56 \text{ mm} \tag{f}$$

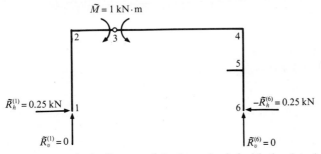

Figure g Free-body diagram of the frame loaded with the virtual loading for computing $\Delta\theta^{(3)}$.

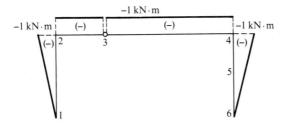

Figure h Moment diagram of the frame loaded with the virtual loading for computing $\Delta\theta^{(3)}$.

Referring to relation (5.66) and using Table A on the inside of the back cover in conjunction with Figs. d and h, we obtain:

$$\Delta\theta^{(3)} = \frac{8 \times 10^9}{EA}(30.6)(0.250) + \frac{10^9}{EI_2}\left\{\tfrac{4}{6}(42.4)(2) + \tfrac{2}{6}[(1)(2)(42.4) + 1(42.4)]\right.$$

$$+ \tfrac{6}{6}[1(112.4) + 1(2)(112.4)] + \tfrac{4}{6}(2)(122.4)$$

$$\left. - \tfrac{2}{6}[1[2(10) + 10] + 0.5[2(10) + 10]\right\} \tag{g}$$

$$= \frac{61.2 \times 10^3}{EA} + \frac{586 \times 10^9}{EI_2}$$

$$= \frac{61.2 \times 10^3}{210,000(1.98 \times 10^4)} + \frac{586 \times 10^9}{(210,000)(5.768 \times 10^8)}$$

$$= 0.00001 + 0.004838 = 0.004848 \text{ rad}$$

The positive sign in the value of $\Delta\theta^{(3)}$ indicates that the sum of the angles $\theta^{(3)R}$ and $\theta^{(3)L}$ shown in Fig. i has the sense of the applied moments of Fig. g.

The first term in the results (f) and (g) represents the effect of axial deformation of the members of the frame on the horizontal component of translation of joint 4 and on the change of slope of its elastic curve at point 3, respectively. It is evident that this term is negligible compared to the second term which represents the effect of bending of the members of the frame.

Referring to the moment diagram of Fig. c and to the results (f) and (g), we sketch the elastic curve of the frame in Fig. i.

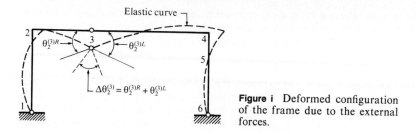

Figure i Deformed configuration of the frame due to the external forces.

2. *For the given change of temperature, compute the horizontal component of translation of joint 4 of the frame and the change of slope of its elastic curve at point 3.* The horizontal component of translation of joint 4 and the change of slope of point 3 due to a temperature difference between the external and internal fibers of the members of the frame and to a change of temperature from that existing during construction may be established from relation (5.66) using the virtual loadings of Figs. e and g, respectively. For a linear temperature distribution, and assuming that the centroid of the cross section of the members of the frame is in the middle of the distance between the extreme fibers, we have

$$T_c = \frac{T_e + T_i}{2} = 25°C \tag{h}$$

hence,
$$\Delta T_c = T_c - T_0 = 25 - 10 = 15°C$$
$$\delta T_3 = T_i - T_e = 40 - 10 = 30°C \tag{i}$$

Thus, referring to Figs. e and g, and applying the method of virtual work (5.66), we get

$$u_h^{(4)} = \sum_{i=1}^{3} \left(\alpha \, \Delta T_c \int_0^L N \, dx_1 + \alpha \, \delta T_3 \int_0^L \frac{\tilde{M}}{h} \, dx_1 \right)^{(i)}$$

$$= (15 \times 10^3)\alpha[\tfrac{1}{2}(4) + \tfrac{1}{4}(8) - \tfrac{1}{2}(4)] + \frac{\alpha(10)^3}{h}(30)[1(2) + 1(1) - 3(\tfrac{6}{2}) - 3(\tfrac{4}{2})]$$

$$= (30 \times 10^3)\alpha - \frac{(30 \times 10^3)\alpha}{h}(12)$$

$$= 0.3 \left(1 - \frac{12 \times 10^3}{400} \right) = 0.3 - 9.0 = -8.7 \text{ mm}$$

(j)

and

$$\Delta\theta^{(3)} = 15\alpha(-0.25)8 + \frac{\alpha(10^3)}{h}(40 - 10)[-1(2) - 1(8) - 1(2)]$$

$$= -30\alpha \left[1 + \frac{12 \times 10^3}{h} \right]$$

$$= -0.0003 \left(1 + \frac{12 \times 10^3}{400} \right) = -0.0003 - 0.0090$$

$$= -0.0093 \text{ rad}$$

(k)

The first term in relations (j) and (k) represents the effect of the increase in temperature of the axis of the frame (from that existing during construction) on the horizontal component of translation of joint 4 and the change of slope of its elastic curve at point 3. The minus sign in relations (j) and (k) indicates that the sense of the horizontal component of translation of joint 4 of the frame and the change of slope of its elastic curve at point 3 is opposite to that of the unit force and the pair of unit moments of Figs. e and g, respectively. Notice that usually the effect of the increase of the temperature of the axis of the frame on the displacement of its cross sections is small. In this example it is about 3 percent of the effect of the difference in temperature between the external and internal fibers of the members of the frame.

The effect of the increase in the temperature of the axis of the frame on the component of the translation $u_h^{(4)}$ and the change of slope $\Delta\theta^{(3)}$ can also be established by considering the deformed configuration of the frame shown in Fig. j. That is, referring to Fig. j, the total change of length of beam 2,4 is equal to

$$\varepsilon + \frac{\varepsilon}{3} = \alpha\,\Delta T_c\,L = \alpha(15)(8)\text{ m} \tag{l}$$

thus
$$\varepsilon = 90\alpha\text{ m} \tag{m}$$

Consequently, referring to Fig. j, we have

$$u_h^{(4)} = \frac{\varepsilon}{3} = 30\alpha\text{ m} \tag{n}$$

$$\Delta\theta^{(3)} = \theta^{(5)} - \theta^{(1)} = -\frac{\varepsilon}{12} - \frac{\varepsilon}{4} = -\frac{\varepsilon}{3} = -30\alpha \tag{o}$$

Notice that the members of a statically determinate structure, such as the frame under consideration, deform freely because of a change of temperature and, consequently, they are not stressed.

3. *Compute the horizontal component of translation of joint 4 of the frame and the change of slope of its elastic curve at point 3 due to a downward settlement and a horizontal movement to the right of support 6.* These quantities may be established from relation (5.66) using the virtual loading of Figs. e and g, respectively. Referring

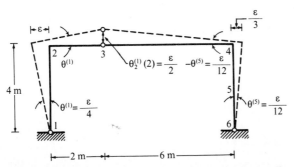

Figure j Deformed configuration of the frame due to the increase of temperature of its axis above that existing during construction.

to Fig. e, the reactions $\tilde{R}_v^{(6)}$ and $\tilde{R}_h^{(6)}$ at support 6 are equal to 0.50 kN upward and 0.75 kN to the right, respectively, while the displacements $\Delta^{(1)}$ and $\Delta^{(2)}$ are downward and to the right, respectively. Consequently, in Eq. (5.66) we use

$$\tilde{R}^{(1)} = -0.50 \text{ kN} \qquad \tilde{R}^{(2)} = 0.75 \text{ kN}$$

Thus, Eq. (5.66) gives

$$u_h^{(4)} - 5\tilde{R}^{(1)} + 10\tilde{R}^{(2)} = u_h^{(4)} - 0.5(5) - 0.75(10) = 0$$

or
$$u_h^{(4)} = 10 \text{ mm} \tag{p}$$

Similarly, referring to Fig. g, we get

$$\tilde{R}^{(1)} = 0 \qquad \tilde{R}^{(2)} = 0.25$$

Thus, Eq. (5.66) gives

$$\Delta\theta^{(3)} = 0.005R^{(1)} + 0.01R^{(2)} = 0.01(0.25) = 0.0025 \text{ rad} \tag{q}$$

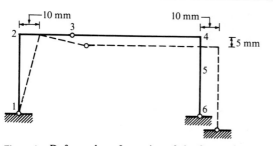

Figure k   Deformed configuration of the frame due to the given movement of support 6.

Notice that the length of the members of a statically determinate structure, such as the frame under consideration, does not change because of the movement of a support. That is, the members of the structure move as rigid bodies and, consequently, are not stressed.

Example 4.   Calculate the deflection of the free end of the beam subjected to the loading shown in Fig. a. The beam is made of steel ($E = 210,000$ MPa) and has a constant cross section ($I = 200 \times 10^6$ mm⁴).

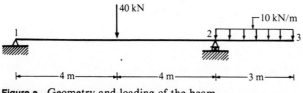

Figure a   Geometry and loading of the beam.

solution We will compute the deflection of point 3 of the beam using relation (5.76).

STEP 1. We plot the M/EI diagram for the beam subjected to the given external actions and compute the areas under it. Referring to Fig. c and using Table B on the inside of the back cover, we have

$$A_M^{(1)} = \frac{57.5}{EI}\left(\frac{4}{2}\right) = \frac{115}{EI}$$

$$\bar{x}_1^{(1)} = \frac{8}{3}$$

$$A_M^{(2)} = \frac{57}{EI}\left(\frac{2.244}{2}\right) = \frac{64.5}{EI}$$

$$\bar{x}_1^{(2)} = 4 + \frac{2.244}{3} = 4.7487$$

(a)

$$A_M^{(3)} = -\frac{45}{EI}\left(\frac{1.756}{2}\right) = -\frac{39.5}{EI}$$

$$\bar{x}_1^{(3)} = 8 - \frac{1.756}{3} = 7.4147$$

$$A_M^{(4)} = -\frac{1}{3}\left(\frac{45}{EI}\right)3 = -\frac{45}{EI}$$

$$\bar{x}_1^{(4)} = 8 + \frac{3}{4} = 8.75$$

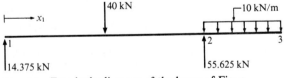

Figure b Free body diagram of the beam of Fig. a.

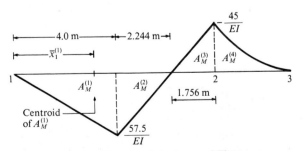

Figure c Moment diagram of the beam of Fig. a.

Figure d   Free-body diagram of the beam subjected to the virtual loading for computing $u_v^{(3)}$.

STEP 2.   We establish the values of the moment at points $\bar{x}_1^{(j)}(j = 1, 2, 3, 4)$ of the beam subjected to the virtual force for computing the deflection of point 3. Referring to Fig. d, we obtain the values of the moment at $x_1 = \bar{x}_1^{(j)}$:

$$\tilde{M} = -1 \text{ kN} \cdot \text{m} \qquad \text{at } x_1 = 8/3$$

$$\tilde{M} = -1.7805 \text{ kN} \cdot \text{m} \qquad \text{at } x_1 = 4.7487$$

$$\tilde{M} = -2.7805 \text{ kN} \cdot \text{m} \qquad \text{at } x_1 = 7.4147$$

(b)

$$\tilde{M} = -2.25 \text{ kN} \cdot \text{m} \qquad \text{at } x_1 = 8.75$$

STEP 3.   We substitute relations (a) and (b) into (5.76) to obtain the deflection of point 3. That is

$$u_v^{(3)} = \sum_{j=1}^{4} A_M^{(j)} \tilde{M}(\bar{x}_1^{(j)}) = -\frac{115}{EI}(1) - \frac{64.52}{EI}(1.785) + \frac{39.5}{EI}(2.7805) + \frac{45}{EI}(2.25)$$

$$= \frac{18.75}{EI} = \frac{18.75}{(210)(200)}$$

$$= 0.000447 \text{ m} = 0.447 \text{ mm } \downarrow$$

It is apparent that this approach is more laborious than that involving the use of Table A on the inside of the back cover.

Example 5.   Consider the cantilever frame shown in Fig. a. The cross sections of the members of the frame are circular tubes of external diameter $d_e = 200$ mm and internal diameter $d_i = 100$ mm. Compute the components of translation of joint 5 when the frame is loaded by a horizontal force $P = 10$ kN at joint 5. Disregard the effect of axial deformation of the members of the frame. The modulus of elasticity and the shear modulus of the material from which the members of the frame are made are $E = 210,000$ MPa and $G = 84,000$ MPa, respectively.

solution   This is a space structure whose members are subjected to bending moments and to torsional moment. For each member of the structure, we choose the local system of axes shown in Fig. b.

The global components of translation $\bar{u}_i^{(5)}(i = 1, 2, 3)$ of joint 5 of the frame will be computed using the method of virtual work (5.66). That is,

$$\bar{u}_i^{(5)} = \sum_{k=1}^{4} \left(\int_0^L \frac{M_1 \tilde{M}_1}{GK} dx_1 + \int_0^L \frac{M_2 \tilde{M}_2}{EI_2} dx_1 + \int_0^L \frac{M_3 \tilde{M}_3}{EI_3} dx_1 \right)^{(k)}$$

(a)

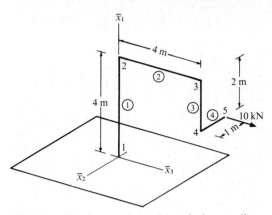

Figure a Geometry and loading of the cantilever frame.

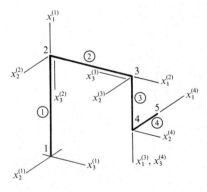

Figure b Local axes for the members of the frame.

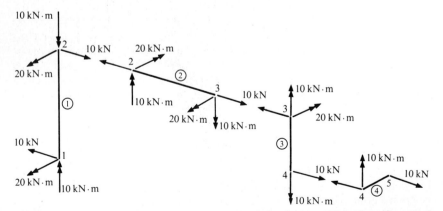

Figure c Free-body diagrams of the members of the frame subjected to the actual loading.

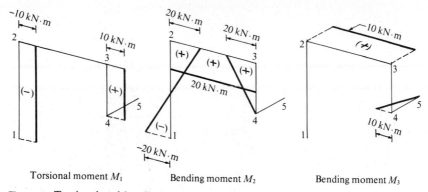

Torsional moment M_1 Bending moment M_2 Bending moment M_3

Figure d Torsional and bending movement diagrams due to the actual loading.

In the above relation, M_1, M_2, and M_3 are the torsional and bending components of moment due to the actual loading. The free-body diagrams of the members of the structure subjected to the actual loading and the corresponding torsional and bending moment diagrams are shown in Figs. c and d, respectively. In relation (a), $\tilde{M}_1$, $\tilde{M}_2$, and $\tilde{M}_3$ are the torsional and bending components of moment due to the virtual loading for finding the global components of translation $\overline{u}_i^{(5)}$. The virtual loadings for finding $\overline{u}_1^{(5)}$, $\overline{u}_2^{(5)}$, and $\overline{u}_3^{(5)}$ are shown in Figs. e, h, and k, respectively. The free-body diagrams of the members of the structure corresponding to the loading of Figs. e and h are shown in Figs. f and i, respectively. The torsional and the bending moment diagrams corresponding to the loadings of Figs. e, h, and k are shown in Figs. g, j, and l, respectively.

The moment of inertia of a circular cross section is equal to

$$I_2 = I_3 = \frac{K}{2} = \frac{\pi}{64}(d_e^4 - d_i^4) = \frac{\pi}{64}[(0.20)^4 - (0.10)^4]$$
$$= 7.36 \times 10^{-5} \text{ m}^4$$

(b)

To compute the components of translation of point 5, we evaluate the integrals of relation (a) by using the formulas of Table A on the inside of the back cover. Referring to the torsional and bending moment diagrams of Fig. d and to those of Figs. g, j, and l, we obtain

$$\overline{u}_1^{(5)} = \int_0^4 \frac{M_2\tilde{M}_2}{EI_2}\,dx|^{(2)} + \int_0^4 \frac{M_2\tilde{M}_2}{EI_2}\,dx|^{(1)}$$

$$= \frac{2 \times 10^3}{6EI_2}[4(20 - 20)] + \frac{(4 \times 10^3)(4)(60)}{6EI_2}$$

$$= \frac{16 \times 10^4}{EI_2} = \frac{16 \times 10^4}{(210,000 \times 10^6)(7.36 \times 10^{-5})} = 0.0104 \text{ m}$$

$$= 10.4 \text{ mm} \quad \text{upwards}$$

(c)

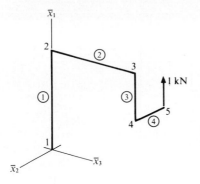

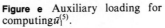

Figure e Auxiliary loading for computing $\tilde{u}_1^{(5)}$.

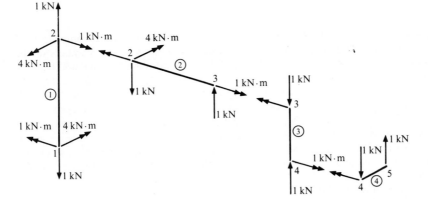

Figure f Free-body diagrams of the members of the frame subjected to the virtual loading for computing $\tilde{u}_1^{(5)}$.

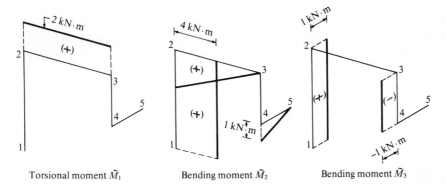

Figure g Torsional and bending moment diagrams due to the virtual loading for computing $\tilde{u}_1^{(5)}$.

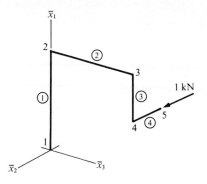

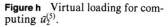

Figure h Virtual loading for computing $\tilde{u}_2^{(5)}$.

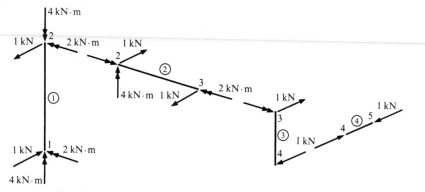

Figure i Free-body diagram of the members of the frame subjected to the virtual loading for computing $\tilde{u}_2^{(5)}$.

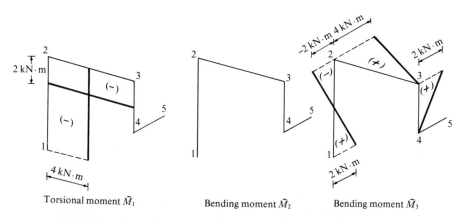

Torsional moment $\tilde{M}_1$ Bending moment $\tilde{M}_2$ Bending moment $\tilde{M}_3$

Figure j Torsional and bending moment diagram, resulting from the virtual loading for computing $\tilde{u}_2^{(5)}$.

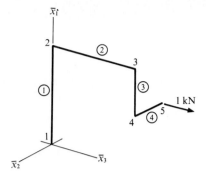

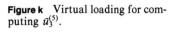

Figure k Virtual loading for computing $\bar{u}_3^{(5)}$.

$$\bar{u}_2^{(5)} = \int_0^4 \frac{M_1 \tilde{M}_1}{GK} \, dx_1^{(1)} + \int_0^4 \frac{M_3 \tilde{M}_3}{EI_3} \, dx_1^{(3)}$$

$$= \frac{4 \times 10^3}{6GK} [4(30) + 4(30)] + \frac{4 \times 10^3}{6EI_3} [4(30)]$$

$$= \frac{16 \times 10^4}{GK} + \frac{8 \times 10^4}{EI_3}$$

$$= \frac{16 \times 10^4}{(84,000 \times 10^6)(14.72 \times 10^{-5})} + \frac{8 \times 10^4}{(210,000 \times 10^6)(7.36 \times 10^{-5})}$$

$$= 0.01294 + 0.05176 = 0.0181 \text{ m} = 18.1 \text{ mm} \tag{d}$$

$$\bar{u}_3^{(5)} = \int_0^4 \frac{M_1 \tilde{M}_1}{GK} \, dx_1^{(1)} + \int_0^2 \frac{M_1 \tilde{M}_1}{GK} \, dx_1^{(3)} + \int_0^4 \frac{M_2 \tilde{M}_2}{EI_2} \, dx_1^{(1)} + \int_0^4 \frac{M_2 \tilde{M}_2}{EI_2} \, dx_1^{(2)}$$

$$+ \int_0^2 \frac{M_2 \tilde{M}_2}{EI_2} \, dx_1^{(3)} + \int_0^4 \frac{M_3 \tilde{M}_3}{EI_3} \, dx_1^{(2)} + \int_0^1 \frac{M_3 \tilde{M}_3}{EI_3} \, dx_1^{(4)}$$

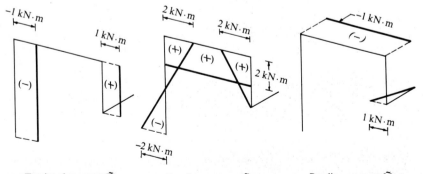

Torsional moment $\tilde{M}_1$ Bending moment $\tilde{M}_2$ Bending moment $\tilde{M}_3$

Figure l Torsional and bending moment diagram, resulting from the virtual loading for computing $\bar{u}_3^{(5)}$.

$$= \frac{4 \times 10^3}{6GK}[10(3) + 10(3)] + \frac{2 \times 10^3}{6GK}[10(3) + 10(3)]$$

$$+ \frac{2 \times 10^3}{6EI_2}[20(4) + 20(4)] + \frac{4 \times 10^3}{6EI_2}[20(6) + 20(6)]$$

$$+ \frac{2 \times 10^3}{6EI_2}(20)4 + \frac{4 \times 10^3}{6EI_3}[10(3) + 10(3)] + \frac{10^3}{6EI_3}(10)2$$

$$= \frac{6 \times 10^4}{GK} + \frac{85 \times 10^4}{3EI_2}$$

$$= \frac{6 \times 10^4}{(84,000 \times 10^6)(14.72 \times 10^{-5})} + \frac{85 \times 10^4}{(210,000 \times 10^6)(7.36 \times 10^{-5})}$$

$$= 0.00485 + 0.01833 = 0.0232 \text{ m} = 23.2 \text{ mm} \tag{e}$$

The plus sign indicates that $\overline{u}_2^{(5)}$ and $\overline{u}_3^{(5)}$ are in the direction of the unit force shown in Figs. h and k, respectively.

5.11 Application of the Method of Virtual Work to Curved Beams

In this section the method of virtual work is employed in establishing the components of displacement of the unsupported end of two curved cantilever beams. The axis of these beams is an arch of a circle. The one beam is subjected to a concentrated force acting in its plane while the other is subjected to a uniformly distributed force acting normal to its plane.

Example 1.   Consider the cantilever curved beam of constant cross sections and radius R shown in Fig. a. Assuming that the thickness-to-radius ratio of the beam is very small, compute the components of translation of point 2.

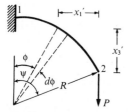

Figure a   Geometry and loading of the curved beam.

solution   Referring to Fig. a, the coordinates (x_1', x_3') of a point of the beam specified by the angle ϕ are

$$x_1' = R \sin \psi - R \sin \phi \qquad x_3' = R(\cos \phi - \cos \psi) \tag{a}$$

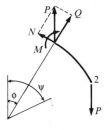

Figure b Free-body diagram of a segment of the curved beam subjected to the actual loading.

From the equilibirum of the segment of the curved beam shown in Fig. b we have

$$M(x_1') = -Px_1' = -PR(\sin \psi - \sin \phi) \qquad N(x_1') = P \sin \phi \qquad \text{(b)}$$

In order to compute the horizontal component of translation of point 2, we use the method of virtual work; the beam is loaded with the virtual load shown in Fig. c. From the equilibrium of the segment of the beam shown in Fig. d, we get

$$\tilde{M} = x_3' = R(\cos \phi - \cos \psi) \qquad \tilde{N} = \cos \phi \qquad \text{(c)}$$

Similarly, in order to compute the vertical component of translation at point 2, we use the method of virtual work; the beam is loaded by the virtual load shown in Fig. e. From the equilibrium of the segment of the beam shown in Fig. f we get

$$\tilde{M} = x_1' = -R (\sin \psi - \sin \phi) \qquad \tilde{N} = \sin \phi \qquad \text{(d)}$$

For the beam of Fig. a the principle of virtual work (5.66) reduces to

$$u_n^{(2)} = \int_0^L \left(\frac{N\tilde{N}}{EA} + \frac{M\tilde{M}}{EI_2} \right) ds = \int_{\phi=0}^{\phi=\psi} \left(\frac{N\tilde{N}}{EA} + \frac{M\tilde{M}}{EI_2} \right) R \, d\phi \qquad \text{(e)}$$

Substituting relations (b) and (c) into (e), we get

$$u_h^{(2)} = -\int_0^\psi \frac{PR^2 (\sin \psi - \sin \phi)(\cos \phi - \cos \psi) R \, d\phi}{EI} + \int_0^\psi \frac{P \sin \phi \cos \phi \, R \, d\phi}{AE}$$

$$= -\frac{PR^3}{EI} \left[\sin^2 \psi + \frac{\cos 2\psi}{4} - \frac{1}{4} - \frac{\psi \sin 2\psi}{2} - \cos^2 \psi + \cos \psi \right]$$

$$+ \frac{PR}{AE} \left[-\frac{\cos 2\psi}{4} + \frac{1}{4} \right] \qquad \text{(f)}$$

Figure c Virtual loading for computing the horizontal component of translation of point 2.

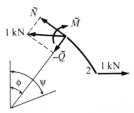

Figure d Free-body diagram of a segment of the curved beam loaded as shown in Fig. c.

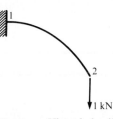

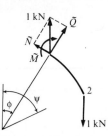

Figure e Virtual loading for computing the vertical component of translation of point 2.

Figure f Free-body diagram of a segment of the curved beam loaded as shown in Fig. e.

Substituting relation (b) and (d) into (e), we get

$$u_v^{(2)} = \int_0^\psi \frac{PR^2 (\sin \psi - \sin \phi)(\sin \psi - \sin \phi) R \, d\phi}{EI} + \int_0^\psi \frac{P \sin^2 \phi R \, d\phi}{AE}$$

$$= \frac{PR^3}{EI} \left[\psi \sin^2 \psi + \frac{3}{2} \cos \psi \sin \psi + \frac{\psi}{2} - 2 \sin \psi \right] + \frac{PR}{2AE} [\psi - \sin \psi \cos \psi]$$

$$\tag{g}$$

The first term in relations (f) and (g) represents the effect of bending while the second term represents the effect of axial force. For a beam with rectangular cross section, we have

$$I = \frac{bh^3}{12} \qquad A = bh \qquad \frac{I}{A} = \frac{h^2}{12} \tag{h}$$

Thus, substituting relations (h) into (f) and (g), we can readily see that for this example the effect of the axial force is of the order of h^2/R^2 as compared to unity. As stated previously, we are limiting our attention to curved beams with small h/R ratios; consequently, h^2/R^2 may be disregarded as compared to unity. Hence, in this example the effect of the axial force is negligible.

For $\psi = 90°$, and disregarding terms of the order of h^2/R^2, relations (f) and (g) yield

$$u_h^{(2)} = -\frac{PR^3}{2EI} \qquad u_v^{(2)} = \frac{PR^3}{2EI} \left(\frac{3\pi}{2} - 4 \right) \tag{i}$$

The negative sign for $u_h^{(2)}$ indicates that its direction is opposite to that of the unit force of Fig. c.

Example 2. Compute the vertical component of translation (deflection) of the free end of the horizontal, semicircular, cantilever bow girder shown in Fig. a when it is subjected to a vertical, uniform load, q kN/m.

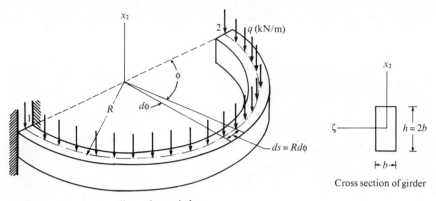

Cross section of girder

Figure a   Circular cantilever bow girder.

---

solution   The loading in this curved girder induces internal torsional moments as well as bending moments. As mentioned in Sec. 5.8, the principle of virtual work (5.66) is valid for both straight and curved (with small h/R ratios) members. The girder under consideration is bent only about the local ζ axis. Consequently, the principle of virtual work (5.66) reduces to

$$u_2^{(2)} = \int_0^\pi \frac{M_\zeta \tilde{M}_\zeta}{EI_\zeta} R \, d\phi + \int_0^\pi \frac{M_t \tilde{M}_t}{GK} R \, d\phi \qquad (a)$$

where M_t is the torsional component of the moment while M_ζ is the bending component of the moment. To compute these moments, we consider the equilibrium of the segment of the girder whose free-body diagram is shown in Fig. b. Notice that the resultant of the distributed forces acting on an element of the girder bounded by the radial lines at α and $\alpha + d\alpha$ is equal to $qR \, d\alpha$. Thus

$$\Sigma F_2 = \int_0^\phi qR \, d\alpha - Q_2 = 0 \quad \text{or} \quad Q_2 = qR\phi \qquad (b)$$

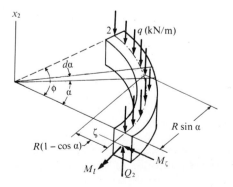

Figure b   Free-body diagram of a segment of the bow girder.

$$\Sigma M_{\varsigma} = \int_0^\phi (qR\, d\alpha)R\sin\alpha + M_{\varsigma} = 0 \quad \text{or} \quad M_{\varsigma} = -qR^2(1 - \cos\phi) \tag{c}$$

$$\Sigma M_t = \int_0^\phi (qR\, d\alpha)R(1 - \cos\alpha) + M_t = 0 \quad \text{or} \quad M_t = -qR^2(\phi - \sin\phi) \tag{d}$$

To compute the vertical component of translation of the free end of the girder, we use the virtual loading shown in Fig. c. The moments corresponding to this loading are

$$\tilde{M}_{\varsigma} = R\sin\phi \qquad \tilde{M}_t = R(1 - \cos\phi) \tag{e}$$

Substituting relations (c), (d), and (e) into (a), we get

$$u_2^{(2)} = -\frac{qR^2}{EI_{\varsigma}} \int_0^\pi (1 - \cos\phi)(R\sin\phi)R\, d\phi$$

$$- \frac{qR^2}{GK} \int_0^\pi (\phi - \sin\phi)R(1 - \cos\phi)R\, d\phi$$

$$= -\frac{2qR^4}{EI_{\varsigma}} - \frac{\pi^2 qR^4}{2GK} \tag{f}$$

Referring to Table A.1 in Sec. A.11.2 for $h/b = 2$ we get $C = 0.229$. Consequently, from relation (A.107), we have

$$K = Cb^3h = 0.229b^3(2b) = 0.458b^4$$

Moreover for $\nu = 0.3$, we obtain

$$G = \frac{E}{2(1 + \nu)} = 0.385E$$

thus,

$$GK = 0.458b^4(0.385E) = 0.176Eb^4 = 0.176(\tfrac{3}{2})EI_{\varsigma} = 0.264EI_{\varsigma} \tag{g}$$

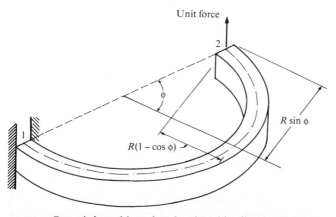

Unit force

2

$R\sin\phi$

ϕ

1

$R(1 - \cos\phi)$

Figure c Bow girder subjected to the virtual loading for computing the vertical component of translation of its free end.

Substituting relation (g) into (f) and integrating, we obtain:

$$u_2^{(2)} = \frac{qR^4}{EI_\zeta} \int_0^\pi \left[\sin \phi - \sin \phi \cos \phi + \frac{1}{0.264} (\phi - \phi \cos \phi - \sin \phi) \right.$$

$$\left. + \sin \phi \cos \phi) \right] d\phi = -\frac{qR^4}{EI_\zeta} (2.0 + 18.68)$$

$$= -20.68 \frac{qR^4}{EI_\zeta} \tag{h}$$

The minus sign indicates that the sense of the translation $u_2^{(2)}$ is opposite to that of the unit force shown in Fig. c. That is, as expected, $u_2^{(2)}$ is downward. Notice, that the first term in the parentheses of relation (h) represents the effect of bending, while the second term represents the effect of torsion. Thus, in this example, the effect of torsion is prevalent.

5.12   Application of the Method of Virtual Work to Members with Variable Cross Section

The moment of inertia of members with variable cross section, is a function of the coordinate x_1, and the integration of the term $\int_0^L (M\tilde{M}/EI) \, dx_1$ could become problematic. In this case, we may use any of the known formulas for numerical integration. These formulas are based on the interpretation of the definite integral $\int_a^b f(x_1) \, dx_1$ as the area under the curve $x_2 = f(x_1)$ between the coordinates $x_1 = a$ and $x_1 = b$. Referring to Fig. 5.26, we subdivide the interval from point a to b into m equal segments by $m + 1$ points $x_1^{(i)}$ ($i = 0$, $1, \ldots, m$), where $x_1^{(0)} = a$ and $x_1^{(m)} = b$. We assume that at these points, the values $f_0, f_1, \ldots, f_m$ of the function $f(x_1)$ can be established. The best known formulas for numerical integration of the integral $\int_a^b f(x_1) \, dx_1$ are as follows.

1. *Simple summation.* The function $f(x_1)$ between any two adjacent points $x_1^{(i)}$ and $x_1^{(i+1)}$ is approximated by its value f_i at $x_1^{(i)}$. That is

$$f(x_1) = f_i \quad \text{for } x_1^{(i)} \leq x_1 \leq x_1^{(i+1)} \quad (i = 0, 1, \ldots, m - 1)$$

The area under the curve from $x_1 = a$ to $x_1 = b$ is approximated by the sum of m rectangles. That is:

$$\int_a^b f(x_1) \, dx_1 \simeq s \sum_{i=0}^{m-1} f_i \tag{5.77}$$

2. *Trapezoidal rule.* The function $f(x_1)$ between any two points $x_1^{(i)}$ and $x_1^{(i+1)}$ is approximated by a straight line. Thus, the area under the curve from

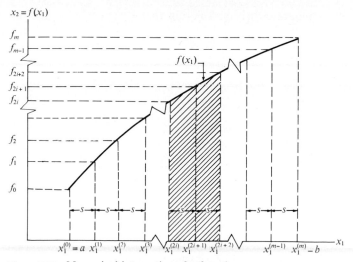

Figure 5.26 Numerical integration of a function.

$x_1 = a$ to $x_1 = b$ is approximated by the sum of the areas of the m trapezoids whose bases have lengths f_0 and f_1, f_1, and $f_2 \ldots , f_{m-1}$ and f_m. That is,

$$\int_a^b f(x_1)\, dx_1 = \sum_{i=0}^{m-1} \Delta A^{(i)} = \sum_{i=0}^{m-1} \left(\frac{f_i + f_{i+1}}{2} \right) s$$

$$= \frac{s}{2}(f_0 + 2f_1 + 2f_2 + \cdots + 2f_{m-1} + f_m)$$

3. *Simpson's rule.* The function $f(x_1)$ between any three points $x_1^{(2i)}$, $x_1^{(2i+1)}$, and $x_1^{(2i+2)}$ [$i = 0, 1, 2, \ldots , (m - 2)/2$ where m is an even number], is approximated by a second-degree parabola passing through these points. That is in the interval $x_1^{(2i)} \le x_1 \le x_1^{(2i+2)}$, we have

$$f(x_1) = A_i x_1^2 + B_i x_1 + C_i \tag{5.79}$$

where the coefficients A_i, B_i, and C_i are obtained by requiring that $f(x_1)$ assumes the values of f_{2i}, f_{2i+1}, and f_{2i+2} at points $x_1^{(2i)}$, $x_1^{(2i+1)}$, and $x_1^{(2i+2)}$, respectively. That is, referring to Fig. 5.26, we have

$$f_{2i} = a_i(x_1^{(2i)})^2 + b_i x_1^{(2i)} + c_i$$

$$f_{2i+1} = a_i(x_1^{(2i+1)})^2 + b_i x_1^{(2i+1)} + c_i \tag{5.79a}$$

$$f_{2i+2} = a_i(x_1^{(2i+2)})^2 + b_i x_1^{(2i+2)} + c_i$$

Solving Eqs. (5.79a) simultaneously and noting that $s = x_1^{(2i+1)} - x_1^{(2i)} = x_1^{(2i+2)} - x_1^{(2i+1)}$, we get

$$
\begin{Bmatrix} a_i \\ b_i \\ c_i \end{Bmatrix} = \frac{1}{2s^2} \times \begin{bmatrix} 1 & -2 & -1 \\ -(2x_1^{(2i)} + s) & 4(x_1^{(2i)} + s) & -(2x_1^{(2i)} + 35) \\ x_1^{(2i)}(x_1^{(2i)} + s) & -2x_1^{(2i)}(x_1^{(2i)} + 2s) & (x_1^{(2i)} + 2)(x_1^{(2i)} + s) \end{bmatrix} \begin{Bmatrix} f_{2i+2} \\ f_{2i+1} \\ f_{2i} \end{Bmatrix}
$$

$$(5.79b)$$

The area A_{2i} under the curve $x_2 = f(x_1)$ from line $x_1 = x_1^{(2i)}$ to line $x_1 = x_1^{(2i+2)}$ is equal to

$$
A_{2i} = \int_{x_1^{(2i)}}^{x_1^{(2i+2)}} f(x_1)\, dx_1 = \int_{x_1^{(2i)}}^{x_1^{(2i+2)}} (a_i x_1^2 + b_i x_i + c_i)\, dx_1
$$

$$
= \tfrac{1}{3} a_i[(x_1^{(2i+2)})^3 - (x_1^{(2i)})^3] + \tfrac{1}{2} b_i[(x_1^{(2i+2)})^2 - (x_1^{(2i)})^2]
$$

$$
+ c_i(x_1^{(2i+2)} - x_1^{(2i)}) \tag{5.79c}
$$

Substituting relations (5.79b) into (5.79c) and simplifying, we obtain

$$
A_{2i} = \int_{x_1^{(2i)}}^{x_1^{(2i+2)}} f(x_1)\, dx_1 = \frac{s}{3}(f_{2i} + 4f_{2i+1} + f_{2i+2}) \tag{5.80}
$$

Applying the above formula for $i = 0, 1, 2, \ldots, (m - 2)/2$, we obtain

$$
\int_a^b f(x_1)\, dx_1 = \sum_{i=1}^{(m-2)/2} A_{2i} = \frac{s}{3}(f_0 + 4f_1 + 2f_2 + 4f_3 + 2f_4
$$

$$
+ \cdots + 2f_{m-2} + 4f_{m-1} + f_m) \tag{5.81}
$$

Example 1. A cantilever beam of rectangular cross section, constant width $b = 60$ mm, and variable depth is subjected to a concentrated force as shown in Fig. a. The modulus of elasticity of the beam is $E = 210{,}000$ MPa. Compute the deflection of the free end of this beam by:

1. Direct integration
2. Numerical integration, by subdividing the beam in eight equal segments and using Simpson's rule

Referring to Fig. a, the height of the beam is equal to

$$
h = 0.75 - 0.1x_1 \text{ m} \tag{a}
$$

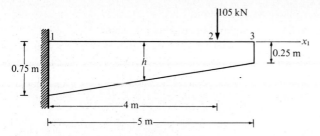

Figure a   Geometry and loading of the beam.

solution   The distribution of moments in the beam subjected to the loading given in Fig. a is

$$M = 105(x_1 - 4) \quad \text{for } 0 \le x_1 \le 4$$
$$M = 0 \quad \text{for } 4 \le x_1 \le 5 \tag{b}$$

In order to compute the deflection of the free end of the beam, we use the virtual loading shown in Fig. b. The distribution of moments corresponding to this loading is

$$\tilde{M} = x_1 - 5 \quad \text{for } 0 \le x_1 \le 5 \tag{c}$$

Using relation (a), we obtain the moment of inertia of the cross sections of the beam as

$$I = \frac{bh^3}{12} = 0.005h^3 = (5 \times 10^{-6})(7.5 - x_1)^3 \quad \text{m}^4 \tag{d}$$

Substituting relations (b), (c), and (d) into the principle of virtual work (5.66), we have

$$u_v^{(3)} = \int_0^4 \frac{M\tilde{M}}{EI} dx_1 = \int_0^4 \frac{(105 \times 10^3)(x_1 - 4)(x_1 - 5) \, dx_1}{(210,000 \times 10^6)(5 \times 10^{-6})(7.5 - x_1)^3}$$
$$= 10^{-1} \int_0^4 \frac{(x_1 - 4)(x_1 - 5)}{(7.5 - x_1)^3} dx_1 \tag{e}$$

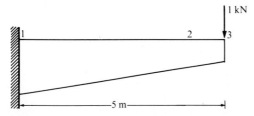

Figure b   Virtual loading for computing the deflection of the free end of the beam.

Part a. Computation of $u_v^{(3)}$ by Direct Integration

In this part, we integrate directly the integral in relation (e). Thus,

$$u_v^{(3)} = 10^{-1} \int_0^4 \frac{x_1^2 - 9x_1 + 20}{(7.5 - x_1)^3} \, dx_1$$

$$= 10^{-1} \int_0^4 \frac{(x_1^2 - 15x_1 + 56.25) - 6(7.5 - x_1) + 8.75}{(7.5 - x_1)^3} \, dx_1$$

$$= 10^{-1} \int_0^4 \left[\frac{1}{7.5 - x_1} - \frac{6}{(7.5 - x_1)^2} + \frac{8.75}{(7.5 - x_1)^3} \right] dx_1 \tag{f}$$

$$= 10^{-1} \left[-\ln(7.5 - x_1) - \frac{6}{7.5 - x_1} + \frac{8.75}{2(7.5 - x_1)^2} \right]_0^4 = 0.012722 \text{ m}$$

or
$$u_v^{(3)} = 12.722 \text{ mm} \downarrow$$

Part b. Computation of $u_v^{(3)}$ by Numerical Integration

In this part, we evaluate the integral (e) by subdividing the interval from $x_1 = 0$ to $x_1 = 4$ into eight equal segments ($x_1 = 0, 0.5, 1.0, 1.5, 2.0, 2.5, 3.0, 3.5, 4.0$) and then using Simpson's rule. Referring to Eqs. (5.81) and (e), we have

$$f(x_1) = \frac{(x_1 - 4)(x_1 - 5)}{(7.5 - x_1)^3} \qquad a = 0, b = 4 \tag{g}$$

The length of the intervals into which the cantilever is subdivided is equal to 0.5 m. The values of the function $f_i = f(x_1^{(i)})$ are tabulated in Table a. Substituting in relation (5.81) the values of f_i given in Table a, we obtain

$$u_v^{(3)} = 10^{-1} \int_0^4 \frac{(x_1 - 4)(x_1 - 5)}{(7.5 - x_1)^3} \, dx_1$$

$$= 10^{-1} \frac{(0.5)}{3} [0.04741 + 0.51260 + 0.20341] \tag{h}$$

$$= 0.012724 \text{ m} = 12.724 \text{ mm}$$

TABLE a

Point (i)	$x_1^{(i)}$	f_i	$4f_i$, i odd	$2f_i$, i even
0	0	0.04741		
1	0.5	0.04592	0.18368	
2	1.0	0.04370	. . . . . .	0.08739
3	1.5	0.04051	0.16204	
4	2.0	0.03606	. . . . . .	0.07212
5	2.5	0.03000	0.12000	
6	3.0	0.02195	. . . . . .	0.04390
7	3.5	0.01172	0.04688	
8	4.0	0.00000		
Total	. .	. . . . . .	0.51260	0.20341

The error in the numerical integration is equal to

$$\text{Percent error} = \frac{12.724 - 12.722}{12.722} \, 100\% = 0.016\% \qquad (i)$$

5.13   Principle of Virtual Work for Framed Structures, Including the Effect of Shear Deformation

In Sec. 5.7 we derived the principle of virtual work for framed structures disregarding the effect of shear deformation of the members of the structure. For most structures of practical interest, this effect is small and it is neglected. In this section we derive the principle of virtual work for framed structures which includes the effect of shear deformation of their members. We establish this effect on the basis of the Timoskenko theory of beams and then on the basis of a more accurate theory.

In the theories of beams (classical and Timoshenko) the shearing components of stress acting on a cross section are established from the shearing components of force acting on this cross section on the basis of relation (A.87); that is

$$\tau_{1n} = \frac{Q_2 Z_3}{I_3 b_s} + \frac{Q_3 Z_2}{I_2 b_s} \qquad (5.82)$$

where, referring to Fig. 5.27, we have defined Z_3 and Z_2 as

$$Z_2 = \iint_{A_n} x_3' \, dA = \bar{x}_{3n} A_n \qquad Z_3 = \iint_{A_n} x_2' \, dA = \bar{x}_{2n} A_n \qquad (5.83)$$

I_2 and I_3 are the moments of inertia of the cross section of the member under consideration about its principal centroidal axes x_2 and x_3 respectively; τ_{1n} is the shearing component of stress at a point of a cross section in the direction of the unit vector $\mathbf{n}$. Referring to Fig. 5.27, A_n is the area of the shaded portion of the cross section while $\bar{x}_{jn}$ ($j = 2$ or 3) is the distance of the centroid of area A_n from the x_i axis ($i = 3$ or 2, $i \neq j$). Moreover b_s is the length of the line AB.

As discussed in Sec. A.9, in general the formula for τ_{1n} gives its average value along line AB (see Fig. 5.27). However, if the variation of τ_{1n} along the direction normal to the unit vector $\mathbf{n}$ (line AB) is negligible, the average value of τ_{1n} is equal to its actual value. This occurs in members having one of the cross sections shown in Fig. 5.27 in the directions indicated in that figure. Moreover in members having cross sections like the ones shown in Fig. 5.27a, b, and c the shearing component of stress normal to τ_{1n} is negligible.

Consider a structure subjected to bending without twisting by given external

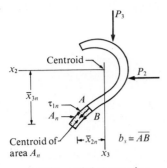

(a) Thin-walled open section

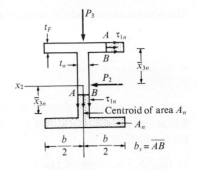

(b) Thin-walled open section

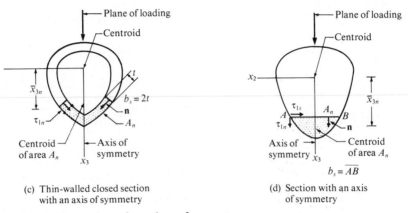

(c) Thin-walled closed section with an axis of symmetry

(d) Section with an axis of symmetry

Figure 5.27 Cross sections of members of a structure.

forces (actual loading). The undeformed and the deformed configurations of an element of length dx_1 of this structure are shown in Fig. 5.28. The geometry of the latter is based on the assumption that plane sections normal to the axis of the element prior to deformation can be considered plane subsequent to deformation but not normal to its axis. As discussed in Sec. 5.7, in this case the increments du_3 and du_2 of the transverse components of translation consist of two parts. One part du_3^r and du_2^r is due to the rotation of the element as a rigid body about the x_2 and x_3 axis, respectively; the other part du_3^s and du_2^s is due to the shear deformation of the element. Referring to Fig. 5.28 and to relations (5.48), (5.50), and (5.6) we have

$$du_3^s = 2e_{13} \, dx_1 = \frac{\lambda_3 Q_3}{GA} \, dx_1 \tag{5.84}$$

$$du_2^s = 2e_{12} \, dx_1 = \frac{\lambda_2 Q_2}{GA} \, dx_1 \tag{5.85}$$

Consider the same structure subjected only to the virtual loading for computing a component of displacement of one of its points (see Sec. 5.8), and denote by $\tilde{Q}_2$ and $\tilde{Q}_3$ statically admissible shearing forces acting on the faces of an element of length dx_1 of a member of the structure. If the structure is subsequently subjected to the actual loading, the shearing forces $\tilde{Q}_2$ and $\tilde{Q}_3$ will perform work due to the additional deformation of the element on which they act. Referring to relation (5.51) this work is equal to

$$dW_d^s = \tilde{Q}_2 \, du_2^s + \tilde{Q}_3 \, du_3^s \tag{5.86}$$

Substituting relations (5.84) and (5.85) into (5.86), we have

$$dW_d^s = \frac{\lambda_3 Q_3 \tilde{Q}_3 \, dx_1}{GA} + \frac{\lambda_2 Q_2 \tilde{Q}_2 \, dx_1}{GA} \tag{5.87}$$

Consequently, the work of shear deformation of the members of a structure on the basis of the Timoshenko theory of beams is equal to

$$W_d^s = \sum_{k=1}^{NM} \left[\lambda_3 \int_0^L \frac{Q_3 \tilde{Q}_3 \, dx_1}{GA} + \lambda_2 \int_0^L \frac{Q_2 \tilde{Q}_2 \, dx_1}{GA} \right]^{(k)} \tag{5.88}$$

Adding W_d^s to the right-hand side of relation (5.66) we obtain the principle of virtual work which includes the effect of shear deformation of the members of a structure on the basis of the Timoshenko theory of beams.

We now establish the work of shear deformation of the members of a structure on the basis of a more accurate theory than the Timoshenko theory of beams. In this theory, the work of shear deformation of a member of a structure is computed from that of an infinitesimal element of the member. In Fig. 5.29 we show statically admissible shearing components of stress $\tilde{\tau}_{1n}$ acting on the faces of an element when the structure is subjected to the virtual loading for computing a component of displacement of one of its points. If the structure

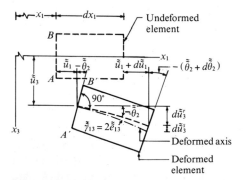

Figure 5.28 Deformation of an element of a member of a planar structure subjected to bending without twisting.

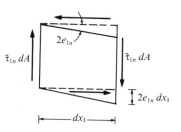

Figure 5.29 Shear deformation of an infinitesimal element and statically admissible shearing components of stress.

is subsequently subjected to the actual loading, the element under consideration will deform. The corresponding shear deformation of the element is shown
in Fig. 5.29. The work of the statically admissible shearing components of
stress $\tilde{\tau}_{1n}$ due to the shear deformation e_{1n} of the element is equal to

$$dW_d^s = 2\tilde{\tau}_{1n}e_{1n}\,dA\,dx_1 \tag{5.89}$$

Using relation (5.82), noting that $2e_{1n} = \tau_{1n}/G$, and integrating relation
(5.89), we obtain

$$W_d^s = \int_0^L \int \int_A \tilde{\tau}_{1n}\,2e_{1n}\,\bigg|_{\text{due to } Q_2}\,dx_1\,dA$$

$$+ \int_0^L \int \int_A \tilde{\tau}_{1n}\,2e_{1n}\,\bigg|_{\text{due to } Q_3}\,dx_1\,dA$$

$$= K_2 \int_0^L \frac{Q_2\tilde{Q}_2}{GA}\,dx_1 + K_3 \int_0^L \frac{Q_3\tilde{Q}_3}{GA}\,dx_1 \tag{5.90}$$

where

$$K_2 = \int \int_A \frac{Z_3^2 A}{I_3^2 b_s^2}\,dA \qquad K_3 = \int \int_A \frac{Z_2^2 A}{I_2^2 b_s^2}\,dA \tag{5.91}$$

The parameters Z_3 and Z_2 are defined by relation (5.83). The factors K_2 and
K_3 depend on the geometry of the cross section of the member and are referred
to as *form factors*.

Adding the effect of shear deformation (5.90) to the principle of virtual work
(5.66) we get

$$d + \sum_{s=1}^{S} \tilde{R}^{(s)}\Delta^{(s)} = \sum_{k=1}^{NM} \left[\int_0^L \left(\frac{N\tilde{N}}{EA} + \frac{M_1\tilde{M}_1}{KG} + \frac{M_2\tilde{M}_2}{EI_2} + \frac{M_3\tilde{M}_3}{EI_3} \right) dx_1 \right.$$

$$+ K_2 \int_0^L \frac{Q_2\tilde{Q}_2}{GA}\,dx_1 + K_3 \int_0^L \frac{Q_3\tilde{Q}_3}{GA}\,dx_1 + \int_0^L \left(\alpha\tilde{N}\,\Delta T_c + \alpha\tilde{M}_2\frac{\delta T_3}{h^{(3)}} \right.$$

$$\left. - \alpha\tilde{M}_3\frac{\delta T_2}{h^{(2)}} \right) dx_1 + \int^{L_0} (\tilde{N}e_{11}^I + \tilde{M}_2 k_2^I + \tilde{M}_3 k_3^I)\,dx_1 \Bigg]^{(k)} \tag{5.92}$$

This is the principle of virtual work for framed structures which includes the
effect of shear deformation of the members of the structure. Notice that we
did not include the effect of $\tau_{1s}e_{1s}$ in relation (5.90) because it is small compared to that of $\tau_{1n}e_{1n}$ (see Sec. A.9). Moreover, notice that the work of shear
deformation (5.90) has the same form as that obtained on the basis of the
Timoshenko theory of beams (5.88). However, the factors K_2 and K_3 are
obtained from Eqs. (5.91) while the factors λ_2 and λ_3 represent the ratio of the

maximum to the average values of the shearing components of stress τ_{12} and τ_{13}, respectively. As shown in the example at the end of this section, for a rectangular cross section $K_2 = K_3 = 1.2$, while as mentioned previously, $\lambda_2 = \lambda_3 = 1.5$. This indicates that the effect of shear deformation on the components of displacement of a point of a structure whose members have rectangular cross sections, obtained on the basis of the Timoshenko theory of beams, is 1.25 times larger than that obtained on the basis of the principle of virtual work (5.92).

In the sequel, we apply the method of virtual work in computing the deflection of a beam, including the effect of shear deformation. It is shown that the effect of shear deformation is negligible when the ratio of the depth to the length of the beam is small.

Example 1. Consider a cantilever beam of rectangular cross section loaded as shown in Fig. a. Compute the deflection of the beam at point 1. Include the effect of shear deformation and establish the range of the parameters characterizing the geometry of the beam for which the effect of shear deformation is negligible.

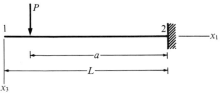

Figure a Geometry and loading of the beam.

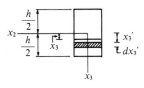

Figure b Cross section of beam.

solution The effect of shear deformation is represented by the second and third terms on the right-hand side of relation (5.92). To compute this effect, we must calculate the value of the form factor K_3 of the beam [see relation (5.91)]. To accomplish this, we need the value of Z_2 defined by relation (5.83). Thus, referring to Fig. b, we have

$$Z_2 = \int_{x_3}^{h/2} x_3' b \, dx_3' = \frac{b}{2}\left(\frac{h^2}{4} - x_3^2\right) \tag{a}$$

and

$$Z_2^2 = \frac{b^2}{4}\left(\frac{h^4}{16} - \frac{h^2 x_3^2}{2} + x_3^4\right) \tag{b}$$

Consequently, the form factor for this beam is equal to

$$K_3 = \int_{-h/2}^{h/2} \frac{A Z_2^2}{I_2^2 b^2} b \, dx_3 = \frac{b^2 h^6}{120 I_2^2} = 1.2 \tag{c}$$

Referring to Figs. a and c, we get

$$
\left.\begin{array}{ll}
Q = 0 & M = 0 \\
\tilde{Q} = -1 & \tilde{M} = x_1
\end{array}\right\} \; 0 \le x_1 \le (L - a)
$$

$$
\left.\begin{array}{ll}
Q = -P & M = P(x_1 - L + a) \\
\tilde{Q} = -1 & \tilde{M} = x_1
\end{array}\right\} \; (L - a) \le x_1 \le L
$$

(d)

Substituting relations (c) and (d) into (5.92), we obtain

$$
\bar{u}_3^{(1)} = \int_{L-a}^{L} \frac{P(x_1 - L + a)x_1 \, dx_1}{EI_2} + 1.2 \int_{L-a}^{L} \frac{P}{GA} \, dx_1
$$

$$
= \frac{Pa}{Ebh} \left[\frac{2a}{h} \left(\frac{3L}{h} - \frac{a}{h} \right) + 2.4(1 + \nu) \right]
$$

(e)

The last term in the bracket represents the effect of shear deformation. It is apparent that the effect of shear deformation depends on the a/h and L/h ratios. For the case $v = 0.3$ and $a = L$, relation (e) reduces to:

$$
\bar{u}_3^{(1)} = \frac{PL}{Ebh} \left[4 \left(\frac{L}{h} \right)^2 + 3.12 \right]
$$

(f)

For $L/h = 2$, the effect of shear deformation is approximately equal to 20 percent, while for $L/h = 6$, it is less than 2 percent and may be disregarded. Thus, it is only for very short and deep beams that the effect of shear deformation is not negligible.

Figure c Beam subjected to the auxiliary loading for the computation of the deflection of point 1.

5.14 Total Strain Energy of Framed Structures

As shown in Sec. A.5.2 [see relation (A.45)], the total strain energy of an elastic body subjected to external actions in an environment of constant temperature is equal to the work performed by the external actions and the reactions as they are applied to the body in order to bring it from its stress-free strain-free state to its deformed state. That is

$$
(U_s)_T = W_{\text{ext. actions}} + W_{\text{reactions}}
$$

(5.93)

Consider a framed structure whose supports do not move (workless supports) composed of NM members (straight or curved with small h/R ratio) made of linearly elastic materials, subjected in an environment of constant temperature to external actions. As these external actions are applied, the structure deforms. We denote the displacement vector at the point of application of the force $\mathbf{P}^{(i)}$ by $\mathbf{u}^{(i)}$ and the rotation vector at the point of application of the moment $\mathbf{M}^{(i)}$ by θ^i. Thus, referring to relations (5.19) and (5.21), we see that the work performed by the external actions as they are applied to the structure is equal to

$$(U_s)_T = \frac{1}{2} \left[\sum_{i=1}^{N} \mathbf{P}^{(i)} \cdot \mathbf{u}^{(i)} + \sum_{i=1}^{M} \mathbf{M}^{(i)} \cdot \theta^{(i)} + \sum_{k=1}^{NM} \left[\int_0^L (\mathbf{q} \cdot \mathbf{u} + \mathbf{m} \cdot \theta)\, dx_1 \right]^{(k)} \right]$$

(5.94)

If we choose the actual displacements of the members of the structure as the geometrically admissible displacement field and the actual internal actions in the members of the structure as the statically admissible internal actions, the principle of virtual work (5.53) gives

$$\sum_{i=1}^{N} \mathbf{P}^{(i)} \cdot \mathbf{u}^{(i)} + \sum_{i=1}^{M} \mathbf{M}^{(i)} \cdot \theta^{(i)} + \sum_{k=1}^{NM} \left[\int_0^L (\mathbf{q} \cdot \mathbf{u} + \mathbf{m} \cdot \theta)\, dx_1 \right]^{(k)}$$

$$= \sum_{k=1}^{NM} \left[\int_0^L \left(N \frac{du_1}{dx_1} + M_1 \frac{d\theta_1}{dx_1} + M_2 \frac{d\theta_2}{dx_1} + M_3 \frac{d\theta_3}{dx_1} \right) dx_1 \right]^{(k)}$$

(5.95)

Comparing relations (5.94) with (5.95) we obtain

$$(U_s)_T = \frac{1}{2} \sum_{k=1}^{NM} \left[\int_0^L \left(N \frac{du_1}{dx_1} + M_1 \frac{d\theta_1}{dx_1} + M_2 \frac{d\theta_2}{dx_1} + M_3 \frac{d\theta_3}{dx_1} \right) dx_1 \right]^{(k)}$$

(5.96)

If the members of the structure are made of isotropic linearly elastic materials relations (5.3) to (5.5) may be substituted into (5.96) to yield the following expression for the total strain energy of a structure consisting of NM members.

$$(U_s)_T = \frac{1}{2} \sum_{K=1}^{NM} \left[\int_0^L \left(\frac{N^2}{EA} + \frac{M_1^2}{GK} + \frac{M_2^2}{EI_2} + \frac{M_3^2}{EI_3} \right) dx_1 \right]^{(k)}$$ (5.97)

5.15 Castigliano's Second Theorem

In this section, we present Castigliano's second theorem, and, using the principle of virtual work, we prove it for framed structures made of isotropic linearly elastic materials. Moreover, we employ this theorem in computing the components of translation and rotation of points of framed structures.

Theorem. Consider a body made of a linearly elastic material, subjected to external actions (N concentrated forces, M concentrated moments, and distributed forces and moments) in an environment of constant temperature, and assume that its supports do not move (workless supports). Moreover, assume that at point A of the body, a concentrated force $P_n^{(A)}$ acts in the direction specified by the unit vector **n**, while at point B of the body a concentrated moment $M_m^{(B)}$ acts in the direction specified by the unit vector **m**. The total strain energy $(U_s)_T$ of the body (see Sec. A.5.2) may be considered a function of the external forces and moments and satisfies the following relations:

$$u_n^{(A)} = \frac{\partial (U_s)_T}{\partial P_n^{(A)}} \qquad \theta_m^{(B)} = \frac{\partial (U_s)_T}{\partial M_m^{(B)}} \tag{5.98}$$

where $u_n^{(A)}$ is the component of translation in the direction specified by the unit vector **n** at point A; $\theta_m^{(B)}$ is the component of rotation in the direction specified by the unit vector **m** at point B. Relations (5.98) were established by Castigliano in 1873 and are referred to as *Castigliano's second theorem.*

5.15.1 Proof of Castigliano's second theorem

As mentioned previously Castigliano's second theorem is valid for bodies made of linearly elastic materials. In this section, however, we prove it only for framed structures whose members are made of isotropic linearly elastic materials.

Consider a framed structure consisting of NM members made of isotropic linearly elastic materials subjected to external actions in an environment of constant temperature and assume that its supports do not move (workless supports). We denote the corresponding internal actions in the members of the structure by N, M_1, M_2, M_3 and the resulting components of displacement by u_1, u_2, u_3, θ_1, θ_2, θ_3. Referring to relation (5.96), we see that the total strain energy of this structure is equal to

$$(U_s)_T = \frac{1}{2} \sum_{k=1}^{NM} \left[\int_0^L \left(N \frac{du_1}{dx_1} + M_1 \frac{d\theta_1}{dx_1} + M_2 \frac{d\theta_2}{dx_1} + M_3 \frac{d\theta_3}{dx_1} \right) dx_1 \right]^{(k)}$$

$$\tag{5.99}$$

If we choose the actual components of displacement (u_1, u_2, u_3, θ_1, θ_2, θ_3) of the members of this structure as the geometrically admissible displacement field and the actual internal actions (N, M_1, M_2, M_3) in the members of this structure as the statically admissible distribution of internal actions, the principle of virtual work (5.53) gives

$$\sum_{i=1}^{N} \mathbf{P}^{(i)} \cdot \mathbf{u}^{(i)} + \sum_{i=1}^{M} \mathbf{M}^{(i)} \cdot \boldsymbol{\theta}^{(i)} + \sum_{k=1}^{NM} \left[\int_0^L (\mathbf{q} \cdot \mathbf{u} + \mathbf{m} \cdot \boldsymbol{\theta})\, dx_1 \right]^{(k)}$$

$$= \sum_{k=1}^{NM} \left[\int_0^L \left(N \frac{du_1}{dx_1} + M_1 \frac{d\theta_1}{dx_1} + M_2 \frac{d\theta_2}{dx_1} + M_3 \frac{d\theta_3}{dx_1} \right) dx_1 \right]^{(k)} \qquad (5.100)$$

Consider the same framed structure subjected to a slightly different loading consisting of the external actions of the previous loading except that the concentrated force $P_n^{(A)}$ has increased to $P_n^{(A)} + \hat{d}P_n^{(A)}$. As a result of this loading, the components of internal actions are $N + \hat{d}N$, $M_1 + \hat{d}M_1$, $M_2 + \hat{d}M_2$, $M_3 + \hat{d}M_3$. Moreover, the components of displacement of the points of the structure are $u_1 + \hat{d}u_1$, $u_2 + \hat{d}u_2$, $u_3 + \hat{d}u_3$, $\theta_1 + \hat{d}\theta_1$, $\theta_2 + \hat{d}\theta_2$, $\theta_3 + \hat{d}\theta_3$. The total strain energy of the structure subjected to this second loading is equal to

$$(U_s)_T + \hat{d}(U_s)_T = \frac{1}{2} \sum_{k=1}^{NM} \left\{ \int_0^L \left[(N + \hat{d}N) \left[\frac{du_1}{dx_1} + \hat{d}\left(\frac{du_1}{dx_1} \right) \right] \right. \right.$$

$$+ (M_1 + \hat{d}M_1) \left[\frac{d\theta_1}{dx_1} + \hat{d}\left(\frac{d\theta_1}{dx_1} \right) \right] + (M_2 + \hat{d}M_2) \left[\frac{d\theta_2}{dx_1} + \hat{d}\left(\frac{d\theta_2}{dx_1} \right) \right]$$

$$\left. \left. + (M_3 + \hat{d}M_3) \left[\frac{d\theta_3}{dx_1} + \hat{d}\left(\frac{d\theta_3}{dx_1} \right) \right] \right] dx_1 \right\}^{(k)} \qquad (5.101)$$

If we choose the components of displacements (u_1, u_2, u_3, θ_1, θ_2, θ_3) of the structure subjected to the first loading as the geometrically admissible displacement field and the internal actions $N + \hat{d}N$, $M_1 + \hat{d}M_1$, $M_2 + \hat{d}M_2$, $M_3 + \hat{d}M_3$ of the structure subjected to the second loading as the statically admissible internal actions, the principle of virtual work (5.53) reduces to

$$\sum_{i=1}^{N=1} \mathbf{P}^{(i)} \cdot \mathbf{u}^{(i)} + \hat{d}P_n^{(A)} u_n^{(A)} + \sum_{i=1}^{M} \mathbf{M} \cdot \boldsymbol{\theta}^{(i)}$$

$$+ \sum_{k=1}^{NM} \left[\int_0^L (\mathbf{q} \cdot \mathbf{u} + \mathbf{m} \cdot \boldsymbol{\theta})\, dx_1 \right]^{(k)} = \sum_{k=1}^{NM} \left\{ \int_0^L \left[(N + \hat{d}N) \frac{du_1}{dx_1} \right. \right.$$

$$\left. \left. + (M_1 + \hat{d}M_1) \frac{d\theta_1}{dx_1} + (M_2 + \hat{d}M_2) \frac{d\theta_2}{dx_1} + (M_3 + \hat{d}M_3) \frac{d\theta_3}{dx_1} \right] dx_1 \right\}^{(k)}$$

$$(5.102)$$

Subtracting relations (5.99) from (5.101), we get

$$\hat{d}(U_s)_T = \frac{1}{2} \sum_{k=1}^{NM} \left\{ \int_0^L \left[\hat{d}N \frac{du_1}{dx_1} + N\hat{d}\left(\frac{du_1}{dx_1}\right) + \hat{d}M_1 \frac{d\theta_1}{dx_1} + M_1\hat{d}\left(\frac{d\theta_1}{dx_1}\right) \right. \right.$$

$$\left. \left. + \hat{d}M_2 \frac{d\theta_2}{dx_1} + M_2\hat{d}\left(\frac{d\theta_2}{dx_1}\right) + \hat{d}M_3 \frac{d\theta_3}{dx_1} + M_3\hat{d}\left(\frac{d\theta_3}{dx_1}\right) \right] dx_1 \right\}^{(k)} \quad (5.103)$$

Referring to relations (5.3) to (5.5) for a structure made of isotropic linearly elastic materials, we have

$$\frac{N}{EA} = \frac{du_1}{dx_1} \qquad \frac{M_2}{EI_2} = \frac{d\theta_2}{dx_1}$$

$$\frac{M_1}{KG} = \frac{d\theta_1}{dx_1} \qquad \frac{M_3}{EI_3} = \frac{d\theta_3}{dx_1} \quad (5.104)$$

Substituting relations (5.104) into (5.103), we obtain

$$\hat{d}(U_s)_T = \sum_{k=1}^{NM} \left[\int_0^L \left(\hat{d}N \frac{du_1}{dx_1} + \hat{d}M_1 \frac{d\theta_1}{dx_1} + \hat{d}M_2 \frac{d\theta_2}{dx_1} + \hat{d}M_3 \frac{d\theta_3}{dx_1} \right) dx_1 \right]^{(k)}$$

$$(5.105)$$

Moreover subtracting relation (5.100) from (5.102), we have

$$dP_n^{(A)} u_n^{(A)} = \sum_{k=1}^{NM} \left[\int_0^L \left(\hat{d}N \frac{du_1}{dx_1} + \hat{d}M_1 \frac{d\theta_1}{dx_1} \right. \right.$$

$$\left. \left. + \hat{d}M_2 \frac{d\theta_2}{dx_1} + \hat{d}M_3 \frac{d\theta_3}{dx_1} \right) dx_1 \right]^{(k)} \quad (5.106)$$

Comparing relation (5.106) and (5.105), we get

$$\hat{d}P_n^{(A)} u_n^{(A)} = \hat{d}(U_s)_T = \frac{\partial(U_s)_T}{\theta P_n^{(A)}} \hat{d}P_n^{(A)} \quad (5.107)$$

Thus
$$u_n^{(A)} = \frac{\partial(U_s)_T}{\partial P^{(A)}} \quad (5.108)$$

Similarly we can prove the validity of the second of relations (5.98).

5.15.2 Application of Castigliano's second theorem in computing components of displacements of framed structures

Using relation (5.97) for a framed structure made of isotropic linearly elastic materials, we may rewrite relations (5.98) as

$$
u_n^{(A)} = \sum_{k=1}^{NM} \left[\int_0^L \left(\frac{N}{EA} \frac{\partial N}{\partial P^{(A)}} + \frac{M_1}{GK} \frac{\partial M_1}{\partial P^{(A)}} + \frac{M_2}{EI_2} \frac{\partial M_2}{\partial P^{(A)}} \right. \right.
$$
$$
\left. \left. + \frac{M_3}{EI_3} \frac{\partial M_3}{\partial P^{(A)}} \right) dx_1 \right]^{(k)}
\tag{5.109}
$$

$$
\theta_m^{(B)} = \sum_{k=1}^{NM} \left[\int_0^L \left(\frac{N}{EA} \frac{\partial N}{\partial M^{(B)}} + \frac{M_1}{GK} \frac{\partial M_1}{\partial M^{(B)}} + \frac{M_2}{EI_2} \frac{\partial M_2}{\partial M^{(B)}} \right. \right.
$$
$$
\left. \left. + \frac{M_3}{EI_3} \frac{\partial M_3}{\partial M^{(B)}} \right) dx_1 \right]^{(k)}
\tag{5.110}
$$

Relations (5.109) or (5.110) can be used to compute a component of translation or a component of rotation at a point A or B of a framed structure in the direction of the unit vector $\mathbf{n}$ or $\mathbf{m}$, respectively, by adhering to the following procedure:

Step 1. The structure is considered subjected to an auxiliary loading consisting of

1. The given actions, except the force (moment) acting at point A (B) in the direction of the unit vector $\mathbf{n}$ ($\mathbf{m}$) if the given actions include such a force (moment)
2. An unknown force $P_n^{(A)}$ [moment $M_m^{(B)}$] acting at point A (B) in the direction of the unit vector $\mathbf{n}$ ($\mathbf{m}$)

Step 2. The internal actions N, M_1, M_2, M_3 in the members of the structure subjected to the auxiliary loading described in step 1 are established as functions of the force $P_n^{(A)}$ (moment $M_m^{(B)}$).

Step 3. The internal actions are differentiated with respect to $P_n^{(A)}$ ($M_m^{(B)}$) and the results are substituted in relation (5.109) or (5.110). This gives the translation $u_n^{(A)}$ (rotation $\theta_m^{(B)}$) of the structure subjected to the auxiliary loading described in step 1.

Step 4. The translation $u_n^{(A)}$ (rotation $\theta_m^{(B)}$) of the structure subjected to the given actions is obtained by setting $P_n^{(A)}$($M_m^{(B)}$) equal to the given force (moment) acting at point A (B) in the direction of the unit vector $\mathbf{n}$ ($\mathbf{m}$). If there is not a force (moment) acting at point A (B) in the direction of the unit vector $\mathbf{n}$ ($\mathbf{m}$), the force $P_n^{(A)}$ (moment $M_m^{(B)}$) is set equal to zero.

In what follows, we illustrate with two examples the use of Castigliano's second theorem in computing components of displacement of points of framed structures made of isotropic linearly elastic materials.

Example 1. Consider the truss shown in Fig. a. The cross-sectional area of its top and bottom chords is equal to 2.4×10^3 mm^2, while the cross-sectional area of its diagonal and vertical members is equal to 1.6×10^3 mm^2. The modulus of elasticity of the material from which its members are made is $E = 200{,}000$ MPa. Compute the vertical component of translation (deflection) of joint 7.

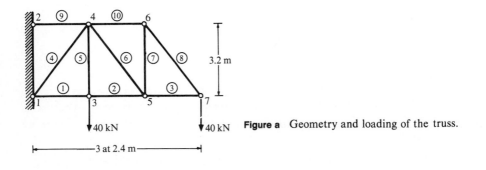

Figure a Geometry and loading of the truss.

---

solution In order to compute the vertical component of translation of joint 7, the vertical force of 40 kN acting at joint 7 is replaced by a vertical force P. The reactions of the truss are computed by referring to Fig. b and considering its equilibrium. Moreover the internal forces in the members of the truss are computed by considering the equilibrium of its joints.

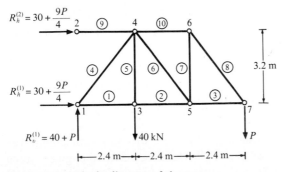

Figure b Free-body diagram of the truss.

Equilibrium of Joint 1 (see Fig. *c*)

$$\Sigma \bar{F}_v = 0 \qquad N^{(4)} = -\tfrac{5}{4}(40 + P)$$

$$\Sigma \bar{F}_h = 0 \qquad N^{(1)} + \tfrac{3}{5}N^{(4)} + 30 + \tfrac{9}{4}P = 0$$

$$N^{(1)} = -\tfrac{3}{2}P$$

Equilibrium of Joint 3 (see Fig. *d*)

$$\Sigma \bar{F}_v = 0 \qquad N^{(5)} = 40 \text{ kN}$$

$$\Sigma \bar{F}_h = 0 \qquad N^{(2)} = -\tfrac{3}{2}P$$

Equilibrium of Joint 4 (see Fig. *e*)

$$\Sigma \bar{F}_v = 0 \qquad \tfrac{4}{5}N^{(6)} - (40 + P) + 40 = 0$$

$$N^{(6)} = \tfrac{5}{4}P$$

$$\Sigma \bar{F}_h = 0 \qquad \tfrac{3}{4}P + \tfrac{3}{4}P = \tfrac{3}{4}(40 + P) - N^{(9)} = 0$$

$$N^{(9)} = 30 + \tfrac{9}{4}P$$

Equilibrium of Joint 7 (see Fig. *f*)

$$\Sigma \bar{F}_v = 0 \qquad N^{(8)} = \tfrac{5}{4}P$$

$$\Sigma \bar{F}_h = 0 \qquad N^{(3)} = -\tfrac{3}{4}P$$

Equilibrium of Joint 6 (see Fig. *g*)

$$\Sigma \bar{F}_v = 0 \qquad N^{(7)} = -P$$

$$\Sigma \bar{F}_h = 0 \qquad N^{(10)} = \tfrac{3}{4}P$$

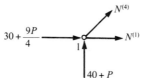

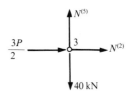

Figure c Free-body diagram of joint 1.

Figure d Free-body diagram of joint 3.

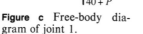

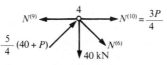

Figure e Free-body diagram of joint 4.

Figure f Free-body diagram of joint 7.

Figure g Free-body diagram of joint 6.

Taking into account that the internal force N is constant in each member of the truss, we find that Castigliano's second theorem (5.109) reduces to

$$u_v^{(7)} = \frac{\partial (U_s)_T}{\partial P} = \sum_{k=1}^{NM} \left(\frac{NL}{EA} \frac{\partial N}{\partial P} \right)^{(k)} \tag{d}$$

The deflection of joint 7 of the truss, subjected to the given loading, may be obtained by substituting in relation (a) the expressions for the internal forces established above, carrying out the differentiation, and setting $P = 40$ kN. This is done in Table a. Thus, the vertical component of translation of joint 7 of the truss is

$$u_v^{(7)} = 6 \text{ mm} \downarrow$$

Table a

Member	$N\big|_{P=40\text{ kN}}$	$A,$ 10^3 mm^2	$L,$ mm	$\dfrac{NL}{EA}\big|_{P=40\text{ kN}}$	$\dfrac{\partial N}{\partial P}\big|_{P=40\text{ kN}}$	$\dfrac{NL}{EA}\dfrac{\partial N}{\partial P}\big|_{P=40\text{ kN}}$
1	−60	2.4	2400	−0.30	−1.5	0.450
2	−60	2.4	2400	−0.30	−1.5	0.450
3	−30	2.4	2400	−0.15	−0.75	0.1125
4	−100	1.6	4000	−1.25	−1.25	1.5625
5	40	1.6	3200	0.40	0	0.000
6	50	1.6	4000	0.625	1.25	0.78125
7	−40	1.6	3200	−0.40	−1.00	0.400
8	50	1.6	4000	0.625	1.25	0.78125
9	120	2.4	2400	0.60	2.25	1.350
10	30	2.4	2400	0.15	0.75	0.1125
Total						6.000

Example 2. Consider a cantilever beam made of a linearly elastic material subjected to a uniform load over part of its span, as shown in Fig. a. Compute the deflection and the slope of the free end of this beam using Castigliano's second theorem.

solution

1. *Computation of the deflection of the free end of the beam.* In this case, we consider the beam loaded with the auxiliary loading shown in Fig. b, consisting of the

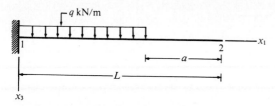

Figure a Geometry and loading of beam.

actual load and a concentrated force P acting at point 2. The bending moment in the beam is given as

$$M = -Px'_1 \qquad\qquad 0 \le x'_1 \le a$$

$$M = -Px'_1 - \frac{q(x'_1 - a)^2}{2} \qquad a \le x'_1 \le L \tag{a}$$

The deflection $u_v^{(2)}$ of point 2 of the beam of Fig. b is obtained from relation (5.109) which, in this case, reduces to

$$u_v^{(2)} = \frac{\partial (U_s)_T}{\partial P} = \int_0^L \frac{M}{EI} \frac{\partial M}{\partial P} \, dx'_1 \tag{b}$$

Substituting relations (a) into (b), we get

$$u_v^{(2)} = \int_0^a \frac{P(x'_1)^2}{EI} \, dx'_1 + \int_a^L \frac{\left[Px'_1 + q\dfrac{(x'_1 - a)^2}{2} \right] x'_1 \, dx'_1}{EI} \tag{c}$$

The deflection of point 2 of the beam of Fig. a may be obtained from relation (c) by setting $P = 0$ and integrating. Thus

$$u_v^{(2)} = \int_a^L \frac{q(x'_1 - a)^2 x'_1 \, dx'_1}{2EI} = \frac{q}{24EI}(3L^4 - 8aL^3 + 8a^2L^2 - a^4) \tag{d}$$

In case $a = 0$, the above relation yields

$$u_v^{(2)} = \frac{qL^4}{8EI} \tag{e}$$

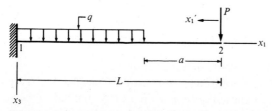

Figure b Beam subjected to the auxiliary loading for the computation of the displacement of point 2.

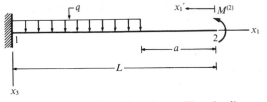

Figure c Beam subjected to the auxiliary loading
for the computation of the rotation at point 2.

2. *Computation of the slope at the free end of the beam.*   In this case, we consider
the beam loaded with the auxiliary loading of Fig. c, consisting of the actual load and
a concentrated moment acting at point 2. The bending moment in this beam is given
as

$$M = M^{(2)} \qquad\qquad\qquad 0 \leq x_1' \leq a \qquad\qquad (f)$$

$$M = M^{(2)} - \frac{q(x_1' - a)^2}{2} \qquad a \leq x_1' \leq L$$

Substituting relations (f) into (5.110), the slope du_v/dx_1 at point 2 of the auxiliary
beam of Fig. c is obtained as

$$\theta^{(2)} = -\left(\frac{du_v}{dx_1}\right)^{(2)} = \frac{\partial(U_s)_T}{\partial M^{(3)}} = \int_0^a \frac{M^{(2)}\,dx_1'}{EI} + \int_a^L \frac{M^{(2)} - q(x_1' - a)^2/2}{EI}\,dx_1'$$

$$(g)$$

The slope du_v/dx_1 at point 2 of the beam of Fig. a may be obtained from relation (g)
by setting $M^{(2)} = 0$ and integrating. Thus

$$\theta^{(2)} = -\left(\frac{du_v}{dx_1}\right)^{(2)} = -\int_0^L \frac{q(x_1' - a)^2}{2EI}\,dx_1' = -\frac{q}{6EI}(L - a)^3 \qquad (h)$$

The minus sign indicates that the rotation $\theta^{(2)}$ is opposite to the moment $M^{(2)}$ shown
in Fig. c; that is $\theta^{(2)}$ is clockwise. If $a = 0$, relation (h) yields

$$\left(\frac{du_v}{dx_1}\right)^{(2)} = \frac{qL^3}{6EI} \qquad\qquad (i)$$

5.16   Betti-Maxwell Reciprocal Theorem

Consider a body made of a linearly elastic material subjected to a set of external actions A in an environment of constant temperature. Consider the same
body subjected to another set of external actions B in an environment of constant temperature. We denote the displacement field of the body resulting from
the set of external actions A and B by $\mathbf{u}^A$ and $\mathbf{u}^B$, respectively. Betti's reciprocal
theorem states that *the work performed by the system of external actions A on*

the displacement field $\mathbf{u}^B$ is equal to the work performed by the system of external action B on the displacement field $\mathbf{u}^A$.

For a framed structure, Betti's reciprocal theorem can be stated as follows.

Theorem. Consider a framed structure consisting of NM members made of a linearly elastic material and subjected to loading in an environment of constant temperature. The loading consists of N concentrated forces $\mathbf{P}^{(i)}$, M concentrated moments $\mathbf{M}^{(i)}$, a distributed force $\mathbf{q}$, and a distributed moment $\mathbf{m}$. We denote the resulting distributions of the translation and rotation vectors by $\mathbf{u}$ and $\boldsymbol{\theta}$, respectively.

Consider the same structure subjected to another loading in an environment of constant temperature. This loading consists of $\hat{N}$ concentrated forces $\hat{\mathbf{P}}^{(i)}$, $\hat{M}$ concentrated moments $\hat{\mathbf{M}}^{(i)}$, a distributed force $\hat{\mathbf{q}}$, and a distributed moment $\mathbf{m}$. We denote the resulting distribution of the translation and rotation vectors by $\hat{\mathbf{u}}$ and $\hat{\boldsymbol{\theta}}$, respectively.

The work performed by the loading $\mathbf{P}^{(i)}$, $\mathbf{M}^{(i)}$, $\mathbf{q}$, and $\mathbf{m}$ on the displacements $\hat{\mathbf{u}}$ and $\hat{\boldsymbol{\theta}}$ is equal to the work performed by the loading $\hat{\mathbf{P}}^{(i)}$, $\hat{\mathbf{M}}^{(i)}$, $\hat{\mathbf{q}}$, and $\hat{\mathbf{m}}$ on the displacements $\mathbf{u}$ and $\boldsymbol{\theta}$. That is,

$$
\sum_{i=1}^{N} \mathbf{P}^{(i)} \cdot \hat{\mathbf{u}}^{(i)} + \sum_{i=1}^{M} \mathbf{M}^{(i)} \cdot \hat{\boldsymbol{\theta}}^{(i)} + \sum_{k=1}^{NM} \left[\int_{0}^{L} (\mathbf{q} \cdot \hat{\mathbf{u}} + \mathbf{m} \cdot \hat{\boldsymbol{\theta}}) \, dx_1 \right]^{(k)}
$$

$$
= \sum_{i=1}^{N} \hat{\mathbf{P}}^{(i)} \cdot \hat{\mathbf{u}}^{(i)} + \sum_{i=1}^{M} \hat{\mathbf{M}}^{(i)} \cdot \boldsymbol{\theta}^{(i)} + \sum_{k=1}^{NM} \left[\int_{0}^{L} (\hat{\mathbf{q}} \cdot \mathbf{u} + \hat{\mathbf{m}} \cdot \boldsymbol{\theta}) \, dx_1 \right]^{(k)}
$$

$$(5.111)$$

Proof. As previously mentioned, Betti's reciprocal theorem is valid for framed structures made of linearly elastic materials. However, in this text, we limit our attention to framed structures made of isotropic, linearly elastic materials. For this reason, we prove Betti's reciprocal theorem only for framed structures made of isotropic, linearly elastic materials.

To prove this theorem, we employ the principle of virtual work for the structure under consideration subjected to the external actions $\mathbf{P}^{(i)}$, $\mathbf{M}^{(i)}$, $\mathbf{q}$, and $\mathbf{m}$. We choose the resulting distribution of internal actions N, M_1, M_2, M_3 in the members of the structure as the statically admissible distribution of internal actions. Moreover, we choose the displacement fields $\hat{\mathbf{u}}$ and $\hat{\boldsymbol{\theta}}$ of the structure subjected to the external actions $\hat{\mathbf{P}}^{(i)}$, $\hat{\mathbf{M}}^{(i)}$, $\hat{\mathbf{q}}$, and $\hat{\mathbf{m}}$ as the geometrically admissible displacement fields. Thus, the principle of virtual work [see relation (5.53)] yields

$$
\sum_{i=1}^{N} \mathbf{P}^{(i)} \cdot \hat{\mathbf{u}}^{(i)} + \sum_{i=1}^{M} \mathbf{M}^{(i)} \cdot \hat{\boldsymbol{\theta}}^{(i)} + \sum_{k=1}^{NM} \left[\int_{0}^{L} (\mathbf{q} \cdot \hat{\mathbf{u}} + \mathbf{m} \cdot \hat{\boldsymbol{\theta}}) \, dx_1 \right]^{(k)}
$$

$$
= \sum_{k=1}^{NM} \left[\int_{0}^{L} \left(N \frac{d\hat{u}_1}{dx_1} + M_1 \frac{d\hat{\theta}_2}{dx_1} + M_2 \frac{d\hat{\theta}_2}{dx_1} + M_3 \frac{d\hat{\theta}_3}{dx_1} \right) dx_1 \right]^{(k)} \quad (5.112)
$$

We also employ the principle to virtual work for the structure under consideration subjected to the external actions $\hat{\mathbf{P}}^{(i)}$, $\hat{\mathbf{M}}^{(i)}$, $\hat{\mathbf{q}}$, and $\mathbf{m}$. We choose the resulting distribution of internal actions $\hat{N}$, $\hat{M}_1$, $\hat{M}_2$, $\hat{M}_3$ as the statically admissible distribution of internal actions. Moreover, we choose the displacement fields $\mathbf{u}$ and $\boldsymbol{\theta}$ of the structure subjected to the external actions $\mathbf{P}^{(i)}$, $\mathbf{M}^{(i)}$, $\mathbf{q}$, and $\mathbf{m}$ as the geometrically admissible displacement fields. Thus, the principle of virtual work [see relation (5.53)] yields

$$
\sum_{i=1}^{N} \hat{\mathbf{P}}^{(i)} \cdot \mathbf{u}^{(i)} + \sum_{i=1}^{M} \hat{\mathbf{M}}^{(i)} \cdot \boldsymbol{\theta}^{(i)} + \sum_{k=1}^{NM} \left[\int_{0}^{L} (\hat{\mathbf{q}} \cdot \mathbf{u} + \hat{\mathbf{m}} \cdot \boldsymbol{\theta}) \, dx_1 \right]^{(k)}
$$

$$
= \sum_{k=1}^{NM} \left[\int_{0}^{L} \left(\hat{N} \frac{du_1}{dx_1} + \hat{M}_1 \frac{d\theta_1}{dx_1} + \hat{M}_2 \frac{d\theta_2}{dx_1} + \hat{M}_3 \frac{d\theta_3}{dx_1} \right) dx_1 \right]^{(k)} \tag{5.113}
$$

Using the relations between the components of internal actions and the corresponding components of displacements for structures made of isotropic linearly elastic materials (i.e., $N/EA = du_1/dx_1$, $M_1/GK = d\theta_1/dx_1$ etc.) [see relations (5.3) to (5.5)], we can convert the integral on the right-hand side of relation (5.113) to that on the right-hand side of relation (5.112). Thus, the left-hand sides of relations (5.112) and (5.113) must be equal. Hence relation (5.111) is valid, and concomitantly Betti's reciprocal theorem has been proven.

We illustrate Betti's theorem by applying it to the beam loaded as shown in Fig. 5.30a. That is,

$$
P\hat{u}_v^{(A)} = \hat{P}u_v^{(B)} \tag{5.114}
$$

This relation is referred to as Maxwell's theorem. If $P = \hat{P}$, relation (5.114), reduces to

$$
\hat{u}_v^{(A)} = u_v^{(B)} \tag{5.115}
$$

Betti's theorem may be applied to the beam loaded as shown in Fig. 5.30b.

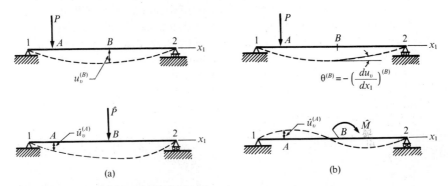

Figure 5.30   Application of Betti's theorem.

That is

$$P\hat{u}_v^{(A)} = \hat{M}\theta^{(B)} = -\hat{M}\left[\frac{du_v}{dx_1}\right]^{(B)} \tag{5.116}$$

5.17 Problems

1 and 2. Using the method of virtual work (dummy-load method), compute the vertical component of translation of joint 3 of the truss subjected to the loads shown in Figs. 5.P1. The members of the truss are made of the steel (E = 210,000 MPa). Repeat with Fig. 5.P2.

3. Using the method of "virtual" work (dummy-load method), compute the vertical component of translation of joint 5 and the rotation of member 5 of the truss subjected to the loads shown in Fig. 5.P3. The members of the truss are made of steel (E = 210,000 MPa).

4. Using the method of virtual work, compute the horizontal and vertical components of translation of joint 4 of the truss shown in Fig. 5.P4. The members of the truss are made of an aluminum alloy (E = 70,000 MPa).

5. Using the method of virtual work, compute the horizontal component of translation of joint 4 of the truss of Fig. 5.P1 resulting from the external actions shown in the figure and from a 20°C increase of temperature of the members of its bottom chord. The members of the truss are made of steel (E = 210,000 MPa, α = $10^{-5}/°C$).

6. Using the method of virtual work, compute the vertical component of translation of joint 4 of the truss of Fig. 5.P2 resulting from the external actions shown in the figure and from a 20°C increase of temperature of the members of its bottom chord. The members of the truss are made of steel (E = 210,000 MPa, α = $10^{-5}/°C$).

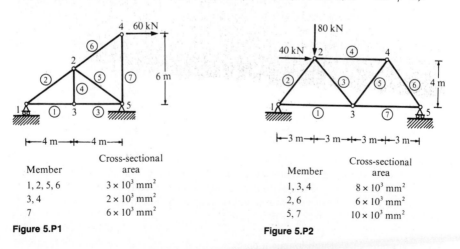

Member	Cross-sectional area
1, 2, 5, 6	3×10^3 mm^2
3, 4	2×10^3 mm^2
7	6×10^3 mm^2

Figure 5.P1

Member	Cross-sectional area
1, 3, 4	8×10^3 mm^2
2, 6	6×10^3 mm^2
5, 7	10×10^3 mm^2

Figure 5.P2

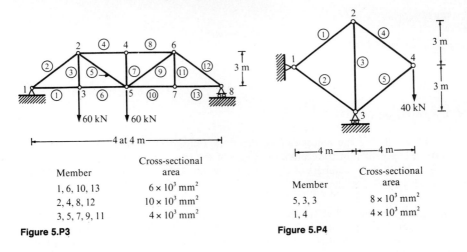

Member	Cross-sectional area
1, 6, 10, 13	6×10^3 mm^2
2, 4, 8, 12	10×10^3 mm^2
3, 5, 7, 9, 11	4×10^3 mm^2

Figure 5.P3

Member	Cross-sectional area
5, 3, 3	8×10^3 mm^2
1, 4	4×10^3 mm^2

Figure 5.P4

7. Using the method of virtual work, compute the vertical component of translation of joint 4 of the truss of Fig. 5.P4 resulting from a 30°C increase of temperature of member 3. The members of the truss are made of an aluminum alloy (E = 70,000 MPa, α = 2.3 $\times$ 10^{-5}/°C).

8. Using the method of virtual work, compute the vertical component of translation of joint 5 of the truss of Fig. 5.P3 resulting from the external actions shown in the figure and from a 20°C decrease of temperature of the members of its bottom chord. The members of the truss are made of steel (E = 210,000 MPa, α = 10^{-5}/°C).

9. Using the method of virtual work, compute the vertical component of translation of joint 3 of the truss of Fig. 5.P1 resulting from an error in the length of member 5. The length of this member is 4.99 m instead of 5 m.

10. Using the method of virtual work, compute the vertical component of translation of joint 4 of the truss of Fig. 5.P4 resulting from an error in the length of member 3. The length of this member is 6.01 m instead of 6 m.

11. Using the method of virtual work, compute the vertical component of translation (deflection) of joint 5 of the truss of Fig. 5.P3, resulting from an error in the length of member 5. The length of this member is 5.01 m instead of 5 m.

12 and 13. Using the method of virtual work, compute for joint 2 of the truss of Fig. 5.P1 the vertical component of translation, due to a 20 mm settlement of the right-hand support. Verify your results by considering the geometry of the deformed truss. Repeat with Fig. 5.P2.

14. Using the method of virtual work, compute for joint 2 of the truss of Fig. 5.P3 the vertical component of translation resulting from the loading shown in the figure and from a 15 mm downward movement of support 8. The truss is made of steel (E = 210,000 MPa).

15. Compute how much longer members 2, 4, 8, and 12 of the truss of Fig. 5.P3 must be made in order to produce a symmetric camber of the bottom chord specified by $u_v^{(3)} = -10$ mm $u_v^{(5)} = -12.5$ mm.

16. Using the method of virtual work, compute for joint 11 of the truss of Fig. 5.P16 the vertical component of translation resulting from the load shown. The members of the truss have the same constant cross section ($A = 4 \times 10^3$ mm^2) and are made of steel ($E = 210,000$ MPa).

17. Using the method of virtual work, compute for joint 4 of the truss of Fig. 5.P17 the horizontal component of translation due to the load shown. The members of the truss have the same constant cross section ($A = 6 \times 10^3$ mm^2) and are made of steel ($E = 210,000$ MPa).

18. Using the method of virtual work, compute for joint 4 of the truss of Fig. 5.P18 the total translation due to the load shown. The members of the truss have the same constant cross section ($A = 5 \times 10^3$ mm^2) and are made of steel ($E = 210,000$ MPa).

19. Using the method of virtual work, compute for joint 9 of the truss of Fig. 5.P19 the vertical component of translation due to the load shown. The members of the truss

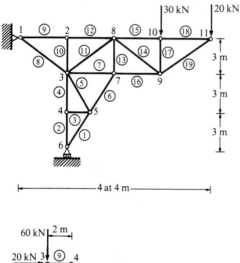

Figure 5.P16

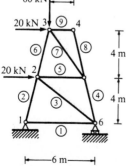

Figure 5.P17

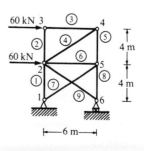

Figure 5.P18

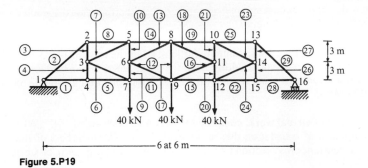

Figure 5.P19

have the same constant cross section ($A = 8 \times 10^3$ mm^2) and are made of steel ($E = 210,000$ MPa).

20. Using the method of virtual work, compute the total translation of joint 2 of the structure loaded as shown in Fig. 5.P20. The members of the structure have the same constant cross section ($A = 4 \times 10^3$ mm^2) and are made of steel ($E = 210,000$ MPa).

21 and 22. Using the method of virtual work, compute the deflection and the rotation of point 2 of the cantilever beam shown in Fig. 5.P21. The beam has a rectangular cross section of 600 mm depth and 200 mm width and it is made of steel ($E = 210,000$ MPa, $G = E/2.6$). Establish what percentage of the total deformation is due to shear. Repeat with Fig. 5.P22.

23. Using the method of virtual work, compute the deflection and the rotation of point 4 of the beam loaded as shown in Fig. 5.P23. The beam is made of steel ($E = 210,000$ MPa) and has a constant cross section (see Fig. 5.P23). Disregard the effect of shear deformation.

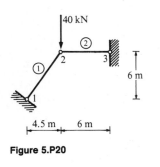

Figure 5.P20

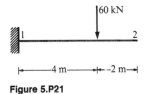

Figure 5.P21

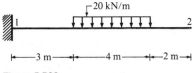

Figure 5.P22

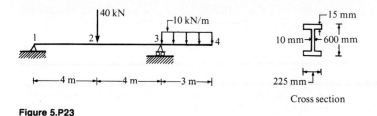

Cross section

Figure 5.P23

24. Using the method of virtual work, compute the deflection and the rotation of point 2 of the beam loaded as shown in Fig. 5.P24. The beam is made of a standard wide-flange steel section ($E = 210,000$ MPa), and it is reinforced at its center portion by steel cover plates. Disregard the effect of shear deformation. The moments of inertia of the three portions of the beam are given in Fig. 5.P24 ($I = 369.7 \times 10^6$ mm^4).

25. Using the method of virtual work, compute the deflection of point 4 of the beam loaded as shown in Fig. 5.P25. Disregard the effect of shear deformation. The beam has a constant cross section ($I = 369.70 \times 10^6$ mm^4), and it is made of steel ($E = 210,000$ MPa).

26. Using the method of virtual work, compute the horizontal and vertical components of translation of point 5 of the structure loaded as shown in Fig. 5.P26. Disregard the effect of shear and axial deformation of the members of the structure. The members of the structure have the same constant cross section ($I = 369.7 \times 10^6$ mm^4) and are made of steel ($E = 210,000$ MPa).

27 and 28. Using the method of virtual work, compute the vertical component of translation of point 3 of the structure loaded as shown in Fig. 5.P27. Disregard the effect of shear and axial deformation of the members of the structure. The members

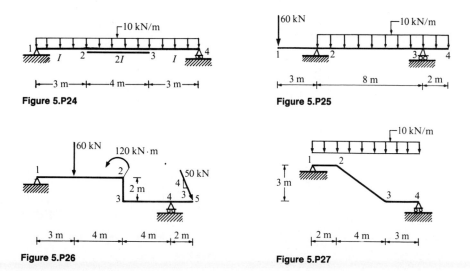

Figure 5.P24

Figure 5.P25

Figure 5.P26

Figure 5.P27

of the structure are made of steel (E = 210,000 MPa) and have the same constant cross section (I = 369.7 × 10^6 mm^4). Repeat with Fig. 5.P28.

29. Using the method of virtual work, compute the horizontal component of translation of point 4 of the structure loaded as shown in Fig. 5.P29. Disregard the effect of axial deformation but consider the effect of shear deformation of the members of the structure. The members of the structure have the same constant rectangular cross section (depth = 600 mm width = 200 mm) and are made of steel (G = $E/2.6$, E = 210,000 MPa). Compute what percentage of the total translation of point 4 represents the effect of shear deformation.

30. Using the method of virtual work, compute the horizontal component of translation of joint 2 of the structure loaded as shown in Fig. 5.P30. Disregard the effect of shear and axial deformation of the members of the structure. The members of the structure have the same constant cross section (I = 369.70 × 10^6 mm^4) and are made of steel (E = 210,000 MPa).

31. Using the method of virtual work, compute the vertical component of translation of point 3, the slope of the elastic curve of member 2,3 at point 3, and the change of the slope of the elastic curve at point 3 of the frame loaded as shown in Fig. 5.P31. The members of the frame have the same constant cross section and are made of the same material (E = 210,000 MPa). Disregard the effect of shear and axial deformation of the members of the frame.

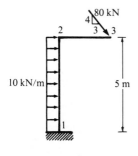

Figure 5.P28

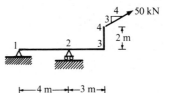

Figure 5.P29

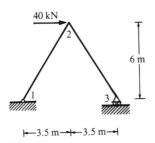

Figure 5.P30

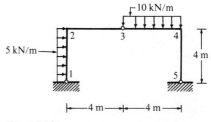

Figure 5.P31

32. Using the method of virtual work, compute the horizontal component of translation of joints 2 and 4 of the frame shown in Fig. 5.P32. Establish what percentage of the total displacement is due to the axial deformation of the members of the frame. Disregard the effect of shear deformation. The members of the frame have the same cross section and are made of the same material. Their area, their moment of inertia, and their elastic constants are $A = 13.2 \times 10^3$ mm^2, $I = 369.7 \times 10^6$ mm^4, $E = 210,000$ MPa, and $G = E/2.6$.

33. Using the method of virtual work, compute the horizontal movement of support 1 of the structure loaded as shown in Fig. 5.P33. The members of the structure have the same constant cross section ($I = 369.7 \times 10^6$ mm^4) and are made of the same material ($E = 210,000$ MPa). Disregard the effect of shear and axial deformation of the members of the structure.

34. Using the method of virtual work, compute the deflection of point 3, the change of slope of the elastic curve at point 2, and the rotation of member 2,3 at point 2 of the structure loaded as shown in Fig. 5.P34. Disregard the effect of shear and axial deformation of the members of the structure. The structure has a constant cross section ($I = 369.70 \times 10^6$ mm^4), and it is made of steel ($E = 210,000$ MPa).

35. Using the method of virtual work, compute the deflection of point 5, the change of slope of the elastic curve at point 3, and the rotation of member 2,3 at point 3 of the beam loaded as shown in Fig. 5.P35. Disregard the effect of shear deformation of the members of the beam. The beam has a constant cross section ($I = 369.70 \times 10^6$ mm^4), and it is made of steel ($E = 210,000$ MPa).

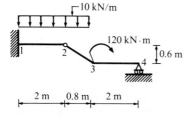

Figure 5.P32

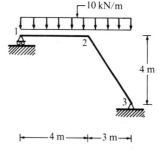

Figure 5.P33

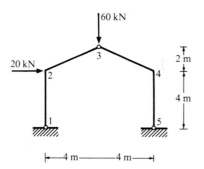

Figure 5.P34

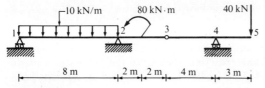

Figure 5.P35

36.   Using the method of virtual work, compute the deflection of point 2, the change of slope of the elastic curve at point 2, and the rotation of member 2,3 at point 2 of the beam loaded as shown in Fig. 5.P36. Disregard the effect of shear deformation of the members of the beam. The beam has a constant cross section ($I = 369.70 \times 10^6$ mm^4) and it is made of steel ($E = 210,000$ MPa).

37 and 38.   Using the method of virtual work, compute the total translation of the free end of the steel ($E = 210,000$ MPa) structure loaded as shown in Fig. 5.P37. Disregard the effect of shear deformation of the members of the structure but include the effect of axial deformation. The area and moment of inertia of the cross sections of the members of the structure are $A = 13.2 \times 10^3$ mm^2 and $I = 369.70 \times 10^6$ mm^4. Repeat with Fig. 5.P38.

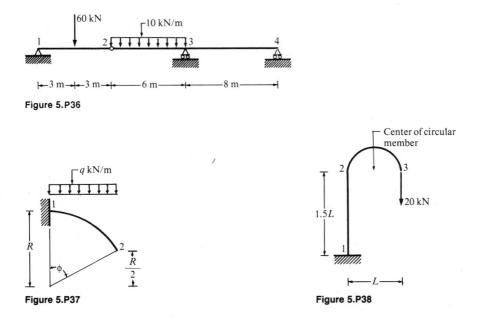

Figure 5.P36

Figure 5.P37

Figure 5.P38

39.   Using the method of virtual work, compute the horizontal movement of support 4 of the frame loaded as shown in Fig. 5.P39. The members of the frame have the same constant cross section ($I = 369.7 \times 10^6$ mm^4) and are made of the same material ($E = 210,000$ MPa). Disregard the effect of shear and axial deformation of the members of the frame.

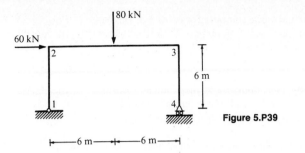

Figure 5.P39

40. Using the method of virtual work, compute for the frame of Fig. 5.P40 the vertical component of translation of point 3, the rotation of member 2,3 at point 3, and
the change of slope of the elastic curve at point 3 due to a temperature differential at
the exterior ($T_e = 30°C$) and interior ($T_i = 0°C$) fibers. The temperature during
construction was 15°C. The members of the frame are made of the same material (α
$- 10^{-5}/°C$) and have the same constant cross section ($A - 16.3 \times 10^3$ mm², $I =$
584.8 × 10⁶ mm⁴, $h = 475$ mm).

41. Using the method of virtual work, compute for points 2, 3, and 4 of the frame
of Fig. 5.P41 the movement due to a temperature differential at the exterior ($T_e =$
$-10°C$) and interior ($T_i = 40°C$) fibers. The temperature during construction was
$T_0 = 5°C$. The members of the frame are made of the same material ($\alpha = 10^{-5}/°C$)
and have the same constant cross section ($A = 13.2 \times 10^3$ mm², $I = 369.7 \times 10^6$
mm⁴, $h = 425$ mm).

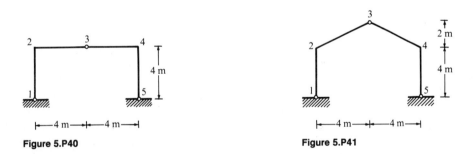

Figure 5.P40   Figure 5.P41

42. Using the method of virtual work, compute for the frame of Fig. 5.P33 the horizontal component of translation of point 1 and the rotation at point 3 due to a temperature differential at the exterior ($T_e = 5°C$) and interior ($T_i = 45°C$) fibers. The
temperature during construction was $T_0 = 5°C$. The members of the frame are made
of the same material ($\alpha = 10^{-5}/°C$) and have the same constant cross section ($A =$
16.3 × 10³ mm², $I = 564.8 \times 10^6$ mm⁴, $h = 475$ mm).

43. Using the method of virtual work, compute for point 4 of the beam of Fig. 5.P25
the deflection due to a 20 mm settlement of support 2. Verify your results by considering the geometry of the deformed beam.

44. Using the method of virtual work compute for point 3 of the beam of Fig. 5.P24 the deflection due to a 25 mm settlement of support 1. Verify your results by considering the geometry of the deformed beam.

45. Using the method of virtual work, compute for the frame of Fig. 5.P31 the slope of the elastic curve of member 2,3 at point 3 and the change of the slope of the elastic curve at point 3 due to a 10 mm horizontal movement of support 1 from left to right. Verify your results by considering the geometry of the deformed frame.

46. Using the method of virtual work, compute for the beam of Fig. 5.P34 the movement of points 2, 3, and 4 due to a 20 mm settlement of support 4.

47. Using the method of virtual work, compute the total translation of point 5 of the frame of Fig. 5.P47. The members of the frame have the same constant tubular cross section (outside diameter 200 mm, thickness 30 mm) and are made from the same material ($G = E/2.6$, $E = 210,000$ MPa). Disregard the effect of the axial and shear deformation of the members of the frame.

48. Using the method of virtual work, compute the total translation of point 3 of the frame of Fig. 5.P48. The members of the frame have the same constant tubular cross section (outside diameter 120 mm, thickness 30 mm) and are made of the same material ($G = E/2.6$, $E = 210,000$ MPa). Disregard the effect of axial and shear deformation of the members of the frame.

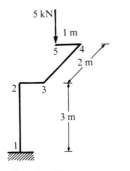

Figure 5.P47

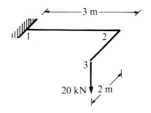

Figure 5.P48

49 and 50. Using the method of virtual work compute the deflection of point 2 of the beam resulting from the load shown in Fig. 5.P49. The cross sections of the beam are rectangular and have a constant width $b = 200$ mm. The beam is made of steel ($E = 210,000$ MPa). Disregard the effect of shear deformation of the beam. Subdivide the length of the beam into eight equal segments, and use Simpson's rule to evaluate the integrals. Repeat with Fig. 5.P50.

51. Using the method of virtual work compute the deflection of point 2 of the beam resulting from the load shown in Fig. 5.P51. The cross sections of the beam are rectangular and have a constant width $b = 300$ mm. The beam is made of steel ($E =$

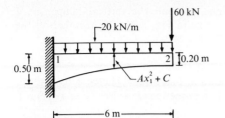

Figure 5.P49

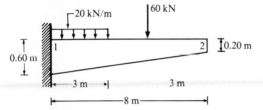

Figure 5.P50

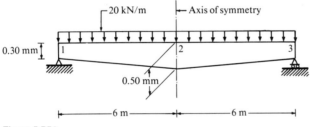

Figure 5.P51

210,000 MPa). Disregard the effect of shear deformation of the beam. Subdivide the half length of the beam into eight equal segments, and use Simpson's rule to evaluate the integrals.

52. Using the second theorem of Castigliano, compute the vertical component of translation of joint 3 of the truss loaded as shown in Fig. 5.P2. The members of the truss are made of steel ($E = 210,000$ MPa).

53. Using the second theorem of Castigliano, compute the horizontal component of translation of joint 4 of the truss loaded as shown in Fig. 5.P1. The members of the truss are made of steel ($E = 210,000$ MPa).

54. Using the second theorem of Castigliano, compute the horizontal component of translation of joint 4 of the truss loaded as shown in Fig. 5.P17. The members of the truss have the same constant cross section ($A = 6 \times 10^3$ mm^2) and are made of steel ($E = 210,000$ MPa).

55.   Using the second theorem of Castigliano, compute the vertical component of translation of joint 2 of the truss loaded as shown in Fig. 5.P4. The members of the truss are made of an aluminum alloy (E = 70,000 MPa).

56 and 57.   Using the second theorem of Castigliano, compute the deflection and rotation of point 4 of the beam loaded as shown in Fig. 5.P25. The beam is made of steel (E = 210,000 MPa) and has a constant cross-section (I = 369.70 $\times$ 10^6 mm^4). Repeat with Fig. 5.P26.

58.   Using the second theorem of Castigliano, compute the vertical component of translation of point 3 and the slope of the elastic curve of member 2,3 at point 3 of the frame loaded as shown in Fig. 5.P31. The members of the frame have the same constant cross section (I = 369.7 $\times$ 10^6 mm^4) and are made of the same material (E = 210,000 MPa).

59.   Using the second theorem of Castigliano, compute the vertical component of translation of point 3 and the rotation of member 2,3 at point 3 of the beam loaded as shown in Fig. 5.P35. The beam has a constant cross section (I = 369.70 $\times$ 10^6 mm^4) and it is made of steel (E = 210,000 MPa).

Photograph and Brief Description of a
Concrete Arch

The Lilac Road overcrossing and Interstate Route 15 are shown in
Fig. 5.31. Its superstructure consists of a post-tensioned prestressed
concrete box girder 695 ft (212 m) long. It is supported at its ends
by abutments and along the central third of its length by a
reinforced concrete cellular arch. The arch rises 69 ft (21 m) above
its abutments which are 505 ft (154 m) apart and rest on rock. The
bridge is skewed to Route 15 at an angle of 20°. To partially
conceal the effect of this skew from those traveling on Route 15, the
angle between the axes of symmetry of the cross sections of the
superstructure and the arch varies along the length of the bridge.
Moreover, the dimensions of the cross sections of the arch vary. The
number of girder stems in the box girder of the superstructure
changes abruptly from five (see Fig. 5.32) in the portion of the
bridge which is not supported on the arch to three at the middle
third of the bridge. Inasmuch as the prestressing tendons are placed

Figure 5.31 The Lilac Road overcrossing and Interstate Route 15 north of Escondido, California. (Courtesy Department of Transportation of the state of California.)

in the stems, this change in the number of stems introduces a discontinuity in the total pre-stressing force, thus requiring careful distribution of forces among the prestressing tendons in order to ensure the correct amount of prestressing force at all locations.

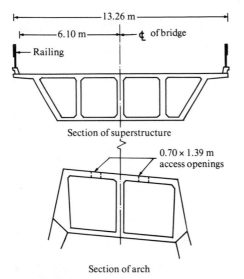

Figure 5.32 Cross section of the bridge. (Courtesy Department of Transportation of the state of California.)

The Conjugate Beam Method

6.1 The Conjugate Beam Method

In this chapter we present the conjugate beam method. This method may be used to establish the components of displacement of a point of a member of a beam or a frame by computing the internal actions at the corresponding point of a member of a conjugate beam or frame subjected to an auxiliary loading. However, in this text we limit its application to beams.

Consider a straight member of a planar structure (beam or frame) subjected to distributed transverse forces $q(x_1)$, to distributed moments $m(x_1)$ given in units of force or moment per unit length, to concentrated forces $P^{(i)}$ ($i = 1, 2, \ldots, N$), and to concentrated moments $M^{(i)}$ ($i = 1, 2, \ldots, M$) (see Fig. 6.1). Consider a segment of this member of length Δx_1 shown in Fig. 6.2. This segment is loaded only by distributed forces $q(x_1)$ and moments $m(x_1)$. The latter are not shown in Fig. 6.1 in order to avoid cluttering it.

Since the segment under consideration is in equilibrium, relations (A.54) to (A.59) are valid. Thus we have

$$q = -\frac{dQ}{dx_1} \tag{6.1}$$

$$Q = \frac{dM}{dx_1} + m \tag{6.2}$$

Moreover, referring to relations (A.71), we have

$$\theta = -\frac{du_3}{dx_1} \tag{6.3}$$

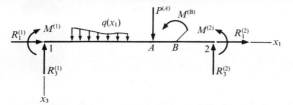

Figure 6.1 Member of a planar structure.

From relation (6.3) and using the relation between transverse component of translation and moment [see relation (A.74b)], we obtain

$$\frac{d\theta}{dx_1} = -\frac{d^2 u_3}{dx_1^2} = \frac{M}{EI} \tag{6.4}$$

Let us consider a conjugate member having the same geometry as the actual member and supported in a manner that we will subsequently establish. The conjugate member is subjected to an auxiliary loading which is related to the internal actions of the real member by the following relation.

$$\bar{q}(x_1) = \frac{M}{EI} = \frac{d\theta}{dx_1} \tag{6.5}$$

It is apparent, that $\bar{q}(x_1)$ represents the change of the angle of rotation per unit length along the axis of the real member. A segment of the conjugate member under consideration is shown in Fig. 6.3. Since this segment is in equilibrium, relations (6.1) and (6.2) are valid. That is,

$$\bar{q} = \frac{M}{EI} = -\frac{d\bar{Q}}{dx_1} \tag{6.6a}$$

$$\bar{Q} = \frac{d\bar{M}}{dx_1} \tag{6.6b}$$

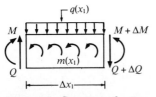

Figure 6.2 Segment of member loaded with the actual loading.

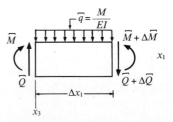

Figure 6.3 Segment of the conjugate member loaded with the auxiliary loading.

From relations (6.4) and (6.6), we obtain

$$\frac{d\theta}{dx_1} = -\frac{d\overline{Q}}{dx_1} \tag{6.7}$$

Furthermore, substituting relation (6.4) into (6.6), we have

$$\overline{q} = -\frac{d\overline{Q}}{dx_1} = -\frac{d^2\overline{M}}{dx_1^2} = \frac{M}{EI} = -\frac{d^2 u_3}{dx_1^2} \tag{6.8}$$

or

$$\frac{d^2 u_3}{dx_1^2} = \frac{d^2\overline{M}}{dx_1^2} \tag{6.9}$$

Equations (6.7) to (6.9) are satisfied by the following relations.

$$\theta(x_1) = -\overline{Q}(x_1) \qquad u_3(x_1) = \overline{M}(x_1) \tag{6.10}$$

The above relations must hold between the displacements of the ends of the real member and the actions at the ends of the conjugate member. That is, if an end of the real member can rotate, the corresponding end of the conjugate member must be subjected to an external force equal to the negative of the rotation of the end of the real member. Moreover, if an end of the real member can translate, the corresponding end of the conjugate member must be subjected to an external moment equal to the translation of the end of the real member. Furthermore, if there is a hinge at some point of a member of a planar structure, the rotation at that point will be discontinuous. Consequently, there must be a discontinuity of the shearing force at the point of the conjugate member corresponding to a hinge of the real member. This discontinuity may be produced by adding a concentrated, external force at the point of the conjugate member corresponding to the point of the real member where the hinge is located. If we denote by θ^L and θ^R the rotation of the cross section to the left and to the right of the hinge from the first of relations (6.10), we have

$$\theta^L = -\overline{Q}^L \qquad \theta^R = -\overline{Q}^R \tag{6.11}$$

$$\Delta\theta = \theta^R - \theta^L = \overline{Q}^L - \overline{Q}^R = \Delta\overline{Q} \tag{6.12}$$

Notice, that is relation (6.12) a positive external force $\Delta\overline{Q}$ acts in the direction of the positive x_3 axis (see Fig. 6.4).

On the basis of the foregoing discussion, the conjugate beam for the simply supported beam shown in Fig. 6.5a is the simply supported beam shown in Fig. 6.5c. That is, since the translations of the ends of the real beam are zero while the rotations are not zero, the moments at the ends of the conjugate beam must vanish while the shearing forces must not. The conjugate beam for the cantilever beam shown in Fig. 6.5d is the cantilever beam shown in Fig. 6.5f. That is, since the rotation θ and the translation u_3 vanish at the fixed end of the real

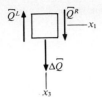

$\overline{Q}^L$ $\overline{Q}^R$

$\Delta\overline{Q}$

Figure 6.4 Segment of a conjugate member, including the point corresponding to a hinge in the real member.

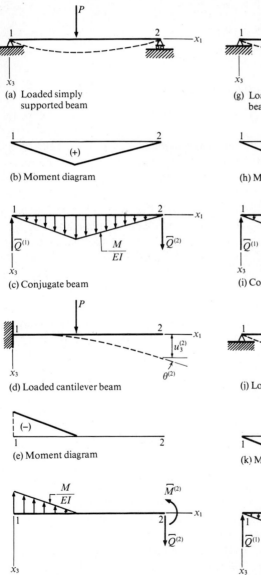

P

(a) Loaded simply
    supported beam

1   2

(+)

(b) Moment diagram

1   2

$\overline{Q}^{(1)}$ $\dfrac{M}{EI}$ $\overline{Q}^{(2)}$

(c) Conjugate beam

P

1   2
$u_3^{(2)}$
$\theta^{(2)}$

(d) Loaded cantilever beam

(−)

1   2

(e) Moment diagram

$\dfrac{M}{EI}$ $\overline{M}^{(2)}$

1   2
$\overline{Q}^{(2)}$

(f) Conjugate beam

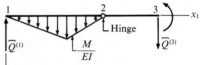

P

(g) Loaded simply supported
    beam with an overhang

1   2   3

(+)

(h) Moment diagram

1   2   3
Hinge
$\overline{Q}^{(1)}$ $\dfrac{M}{EI}$ $\overline{Q}^{(3)}$

(i) Conjugate beam

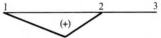

P Hinge

1   2   3   4

(j) Loaded compound beam

(−)

(+)   2   3   4
1

(k) Moment diagram

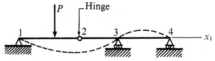

$\dfrac{M}{EI}$

1   2   3   4
Hinge
$\overline{Q}^{(1)}$ $\Delta\overline{Q}^{(2)}$ $\overline{Q}^{(4)}$

(l) Conjugate beam

Figure 6.5 Real and corresponding conjugate beams.

350

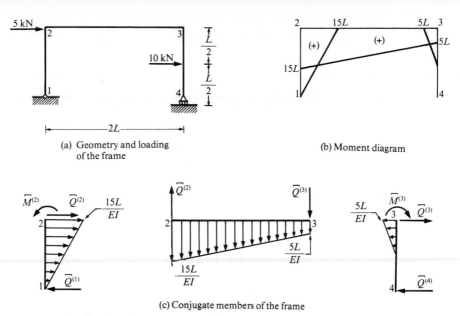

(a) Geometry and loading of the frame

(b) Moment diagram

(c) Conjugate members of the frame

Figure 6.6 Application of the conjugate beam method to a frame.

beam, there should not be a force or a moment at the corresponding end of the conjugate beam. Moreover, since the rotation θ and the translation u_3 are different than zero at the free end of the real beam, there must be a force and a moment at this end of the conjugate beam.

The conjugate beam for the simply supported beam with an overhang, shown in Fig. 6.5g is shown in Fig. 6.5i. Notice, that at support 2, the component of displacement u_3 vanishes, while the component of rotation θ does not vanish. Thus, at point 2 of the conjugate beam the moment must vanish, while the shearing force does not necessarily vanish. That is, a hinge must be placed at the point of the conjugate beam corresponding to support 2 of the real beam.

The conjugate beam for the compound beam of Fig. 6.5j is shown in Fig. 6.5m. As discussed previously, an external force is required at the point of the conjugate beam corresponding to a hinge of the real beam.

The conjugate beam method could also be applied to frames as shown in Fig. 6.6. Notice that in Fig. 6.6c the effect of axial deformation of the members of the frame is disregarded. That is, the ends of the horizontal member of the frame do not translate vertically; consequently, there is not a moment at the ends of its conjugate member.

In what follows, we solve three examples to illustrate the conjugate beam method.

Example 1. Using the conjugate beam method, compute the deflection and rotation at the free end of the cantilever beam of constant cross section, loaded as shown in Fig. a.

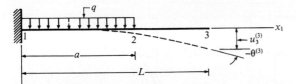

Figure a Geometry and loading of the beam.

solution The moment diagram for the beam is drawn in Fig. b. The conjugate beam is loaded with the M/EI diagram as shown in Fig. c. At point 1 of the actual beam, the deflection $u_3^{(1)}$ and the rotation $\theta^{(1)}$ vanish. Consequently, there must be neither a moment nor a reaction at the corresponding point of the conjugate beam. At point 3 of the actual beam, the deflection $u_3^{(3)}$ and the rotation $\theta^{(3)}$ do not vanish. Consequently, there must be a moment $\overline{M}^{(3)}$ and a reacting force $\overline{Q}^{(3)}$ at the corresponding point of the conjugate beam. In Fig. c the moment $\overline{M}^{(3)}$ and the reaction $\overline{Q}^{(3)}$ are shown as assumed positive. Moreover the external forces are acting upward because the moment is negative. Referring to Fig. c from the equilibrium of the conjugate beam, we have

$$\Sigma \overline{F}_3 = 0 \qquad \overline{Q}^{(3)} = \frac{qa^3}{6EI} = -\theta^{(3)} \tag{a}$$

$$\Sigma \overline{M}^{(3)} = 0 \qquad \overline{M}^{(3)} = \frac{qa^3}{6EI}\left(L - \frac{a}{4}\right) = u_3^{(3)} \tag{b}$$

Thus
$$u_3^{(3)} = \frac{qa^3}{6EI}\left(L - \frac{a}{4}\right) \qquad \text{downward} \tag{c}$$

$$\theta^{(3)} = \frac{qa^3}{6EI} \qquad \text{clockwise} \tag{d}$$

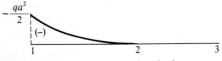

Figure b Moment diagram for the beam.

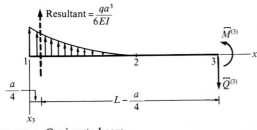

Figure c Conjugate beam.

Example 2.   Using the conjugate beam method, determine the deflection at the center of the simply supported beam loaded as shown in Fig. a. The beam is made of steel (E = 210,000 MPa). The moments of inertia of the members of the beam are given in Fig. a where I = 250 $\times$ 10^6 mm^4.

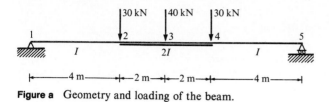

Figure a   Geometry and loading of the beam.

solution   The moment diagram for the beam is first determined (see Fig. b). The conjugate beam is shown in Fig. c. Because of the symmetry of the loading, each reaction of the conjugate beam is equal to half of the loading. Thus,

$$\overline{Q}^{(1)} = -\overline{Q}^{(5)} = \frac{(200)(2)}{EI} + \left(\frac{100 + 120}{2EI}\right)2 = \frac{620}{EI} \tag{a}$$

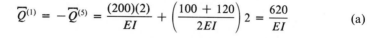

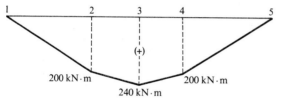

Figure b   Moment diagram of the beam.

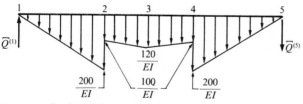

Figure c   Conjugate beam.

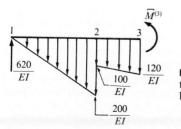

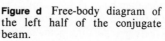

Figure d   Free-body diagram of the left half of the conjugate beam.

The deflection of the center of the beam of Fig. a is equal to the moment at the center of the beam of Fig. c. Thus, referring to Fig. d and setting the sum of the moments about point 3 equal to zero, we obtain

$$\overline{M}^{(3)} = \frac{620(6)}{EI} - \frac{200(2)}{EI}\left(2 + \frac{4}{3}\right) - \frac{100(2)(1)}{EI} - \frac{20(1)}{EI}\left(\frac{2}{3}\right)$$

$$= \frac{2174}{EI} = 0.0414 \text{ m}$$

$$u_3^{(3)} = \overline{M}^{(3)} = 41.4 \text{ mm downward}$$

Example 3. Using the conjugate beam method, compute the deflection of point 4 of the beam loaded as shown in Fig. a. The beam is made of steel ($E = 210,000$ MPa) and has a constant cross section ($I = 200 \times 10^6$ mm^4).

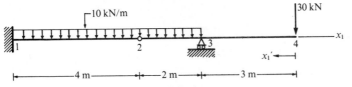

Figure a Geometry and loading of the beam.

solution Referring to Fig. b, the moment of the beam from point 1 to point 3 is equal to

$$M(x_1) = 60 + 5x_1 - \frac{10x_1^2}{2} \qquad 0 \le x_1 \le 6 \tag{a}$$

For the overhang, it is convenient to express the moment as a function of the coordinate x_1' measured from point 4. Thus

$$M(x_1') = -30x_1' \qquad 0 \le x_1' \le 3 \tag{b}$$

In Fig. c, we plot each of the erms of Eqs. (a) and (b) separately. The resulting moment diagram is called the moment diagram by parts. Referring to Table B on the inside of the back cover, we can see that the area under the moment diagram for the beam and the moment of this area about any point can be readily established if the moment diagram of the beam is plotted by parts. The conjugate beam loaded with the M/EI diagram is shown in Fig. d. At point 1 of the actual beam, the deflection $u_3^{(1)}$ and the rotation $\theta^{(1)}$ vanish. Consequently, there must be neither a moment nor a reaction at the corresponding point of the conjugate beam. At the hinge at point 2 of the actual beam, the rotation is not continuous. Consequently, the shearing force must be discontinuous at the corresponding point of the conjugate beam. That is, there must be an external force $\Delta\overline{Q}$ at point 2 of the conjugate beam. At point 3 of the actual

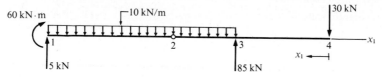

Figure b Free-body diagram of the beam.

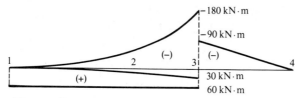

Figure c Moment diagram.

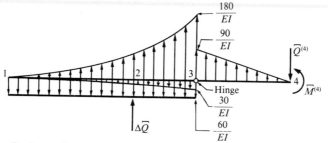

Figure d Conjugate beam.

beam, the deflection vanishes while the rotation does not. Therefore, a hinge is placed at point 3 of the conjugate beam. Finally, at point 4 of the conjugate beam, the deflection $u_3^{(4)}$ and the rotation $\theta^{(4)}$ do not vanish. Consequently, there must be a moment $\overline{M}^{(4)}$ and a shearing force $\overline{Q}^{(4)}$ at point 4 of the conjugate beam. Referring to Fig. e and using Table B on the inside of the back cover, from the equilibrium of the left

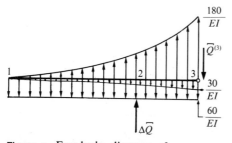

Figure e Free-body diagram of the left part of the conjugate beam.

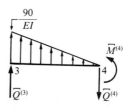

Figure f Free-body diagram of the right part of the conjugate beam.

part of the conjugate beam we have

$$\Sigma \overline{M}^{(3)} = 0 \qquad \frac{60}{EI}(6)(3) + \frac{30}{EI}(3)(2) - \frac{180}{EI}\left(\frac{6}{3}\right)\left(\frac{6}{4}\right) - \Delta\overline{Q}(2) = 0$$

Hence

$$\Delta\overline{Q} = \frac{360}{EI}$$

and $\qquad \Sigma\overline{F}_3 = 0 \qquad \dfrac{60(6)}{EI} + \dfrac{30(3)}{EI} - \dfrac{180}{EI}\left(\dfrac{6}{3}\right) - \dfrac{360}{EI} + \overline{Q}^{(3)} = 0$

Hence

$$\overline{Q}^{(3)} = \frac{270}{EI}$$

Moreover, referring to Fig. e from the equilibrium of the right part of the conjugate beam, we have

$$\Sigma\overline{F}_3 = 0 \qquad \frac{270}{EI} + \frac{90}{EI}\left(\frac{3}{2}\right) - \overline{Q}^{(4)} = 0$$

$$\Sigma\overline{M}^{(4)} = 0 \qquad \frac{270}{EI}(3) + \frac{90}{EI}\left(\frac{3}{2}\right)(2) - \overline{M}^{(4)} = 0$$

or

$$\overline{Q}^{(4)} = \frac{405}{EI} \qquad \overline{M}^{(4)} = \frac{1080}{EI}$$

Thus $\qquad \theta^{(4)} = -\overline{Q}^{(4)} = -\dfrac{405}{EI} = \dfrac{-405}{(210 \times 10^6)(2 \times 10^{-4})} = -0.00964$ rad

$$u_3^{(4)} = \overline{M}^{(4)} = \frac{1080}{EI} = \frac{1080}{(210 \times 10^6)(2 \times 10^{-4})} = 0.0257 \text{ m}$$

$$= 25.7 \text{ mm downward}$$

The minus sign indicates that the rotation $\theta^{(4)}$ is clockwise.

6.2   Discretization of the Forces Acting on a Conjugate Beam

As a result of a complicated distribution of the actions acting on certain members of the real structure or of a complicated variation of the cross section of some members of a structure along their length, it is possible that the distribution of the external forces on some conjugate members is complex. The external forces on these conjugate members may be approximated in one of the following ways.

1. By replacing the actual force distribution in a member by forces which are uniformly distributed over small segments of its length. In this method the

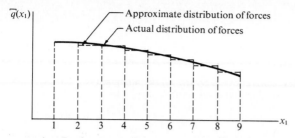

Figure 6.7 Replacement of the actual force distribution on a member by forces which are uniformly distributed over small segments of its length.

loaded part of the member is subdivided into segments and the actual forces acting on each segment are replaced by uniformly distributed forces whose magnitude is equal to the average of the values of the actual forces at the ends of the segment (see Fig. 6.7).

2. By replacing the actual force distribution on a member by a set of concentrated forces (discretization of forces) (see Fig. 6.8). In this method, the loaded part of the member is subdivided into $n - 1$ panels by n panel points. It is assumed that the load is not applied directly to the member but on a series of simply supported stringers. Each one of these stringers is supported at two adjacent panel points of the member. Thus, the distributed forces acting on the member are replaced by the reactions of the imaginary stringers. Consider a

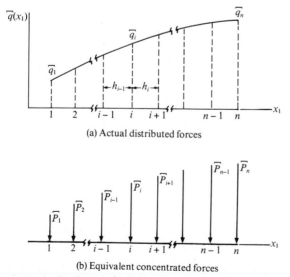

(a) Actual distributed forces

(b) Equivalent concentrated forces

Figure 6.8 Replacement of distributed forces by equivalent concentrated forces.

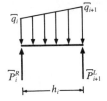

Figure 6.9 Free-body diagram of the imaginary stringer extending between points i and $i+1$.

typical panel of length h_i between points i and $i + 1$ and denote the magnitude of the external force at the left and at the right end of the panel, by $\bar{q}_i$ and $\bar{q}_{i+1}$ respectively. The reactions $\bar{P}_i^R$ and $\bar{P}_{i+1}^L$ of the imaginary stringer extending from point i to point $i + 1$ can be obtained by referring to Fig. 6.9 and considering the equilibrium of the imaginary stringer. That is

$$\Sigma \bar{M}^{(i+1)} = 0 \qquad h_i \bar{P}_i^R = \bar{q}_i h_i \frac{h_i}{2} + (\bar{q}_{i+1} - \bar{q}_i) \frac{h_i}{2} \frac{h_i}{3} \qquad (6.13)$$

or   or $$\bar{P}_i^R = \frac{h_i}{6} (2\bar{q}_i + \bar{q}_{i+1}) \qquad (6.14)$$

$$\Sigma \bar{M}^i = 0 \qquad \bar{P}_{i+1}^L = \frac{h_i}{6} (2\bar{q}_{i+1} + \bar{q}_i) \qquad (6.15)$$

Similarly, the reactions $\bar{P}_i^L$ and $\bar{P}_{i-1}^R$ of the imaginary stringer of length h_{i-1} extending from points $i - 1$ to points i, are

$$\bar{P}_{i-1}^R = \frac{h_{i-1}}{6} (2\bar{q}_{i-1} + \bar{q}_i)$$

$$\bar{P}_i^L = \frac{h_{i-1}}{6} (2\bar{q}_i + \bar{q}_{i-1}) \tag{6.16}$$

Hence, the equivalent concentrated force $\bar{P}_i$ at the typical interior panel point i of the member is equal to

$$\bar{P}_i = \bar{P}_i^L + \bar{P}_i^R = \frac{h_{i-1}}{6} (\bar{q}_{i-1} + 2\bar{q}_i) + \frac{h_i}{6} (2\bar{q}_i + \bar{q}_{i+1}) \qquad (6.17)$$

Usually, the loaded portion of the member is divided into panels of equal length. In this case, relation (6.17) reduces to

$$\bar{P}_i = \frac{h}{6} (\bar{q}_{i-1} + 4\bar{q}_i + \bar{q}_{i+1}) \qquad (6.18)$$

Referring to relations (6.14) and (6.16), it is apparent that for the end panel points 1 and n of the member, we have

$$\overline{P}_1 = \overline{P}_1^R = \frac{h}{6}(2\overline{q}_1 + \overline{q}_2)$$

(6.19)

$$\overline{P}_n = \overline{P}_n^L = \frac{h}{6}(\overline{q}_{n-1} + 2\overline{q}_n)$$

Example 1. Compute the deflection of point C of the beam of Fig. a using the conjugate beam method. Subdivide the distance between points A and C of the conjugate beam into 15 equal spaces by 16 points and replace the load of the conjugate beam by approximately equivalent concentrated forces acting at these points. The beam has a constant cross section.

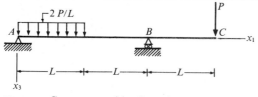

Figure a   Geometry and loading of the beam.

solution   The moment diagram of the beam is shown in Fig. b, while the conjugate beam is shown in Fig. c. We subdivide the distance from point A to C of the conjugate

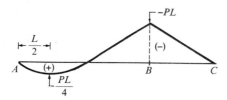

Figure b   Moment diagram of the beam.

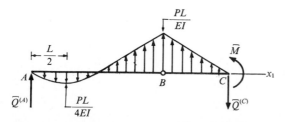

Figure c   Free body diagram of the conjugate beam.

TABLE a   Values of Approximately Equivalent Concentrated Forces and Their Moments about Points _B_ and _C_

Point	$EI\overline{q}/PL$	$EI\overline{P_i}/PL^2$	$EI\overline{P_i}(15 - i)/PL^2$	$EI\overline{P_i}(10 - i)/PL^2$
$0 \equiv A$	0	0.0053	0.0795	0.0530
1	0.16	0.0293	0.4102	0.2637
2	0.24	0.0453	0.5889	0.3624
3	0.24	0.0453	0.5436	0.3171
4	0.16	0.0293	0.3223	0.1758
5	0	−0.0013	−0.0130	−0.0065
6	−0.20	−0.0400	−0.3600	−0.1600
7	−0.40	−0.0800	−0.6400	−0.2400
8	−0.60	−0.1200	−0.8400	−0.2400
9	−0.80	−0.1600	−0.9600	−0.1600
10 = B	−1.00	−0.1867	−0.9335	0
11	−0.80	−0.1600	−0.6400	
12	−0.60	−0.1200	−0.3600	
13	−0.40	−0.0800	−0.1600	
14	−0.20	−0.0400	−0.0400	
15 = C	0.00	−0.0067	0	
Total	. . . . .	. . . .	−3.0020	0.3655

beam with 16 equally spaced points. Consequently, the spacing of these points is equal to $L/5$. We compute the concentrated force which must be placed at each point of the conjugate beam using Eqs. (6.18) and (6.19). The results are shown in Table a.

The free-body diagram of the conjugate beam subjected to the equivalent concentrated forces is shown in Fig. d. From the equilibrium of the conjugate beam of Fig. d and using the results of Table a, we obtain

$$\Sigma \overline{M}^{(B)} = 0 - \sum_{i=1}^{9} \overline{P}_i(10 - i)\left(\frac{L}{5}\right) + 2L\overline{Q}^{(A)}$$

or

$$\overline{Q}^{(A)} = \frac{0.0366 PL^2}{EI} \tag{a}$$

$$\Sigma \overline{M}^{(C)} = 0 = \sum_{i=1}^{15} \overline{P}_i(15 - i)\left(\frac{L}{5}\right) - 3L\overline{Q}^{(A)} + \overline{M}^{(C)} \tag{b}$$

or

$$\overline{M}^{(C)} = \frac{0.7101 PL^3}{EI}$$

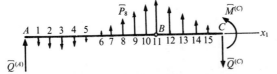

Figure d Free-body diagram of the conjugate beam loaded with approximately equivalent concentrated forces.

Thus
$$u_3^{(C)} = \overline{M}^{(C)} = \frac{0.7101\,PL^3}{EI} \text{ downward} \qquad (c)$$

We may compute the exact value of the deflection u_3^C by considering the equilibrium of the conjugate beam subjected to the distributed load. Referring to Fig. c, we have

$$\Sigma \overline{M}^{(B)} = 0 = 2L\overline{Q}^{(A)} - \frac{PL}{4EI}\left(\frac{2L}{3}\right)\left(\frac{3L}{2}\right) + \frac{PL}{EI}\left(\frac{L}{2}\right)\left(\frac{L}{3}\right) \qquad (d)$$

$$\Sigma \overline{M}^{(C)} = 0 = \overline{M}^{(C)} - 3L\overline{Q}^{(A)} + \frac{PL}{4EI}\left(\frac{2L}{3}\right)\left(\frac{5L}{2}\right) - \frac{PL}{EI}(L)\,L \qquad (e)$$

thus,
$$\overline{Q}^{(A)} = \frac{PL^2}{24EI} = \frac{0.0417\,PL^2}{EI} \qquad (f)$$

$$u_3^{(C)} = \overline{M}^{(c)} = \frac{17PL^3}{24EI} = \frac{0.7083\,PL^3}{EI} \text{ downward} \qquad (g)$$

Comparing the value of the deflections established by relations (c) and (g), it is apparent that the error in the results of the approximate method is about 0.25 percent.

6.3 Problems

1 and 2. Using the conjugate beam method, compute the deflection and the rotation of point 2 of the beam loaded as shown in Fig. 6.P1. The beam is made of steel ($E = 210{,}000$ MPa) and has a constant cross section ($I = 117.7 \times 10^6$ mm^4). Repeat for Fig. 6.P2

3. Using the conjugate beam method, compute the deflection of point 4 of the beam loaded as shown in Fig. 6.P3. The beam is made of steel ($E = 210{,}000$ MPa) and has a constant cross section ($I = 369.7 \times 10^6$ mm^4).

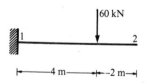

Figure 6.P1

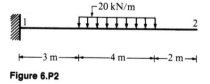

Figure 6.P2

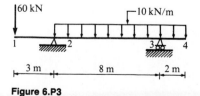

Figure 6.P3

4 and 5. Using the conjugate beam method, compute the deflection at the middle point of the beam loaded as shown in Fig. 6.P4. The beam is made of steel (E = 210,000 MPa) and has a constant cross section (I = 369.7 × 10^6 mm⁴). Repeat for Fig. 6.P5.

Figure 6.P4   Figure 6.P5

6. Using the conjugate beam method, compute the deflection and the rotation of point 2 of the beam loaded as shown in Fig. 6.P6. The beam is made of a standard wide-flange section (E = 210,000 MPa), and it is reinforced at its center portion by cover plates. The moments of inertia of the three portions of the beam are given in Fig. 6.P6 (I = 300 × 10^6 mm⁴).

7. Using the conjugate beam method, compute the deflection of point 4 of the beam loaded as shown in Fig. 6.P7. The members of the beam have the same constant cross section (I = 369.7 × 10^6 mm⁴) and are made of the same material (E = 210,000 MPa).

8. Using the conjugate beam method, compute the deflection of point 5 of the beam loaded as shown in Fig. 6.P8. The members of the beam have the same constant cross section (I = 369.7 × 10^6 mm⁴) and are made of the same material (E = 210,000 MPa).

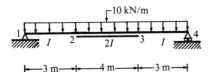

Figure 6.P6

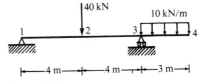

Figure 6.P7

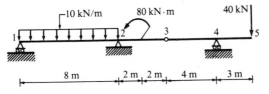

Figure 6.P8

9. Using the conjugate beam method, compute the deflection and the rotation of point 2 of the beam loaded as shown in Fig. 6.P9. The beam is made of a standard wide-flange section (E = 210,000 MPa), and it is reinforced by cover plates. The moments of inertia of the two portions of the beam are given in Fig. 6.P9 (I = 300 $\times$ 10^6 mm^4).

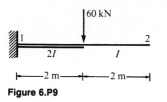

Figure 6.P9

10 and 11. Using the conjugate beam method and discretizing the forces acting on the conjugate beam, establish the deflection and rotation of point 4 of the beam of Fig. 6.P3. Subdivide the conjugate beam into 10 equal intervals. The beam has constant cross section (I = 369.7 $\times$ 10^6 mm^4) and is made of steel (I = 210,000 MPa). Repeat for Fig. 6.P7.

Photographs and Brief Descriptions of Steel Frame Buildings

The Waldorf Astoria Hotel (Fig. 6.10) is an impressive building with a steel skeleton. Tall buildings usually have a steel skeleton which carries the loads acting on them. The floors of buildings have considerable bending stiffness; thus, the rotation of their edges is relatively small. Consequently, the analysis of the skeleton may be simplified considerably by assuming that the rotation of its columns at the floor levels is zero.

Figure 6.10 The Waldorf Astoria Hotel in New York City. (*Courtesy of Bethlehem Steel Corp., Bethlehem, Pa.*)

In the Alcoa Company Tower shown in Fig. 6.11 the diamond-patterned diagonal members have been placed in order to increase the ability of the building to resist earthquakes. The steel frame of this tower will not be covered. It will enclose its external glass walls.

Figure 6.11 Steel skeleton of the tower of Alcoa Co. in San Francisco, Calif. (*Courtesy of Bethlehem Steel Corp., Bethlehem, Pa.*)

The perched 16-story office building shown in Figs. 6.12 and 6.13 towers 300 ft (91 m) above a five-level city parking garage on which it is supported by V-shaped stilts. It was completed in 1981.

Figure 6.12 Office building in Hartford, Conn. (*Courtesy of Bethlehem Steel Corp., Bethlehem, Pa.*)

Figure 6.13 Steel skeleton of the office building in Hartford, Conn. (*Courtesy of Bethlehem Steel Corp., Bethlehem, Pa.*)

The tower in Fig. 6.14 consists of 34 floors and rises 560 ft (161.54 m) above the ground level. The picture was taken when the outside dressing of the steel skeleton was in progress.

The Brendan Byrne Arena (Figs. 6.15 and 6.16) was completed in 1980. Four arched steel box trusses, supported on eight steel-framed towers, allow up to 21,000 spectators an unobstructed view of the arena floor.

Figure 6.14   Steel skeleton of the tower of the Pacific Gas and Electric Co. in San Francisco, Calif. (*Courtesy of Bethlehem Steel Corp., Bethlehem, Pa.*)

Figure 6.15 Steel skeleton of the Brendan Byrne Arena in East Rutherford, N.J. (*Courtesy of Bethlehem Steel Corp., Bethlehem, Pa.*)

Figure 6.16 The Brendan Byrne Arena in East Rutherford, N.J. (*Courtesy of Beth·lehem Steel Corp., Bethlehem, Pa.*)

Analysis of Statically Indeterminate Structures

The Force or Flexibility Methods of Structural Analysis

7.1 Introduction

The reactions and internal actions of statically determinate structures can be determined from the given loads using the equations of equilibrium alone. However, as discussed in Sec. 1.11, the reactions and/or internal actions of statically indeterminate structures cannot be determined using only the equations of equilibrium. Additional equations are required which are established by imposing the requirement that the deformed configuration of the structure be continuous and compatible with the constraints imposed by its supports.

The methods for analyzing statically indeterminate structures are grouped into two categories: *the force or flexibility methods and the displacement or stiffness methods.* In the force methods, reactions and/or internal actions of the structure are chosen as the unknowns. Moreover, the equations from which the unknowns are established represent conditions of continuity of components of displacements. In the displacement methods, the displacements of the joints of the structure are chosen as the unknowns. Furthermore, the equations from which the unknowns are established represent equations of equilibrium of the nodes of the structure.

Once the internal actions in the members of a structure (statically determinate or indeterminate) are established, the displacements of its joints can then be found using, for instance, the method of virtual work (Chap. 5). Moreover, as becomes apparent in Chap. 8, if the components of displacements of the joints of a structure are known, the internal actions in its members can be established.

This part is subdivided into three chapters. In Chap. 7, the classical (systems) approach to the basic force or flexibility method, also known as the *method of consistent deformation,* and a number of other versions of the force or flexibility method are presented. They are applied to simple planar and space structures. These methods have been employed extensively in analyzing simple

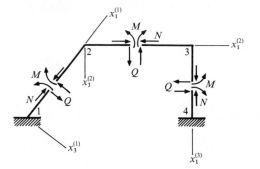

Figure 7.1 Positive internal actions in the members of a planar frame.

structures by hand or with a desk calculator. However, they are not suitable for programming the analysis of a group of structures on a computer.

In the basic force method, pertinent relations are presented in matrix form resulting in a more compact and systematic presentation which enables the reader to concentrate on the basic concepts without being distracted by algebraic details. Moreover, this presentation permits the use of existing subroutines for performing the required matrix operations on a digital computer. However, the pertinent matrixes are generated by hand.

In Chap. 8, two classical displacement methods for analyzing statically indeterminate planar beams and frames are presented—the slope deflection method and the displacement method with moment distribution. These methods are suitable for analyzing planar beams and frames by hand or with a desk calculator when the effect of axial deformation of their members on their internal actions is negligible.

In Chap. 9, the Müller-Breslau principle is applied in plotting the influence lines for statically indeterminate structures.

In this chapter, the sign convention specified in Fig. 1.33a is employed for the internal actions of the members of a structure. Consequently, an axial force inducing a tensile stress distribution in the fibers of the member on which it acts is considered as positive. Moreover, the local axis x_2 of the members of planar structures is chosen normal to their plane and directed toward the viewer. Thus, the shearing forces and bending moments shown in Fig. 7.1 are considered positive.

As mentioned in Sec. 5.13, the components of displacement of a point of a framed structure due to the shearing deformation of its members are of a smaller order of magnitude than the corresponding components of displacement of this point due to the bending deformation of the members of the structure and, consequently, are disregarded in this part.

7.2   The Basic Force or Flexibility Method

In this section, the systems approach to the basic force or flexibility method is described. This is a very general method, and it can be applied in analyzing any statically indeterminate structure.

Consider the beam of Fig. 7.2a. It is apparent that if its support at point 1 is removed, the resulting structure is not a mechanism (it cannot move without deforming) and, moreover, it is statically determinate; that is, its reactions can be computed using only the equations of equilibrium. It is apparent that by removing the support of the beam at point 1, one of its unknown reactions has been eliminated. This reaction is in excess of the minimum number of reactions required in order that the structure not be a mechanism. Thus the beam of Fig. 7.2a is statically indeterminate to the first degree. Consequently, its reactions cannot be computed by considering only its equilibrium. For instance, referring to Fig. 7.2b, from the equilibrium of the beam we obtain

$$\Sigma \bar{F}_h = 0 \qquad R_h^{(2)} = 0$$

$$\Sigma \bar{F}_v = 0 \qquad R_v^{(2)} = 40 - R_v^{(1)} \qquad (7.1)$$

$$\Sigma \bar{M}^{(2)} = 0 \qquad M^{(2)} = 5 R_v^{(1)} - 80$$

It is apparent that for any assumed value of the reaction $R_v^{(1)}$, a set of values can be found for the reactions, $R_v^{(2)}$ and $M^{(2)}$, which satisfy the equations of equilibrium (7.1). As mentioned in Sec. 5.5, any set of values of the reactions of a structure which satisfy the equations of equilibrium is referred to as *statically admissible*. Moreover, the distribution of the internal actions corresponding to a statically admissible set of reactions is referred to as *statically admissible*.

In order to establish the reactions of the beam, we must establish an additional equation by considering its deformed configuration. The equations of the elastic curve of the beam may be obtained by integrating its moment-displacement relation, which referring to relation (A.74b) is

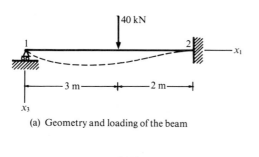

(a) Geometry and loading of the beam

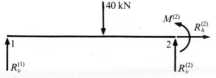

(b) Free-body diagram of the beam

Figure 7.2 Statically indeterminate beam.

$$EI\frac{d^2u_v}{dx_1^2} = -M \tag{7.2}$$

The component of translation (deflection) u_v of the beam is considered positive when its sense is that of the positive axis x_3. In general, the elastic curve of the beam obtained by using a statically admissible distribution of the internal moment and integrating relation (7.2) does not satisfy all the conditions imposed by the supports of the beam and by the requirements for continuity of the beam. This may be illustrated by determining the expressions for the moment of the beam, referring to Fig. 7.2b, substituting them into relation (7.2) and integrating the resulting relations. Thus,

$0 \leq x_1 \leq 3$ $\qquad\qquad\qquad\qquad$ $3 \leq x_1 \leq 5$

$$EI\frac{d^2u_v}{dx_1^2} = -R_v^{(1)}x_1 \qquad\qquad EI\frac{d^2u_v}{dx_1'^2} = -R_v^{(1)}x_1 + 40(x_1 - 3)$$

$$EI\frac{du_v}{dx_1} = -\frac{R_v^{(1)}x_1^2}{2} + C_1 \qquad EI\frac{du_v}{dx_1} = -\frac{R_v^{(1)}x_1^2}{2} + \frac{40(x_1 - 3)^2}{2} + C_2$$

$$EIu_v = -\frac{R_v^{(1)}x_1^3}{6} + C_1x_1 + C_3 \qquad EIu_v = -\frac{R_v^{(1)}x_1^3}{6} + \frac{40(x_1 - 3)^3}{6}$$

$$+ C_2x_1 + C_4 \tag{7.3}$$

Since the beam is continuous at point 2, it is necessary that the slope and the deflection at this point ($x_1 = 3$ m) obtained from the above two sets of equations be identical. Hence, we obtain

$$C_1 = C_2 \qquad C_3 = C_4 \tag{7.4}$$

Moreover, the conditions at support 2 (at $x_1 = 5$, $du_v/dx_1 = u_v = 0$) impose the following relations.

$$C_1 = C_2 = -80 + 12.5R_v^{(1)}$$
$$C_3 = C_4 = 346.67 - 41.667R_v^{(1)} \tag{7.5}$$

Using the above values of the constants in relations (7.3), we obtain the following equations for the elastic curve of the beam of Fig. 7.2a.

$0 \leq x_1 \leq 3$

$$EIu_v = -\frac{R_v^{(1)}x_1^3}{6} - 60x_1 + 12.5R_v^{(1)}x_1 + 346.67 - 41.667R_v^{(1)}$$

$3 \leq x_1 \leq 5$

$$EIu_v = -\frac{R_v^{(1)}x_1^3}{6} + \frac{40(x_1 - 3)^3}{6} - 60x_1 + 12.5R_v^{(1)}x_1$$

$$+ 346.67 - 41.667R_v^{(1)} \tag{7.6}$$

The elastic curve specified by Eqs. (7.6) is continuous and, moreover, satisfies the requirements imposed by the support of the beam at point 2. It is apparent that in order for the elastic curve also to satisfy the condition at support 1, the reaction $R_v^{(1)}$ must assume the value required in order that the displacement at point 1 be equal to zero. That is, referring to the first of Eqs. (7.6), we have

$$u_v(0) = 0 \qquad R_v^{(1)} = \frac{346.67}{41.667} = 8.32 \text{ kN} \tag{7.7}$$

Substituting the above result in relations (7.1), we obtain

$$R_v^{(2)} = 40 - 8.32 = 31.68 \text{ kN}$$

$$M^{(2)} = 5(8.32) - 80 = 38.40 \text{ kN} \cdot \text{m}$$

On the basis of the foregoing discussion, it is evident that the requirements of continuity of the beam and the conditions at its supports are satisfied only by one set of statically admissible reactions which is the actual set of reactions of the structure.

7.2.1 Analysis of statically indeterminate structures to the first degree subjected to external actions

In this section, we limit our attention to statically indeterminate structures to the first degree, subjected only to external actions so as to emphasize the basic steps involved in the analysis of statically indeterminate structures using the basic force or flexibility method.

Let us consider the beam of Fig. 7.3a, and let us choose one of the reactions $R_v^{(1)}$, $R_v^{(2)}$, $M^{(2)}$, or an internal action (say the moment at $x_1 = 3$ m) as *the redundant*. The statically determinate beam resulting from the actual beam by removing the constraint which induces the chosen redundant is referred to as the *primary structure*. If the reaction $R_v^{(1)}$ is chosen as the redundant, the primary structure is the cantilever beam shown in Fig. 7.3b. If the moment $M^{(2)}$ is chosen as the redundant, the primary structure is the simply supported beam shown in Fig. 7.3c. Moreover, if the moment at $x_1 = 3$ m is chosen as the redundant, the primary structure is the beam of Fig. 7.3c. Notice that the reactions of a primary structure represent a set of statically admissible reactions of the actual structure corresponding to a value of zero for the chosen redundant.

Referring to Fig. 7.4, we can readily see that the internal forces, bending

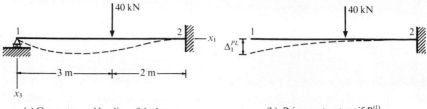

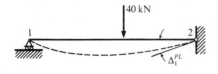

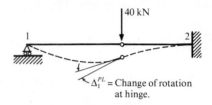

(a) Geometry and loading of the beam

(b) Primary structure if $R_v^{(1)}$ is chosen as the redundant

(c) Primary structure if $M^{(2)}$ is chosen as the redundant

(d) Primary structure if the moment at $x_1 = 3$ m is chosen as the redundant

Figure 7.3   Primary structures.

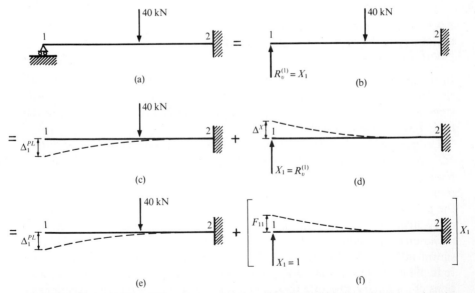

Figure 7.4   Superposition of the primary structure subjected to the given loading and to the redundant.

moments, and the translation (deflection) at any point of the beam can be obtained by superimposing the corresponding quantities of the primary structure subjected to the given loading and those of the primary structure subjected only to the redundant. Thus, for the beam of Fig. 7.4a, we may write that the translation (deflection) $\Delta_1^{(s)}$ at point 1 of the actual structure whose magnitude we know (it is zero) is equal to the sum of the translation (deflection) Δ_1^{PL} of point 1 of the primary structure, subjected to the given loading, and the deflection Δ_1^{PX} of point 1 of the primary structure subjected to the redundant. Hence

$$\Delta_1^{(s)} = 0 = \Delta_1^{PL} + \Delta_1^{PX} \tag{7.8}$$

Moreover, as shown in Fig. 7.4f, the internal actions and the deflections of the primary structure subjected to the redundant are equal to those of the primary structure subjected to a unit value of the redundant multiplied by the value of the unknown redundant. Thus, relation (7.8) can be rewritten as

$$\Delta_1^{(s)} = 0 = \Delta_1^{PL} + X_1 F_{11} \tag{7.9}$$

where F_{11} is the deflection of point 1 of the primary structure subjected to $X_1 = 1$. It is called the flexibility coefficient corresponding to the redundant X_1. Equation (7.9) is called *the compatibility equation* of the beam. The deflections Δ_1^{PL} and F_{11} of point 1 of the primary structure can be computed using the methods presented in Chaps. 5 or 6. Using the values of these deflections, we obtain the redundant X_1 from relation (7.9).

An internal action A in the structure is equal to the sum of the corresponding internal actions A^{PL} in the primary structure, subjected to the actual loading, and the corresponding internal action of the primary structure, subjected to the redundant. That is,

$$A = A^{PL} + A^{PX} X_1 \tag{7.10}$$

where A^{PX} represents the internal action in the primary structure subjected to $X_1 = 1$ corresponding to A. Thus, when the redundant is computed, any internal action in the structure can be determined using relation (7.10).

On the basis of the foregoing discussion, in order to analyze a statically indeterminate structure to the first degree subjected to external actions, we adhere to the following procedure.

1. We select the redundant, and we form the primary structure by removing the constraint which induces the redundant.

2. We compute the displacement Δ_1^{PL} of the primary structure subjected to the given loading, using the method of virtual work (5.66).

3. We compute the displacement F_{11} of the primary structure subjected to $X_1 = 1$ using the method of virtual work (5.66).

4. We compute the redundant using the compatibility equation (7.9).

5. We compute the internal actions in the structure either by using relation (7.10) or by considering the equilibrium of appropriate parts of the structure.

In the sequel, the basic force method is applied to the following four examples.

Example 1. We establish the internal forces in the members of a truss that is internally statically indeterminate to the first degree. The truss is subjected to an external force.

Example 2. We establish the internal actions in the members of a beam that is statically indeterminate to the first degree. One of the supports of the beam is elastic. The beam is subjected to external forces. The shear and moment diagrams for the beam are plotted.

Example 3. We establish the internal actions in the members of a frame that is statistically indeterminate to the first degree. The frame consists of two vertical columns and a horizontal truss, and it is subjected to external forces. The shear and moment diagrams for the columns are plotted.

Example 4. We establish the internal actions in the members of a beam of variable cross section that is statically indeterminate to the first degree. The beam is subjected to external actions. The shear and moment diagrams are plotted.

Example 1. Compute the internal forces $N^{(i)}$ in the members of the statically indeterminate truss shown in Fig. a. The members of the truss have the same constant cross section and are made from the same material.

Notice that the internal forces in the members of the truss will be the same whether or not members 3 and 6 are joined together at point 5. This becomes apparent by assuming that members 3 and 6 are joined at point 5 and considering the equilibrium of forces at joint 5. The force in member 1,5 must be equal to that of member 5,3 while the force in member 2,5 must be equal to that in member 5,4.

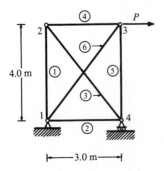

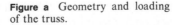

Figure a Geometry and loading of the truss.

solution

STEP 1. The truss under consideration is statically indeterminate to the first degree. We choose the force in member 6 as the redundant ($N^{(6)} = X_1$) and form the primary structure by cutting member 6.

STEP 2. We compute the relative movement Δ_1^{PL} of the ends of the cut of member 6 of the primary structure subjected to the given loading (see Fig. b). For this purpose the following quantities must be computed.

1. The internal forces N^{PLi} ($i = 1, 2, \ldots, 6$) in the members of the primary structure subjected to the given loading (see Fig. b).
2. The internal forces $\tilde{N}^{(i)}$ ($i = 1, 2, \ldots, 6$) in the members of the primary structure subjected to a pair of equal and opposite unit forces $X_1 = 1$ (see Fig. c).

The results are given in Table a.

Using relation (5.67) and, referring to Table a, we have

$$\Delta_1^{PL} = \frac{1}{EA} \sum_{i=1}^{6} [N^{PLi} \tilde{N}^{(i)} L_i] = -\frac{16.2P}{EA}$$

STEP 3. We compute the flexibility coefficient F_{11} of the truss corresponding to the chosen redundant. The flexibility coefficient is equal to the relative movement of the cut ends of member 6 of the primary structure subjected to $X_1 = 1$ (see Fig. c). Using relation (5.67) and referring to the Table a, we have

$$F_{11} = \frac{1}{EA} \sum_{i=1}^{6} [\tilde{N}^{(i)} \tilde{N}^{(i)} L_i] = \frac{17.28}{EA}$$

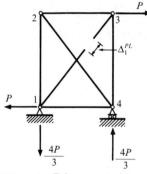

Figure b Primary structure subjected to given loading.

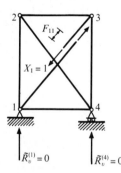

Figure c Primary structure subjected to a pair of unit forces.

TABLE a Computation of the Internal Forces in the Members of the Truss

Member	L_i	N^{PLi}	$\tilde{N}^{(i)}$	$N^{PLi}\tilde{N}^{(i)}L_i$	$\tilde{N}^{(i)}\tilde{N}^{(i)}L_i$	$N^{(i)} = N^{PLi} + \tilde{N}^{(i)}X_1$
1	4.0	$1.3333P$	$_s0.8$	$-4.2667P$	2.56	$0.5833P$
2	3.0	$1.0000P$	$_s0.6$	$-1.8000P$	1.08	$0.4375P$
3	5.0	$-1.6667P$	1.0	$-8.3333P$	5.00	$-0.7292P$
4	3.0	$1.0000P$	$_s0.6$	$-1.8000P$	1.08	$0.4375P$
5	4.0	0	$_s0.8$	0	2.56	$-0.7500P$
6	5.0	0	1.0	0	5.00	$0.9375P$
Total				$-16.2P$	17.28	

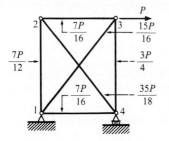

Figure d Internal forces in the members of the truss subjected to the given force.

STEP 4.   We compute the redundant by superimposing the results of steps 2 and 3. Since in the actual structure, the relative movement of the cut ends of member 6 is zero ($\Delta_1^s = 0$), from relation (7.9) we get

$$X_1 = -\frac{\Delta_1^{PL}}{F_{11}} = \frac{16.20P}{17.28} = 0.9375P$$

STEP 5.   We compute the internal forces in the members of the truss using relation (7.10). That is

$$N^{(i)} = N^{PLi} + \tilde{N}^{(i)}X_1$$

The results are tabulated in Table a and in Fig. d.

Example 2.   Support 2 of the continuous beam of Fig. a is elastic with a flexibility coefficient $f = L^3/24EI$ m/kN, where E is the modulus of elasticity (in kilonewtons per square meter) of the material from which the beam is made and I is the moment of inertia of the cross section of the beam. Plot the shear and moment diagrams for the beam subjected to the external actions shown in Fig. a.

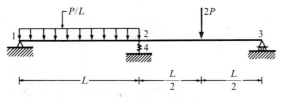

Figure a   Geometry and loading of the beam.

solution

STEP 1.   This is a statically indeterminate structure to the first degree. We choose the reaction at support 4 as the redundant X_1 and form the primary structure by disconnecting the spring from support 4.

STEP 2.   We compute the vertical displacement Δ_1^{PL} of the disconnected end of the spring of the primary structure subjected to the given loading (see Fig. b). For this purpose, the following quantities must be computed.

1. The internal moments in the primary structure subjected to the given loading
2. The internal moments in the primary structure subjected to $X_1 = 1$

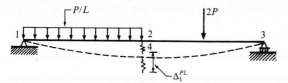

Figure b   Primary structure subjected to the given loading.

The free-body diagram of the primary structure subjected to the given external actions is shown in Fig. c. Referring to this figure, we see that the distribution of the internal moment in the primary structure is

$$M = \frac{5Px_1}{4} - \frac{Px_1^2}{2L} \qquad\qquad \text{for } 0 \le x_1 \le L$$

$$M = \frac{7Px_1'}{4} \qquad\qquad \text{for } 0 \le x_1' \le \frac{L}{2} \qquad\qquad \text{(a)}$$

$$M = \frac{7Px_1'}{4} - 2P\left(x_1' - \frac{L}{2}\right) \qquad \text{for } \frac{L}{2} \le x_1' \le L$$

The primary structure subjected to $X_1 = 1$ is shown in Fig. d. The displacement of point 4 of this structure is equal to the sum of the displacement of point 2 of the beam and the deformation of the spring. Referring to Fig. d, we see that the internal moment in the primary structure subjected to a unit value of the redundant is equal to

$$M = -\frac{x_1}{2} \qquad \text{for } 0 \le x_1 \le L$$

$$\qquad\qquad\qquad\qquad\qquad\qquad\qquad\qquad\qquad\qquad\qquad \text{(b)}$$

$$M = -\frac{x_1'}{2} \qquad \text{for } 0 \le x_1' \le L$$

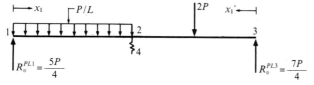

Figure c   Free-body diagram of the primary structure subjected to the given loading.

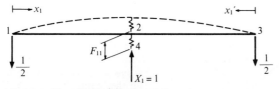

Figure d   Free-body diagram of the primary structure subjected to a unit force.

Substituting relations (a) and (b) into (5.66) and integrating, we have

$$\Delta_1^{PL} = -\frac{1}{EI}\int_0^L \frac{x_1}{2}\left(\frac{5Px_1}{4} - \frac{Px_1^2}{2L}\right)dx_1 - \frac{1}{EI}\int_0^{L/2}\frac{x_1'}{2}\frac{7Px_1'}{4}dx_1'$$

$$- \frac{1}{EI}\int_{L/2}^L \frac{x_1'}{2}\left[\frac{7Px_1'}{4} - 2P\left(x_1' - \frac{L}{2}\right)\right]dx_1' = -\frac{PL^3}{3EI} \qquad (c)$$

STEP 3.   We compute the flexibility coefficient F_{11} of the primary structure. As shown in Fig. d, the flexibility coefficient F_{11} is equal to the vertical displacement of the disconnected end of the spring of the primary structure subjected to $X_1 = 1$. Substituting relations (b) into (5.66) and integrating, we get

$$F_{11} = \frac{2}{EI}\int_0^L \left(\frac{x_1}{2}\right)^2 dx_1 + f = \frac{L^3}{6EI} + \frac{L^3}{24EI} = \frac{5L^3}{24EI} \qquad (d)$$

STEP 4.   We compute the redundant X_1 by superimposing the results of steps 2 and 3. Since the cut end of the spring of the actual beam does not move ($\Delta_1^s = 0$), substituting relations (c) and (d) into (7.9), we get

$$X_1 = -\frac{\Delta_1^{PL}}{F_{11}} = \frac{PL^3/3EI}{5L^3/24EI} = \frac{8P}{5}$$

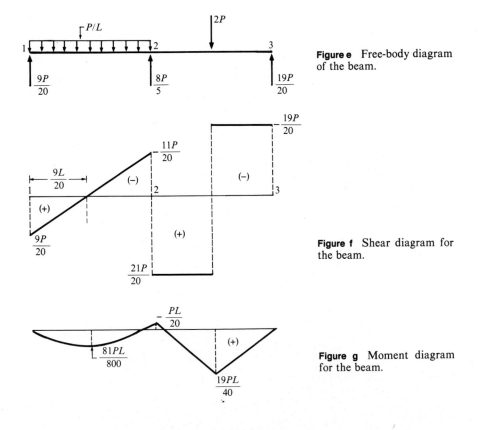

Figure e Free-body diagram of the beam.

Figure f Shear diagram for the beam.

Figure g Moment diagram for the beam.

STEP 5.   We compute the reactions at supports 1 and 3 using relation (7.10). Thus, referring to Fig. c and using the value of X_1 established in step 4 we obtain

$$R_v^{(1)} = R_v^{PL1} - \frac{1}{2} X_1 = \frac{5P}{4} - \frac{4P}{5} = \frac{9P}{20}$$

$$R_v^{(3)} = R_v^{PL3} - \frac{1}{2} X_1 = \frac{7P}{4} - \frac{4P}{5} = \frac{19P}{20}$$

The results are shown in Fig. e. The shear and moment diagrams are plotted in Figs. f and g, respectively.

Example 3. The ratio of the moment of inertia of the columns of the structure of Fig. a to the area of the truss members is $I/A = KL_p^2$. All the members of the structure are made from the same material. Compute the reactions of the structure and the internal forces in the members of the truss when $K = 1$. Disregard the effect of the axial deformation of the columns.

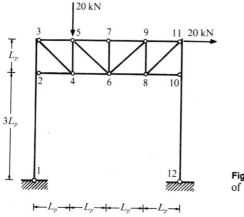

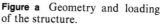

Figure a Geometry and loading of the structure.

solution

STEP 1.   This is a statically indeterminate structure of the first degree. We choose the axial force in member 7,9 of the truss as the redundant.

STEP 2.   We compute the relative movement Δ_1^{PL} of the ends of the cut of member 7,9 of the primary structure subjected to the given forces (see Fig. b).

Referring to Fig. b, we see that the vertical reactions at the supports of the structure may be obtained as follows.

$$\Sigma \overline{M}^{(12)} = 0 = R_v^{(1)}(4L_p) - 20(3L_p) + 20(4L_p)$$

$$\Sigma \overline{M}^{(1)} = 0 = R_v^{(12)}(4L_p) - 20(L_p) - 20(4L_p)$$

or

$$R_v^{(1)} = -5 \text{ kN} = 5 \text{ kN downward}$$

$$R_v^{(12)} = 25 \text{ kN upward}$$

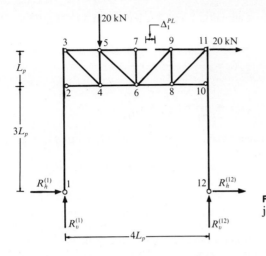
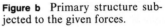

Figure b Primary structure subjected to the given forces.

$$\Sigma \bar{F}_v = 0 \qquad R_v^{(1)} + R_v^{(12)} = 20 \text{ kN} \qquad check$$

Moreover $\qquad \Sigma \bar{F}_h = 0 \qquad R_h^{(1)} + R_h^{(12)} + 20 = 0 \qquad$ (a)

Referring to the free-body diagram of the left portion of the structure shown in Fig. c, we have

$$\Sigma \overline{M}^{(6)} = 0 = 5(2L_p) + R_h^{(1)}(3L_p) + 20L_p$$

$$\Sigma \bar{F}_h = 0 = R_h^{(1)} + H^{(6)} \qquad \Sigma \bar{F}_v = 0 = 5 + 20 - V^{(6)}$$

or $\qquad R_h^{(1)} = -10 \text{ kN} = 10 \text{ kN to the left}$

$$H^{(6)} = 10 \text{ kN} \qquad V^{(6)} = 25 \text{ kN}$$

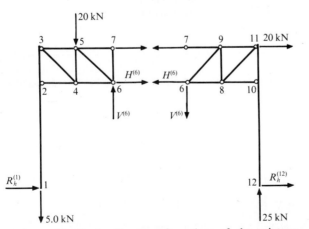

Figure c Free-body diagram of portions of the primary structure.

Using the value of $R_h^{(1)}$ obtained above in relation (a), we get

$$R_h^{(12)} = -10 \text{ kN or } 10 \text{ kN to the left}$$

The internal actions in the members of the truss may be established from the equilibrium of its joints. The results are shown in Fig. d. Referring to this figure, we see that the moment in the columns is

Column 1,2,3

$$0 \leq x_1 \leq 3L_p \qquad M = 10x_1$$

$$3L_p \leq x_1 \leq 4L_p \qquad M = 10x_1 - 40(x_1 - 3L_p) \tag{b}$$

$$= -30x_1 + 120L_p$$

Column 12,10,11

$$0 \leq x_1 \leq 3L_p \qquad M = -10x_1$$

$$3L_p \leq x_1 \leq 4L_p \qquad M = -10x_1 + 40(x_1 - 3L_p) \tag{c}$$

$$= 30x_1 - 120L_p$$

The free-body diagram of the primary structure subjected to $X_1 = 1$ is shown in Fig. e. In this figure the internal forces acting in the members of the truss are also shown. Referring to this figure, the moment in the columns is

$$0 \leq x_1 \leq 3L_p \qquad \tilde{M} = -\frac{x_1}{3}$$

$$3L_p \leq x_1 \leq 4L_p \qquad \tilde{M} = -\frac{x_1}{3} + \frac{4}{3}(x_1 - 3L_p) = x_1 - 4L_p \tag{d}$$

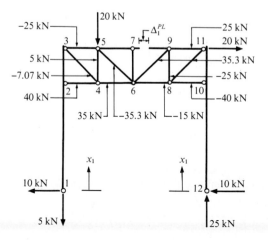

Figure d Internal forces and reactions of the primary structure.

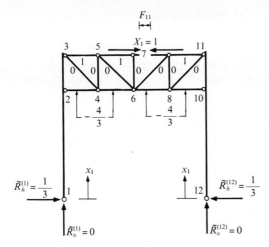

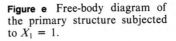

Figure e Free-body diagram of the primary structure subjected to $X_1 = 1$.

Referring to Fig. d and e and substituting relations (b), (c), and (d) into (5.66), we obtain

$$\Delta_1^{PL} = \frac{1}{EI}\left\{-\int_0^{3L_p} 10x_1 \frac{x_1}{3} dx_1 + \int_{3L_p}^{4L_p}(-30x_1 + 120L_p)(x_1 - 4L_p)\,dx_1\right.$$

$$-\int_0^{3L_p} -10x_1 \frac{x_1}{3} dx_1 + \left.\int_{3L_p}^{4L_p}(30x_1 - 120L_p)(x_1 - 4L_p)\,dx_1\right\}$$

$$+ \frac{L_p}{AE}\left[40\left(-\frac{4}{3}\right) + (-25)(1) + 35\left(-\frac{4}{3}\right) + (-15)\left(-\frac{4}{3}\right) - 40\left(-\frac{4}{3}\right)\right.$$

$$+ (25)(1)\bigg] = -\frac{80L_p}{3AE} = -26.67K\frac{L_p^3}{EI} \qquad (e)$$

STEP 3. We compute the flexibility coefficient F_{11} of the primary structure corresponding to the chosen redundant. As shown in Fig. e, F_{11} is equal to the relative movement of the ends of the cut member 7,9 of the primary structure subjected to $X_1 = 1$. Referring to Fig. e and to relations (d) and using relation (5.66), we get

$$F_{11} = \frac{2}{EI}\left[\int_0^{3L_p}\left(\frac{x_1}{3}\right)^2 dx_1 + \int_{3L_p}^{4L_p}(x_1 - 4L_p)^2\,dx_1\right] + \frac{L_p}{AE}\left[4(1)^2 + 4\left(\frac{4}{3}\right)^2\right]$$

$$= 2.66\frac{L_p^3}{EI} + \frac{11.11L_p}{AE} = (2.66 + 11.11K)\frac{L_p^3}{EI} \qquad (f)$$

STEP 4. We compute the redundant by superimposing the results of steps 2 and 3. Substituting relations (e) and (f) into the compatibility relation (7.9), we have

$$\Delta_1^{PL} + F_{11}X_1 = -26.67K\frac{L_p^3}{EI} + (2.66 + 11.11K)\frac{L_p^3}{EI}X_1 = 0$$

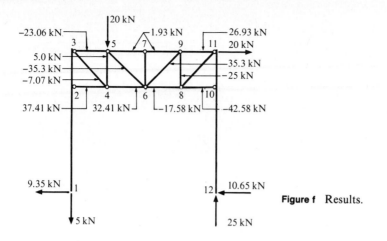

Figure f Results.

Thus
$$X_1 = \frac{26.67K}{2.66 + 11.11K}$$

STEP 5. We compute the reactions of the structure and the internal actions in the members of the truss using relation (7.10). That is

$$A = A_{\text{Fig. d}} + (A_{\text{Fig. e}})X_1$$

The results for $K = 1$ are shown in Fig. f.

Example 4. Plot the shear and moment diagrams of the beam shown in Fig. a. The cross sections of the beam are rectangular of constant width b.

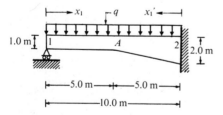

Figure a Geometry and loading of the beam.

solution

STEP 1. This is a statically indeterminate structure to the first degree. We choose the moment at point 2 as the redundant. Thus, the primary structure is a simply supported beam.

STEPS 2 AND 3. We compute the rotation Δ_1^{PL} of end 2 of the primary structure subjected to the given loads (see Fig. b) and the flexibility coefficient F_{11} corresponding to the chosen redundant (see Fig. c). Referring to Fig. b, we see that the moment in the primary structure subjected to the given load is equal to

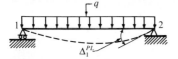

Figure b Primary structure subjected to the given load.

Figure c Primary structure subjected to $X_1 = 1$.

$$M = \frac{qLx_1}{2} - \frac{qx_1^2}{2} = \frac{qLx_1'}{2} - \frac{q(x_1')^2}{2} = 5qx_1' - \frac{q(x_1')^2}{2} \tag{a}$$

Referring to Fig. c, the moment in the primary structure subjected to a unit value of the redundant is equal to

$$\tilde{M} = \frac{x_1}{10} = \frac{10 - x_1'}{10} \tag{b}$$

The moments of inertia of the cross sections of the beam are

$$I_{1-A} = \frac{b}{12} = I \qquad\qquad 0 \le x_1 \le 5 \tag{c}$$

$$I_{A-2} = \frac{b}{12}\left[1 + \left(\frac{5 - x_1'}{5}\right)\right]^3 = \frac{b}{12}\left(2 - \frac{x_1'}{5}\right)^3 = I\rho \qquad 0 \le x_1' \le 5$$

where

$$\rho = \left(2 - \frac{x_1'}{5}\right)^3$$

The displacement Δ_1^{PL} and the flexibility coefficient F_{11} are established by substituting relations (a), (b), and (c) into relation (5.66). Thus

$$\Delta_1^{PL} = \int_0^5 \frac{M\tilde{M}}{EI}\,dx_1 + \int_0^5 \frac{M\tilde{M}}{EI\rho}\,dx_1'$$

$$= \frac{1}{EI}\left[\int_0^5 \left(5qx_1 - \frac{qx_1^2}{2}\right)\left(\frac{x_1}{10}\right)dx_1 + \int_0^5 \frac{M\tilde{M}}{\rho}\,dx_1'\right] \tag{d}$$

$$= \frac{1}{EI}\left[\left[\frac{qx_1^3}{6} - \frac{qx_1^4}{80}\right]\Bigg|_0^5 + \int_0^5 \frac{M\tilde{M}}{\rho}\,dx_1'\right] = \frac{1}{EI}\left[13.021q + \int_0^5 \frac{M\tilde{M}}{\rho}\,dx_1'\right]$$

$$F_{11} = \int_0^5 \frac{\tilde{M}^2}{EI}\,dx_1 + \int_0^5 \frac{\tilde{M}^2}{EI\rho}\,dx_1' = \frac{1}{EI}\left[\int_0^5 \frac{x_1^2}{100}\,dx_1 + \int_0^5 \frac{\tilde{M}^2}{\rho}\,dx_1'\right]$$

$$= \frac{1}{EI}\left[\frac{x_1^3}{300}\Bigg|_0^5 + \int_0^5 \frac{\tilde{M}^2}{\rho}\,dx_1'\right] = \frac{1}{EI}\left[0.416 + \int_0^5 \frac{\tilde{M}^2}{\rho}\,dx_1'\right] \tag{e}$$

We evaluate numerically the integrals in relations (d) and (e) by using Simpson's rule. We subdivide the interval from point A to 2 into eight equal segments (see Fig. d). Referring to relation (5.81), we may rewrite relations (d) and (e) as

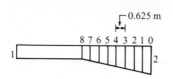

Figure d Beam subdivided into segments.

$$\Delta_I^{PL} = \frac{1}{EI} \left[13.021q + \frac{h}{3} (f_0 + 4f_1 + 2f_2 + 4f_3 + 2f_4 \right.$$

$$\left. + 4f_5 + 2f_6 + 4f_7 + f_8) \right] \tag{f}$$

$$F_{11} = \frac{1}{EI} \left[0.416 + \frac{h}{3} (F_0 + 4F_1 + 2F_2 + 4F_3 + 2F_4 \right.$$

$$\left. + 4F_5 + 2F_6 + 4F_7 + F_8) \right]$$

where

$$f(x_1') = \frac{M\tilde{M}}{\rho} \quad \text{and} \quad F(x_1') = \frac{(\tilde{M})^2}{\rho} \tag{g}$$

Substituting the values of f_i and F_i given in Table a into relations (f) and noting that the length h of each segment is equal to 0.625 m, we see that the integrals in relations (d) and (e) are equal to

$$\int_0^5 f(x_1') \, dx_1' = \frac{0.625}{3} [6.25 + 4(9.57) + 2(6.725)] q = 12.079q$$

$$\int_0^5 F(x_1') \, dx_1' = \frac{0.625}{3} [0.375 + 4(0.69) + 2(0.51)] = 0.865 \tag{h}$$

TABLE a   Evaluation of the Integrals in Relations (d) and (e)

Point	ρ	M	$\tilde{M}$	f_0, f_8	f_i	f_i	F_0, F_8	F_i	F_i
0	8.00	0	1.000	0			0.125		
1	6.59	2.93q	0.937		0.416q			0.133	
2	5.36	5.47q	0.875			0.893q			0.143
3	4.29	7.61q	0.812		1.44q			0.153	
4	3.375	9.37q	0.750			2.082q			0.167
5	2.60	10.74q	0.687		2.84q			0.182	
6	1.95	11.72q	0.625			3.750q			0.200
7	1.42	12.30q	0.562		4.87q			0.222	
8	1.00	12.50q	0.500	6.25q			0.250		
Total	....	.....	....	6.25q	9.57q	6.725q	0.375	0.690	0.510

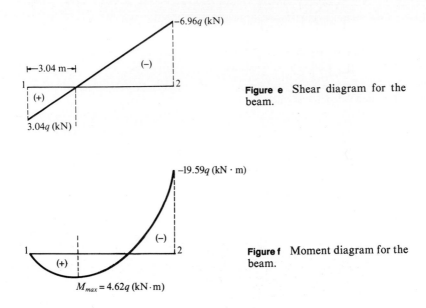

Figure e Shear diagram for the beam.

Figure f Moment diagram for the beam.

Substituting relations (h) into (d) and (e), we get

$$\Delta_1^{PL} = \frac{1}{EI}[13.021q + 12.079q] = \frac{25.10q}{EI} \tag{i}$$

$$F_{11} = \frac{1}{EI}(0.416 + 0.865) = \frac{1.281}{EI} \tag{j}$$

STEP 4.   We compute the redundant X_1 by substituting results (i) and (j) into the compatibility equation (7.9). Thus,

$$X_1 = -\frac{\Delta_1^{PL}}{F_{11}} = \frac{25.10q}{1.281} = -19.59q \tag{k}$$

STEP 5.   The shear and moment diagrams are plotted in Figs. e and f, respectively.

7.2.2 Analysis of statically indeterminate structures subjected to a given loading

Consider a statically indeterminate structure to the nth degree, subjected to a given loading. This structure has n reactions and/or internal actions in excess of those which can be established by considering the equilibrium of its parts. The loading could consist of external actions, change of temperature, initial stress, and movement of its supports. When analyzing this structure using the basic force or flexibility method, we adhere to the following steps.

Step 1. We choose n reactions and/or internal actions as the unknown quantities. We term these reactions and/or internal actions the *redundants* of the structure and denote them by X_i ($i = 1, 2, \ldots, n$). We form an auxiliary, statically determinate structure by removing the constraints of the actual structure which induce the chosen redundants. This auxiliary structure is referred to as the *primary structure*. The choice of the redundants must be such that the primary structure is not a mechanism (see Sec. 1.10.3). For instance, let us consider the continuous beam shown in Fig. 7.5a, which is statically indeterminate to the second degree. Two of the most convenient choices of redundants are shown in Fig. 7.5b. At the left-hand side of Fig. 7.5b, the reactions at supports 2 and 3 of the beam are chosen as the redundants; consequently, the primary structure is a simply supported beam. On the right-hand side of Fig. 7.5b, the internal moments at supports 2 and 3 of the beam are chosen as the redundants; consequently, the primary structure is a sequence of three simply supported beams obtained by introducing hinges at the two intermediate supports of the actual beam.

When the number of redundants becomes greater than two or three, it is necessary to organize systematically the analysis of statically indeterminate structures. This may be accomplished by using a matrix formulation which is also adaptable to programming the solution of the resulting linear algebraic equations on a digital computer. With this in mind, we form the matrix of the redundants as

$$\{X\} = \left\{ \begin{array}{c} X_1 \\ X_2 \\ X_3 \\ \cdots \\ X_n \end{array} \right\} \tag{7.11}$$

Step 2. We consider the primary structure subjected to the loading acting on the real structure, including the external actions, the change of temperature, the initial stress in its members, and the translation or rotation of its supports,† which does not correspond to the chosen redundants. For instance, suppose that support 2 of the beam of Fig. 7.5a settles by a given amount Δ_1^s. If, as on the left-hand side of Fig. 7.5b, the reaction at support 2 is chosen as a redundant, the effect of this settlement is not taken into account in this step. However, if the internal moments at supports 2 and 3 are chosen as the redundants, the

---

† Since the primary structure is statically determinate, its members are free to expand or contract when subjected to a change of temperature and, moreover, the structure moves as a rigid body when one or more of its supports translate or rotate. Thus, under this loading, the members of the primary structure are not subjected to any internal actions. However, the configuration of the primary structure changes.

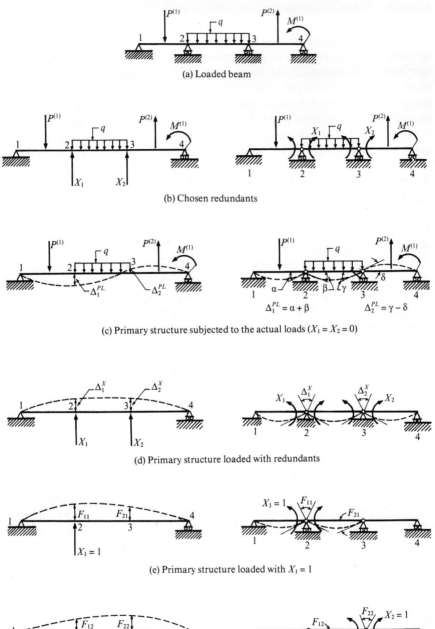

(a) Loaded beam

(b) Chosen redundants

(c) Primary structure subjected to the actual loads ($X_1 = X_2 = 0$)

(d) Primary structure loaded with redundants

(e) Primary structure loaded with $X_1 = 1$

(f) Primary structure loaded with $X_2 = 1$

Figure 7.5   Example of analysis of an indeterminate beam using the basic force method.

effect of the previously mentioned settlement must be taken into account in this step.

We compute the components of displacements (translations and rotations) Δ_i^{PL} $(i = 1, 2, \ldots, n)$ of the points of the primary structure where the redundants are applied and in the direction of the redundants, using one of the methods presented in Chaps. 5 and 6 (i.e., the method of virtual work). We then form the matrix of the components of displacement of the primary structure as

$$[\Delta^{PL}] = \left\{ \begin{array}{c} \Delta_1^{PL} \\ \Delta_2^{PL} \\ \cdots \\ \Delta_n^{PL} \end{array} \right\} \tag{7.12}$$

The displacements Δ_i^{PL} *are taken as positive in the assumed direction of the corresponding redundants* X_i. Consequently, when we employ the method of virtual work, we use $X_i = 1$ as the "virtual" loading, in order to compute the displacement Δ_i^{PL}.

After establishing the redundants of a statically indeterminate structure, we can compute the internal actions at any cross section of its members by considering the equilibrium of appropriate segments of the structure. However, it is more convenient to compute the desired internal actions in the members of a statically indeterminate structure from the corresponding internal actions in the members of the primary structure. In such an eventuality, at this step, we store in the matrix $\{A^{PL}\}$ the internal actions of the ends of the members of the primary structure, subjected to the given loading except the given settlement of supports of the structure corresponding to the chosen redundants. Thus,

$$\{A^{PL}\} = \left\{ \begin{array}{c} A_1^{PL} \\ A_2^{PL} \\ \cdots \\ A_q^{PL} \end{array} \right\} \tag{7.13}$$

Step 3. We compute the flexibility matrix of the primary structure corresponding to the chosen redundants. We subject the primary structure to a unit value of one of the redundants $X_j = 1$ and compute the displacements F_{ij} $(i = 1, 2, \ldots, n)$ corresponding to the redundants X_i $(i = 1, 2, \ldots, n)$. We repeat this process for $j = 1, 2, \ldots, n$. The displacements F_{ij} are referred to as the *flexibility coefficients of the primary structure corresponding to the redundants* $\{X\}$ (see Sec. 5.3).

We denote by Δ_i^{PX} the displacement of the primary structure which corresponds to the redundant X_i when the primary structure is loaded only with the redundants. The displacement Δ_i^{PX} may be expressed as a linear combination of the redundants, since, as discussed in Sec. 1.9 for the structures which we are considering, the relations between the loading acting on them and the resulting displacements are linear. Thus

$$\Delta_i^{PX} = \sum_{j=1}^{n} F_{ij}X_j \qquad i = 1, 2, \ldots, n$$

(7.14)

or $$\{\Delta^{PX}\} = [F]\{X\}$$

It is apparent that the term F_{ij} of the matrix $[F]$ is a flexibility coefficient of the primary structure. It represents the displacement corresponding to the redundant X_i when the structure is subjected only to a unit value of the redundant X_j. The matrix $[F]$ is referred to as *the flexibility matrix of the primary structure corresponding to the redundants* $\{X\}$. Notice that as shown in Sec. 5.3, the flexibility matrix of a framed structure is symmetric. That is,

$$F_{ij} = F_{ji}$$

(7.15)

The displacements Δ_1^{PX} and Δ_2^{PX} of the primary structure of the continuous beam of Fig. 7.5a are shown in Fig. 7.5d. For this beam, relation (7.14) becomes

$$\Delta_1^{PX} = F_{11}X_1 + F_{12}X_2 \qquad \Delta_2^{PX} = F_{21}X_1 + F_{22}X_2$$

(7.16)

The flexibility coefficients of the primary structure of the continuous beam of Fig. 7.5a are shown in Fig. 7.5e and f.

In order to compute, in a subsequent step, the internal actions at the ends of the members of the structure which we are analyzing, we form in this step the matrix $[A^{PX}]$ of the internal actions acting at the ends of the members of the primary structure, subjected to a unit value of one of the redundants at a time. That is,

$$[A^{PX}] = \begin{bmatrix} A_1^{PX_1} & A_1^{PX_2} & \cdots & A_1^{PX_n} \\ A_2^{PX_1} & A_2^{PX_2} & \cdots & A_2^{PX_n} \\ \cdots & \cdots & \cdots & \cdot \\ A_q^{PX_1} & A_q^{PX_2} & \cdots & A_q^{PX_n} \end{bmatrix}$$

(7.17)

The elements of the ith column of the matrix $[A^{PX}]$ represent the internal actions acting at the ends of the members of the structure subjected to $X_i = 1$.

Step 4.   We compute the redundants by superimposing the results of steps 2 and 3. That is, from physical intuition, it is apparent that a component of displacement of a point of the given structure is equal to the sum of the corresponding components of displacements of the primary structure, subjected separately to the following loadings.

1. To the given loading of the structure except the displacements of supports which correspond to redundants

2. To all the redundants

Thus, the component of the displacement (translation or rotation) Δ_i^s in the direction of the redundant X_i of the point of the structure where the redundant X_i is applied is equal to

$$\Delta_i^s = \Delta_i^{PL} + \Delta_i^{PX} = \Delta_i^{PL} + \sum_{j=1}^{n} F_{ij} X_j \qquad i = 1, 2, \ldots, n \qquad (7.18)$$

In the above relation the component of displacement Δ_i^s is considered positive when its sense is the same as that of the corresponding redundant. Relations (7.18) can be written in matrix form as

$$\begin{Bmatrix} \Delta_1^s \\ \Delta_2^s \\ \cdots \\ \Delta_n^s \end{Bmatrix} = \begin{Bmatrix} \Delta_1^{PL} \\ \Delta_2^{PL} \\ \cdots \\ \Delta_n^{PL} \end{Bmatrix} + \begin{bmatrix} F_{11} & F_{12} & \cdots & F_{1n} \\ F_{21} & F_{22} & \cdots & F_{2n} \\ \cdots & \cdots & \cdots & \cdots \\ F_{n1} & F_{n2} & \cdots & F_{nn} \end{bmatrix} \begin{Bmatrix} X_1 \\ X_2 \\ \cdots \\ X_n \end{Bmatrix}$$

$$(7.19)$$

or
$$\{\Delta^s\} = \{\Delta^{PL}\} + [F]\{X\} \qquad (7.20)$$

Equation (7.20) is *the equation of compatibility of the structure*. Also, it is often referred to as the *Maxwell-Mohr equation*. It represents the restrictions which the redundants X_i must satisfy in order that the deformed configuration of the structure be continuous and compatible with the given displacement conditions at its supports. For the continuous beam of Fig. 7.5a, relation (7.20) reduces to

$$\Delta_1^s = \Delta_1^{PL} + F_{11} X_1 + F_{12} X_2$$
$$\Delta_2^s = \Delta_2^{PL} + F_{21} X_1 + F_{22} X_2 \qquad (7.21)$$

Notice, that if we choose as the primary structure the simply supported beam shown on the left-hand side of Fig. 7.5c, the displacements Δ_1^s and Δ_2^s are equal to minus the given settlement of supports 2 and 3, respectively. If we choose as the primary structure the row of three simply supported beams shown on the right-hand side of Fig. 7.5c, the displacements Δ_1^s and Δ_2^s vanish. In this case, the effect of the settlement of a support must be considered in step 2 because it affects the displacements Δ_1^{PL} and Δ_2^{PL} of the primary structure.

We solve the system of linear algebraic equations (7.20) and obtain

$$\{X\} = [F]^{-1}[\{\Delta^s\} - \{\Delta^{PL}\}] \qquad (7.22)$$

Step 5. We determine the support reactions and the internal actions of the structure which were not evaluated in step 4.

When the redundants have been determined, any reaction or internal action of the actual structure which we are interested in computing can be obtained either by considering the equilibrium of appropriate parts of the structure (see Example 1 at the end of this section) or by superimposing the corresponding quantities of the primary structure obtained in steps 2 and 3 (see Example 2 at the end of this section). Denoting a reaction or an internal action of the actual structure which we are interested in computing by A_k, the corresponding quantity of the primary structure subjected to the given loading† by A_k^{PL}, and the corresponding quantity of the primary structure subjected to $X_i = 1$ by $A_k^{PX_i}$, we have

$$A_k = A_k^{PL} + \sum_{i=1}^{n} (A_k^{PX_i})X_i \qquad (7.23)$$

Thus, if we are interested in computing q quantities A_k, we have

$$\begin{Bmatrix} A_1 \\ A_2 \\ \cdots \\ A_q \end{Bmatrix} = \begin{Bmatrix} A_1^{PL} \\ A_2^{PL} \\ \cdots \\ A_q^{PL} \end{Bmatrix} + \begin{bmatrix} A_1^{PX_1} & A_1^{PX_2} & \cdots & A_1^{PX_n} \\ A_2^{PX_1} & A_2^{PX_2} & \cdots & A_2^{PX_n} \\ \cdots & \cdots & \cdots & \cdots \\ A_q^{PX_1} & A_q^{PX_2} & \cdots & A_q^{PX_n} \end{bmatrix} \begin{Bmatrix} X_1 \\ X_2 \\ \cdots \\ X_n \end{Bmatrix}$$

$$(7.24)$$

or
$$\{A\} = \{A^{PL}\} + [A^{PX}]\{X\} \qquad (7.25)$$

We now illustrate the use of the basic force or flexibility method by the following examples.

Example 1. We establish the internal actions in a girder of a cable-stayed bridge. The girder is statically indeterminate to the second degree, and it is subjected to uniformly distributed forces. The shear and moment diagrams for the girder are plotted.

Example 2. We establish the internal actions in the members of a beam that is statically indeterminate to the second degree. The beam is subjected to external actions to a settlement of a support and to a rotation of another support. The shear and moment diagrams for the beam are plotted.

Example 3. We establish the internal actions in the members of a grid that is statically indeterminate to the second degree.

---

† The loading can be external actions, change of temperature, initial stress of the members of the structure and given movement of its supports which do not correspond to the chosen redundants.

Example 1. In the two-span cable-stayed bridge (see Sec. 1.4.1) of Fig. a, the cable is anchored to the ground at point 1 and is continuous at point 3 where it rests on a frictionless cylinder. The area of the cable is $A = 855.3$ mm^2. The steel girder is continuous at the tower on which it is supported on a fixed bearing. The moments of inertia of members 1, 2 and 2, 6 of the girder are 330.90×10^6 mm^4 and 576.80×10^6 mm^4, respectively. The modulus of elasticity of steel is $E = 210,000$ MPa.

Plot the shear and moment diagrams for the girder of the bridge of Fig. a. Neglect the effect of axial deformation of the girder and tower.

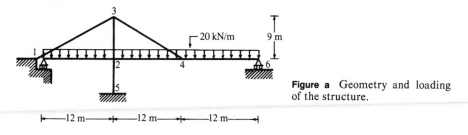

Figure a   Geometry and loading of the structure.

solution

STEP 1.   The structure of Fig. a is statically indeterminate to the second degree. We choose as redundants the moment in the girder at point 2 and the axial force in the cable. Thus, the primary structure is obtained from the actual structure by cutting the cable and by inserting a hinge in the girder at point 2 (see Fig. b).

STEP 2.   We consider the primary structure, subjected to the given loading, and we compute the displacements Δ_1^{PL} and Δ_2^{PL} corresponding to the redundants X_1 and X_2 (see Fig. c) using relation (5.66). The displacement Δ_1^{PL} is obtained using the moment

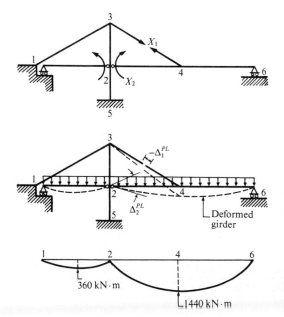

Figure b   Primary structure and redundants.

Figure c   Primary structure subjected to the given loading.

Figure d   Moment diagram for the girder of the primary structure subjected to the given loading.

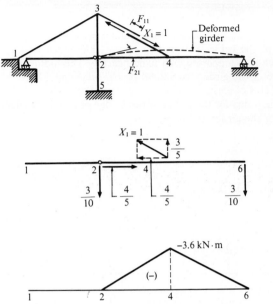

Figure e Primary structure subjected to $X_1 = 1$.

Figure f Free-body diagram of the girder of the primary structure subjected to $X_1 = 1$.

Figure g Moment diagram for the girder of the primary structure subjected to $X_1 = 1$.

diagrams of Figs. d and g for M and $\tilde{M}$, respectively; while the displacement Δ_2^{PL} is obtained using the moment diagrams of Figs. d and j for M and $\tilde{M}$, respectively. Thus, referring to Table A on the inside of the back cover and denoting by I the moment of inertia of member 2, 6 of the girder, we get

$$\Delta_1^{PL} = \int \frac{M\tilde{M}}{EI}\, dx_1 = -\frac{24}{3EI}(1440)(3.6)\left(1 + \frac{1}{4}\right) = -\frac{51,840}{EI}$$

$$\Delta_2^{PL} = \int \frac{M\tilde{M}}{EI}\, dx_1 = \frac{12(360)}{3EI}(0 + 1)\left(\frac{5768}{3309}\right) + \frac{24}{3EI}(1440)(1 + 0) = \frac{14,030.1}{EI}$$

thus

$$\{\Delta^{PL}\} = \frac{1}{EI}\begin{Bmatrix} -51,840.0 \\ 14,030.1 \end{Bmatrix} \tag{a}$$

STEP 3.   We evaluate the flexibility matrix of the structure. Using the moment diagram of Fig. g for M and $\tilde{M}$ and referring to Table A on the inside of the back cover, we get

$$F_{11} = \int \frac{M\tilde{M}}{EI}\, dx_1 + \text{(the effect of elongation of the cable)}$$

$$= \frac{24}{6EI}(3.6)^2(2) + \frac{(2)(1)(15)}{EA}$$

$$= \frac{103.68}{EI} + \frac{30(0.5768)}{EI(0.8553)} = \frac{123.91}{EI}$$

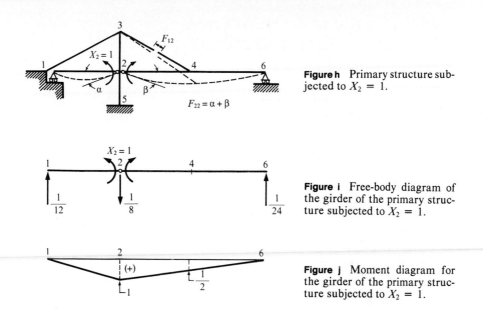

Figure h Primary structure subjected to $X_2 = 1$.

Figure i Free-body diagram of the girder of the primary structure subjected to $X_2 = 1$.

Figure j Moment diagram for the girder of the primary structure subjected to $X_2 = 1$.

Using the moment diagrams of Figs. g and j for M and $\tilde{M}$, respectively, and referring to Table A on the inside of the back cover, we obtain

$$F_{21} = F_{12} = -\frac{24}{6EI}(3.6)(1)\left(1 + \frac{1}{2}\right) = -\frac{21.60}{EI}$$

Using the moment diagram of Fig. j for M and $\tilde{M}$ and referring to Table A on the inside of the back cover, we get

$$F_{22} = \frac{24}{6EI}(1)(2) + \frac{12}{6EI}(1)(2)\frac{5768}{3309} = \frac{14.97}{EI}$$

Thus the flexibility matrix of the structure is

$$[F] = \frac{1}{EI}\begin{bmatrix} 123.91 & -21.60 \\ -21.60 & 14.97 \end{bmatrix} \tag{b}$$

STEP 4.   We compute the redundants by substituting relations (a) and (b) into the compatibility equation (7.20). Thus

$$\frac{1}{EI}\begin{Bmatrix} -51,840.00 \\ 14,030.10 \end{Bmatrix} + \frac{1}{EI}\begin{bmatrix} 123.91 & -21.60 \\ -21.60 & 14.97 \end{bmatrix}\begin{Bmatrix} X_1 \\ X_2 \end{Bmatrix} = \begin{Bmatrix} 0 \\ 0 \end{Bmatrix}$$

thus $$\begin{Bmatrix} X_1 \\ X_2 \end{Bmatrix} = -\frac{1}{1388.373}\begin{bmatrix} 14.97 & 21.60 \\ 21.60 & 123.91 \end{bmatrix}\begin{Bmatrix} -51,840.0 \\ 14,030.1 \end{Bmatrix} = \begin{Bmatrix} 340.68 \\ -445.65 \end{Bmatrix} \tag{c}$$

STEP 5.   Using the values of the redundants computed in step 4, the free-body dia-

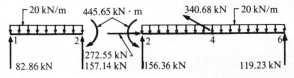

Figure k Free-body diagrams of the two members of the girder subjected to the given loads.

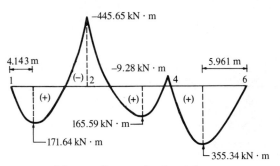

Figure l Shear diagram for the girder.

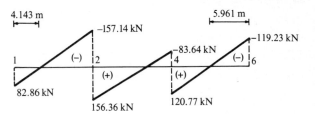

Figure m Moment diagram for the girder.

grams of the members of the girder are shown in Fig. k. Referring to this figure, we plot the shear and moment diagrams in Figs. l and m, respectively.

Example 2. In addition to the external actions indicated in Fig. a, the beam is subjected to a settlement of support 2 of magnitude 0.0256 m and a rotation of support 3, of magnitude 0.0016 rad. The members of the beam are made of steel ($E = 200,000$ MPa) and have constant cross sections whose moments of inertia are given in Fig. a ($I = 400 \times 10^6$ mm^4). Plot the shear and moment diagram for the beam.

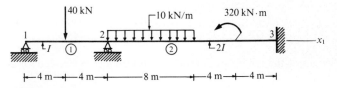

Figure a Geometry and loading of the beam.

solution

STEP 1.   This beam is statically indeterminate to the second degree. We choose the internal moments at joints 2 and 3 as the redundants (see Fig. b). Thus, the primary structure is a series of two simply supported beams.

STEP 2.   We consider the primary structure subjected to the given loading and compute the displacements Δ_1^{PL} and Δ_2^{PL} corresponding to the redundants X_1 and X_2 (see Fig. c). These displacements include the effect of the given external actions and of the settlement of support 2. However, they do not include the effect of the rotation of support 3. This is included as Δ_2^s in the compatibility equation (7.20). Referring to Fig. d, the moment in the primary structure subjected to the given external actions is

$$M = 20x_1 \qquad\qquad \text{for } 0 < x_1 < 4$$

$$M = 20x_1 - 40(x_1 - 4) \qquad \text{for } 4 < x_1 < 8$$

$$M = 40x_1'\left(2 - \frac{x_1'}{8}\right) \qquad \text{for } 0 < x_1' < 8 \tag{a}$$

$$M = 320 \qquad\qquad \text{for } 8 < x_1' < 12$$

$$M = 0 \qquad\qquad \text{for } 12 < x_1' < 16$$

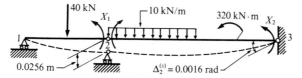

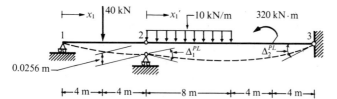

Figure b Primary structure subjected to the given loading and to the redundants.

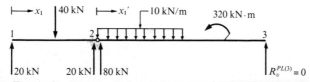

Figure c Primary structure subjected to the given actions and settlement of support 2.

Figure d Free-body diagram of the primary structure subjected to the given actions.

Moreover, referring to Fig. e the moment in the primary structure subjected to $X_1 = 1$ is

$$\tilde{M} = \frac{x_1}{8} \qquad \text{for } 0 < x_1 < 8$$

$$\tilde{M} = 1 - \frac{x_1'}{16} \qquad \text{for } 0 < x_1' < 16$$

(b)

Furthermore, referring to Fig. f the moment in the primary structure subjected to $X_2 = 1$ is

$$\tilde{M} = 0 \qquad \text{for } 0 < x_1 < 8$$

$$\tilde{M} = \frac{x_1'}{16} \qquad \text{for } 0 < x_1' < 16$$

(c)

Substituting the moments from relations (a) and (b) into relation (5.66), we obtain

$$\Delta_1^{PL} = - \sum_{s=1}^{S} \tilde{R}^{(s)} \Delta^{(s)} + \int_0^8 \frac{M\tilde{M}}{EI} dx_1 + \int_0^{16} \frac{M\tilde{M}}{2EI} dx_1' = - \left(\frac{3}{16}\right)(0.0252)$$

$$+ \frac{1}{EI}\left[\int_0^4 20x_1 \left(\frac{x_1}{8}\right) dx_1 + \int_4^8 [20x_1 - 40(x_1 - 4)]\frac{x_1}{8} dx_1 \right.$$

$$+ \frac{1}{2}\int_0^8 40x_1' \left(2 - \frac{x_1'}{8}\right)\left(1 - \frac{x_1'}{16}\right) dx_1'$$

$$\left. + \frac{1}{2}\int_8^{12} 320\left(1 - \frac{x_1'}{16}\right) dx_1 \right] = 0.007533 \text{ m}$$

Substituting the moments from relations (a) and (c) into relation (5.66), we get

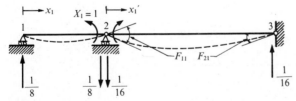

Figure e Primary structure subjected to a pair of unit moments $X_1 = 1$.

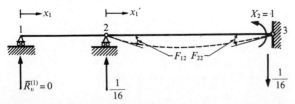

Figure f Primary structure subjected to a unit moment $X_2 = 1$.

$$\Delta_2^{PL} = -\sum_{s=1}^{S} \tilde{R}^{(s)}\Delta^{(s)} + \int_0^{12} \frac{M\tilde{M}}{2EI}\,dx_1' = -\left(-\frac{1}{16}\right)(0.0256)$$

$$+ \frac{1}{2EI}\left[\int_0^8 40x_1'\left(2 - \frac{x_1'}{8}\right)\left(\frac{x_1'}{16}\right)dx_1' + \int_8^{12} 320\frac{x_1'}{16}\,dx_1\right] = 0.009934 \text{ m}$$

thus
$$[\Delta^{PL}] = \begin{Bmatrix} 0.007533 \\ 0.009934 \end{Bmatrix} \tag{d}$$

STEP 3.   We determine the flexibility matrix of the structure. Using relation (5.66) and referring to relation (b), we have

$$F_{11} = \int_0^8 \frac{M\tilde{M}}{EI}\,dx_1 + \int_0^{16} \frac{M\tilde{M}}{2EI}\,dx_1'$$

$$= \frac{1}{EI}\int_0^8 \frac{x_1^2}{64}\,dx_1 + \frac{1}{2EI}\int_0^{16}\left(1 - \frac{x_1'}{16}\right)^2 dx_1' = \frac{16}{3EI}$$

Using relation (5.66) and referring to relations (b) and (c), we get

$$F_{21} = \int_0^{16} \frac{M\tilde{M}}{2EI}\,dx_1' = \frac{1}{2EI}\int_0^{16}\left(1 - \frac{x_1'}{16}\right)\left(\frac{x_1'}{16}\right)dx_1' = \frac{8}{6EI}$$

Using relation (5.66) and referring to relation (c), we obtain

$$F_{22} = \int_0^{16} \frac{M\tilde{M}}{2EI}\,dx_1' = \frac{1}{2EI}\int_0^{16}\left(\frac{x_1'}{16}\right)^2 dx_1' = \frac{8}{3EI}$$

hence
$$[F] = \frac{8}{6EI}\begin{bmatrix} 4 & 1 \\ 1 & 2 \end{bmatrix} \tag{e}$$

STEP 4.   We compute the redundants by superimposing the results of steps 2 and 3. Rotation of support 3 is equal to 0.0016 rad in the counterclockwise direction. Thus,

$$\Delta_2^s = 0.0016 \text{ rad} \tag{f}$$

Substituting relations (d), (e), and (f) into compatibility equation (7.20), we get

$$\begin{Bmatrix} 0 \\ 0.0016 \end{Bmatrix} = \begin{Bmatrix} 0.007533 \\ 0.009934 \end{Bmatrix} + \frac{8}{6EI}\begin{bmatrix} 4 & 1 \\ 1 & 2 \end{bmatrix}\begin{Bmatrix} X_1 \\ X_2 \end{Bmatrix}$$

thus $$\begin{Bmatrix} X_1 \\ X_2 \end{Bmatrix} = \frac{3EI}{4}\begin{bmatrix} 4 & 1 \\ 1 & 2 \end{bmatrix}^{-1}\left\{\begin{Bmatrix} 0 \\ 0.0016 \end{Bmatrix} - \begin{Bmatrix} 0.007533 \\ 0.009934 \end{Bmatrix}\right\} = \begin{Bmatrix} -57.70 \\ -221.12 \end{Bmatrix} \tag{g}$$

STEP 5.   We compute the reactions of the beam referring to Figs. d, e, and f and using relations (g) and (7.25). Thus

$$\begin{Bmatrix} R_v^{(1)} \\ R_v^{(2)} \\ R_v^{(3)} \end{Bmatrix} = \begin{Bmatrix} 20 \\ 100 \\ 0 \end{Bmatrix} + \frac{1}{8}\begin{bmatrix} 1 & 0 \\ -1.5 & 0.5 \\ 0.5 & -0.5 \end{bmatrix}\begin{Bmatrix} -57.70 \\ -221.12 \end{Bmatrix} = \begin{Bmatrix} 12.79 \\ 97.00 \\ 10.21 \end{Bmatrix}$$

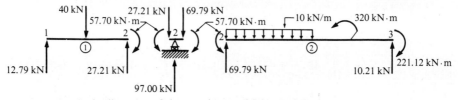

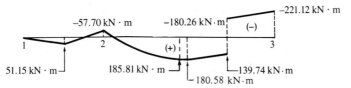

Figure g   Free-body diagrams of the members and joint 2 of the beam.

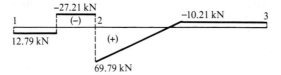

Figure h   Shear diagram for the beam.

Figure i   Moment diagram for the beam.

The free-body diagrams of the members and joint 2 of the beam are shown in Fig. g. From these, the shear and moment diagrams are plotted in Figs. h and i, respectively.

Example 3.   The members of the grid shown in Fig. a are tubular. The polar moment of inertia of member 2 is half that of the other members. The modulus of elasticity of the material from which the tubes are made is 2.6 times larger than its shear modulus. Plot the shear and moment diagrams for the members of the grid loaded as shown in Fig. a.

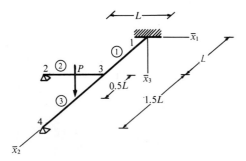

Figure a   Geometry and loading of the grid.

solution

STEP 1.   This is a statically indeterminate structure to the second degree. We choose the reactions at supports 2 and 4 as the redundants. (See Fig. b.)

STEP 2.   We compute the displacements Δ_1^{PL} and Δ_2^{PL} of points 2 and 4 of the primary structure subjected to the given force (see Fig. c) using relation (5.66). The moment diagram of the primary structure subjected to the given force is shown in Fig. d. The primary structure subjected to the virtual loadings for computing the displacements Δ_1^{PL} and Δ_2^{PL} are shown in Figs. e and f, respectively. The corresponding moment diagrams are shown in Figs. g and h. The displacement Δ_1^{PL} is obtained using the moment diagram of Fig. d for M_2 and of Fig. g for $\bar{M}_2$. The displacement Δ_2^{PL} is obtained using the moment diagram of Fig. d for M_2 and that of Fig. h for $\tilde{M}_2$. Thus, denoting the moment of inertia of members 3 and 1 by $2I$ and referring to Table A on the inside

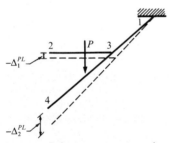

Figure b Primary structure subjected to the given loading and to the redundants.

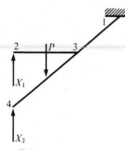

Figure c Primary structure subjected to the given loading.

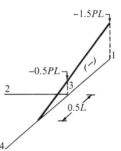

Figure d Moment diagram (M_2) for the primary structure subjected to the given loading.

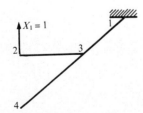

Figure e Primary structure subjected to $X_1 = 1$.

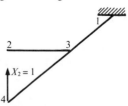

Figure f Primary structure subjected to $X_2 = 1$.

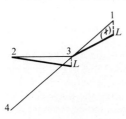

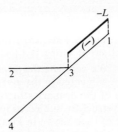

(a) Bending moment M_2 (b) Torsional moment M_1

Figure g Moment diagrams for the primary structure loaded as shown in Fig. e.

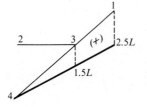

Figure h Moment diagram (M_2) for the primary structure loaded as shown in Fig. f.

of the back cover, we get

$$\Delta_1^{PL} = \int_{pt.1}^{pt.3} \frac{M_2 \tilde{M}_2 \, dx_1}{E(2I)} = -\frac{L(L)}{6E(2I)}(0.5PL + 3PL) = -\frac{3.5PL^3}{12EI}$$

$$\Delta_2^{PL} = \int_{pt.1}^{pt.4} \frac{M_2 \tilde{M}_2 \, dx_1}{E(2I)} = -\frac{(1.5L)}{6E(2I)}(1.5PL)(L + 5L) = -\frac{13.5PL^3}{12EI}$$

thus
$$[\Delta^{PL}] = -\frac{PL^3}{12EI}\begin{Bmatrix} 3.5 \\ 13.5 \end{Bmatrix} \tag{a}$$

STEP 3.   We compute the flexibility matrix of the primary structure using relation (5.66). The flexibility coefficient F_{11} may be established by using the moment diagrams of Fig. g for $M_2 = \tilde{M}_2$ and $M_1 = \tilde{M}_1$. Thus, denoting by J the polar moment or inertia of members 3 and 1 and referring to Table A on the inside of the back cover, we obtain

$$F_{11} = \int_{pt.2}^{pt.3} \frac{M_2 \tilde{M}_2 \, dx_1}{EI} + \int_{pt.1}^{pt.3} \frac{M_2 \tilde{M}_2 \, dx_1}{E(2I)} + \int_{pt.1}^{pt.3} \frac{M_1 \tilde{M}_1 \, dx_1}{GJ}$$

$$= \frac{L}{6(EI)}L(2L) + \frac{L}{6(2EI)}L(2L) + \frac{(L)(L)(L)}{G(J)} = \frac{L^3}{2EI} + \frac{L^3}{GJ}$$

Notice that for a member with circular or tubular cross section, the polar moment of inertia is equal to $J = 2I$. Thus, for $G = E/2.6$, we get

$$F_{11} = \frac{L^3}{2EI}(1 + 2.6) = \frac{1.8L^3}{EI}$$

The flexibility coefficient F_{12} may be established by using the moment diagrams of

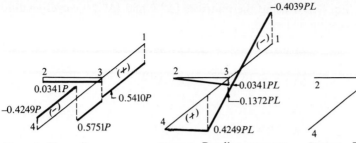

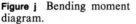

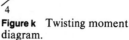

Figure i Shear diagram.   **Figure j** Bending moment diagram.   **Figure k** Twisting moment diagram.

Figs. g and h for M_2 and $\tilde{M}_2$, respectively. Thus, referring to Table A on the inside of the back cover, we obtain

$$F_{12} = F_{21} = \int_{\text{pt.1}}^{\text{pt.3}} \frac{M_2 \tilde{M}_2 \, dx_1}{E(2I)} = \frac{L}{6E(2I)} (L)(1.5L + 5L) = 0.5417 \frac{L^3}{EI}$$

The flexibility coefficient F_{22} is obtained by using the moment diagram of Fig. h for M_2 and $\tilde{M}_2$. Thus, referring to Table A on the inside of the back cover, we get

$$F_{22} = \int_{\text{pt.1}}^{\text{pt.4}} \frac{M_2 \tilde{M}_2 \, dx_1}{E(2I)} = \frac{2.5L}{3(2I)} (2.5L)^2 = 2.6041 \frac{L^3}{EI}$$

Hence, the stiffness matrix of the primary structure is

$$[F] = \frac{L^3}{EI} \begin{bmatrix} 1.8000 & 0.5417 \\ 0.5417 & 2.6041 \end{bmatrix} \tag{b}$$

STEP 4. We compute the redundants. Substituting results (a) and (b) in the compatibility equation (7.20), we obtain

$$-\frac{PL^3}{12EI} \begin{Bmatrix} 3.5 \\ 13.5 \end{Bmatrix} + \frac{L^3}{EI} \begin{bmatrix} 1.8000 & 0.5417 \\ 0.5417 & 2.6041 \end{bmatrix} \begin{Bmatrix} X_1 \\ X_2 \end{Bmatrix} = 0$$

Thus

$$\begin{Bmatrix} X_1 \\ X_2 \end{Bmatrix} = P \begin{Bmatrix} -0.0341 \\ 0.4249 \end{Bmatrix}$$

STEP 5. Using the values of the redundants established in step 4, the shear, the bending moment, and the twisting moment diagrams are plotted in Figs. i, j, and k, respectively.

7.3 Analysis of Structures Subjected to Multiple Loading

Frequently, we are interested in establishing the internal actions in the members of a statically indeterminate structure to the nth degree, subjected to m different loading conditions (multiple loading). The values of the redundants $\{X\}^{(j)}$ corresponding to the jth loading condition ($j = 1, 2, \ldots, m$) may be obtained from Eq. (7.20) using the matrices $\{\Delta^s\}^{(j)}$ and $\{\Delta^{PL}\}^{(j)}$ corresponding to this loading condition. That is,

$$\{\Delta^s\}^{(1)} = \{\Delta^{PL}\}^{(1)} + [F]\{X\}^{(1)}$$

$$\{\Delta^s\}^{(2)} = \{\Delta^{PL}\}^{(2)} + [F]\{X\}^{(2)} \tag{7.26}$$

$$\cdots\cdots\cdots\cdots\cdots\cdots\cdots$$

$$\{\Delta^s\}^{(m)} = \{\Delta^{PL}\}^{(m)} + [F]\{X\}^{(m)}$$

These relations may be rewritten as follows:

$$[\{\Delta^s\}^{(1)}\ \{\Delta^s\}^{(2)}\ \cdots\ \{\Delta^s\}^{(m)}] = [\{\Delta^{PL}\}^{(1)}\ \{\Delta^{PL}\}^{(2)}\ \cdots\ \{\Delta^{PL}\}^{(m)}] \tag{7.27}$$

$$+ [F][\{X\}^{(1)}\ \{X\}^{(2)}\ \cdots\ \{X\}^{(m)}] \tag{7.28}$$

or

$$[\Delta^s] = [\Delta^{PL}] + [F][X]$$

where $[\Delta^s]$, $[\Delta^{PL}]$, and $[X]$ are the following $n \times m$ matrices:

$$[\Delta^{PL}] = \begin{bmatrix} \Delta_1^{PL(1)} & \Delta_1^{PL(2)} & \cdots & \Delta_1^{PL(m)} \\ \Delta_2^{PL(1)} & \Delta_2^{PL(2)} & \cdots & \Delta_2^{PL(m)} \\ \cdots & \cdots & \cdots & \cdots \\ \Delta_n^{PL(1)} & \Delta_n^{PL(2)} & \cdots & \Delta_n^{PL(m)} \end{bmatrix} \tag{7.29a}$$

$$[\Delta^s] = \begin{bmatrix} \Delta_1^{(1)} & \Delta_1^{(2)} & \cdots & \Delta_1^{(m)} \\ \Delta_2^{(1)} & \Delta_2^{(2)} & \cdots & \Delta_2^{(m)} \\ \cdots & \cdots & \cdots & \cdots \\ \Delta_n^{(1)} & \Delta_n^{(2)} & \cdots & \Delta_n^{(m)} \end{bmatrix} \tag{7.29b}$$

$$[X] = \begin{bmatrix} X_1^{(1)} & X_1^{(2)} & \cdots & X_1^{(m)} \\ X_2^{(1)} & X_2^{(2)} & \cdots & X_2^{(m)} \\ \cdots & \cdots & \cdots & \cdots \\ X_n^{(1)} & X_n^{(2)} & \cdots & X_n^{(m)} \end{bmatrix} \tag{7.29c}$$

Similarly, the internal actions $\{A\}^{(j)}$ in the members of the structure, subjected to the jth loading condition ($j = 1, 2, \ldots, m$) may be obtained from relation (7.25) using the matrices $\{A^{PL}\}^{(j)}$ and $\{X\}^{(j)}$ corresponding to this loading condition. That is,

$$\{A\}^{(1)} = \{A^{PL}\}^{(1)} + [A^{PX}]\{X\}^{(1)}$$

$$\{A\}^{(2)} = \{A^{PL}\}^{(2)} + [A^{PX}]\{X\}^{(2)}$$

$$\cdots\cdots\cdots\cdots\cdots\cdots\cdots$$

$$\{A\}^{(m)} = \{A^{PL}\}^m + [A^{PX}]\{X\}^{(m)}$$

The above relations may be rewritten as

$$[A] = [A^{PL}] + [A^{PX}][X] \qquad (7.30)$$

where the matrix $[A^{PX}]$ is defined by relation (7.17), the matrix $[X]$ is defined by relation (7.29c), while the matrices $[A]$ and $[A^{PL}]$ are equal to

$$[A] = \begin{bmatrix} A_1^{(1)} & A_1^{(2)} & \cdots & A_1^{(m)} \\ A_2^{(1)} & A_2^{(2)} & \cdots & A_2^{(m)} \\ \cdots & \cdots & \cdots & \cdots \\ A_q^{(1)} & A_q^{(2)} & \cdots & A_q^{(m)} \end{bmatrix}$$

$$[A^{PL}] = \begin{bmatrix} A_1^{PL(1)} & A_1^{PL(2)} & \cdots & A_1^{PL(m)} \\ A_2^{PL(1)} & A_2^{PL(2)} & \cdots & A_2^{PL(m)} \\ \cdots & \cdots & \cdots & \cdots \\ A_q^{PL(1)} & A_q^{PL(2)} & \cdots & A_q^{PL(m)} \end{bmatrix}$$

(7.31)

7.4 Analysis of Symmetric Structures Subjected to a General Loading

The analysis of a statically indeterminate structure of the nth degree by the force method results in a set of n linear algebraic equations of compatibility (7.20) involving the n redundants. That is,

$$\{\Delta^s\} = \{\Delta^{PL}\} + [F]\{X\} \qquad (7.32)$$

When a digital computer is available, the solution of the system of linear algebraic equations (7.32), may be easily obtained. However, if a computer is not available, its solution for a highly redundant structure could be very laborious. For symmetric structures, the work involved in solving the system of equations

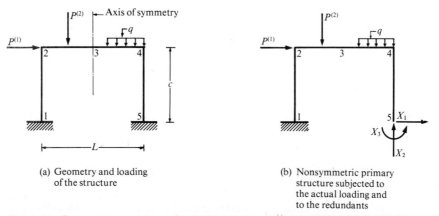

(a) Geometry and loading
of the structure

(b) Nonsymmetric primary
structure subjected to
the actual loading and
to the redundants

Figure 7.6 Symmetric structure subjected to nonsymmetric loading.

(7.32) can be reduced considerably by choosing a symmetric primary structure and symmetric or antisymmetric pairs of redundants so that the moment distribution in the primary structure subjected to a pair of redundants is either symmetric or antisymmetric. The flexibility coefficients are equal to integrals of products of two of these moment distributions. Since the integral of the product of a symmetric and an antisymmetric function vanishes, it is apparent that for the previously described choice of primary structure and redundants, some of the flexibility coefficients vanish, resulting in a simpler set of linear algebraic equations for the redundants X_i. As an example, consider the symmetric frame loaded as shown in Fig. 7.6a. If we choose the nonsymmetric primary structure shown in Fig. 7.6b, none of the flexibility coefficients F_{ij} vanish and consequently, the compatibility equations (7.32) are

$$0 = \Delta_1^{PL} + F_{11}X_1 + F_{12}X_2 + F_{13}X_3$$

$$0 = \Delta_2^{PL} + F_{21}X_1 + F_{22}X_2 + F_{23}X_3 \qquad (7.33)$$

$$0 = \Delta_3^{PL} + F_{31}X_1 + F_{32}X_2 + F_{33}X_3$$

However, if, as the primary structure, we choose the symmetric structure obtained by cutting the frame by the plane normal to it and containing its axis of symmetry, then, as shown in Fig. 7.7, when the primary structure is sub-

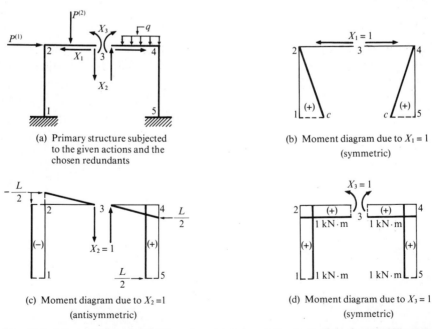

(a) Primary structure subjected
to the given actions and the
chosen redundants

(b) Moment diagram due to $X_1 = 1$
(symmetric)

(c) Moment diagram due to $X_2 = 1$
(antisymmetric)

(d) Moment diagram due to $X_3 = 1$
(symmetric)

Figure 7.7   Moment diagrams for the symmetric primary structure of the frame of Fig. 7.6a subjected to a unit value of a redundant.

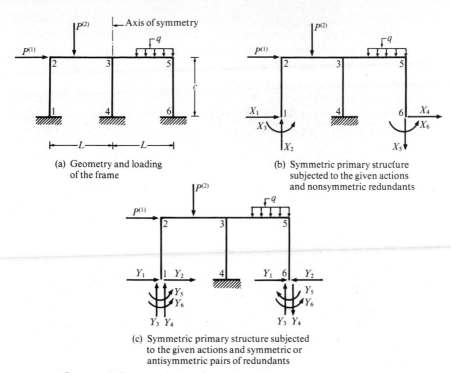

(a) Geometry and loading
of the frame

(b) Symmetric primary structure
subjected to the given actions
and nonsymmetric redundants

(c) Symmetric primary structure subjected
to the given actions and symmetric or
antisymmetric pairs of redundants

Figure 7.8 Symmetric frame and redundants.

jected to $X_1 = 1$ and $X_3 = 1$, the moment diagram is symmetric; whereas, when the primary structure is subjected to $X_2 = 1$, the moment diagram is antisymmetric. Thus,

$$F_{12} = 0 \qquad F_{23} = 0 \qquad (7.34)$$

Consequently, the equations of compatibility (7.32) reduce to

$$0 = \Delta_1^{PL} + F_{11}X_1 + F_{13}X_3$$

$$0 = \Delta_2^{PL} + F_{22}X_2 \qquad (7.35)$$

$$0 = \Delta_3^{PL} + F_{31}X_1 + F_{33}X_3$$

Thus, the analysis of the symmetric structure under consideration is simplified with the choice of a symmetric primary structure.

As a second example, consider the symmetric, two-bay frame loaded as shown in Fig. 7.8a. This is an indeterminate structure to the sixth degree. If we choose the symmetric primary structure and the redundants shown in Fig. 7.8b, none of the flexibility coefficients vanish. That is, for this choice of redundants, the compatibility equations are a set of six linear algebraic equations, each involving the six redundants X_i ($i = 1, 2, \ldots, 6$). In order to take advantage of the symmetry of the structure, we must choose symmetric and antisymmetric pairs of redundants so that the moment diagrams of the primary structure subjected to one pair of redundants at a time are either symmetric or antisymmetric. In this example, we can choose the symmetric and antisymmetric pairs or redundants Y_i ($i = 1, 2, \ldots, 6$) shown in Fig. 7.8c. Referring to Fig. 7.8b and c, we see that the following relations exist between the redundants X_i and Y_i.

$$X_1 = Y_1 + Y_2 \qquad X_4 = Y_2 - Y_1$$

$$X_2 = Y_3 + Y_4 \qquad X_5 = Y_4 - Y_3 \qquad (7.36)$$

$$X_3 = Y_5 + Y_6 \qquad X_6 = Y_6 - Y_5$$

When the primary structure is subjected to the symmetric pairs of redundants $Y_1 = 1$ or $Y_3 = 1$ or $Y_5 = 1$, as shown in Fig. 7.9a, c, and e, the corresponding moment diagrams are symmetric. On the other hand, when the primary structure is subjected to the antisymmetric pairs of redundants $Y_2 = 1$, $Y_4 = 1$, or $Y_6 = 1$, as shown in Fig. 7.9b, d, and f, the corresponding moment diagrams are antisymmetric. Consequently, the flexibility coefficients F_{12}, F_{14}, F_{16}, F_{32}, F_{34}, F_{36}, F_{52}, F_{54}, and F_{56} must vanish and the compatibility relations (7.32) reduce to the following two systems of linear algebraic equations.

$$0 = \Delta_1^{PL} + F_{11}Y_1 + F_{13}Y_3 + F_{15}Y_5$$

$$0 = \Delta_3^{PL} + F_{31}Y_1 + F_{33}Y_3 + F_{35}Y_5 \qquad (7.37)$$

$$0 = \Delta_5^{PL} + F_{51}Y_1 + F_{53}Y_3 + F_{55}Y_5$$

and

$$0 = \Delta_2^{PL} + F_{22}Y_2 + F_{24}Y_4 + F_{26}Y_6$$

$$0 = \Delta_4^{PL} + F_{42}Y_2 + F_{44}Y_4 + F_{46}Y_6 \qquad (7.38)$$

$$0 = \Delta_6^{PL} + F_{62}Y_2 + F_{64}Y_4 + F_{66}Y_6$$

Δ_1^{PL} and Δ_4^{PL} are the relative horizontal and vertical components of translation, respectively, while Δ_5^{PL} is the relative rotation of points 1 and 6 of the primary structure subjected to the given external actions. Δ_2^{PL} and Δ_3^{PL} are, respectively, the sum of the horizontal and vertical components of translation while Δ_6^{PL} is the sum of the rotation of points 1 and 6 of the primary structure subjected to the given external actions. The displacements Δ_i^{PL} ($i = 1, 2, \ldots, 6$) may be established using the method of virtual work (5.66) and the virtual loadings of Fig. 7.9. For instance, the relative horizontal translation Δ_1^{PL} of points 1 and 6

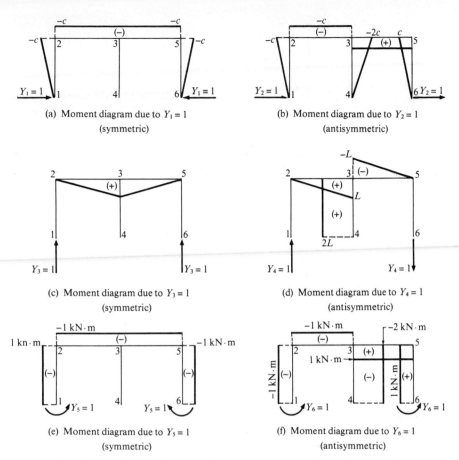

(a) Moment diagram due to $Y_1 = 1$
(symmetric)

(b) Moment diagram due to $Y_2 = 1$
(antisymmetric)

(c) Moment diagram due to $Y_3 = 1$
(symmetric)

(d) Moment diagram due to $Y_4 = 1$
(antisymmetric)

(e) Moment diagram due to $Y_5 = 1$
(symmetric)

(f) Moment diagram due to $Y_6 = 1$
(antisymmetric)

Figure 7.9 Moment diagrams for the primary structure of the frame of Fig. 7.8a subjected to unit values of symmetric and antisymmetric pairs of redundants.

of the primary structure subjected to the given external actions is

$$\Delta_1^{PL} = \int \frac{M^{PL} M^{Y_1}}{EI} \, dx_1 \tag{7.39}$$

where M^{PL} and M^{Y_1} are the bending moment in the members of the primary structure due to the given external actions and the symmetric pair of unit forces $Y_1 = 1$, respectively (see Fig. 7.9a).

On the basis of the foregoing discussion, it is apparent that in the example under consideration, by choosing the symmetric and antisymmetric pairs of reactions shown in Fig. 7.8c, as redundants, the compatibility equations have been reduced to two uncoupled systems of equations. The one system contains

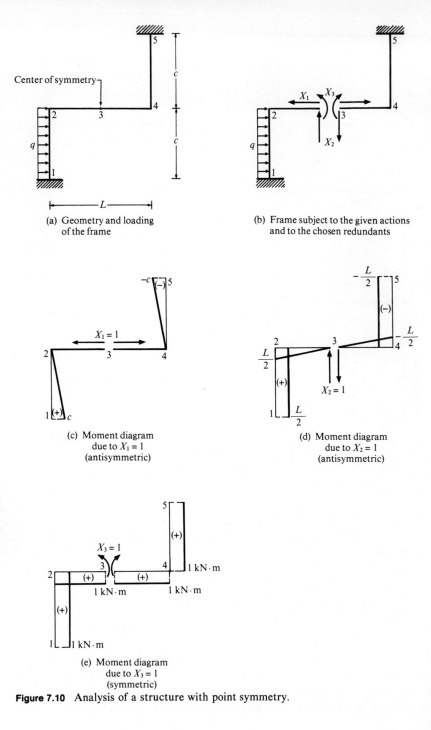

(a) Geometry and loading
of the frame

(b) Frame subject to the given actions
and to the chosen redundants

(c) Moment diagram
due to $X_1 = 1$
(antisymmetric)

(d) Moment diagram
due to $X_2 = 1$
(antisymmetric)

(e) Moment diagram
due to $X_3 = 1$
(symmetric)

Figure 7.10 Analysis of a structure with point symmetry.

three linear, algebraic equations for the three pairs of symmetric redundants Y_1, Y_3, and Y_5, while the other contains three linear algebraic equations for the three pairs of antisymmetric redundants Y_2, Y_4, and Y_6. When a computer is not available, the work required to solve these two systems is less than that required to solve one system of six linear algebraic equations, each containing the six unknown redundants.

As a third example, let us consider the planar structure shown in Fig. 7.10a, which is statically indeterminate to the third degree and is symmetric with respect to point 3. If we choose the internal actions at point 3 as redundants, the moment diagrams for the primary structure, subjected to a unit value of one of the redundants, are either symmetric or antisymmetric (see Fig. 7.10c, d, and e). Referring to Fig. 7.10c, d, and e, it can be deduced that the flexibility coefficients F_{13} and F_{23} vanish. Thus, the equations of compatibility (7.32) reduce to

$$0 = \Delta_1^{PL} + F_{11}X_1 + F_{12}X_2$$

$$0 = \Delta_2^{PL} + F_{21}X_1 + F_{22}X_2 \tag{7.40}$$

$$0 = \Delta_3^{PL} + F_{33}X_3$$

7.5 Analysis of Symmetric Structures Subjected to Symmetric or Antisymmetric Loading

In the previous section, we considered symmetric structures subjected to a general loading, and we established ways to take advantage of their symmetry in order to simplify their analysis. In this section, we demonstrate by six examples the additional simplifications in the analysis of symmetric sructures which can be attained when the applied loads are symmetric or antisymmetric with respect to the axis, the point, or the plane of symmetry of the structure.

Example 1. We establish the internal actions in the members of a statically indeterminate to the first degree planar frame having an axis of symmetry. The frame is subjected to an external antisymmetric force. Because of the symmetry of the frame and the antisymmetry of the loading, the reactions of the frame can be established without the use of a compatibility equation.

Example 2. We establish the internal actions in the members of a planar frame having an axis of symmetry. The frame is subjected to a horizontal force. This nonsymmetric loading is replaced by a symmetric and an antisymmetric component. For this example, the analysis of the structure subjected separately to the symmetric and to the antisymmetric component of the loading is preferable to the analysis of the structure subjected directly to the given nonsymmetric loading. This conclusion is not valid for every type of nonsymmetric loading.

Example 3. We establish the internal actions in the members of a planar frame having a center of symmetry. The frame is subjected to an external (antisymmetric) force.

Example 4. We establish the internal actions in the members of a planar frame having an axis of symmetry. The frame is subjected to different temperature at its internal and external fibers.

Example 5. We establish the internal actions in the members of a circular ring. The ring is subjected to two external radial, collinear forces. This loading has two mutually perpendicular axes of symmetry.

Example 6. We establish the internal actions in the members of a symmetric grid. The grid has a plane and an axis of symmetry and is subjected to an external moment which is antisymmetric with respect to the plane of symmetry of the grid and symmetric with respect to its axis of symmetry.

Example 1. Consider the frame loaded as shown in Fig. a. The members of the frame have the same constant cross sections and are made from the same material. Compute the reactions at the supports of the frame, plot the shear, the moment, and the axial force diagrams, and sketch the elastic curve of the frame.

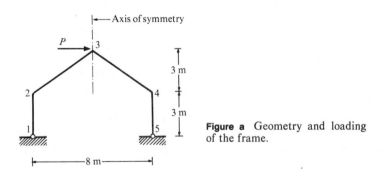

Figure a Geometry and loading of the frame.

---

solution   This structure has an axis of symmetry. Moreover, its loading is antisymmetric with respect to this axis. Thus, the reactions at the supports of the frame are antisymmetric as shown in Fig. b. Notice, that the structure under consideration is indeterminate to the first degree. However, because of its symmetry and the antisymmetry of the loading, its reactions can be found without employing a compatibility relation. This indicates that the antisymmetry of the loading and of the reactions of this symmetric structure has imposed upon its response the restriction which is

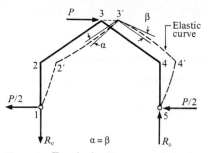

Figure b Free-body diagram and elastic curve of the frame.

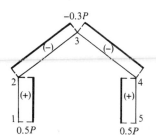

Figure   c Shear   diagram (symmetric).

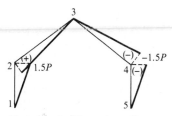

Figure   d Moment   diagram (antisymmetric).

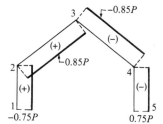

Figure e Axial force diagram (antisymmetric).

required for the compatibility of its deformation. Referring to Fig. b and taking moments about point 1, we have

$$R_v = \tfrac{3}{4}P$$

The shear, moment, and axial force diagrams and the elastic curve of the frame are shown in Figs. c, d, e and b, respectively.

Example 2.   Consider the frame loaded with a horizontal load as shown in Fig. a. The members of the frame are made from the same material and have the same constant

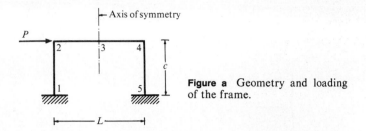

Figure a Geometry and loading of the frame.

cross section. Compute the reactions at the supports of the frame. Plot the moment diagram and sketch the elastic curve of the frame.

---

solution This structure has an axis of symmetry. However, its loading is not symmetric. In order to employ the properties of symmetric structures, subjected to symmetric or antisymmetric loads, the given loading is replaced by the sum of a symmetric and an antisymmetric loading as shown in Fig. b.

It can be seen that since we are disregarding the influence of axial deformation, the symmetric load induces only an axial compressive force equal to $P/2$ in the horizontal member. Thus, we have to analyze only the structure subjected to the antisymmetric loading. Therefore, in this case, the analysis of the structure subjected separately to the symmetric and antisymmetric components of the given loading is preferable to the analysis of the structure subjected directly to the given loading.

As a result of the symmetry of the structure and the antisymmetry of the loads, the moment and axial force at point 3 must vanish. Thus, we choose the shearing force at point 3 as the redundant. Referring to Fig. c, we see that the moment in the primary structure, when subjected to the actual loading, is given by

$$M = -\frac{Px_1'}{2} \qquad 0 \le x_1' \le c$$

Referring to Fig. c, we see that the moment in the primary structure when subjected to $X_1 = 1$ is given as

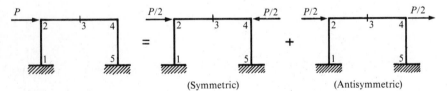

(Symmetric)   (Antisymmetric)

Figure b Replacement of nonsymmetric loading by the sum of symmetric and antisymmetric loadings.

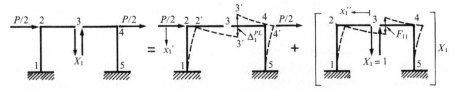

Figure c Analysis of the frame subjected to the antisymmetric loading.

$$\tilde{M} = -\frac{L}{2}, \qquad 0 \le x_1' \le c$$

$$\tilde{M} = x_1'', \qquad 0 \le x_1'' \le \frac{L}{2}$$

Thus, using relation (5.66), we see that the displacement Δ_1^{PL} is equal to

$$\Delta_1^{PL} = 2 \int_0^c \frac{M\tilde{M}}{EI} dx_1' = 2 \int_0^c \frac{PLx_1'}{4EI} dx_1' = \frac{PLc^2}{4EI} \qquad \text{(a)}$$

The flexibility coefficient F_{11} is

$$F_{11} = 2 \int_0^c \left(\frac{L}{2}\right)^2 \frac{dx_1'}{EI} + 2 \int_0^{L/2} \frac{(x_1'')^2 dx_1''}{EI}$$

$$= \frac{L^2}{2EI}\left(c + \frac{L}{6}\right) \qquad \text{(b)}$$

Substituting relations (a) and (b) into (7.20), we get

$$X_1 = -\frac{3Pc^2}{L(6c + L)}$$

The minus sign indicates that X_1 acts opposite to the direction assumed in Fig. c.

The free-body diagram of the left half of the frame is shown in Fig. d, while the moment diagram for the frame is shown in Fig. e and the elastic curve of the frame is shown in Fig. f.

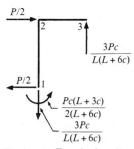

Figure d Free-body diagram of the left half of the frame.

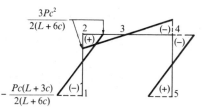

Figure e Moment diagram for the frame.

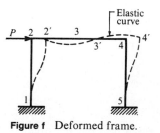

Figure f Deformed frame.

Example 3.   Consider the structure loaded as shown in Fig. a. The members of this structure are made from the same material and have the same constant cross section. Compute the reactions at its supports, plot the moment diagram, and draw the elastic curve of the structure.

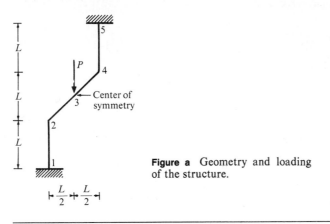

Figure a   Geometry and loading of the structure.

solution

STEP 1.   This structure is symmetric with respect to point 3. Moreover, the loading is antisymmetric with respect to this point (see Chap. 2). Thus, the reactions at the supports of the structure assume the values shown in Fig. b. Notice that the structure under consideration is indeterminate to the third degree. However, because of its symmetry and the antisymmetry of the loading, only one pair of equal moments is not known. Thus, we need only one compatibility equation to establish the unknown pair of moments.

STEP 2.   We compute the displacement Δ_1^{PL} of the primary structure. This is the sum

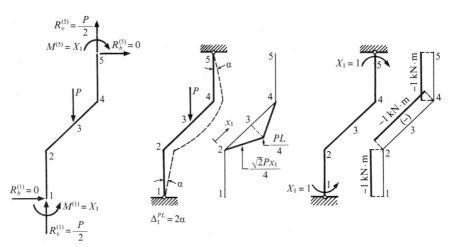

Figure b   Free-body diagram of the structure.

Figure c   Primary structure subjected to the actual load and corresponding moment diagram.

Figure d   Primary structure subjected to $X_1 = 1$ and corresponding moment diagram.

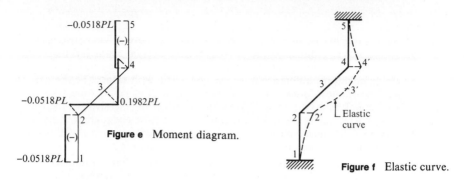

Figure e Moment diagram.

Figure f Elastic curve.

of the rotations of points 1 and 5 of the primary structure. The primary structure loaded with the actual load is shown in Fig. c. Referring to this figure and using the method of virtual work (5.66) with the virtual loading of Fig. d, we obtain

$$\Delta_1^{PL} = 2 \int_0^{L\sqrt{2}/2} (-1) \frac{\sqrt{2}}{4} Px_1 \, dx_1 = -\frac{\sqrt{2}}{8} PL^2 \qquad \text{(a)}$$

STEP 3.   We compute the flexibility coefficient F_{11} of the structure corresponding to our choice of redundant. This is equal to the sum of the rotations of points 1 and 5 of the primary structure when subjected to a unit value of the redundant. Referring to Fig. d, and employing relation (5.66), we get

$$F_{11} = 2 \int_0^L dx_1 + \int_0^{L\sqrt{2}} dx_1 = L(\sqrt{2} + 2) \qquad \text{(b)}$$

STEP 4.   We compute the redundant. Substituting relations (a) and (b) into (7.20), we obtain

$$X_1 = -\frac{\Delta_1^{PL}}{F_{11}} = \frac{\sqrt{2}}{8(2 + \sqrt{2})} PL = 0.0518PL$$

STEP 5.   For moment diagram see Fig. e; for elastic curve see Fig. f.

Example 4.   The members of the frame of Fig. a are made of the same material and have the same constant cross section of moment of inertia I and depth h. Plot the moment diagram for the frame when its members are subjected to a temperature T_i at their internal fibers and to a temperature T_e at their external fibers.

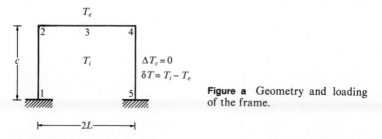

Figure a Geometry and loading of the frame.

solution

STEP 1. This frame has an axis of symmetry. In order to have a symmetric primary structure, we choose as redundants the internal actions of point 3 which is located on the axis of symmetry of the frame. From physical intuition we can deduce that the effect of the temperature difference must be symmetric with respect to the axis of symmetry of the frame. Consequently, the shearing force at point 3 must be zero. Moreover, inasmuch as ΔT_c is equal to zero, the length of the members of the frame does not change. Thus joints 2 and 4 of the frame do not move and moreover the axial force in member 2,3,4 must vanish. Hence, only the moment at point 3 of this statically interdeterminate to the third degree structure is not known.

STEP 2. We compute the displacement Δ_1^{PL} of the primary structure subjected to the given temperature difference (see Fig. b) using the method of virtual work. Thus refer-

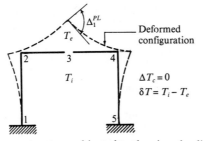

Figure b Primary structure subjected to the given loading.

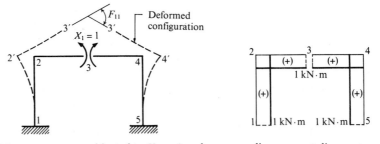

Figure c Primary structure subjected to $X_1 = 1$ and corresponding moment diagram.

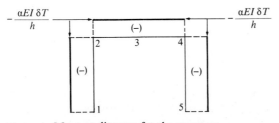

Figure d Moment diagram for the structure.

ring to relation (5.66) and using the virtual loading shown in Fig. c we obtain

$$\Delta_1^{PL} = \frac{2\alpha \, \delta T}{h} \left(\int_0^L dx_1 + \int_0^c dx_1 \right) = \frac{2a \, \delta T \, (L + c)}{h} \tag{a}$$

STEP 3.   We compute the flexibility coefficient F_{11} of the structure corresponding to the redundant. Thus using relation (5.66) and referring to Fig. c we get:

$$F_{11} = 2 \int_0^L \frac{dx_1}{EI} + 2 \int_0^c \frac{dx_1}{EI} = \frac{2(L + c)}{EI} \tag{b}$$

STEP 4.   Substituting relations (a) and (b) into the compatibility equation (7.20), we obtain:

$$X_1 = -\frac{\Delta_1^{PL}}{F_{11}} = -\frac{\alpha \, EI \, \delta T}{h}$$

STEP 5.   Using the above results, we plot the moment diagram in Fig. d. As can be seen from App. C, the calculated moment in any member of the frame due to the given change of temperature is equal to the corresponding moment in this member when subjected to the given change of temperature with its ends fixed. Thus it is apparent that the joints of the frame do not rotate.

Example 5.   Determine the internal actions, plot the moment diagram, and sketch the elastic curve of the ring loaded as shown in Fig. a. The ring has a constant cross section.

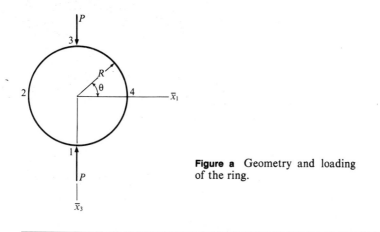

Figure a Geometry and loading of the ring.

solution

STEP 1.   This structure is symmetric with respect to any axis in its plane passing through its geometric center. Moreover, the loading is symmetric with respect to the $\bar{x}_1$ and the $\bar{x}_3$ axes. Because of the symmetry of the ring and the symmetry of the loading with respect to the $\bar{x}_1$ and $\bar{x}_3$ axes, the internal actions in the ring must be

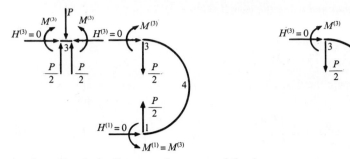

Figure b   Free-body diagrams of segments of the ring.

symmetric with respect to these axes. Thus, referring to Fig. b, we can readily see that from the equilibrium of point 3, the radial component of the internal force to the left or to the right of this point is equal to $P/2$. Moreover, from the equilibrium of portion $\widehat{341}$ of the ring, it is apparent that the tangential component of the internal force at point 3 vanishes. Consequently, we have only one unknown internal action. We choose the two equal pairs of moments at points 2 and 4 of the structure as the redundant X_1 (see Fig. c). Thus, the primary structure is obtained from the actual structure by introducing hinges at points 2 and 4. This choice of redundant results in a primary structure which is symmetric with respect to the $\bar{x}_1$ and $\bar{x}_3$ axes (see Fig. c).

STEP 2.   We compute Δ_1^{PL}. The primary structure loaded with the given external forces is shown in Fig. d. Referring to Fig. e, we see that the moment in the primary structure when loaded with the given forces is equal to

$$M^{PL} = \frac{P}{2}(R - \bar{x}_1) = \frac{PR}{2}(1 - \cos\theta) \qquad 0 \le \theta < \frac{\pi}{2} \qquad \text{(a)}$$

When the primary structure is subjected to the given loading, the tangents to the elastic curve to the left and the right of the hinge at points 2 and 4 are not parallel to the $\bar{x}_3$ axis but form an angle α with it (see Fig. d). However, when the actual structure is subjected to the given loading, the tangents at points 2 and 4 are parallel to the $\bar{x}_3$ axis. Thus, when the primary structure is subjected to the given loading and to

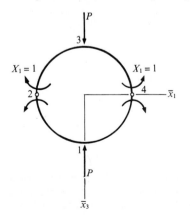

Figure c   Primary structure subjected to the given loading and to the redundant.

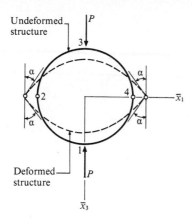

Undeformed structure

Deformed structure

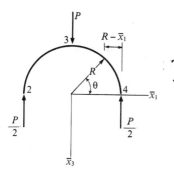

Figure d Primary structure subjected to the given loading.

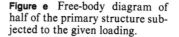

Figure e Free-body diagram of half of the primary structure subjected to the given loading.

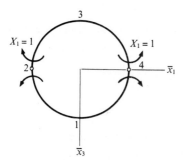

Figure f Primary structure subjected to $X_1 = 1$.

the redundants, the angle α must vanish. Therefore, we would compute the angles $\Delta_1^{PL} = 4\alpha$ of the primary structure subjected to the given forces and the corresponding angles $F_{11}X_1$ of the primary structure subjected to the two pairs of moments X_1 and set their sum equal to zero. The displacement Δ_1^{PL} may be obtained using the method of virtual work (5.66) with the virtual loading shown in Fig. f. Because of this loading,

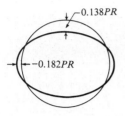

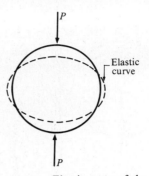

Figure g   Moment diagram for the ring.   **Figure h**   Elastic curve of the ring.

the moment at any cross section of the primary structure is equal to unity. Thus

$$\tilde{M} = 1 \tag{b}$$

Substituting relations (a) and (b) into (5.66) we get

$$\Delta_1^{PL} = \int \frac{M^{PL}\tilde{M}\, ds}{EI} = 4 \int_0^{\pi/2} \frac{PR^2(1 - \cos\theta)}{2EI}\, d\theta$$

$$= 2\frac{PR^2}{EI}\left[\theta - \sin\theta \right]_0^{\pi/2} = \frac{PR^2}{EI}(\pi - 2) \tag{c}$$

STEP 3.   We compute the flexibility coefficient of the structure corresponding to our choice of redundant. Substituting relation (b) into (5.66), we obtain

$$F_{11} = \int \frac{\tilde{M}\tilde{M}\, ds}{EI} = 4 \int_0^{\pi/2} \frac{(1)(1)\, R\, d\theta}{EI} = \frac{2R\pi}{EI} \tag{d}$$

STEP 4.   We compute the redundant. Substituting relations (c) and (d) into the compatibility relation (7.20), we have

$$X_1 = -\frac{\Delta_1^{PL}}{F_{11}} = -\frac{PR(\pi - 2)}{\pi} = -0.182PR$$

STEP 5.   The moment diagram and the elastic curve of the ring are shown in Figs. g and h, respectively.

Example 6.   Plot the shear and moment (bending and twisting) diagrams for the grid subjected to the loading shown in Fig. a. The members of the grid are circular steel tubes of constant cross section ($G = E/2.6$).

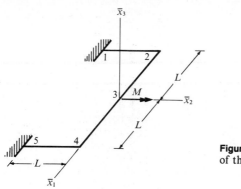

Figure a Geometry and loading of the grid.

---

solution   By definition a grid is a structure whose members lie in one plane. Moreover, the external forces act normal to this plane while the vector of the external moments lies in this plane. Referring to Figs. a and b, we see that the members of the grid are subjected only to a shearing force Q_3, to a bending moment M_2, and to a twisting moment M_1. Moreover, the members of the grid do not rotate about the $\bar{x}_3$ axis ($\theta_3^{(3)} = 0$), and they translate only in the direction of the $\bar{x}_3$ axis. The grid under consideration has a plane of symmetry (the $\bar{x}_2\bar{x}_3$ plane) and an axis of symmetry (the $\bar{x}_2$ axis).

STEP 1.   In this grid, the loading is antisymmetric with respect to the plane x_2x_3 and symmetric with respect to the axis x_2. Thus, as discussed in Sec. 2.4, as a result of the symmetry of the structure and the antisymmetry of loading with respect to the plane x_2x_3, the component of translation $u_3^{(3)}$ and the component of rotation $\theta_1^{(3)}$ vanish at point 3, while on each side of point 3, the magnitude of the component of moment $M_2^{(3)}$ is equal to half the given external moment M. Moreover, as a result of the symmetry of the structure and the loading with respect to the axis x_2, the component of translation $u_3^{(3)}$ and the component of rotation $\theta_3^{(3)}$ vanish at point 3, while on each side of point 3 the magnitude of the moment $M_2^{(3)}$ is equal to $M/2$. Consequently, if we cut the structure to the left and to the right of point 3 as shown in Fig. b, a known bending moment $M_2^{(3)}$, an unknown twisting moment $M_1^{(3)}$, and an unknown shearing

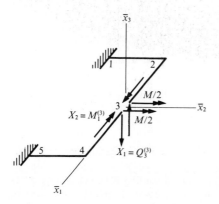

Figure b Primary structure subjected to the given loading and to the redundants.

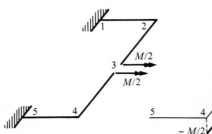

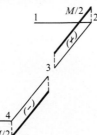

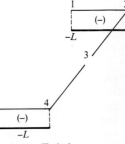

Figure c Primary structure subjected to the given loading.

Figure d Bending moment diagram for the structure of Fig. c.

Figure e Twisting moment diagram for the structure of Fig. c.

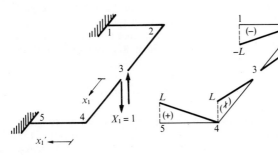

Figure f Primary structure subjected to $X_1 = 1$.

Figure g Bending moment diagram for the structure of Fig. f.

Figure h Twisting moment diagram for the structure of Fig. f.

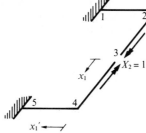

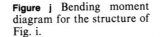

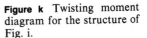

Figure i Primary structure subjected to $X_2 = 1$.

Figure j Bending moment diagram for the structure of Fig. i.

Figure k Twisting moment diagram for the structure of Fig. i.

force $Q_3^{(3)}$ act on the cross sections at the cut. We choose the shearing force $Q_3^{(3)}$ and the twisting moment $M_1^{(3)}$ as the redundants.

STEP 2.   The primary structure subjected to the given loading is shown in Fig. c. Its displacements Δ_1^{PL} and Δ_2^{PL} are computed using the method of virtual work (5.6) in conjunction with the virtual loading shown in Figs. f and i, respectively. Thus, referring to Figs. d and g and to Figs. e and h and noting that $J = 2I$ we get

$$\Delta_1^{PL} = 2 \int_{pt.3}^{pt.4} \frac{M_2 \tilde{M}_2 \, dx_1}{EI} + 2 \int_{pt.4}^{pt.5} \frac{M_1 \tilde{M}_1 \, dx_1'}{GJ}$$

$$= 2 \int_0^L \left(-\frac{M}{2}\right) \frac{x_1 \, dx_1}{EI} - 2 \int_0^L \left(\frac{M}{2}\right) \frac{L \, dx_1'}{GJ}$$

$$= -\frac{ML^2}{2EI} - \frac{ML^2}{GJ} = -\frac{ML^2}{2EI}(1 + 2.6) = -1.8 \frac{ML^2}{EI} \qquad (a)$$

Moreover, referring to Figs. d and j and to Figs. e and k, we obtain

$$\Delta_2^{PL} = 0 \qquad (b)$$

STEP 3.   The flexibility coefficients are computed using relation (5.66). Thus, referring to Figs. g and h, we have

$$F_{11} = 2 \int_{pt.3}^{pt.4} \frac{(M_2)^2 \, dx_1}{EI} + 2 \int_{pt.4}^{pt.5} \frac{(M_2)^2 \, dx'}{EI} \frac{1}{} + 2 \int_{pt.4}^{pt.5} \frac{(M_1)^2 \, dx_1'}{GJ}$$

$$= 2 \int_0^L \frac{(x_1)^2 \, dx_1}{EI} + 2 \int_0^L \frac{(x_1')^2 \, dx_1'}{EI} + 2 \int_0^L \frac{L^2 \, dx_1'}{GJ} = \frac{4}{3} \frac{L^3}{EI} + \frac{2L^3}{GJ}$$

$$= \frac{L^3}{EI}(\frac{4}{3} + 2.6) = 3.933 \frac{L^3}{EI} \qquad (c)$$

Referring to Figs. g and j and to Figs. h and k, we obtain

$$F_{12} = F_{21} = 2 \int_{pt.4}^{pt.5} \frac{M_2 \tilde{M}_2}{EI} \, dx_1 = 2 \int_0^L \frac{(x_1)(1) \, dx_1}{EI} = \frac{L^2}{EI} \qquad (d)$$

Referring to Figs. j and k we get

$$F_{22} = 2 \int_{pt.4}^{pt.5} \frac{(M_2)^2 \, dx_1}{EI} + 2 \int_{pt.3}^{pt.4} \frac{(M_1)^2 \, dx_1}{GJ} = 2 \int_0^L \frac{dx_1}{EI} + 2 \int_0^L \frac{dx_1}{GJ}$$

$$= \frac{2L}{EI} + \frac{2L}{GJ} = \frac{2L}{EI}(1 + 1.3) = \frac{4.6L}{EI} \qquad (e)$$

STEP 4.   The redundants are computed by substituting relations (a) to (e) into (7.20). Thus

$$\begin{Bmatrix} 0 \\ 0 \end{Bmatrix} = \frac{ML^2}{EI} \begin{Bmatrix} -1.8 \\ 0 \end{Bmatrix} + \frac{L}{EI} \begin{bmatrix} 3.933L^2 & L \\ L & 4.6 \end{bmatrix} \begin{Bmatrix} X_1 \\ X_2 \end{Bmatrix}$$

or

$$\begin{Bmatrix} X_1 \\ X_2 \end{Bmatrix} = M \begin{bmatrix} 0.269L^{-1} & -0.058 \\ -0.058 & 0.230L \end{bmatrix} \begin{Bmatrix} 1.8 \\ 0 \end{Bmatrix} = M \begin{Bmatrix} 0.48L^{-1} \\ -0.104 \end{Bmatrix}$$

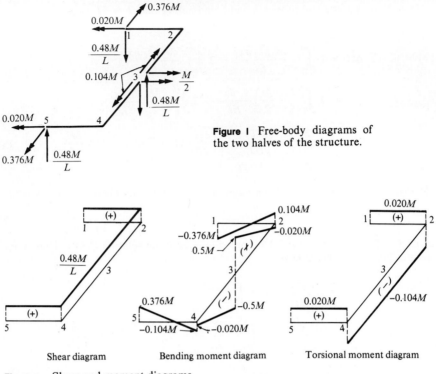

Figure l Free-body diagrams of the two halves of the structure.

Shear diagram   Bending moment diagram   Torsional moment diagram

Figure m Shear and moment diagrams.

STEP 5.   The free-body diagrams of the two halves of the structure are shown in Fig. l. Referring to this figure, the shear and moments diagrams are plotted in Fig. m.

7.6   Analysis of Statically Indeterminate Arches

When a vertical load acts on a horizontal beam, it produces only vertical reactions at its supports. However, when a vertical load acts on an arch, it produces both horizontal and vertical reactions at its supports. The horizontal reactions produce a moment which tends to decrease the moment due to the vertical reactions and, consequently, the resulting moment at any cross section of the arch is less than that in the corresponding cross section of a straight beam spanning the same length as the arch (see Sec. 3.4).

An arch may be three hinged, hinged at both ends, fixed at one end and hinged at the other end, or fixed at both ends. Steel arches are usually constructed as two- or three-hinged arches because of the difficulty of ensuring full fixity at the supports. Concrete arches are usually constructed as fixed-end arches.

Many constructed arches are symmetric because they are more aesthetic than the nonsymmetric arches. Therefore, we will limit our attention to symmetric arches. It should be emphasized, however, that the analysis of nonsymmetric arches is not more difficult than the analysis of symmetric arches except that the calculations are lengthier.

The axis of an arch could have any geometry. However, the most commonly used arches are circular (see Fig. 7.11a) and parabolic (see Fig. 7.11b). The moment of inertia of the cross sections of parabolic arches varies from a minimum I_c at the crown to a maximum at the supports. The moment of inertia I of any cross section of a parabolic arch is usually given as

$$I = I_c \sec \phi \tag{7.41}$$

We limit our attention to arches whose deformation is in the range of validity of the theory of small deformation and whose radius of curvature is large compared to the depth of their cross section. Thus, as discussed in Sec. 5.8, we can apply to these arches the principle of virtual work in the form (5.66) derived for structures made of straight members.

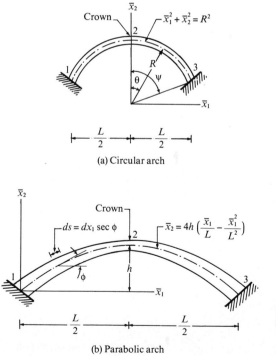

(a) Circular arch

(b) Parabolic arch

Figure 7.11 Circular and parabolic arches.

Simply supported arches having very long spans (over 120 m) deflect considerably; thus, the effect of the change of their geometry, due to their deformation, on their internal forces could result in significant errors.[1] Consequently, their analysis on the basis of the theory of small deformation is not satisfactory. Arches fixed at both ends are stiffer than simply supported arches; thus their analysis on the basis of the theory of small deformation is satisfactory.

Arches hinged at both ends are indeterminate structures to the first degree. They can be analyzed using the basic force method and choosing the horizontal reaction at the supports as the redundant. Arches fixed at both ends are indeterminate structures to the third degree. They may be analyzed using the basic force method of structural analysis taking as redundants either the reactions at the one support (see Fig. 7.12b) or the pairs of internal actions (axial forces, shearing forces, and bending moments) at the crown of the arch (see Fig. 7.12c).

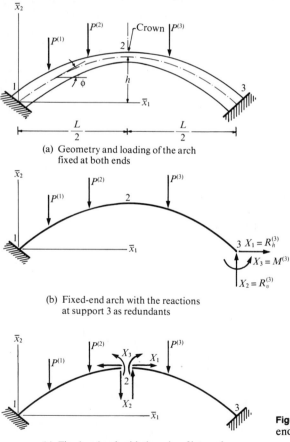

(a) Geometry and loading of the arch fixed at both ends

(b) Fixed-end arch with the reactions at support 3 as redundants

(c) Fixed-end arch with the pairs of internal actions at the crown as redundants

Figure 7.12 Analysis of a fixed-end arch by the force method.

Below we present the following three examples.

Example 1.   The horizontal reaction at the supports of a symmetric, simply supported, parabolic arch is computed using the basic force method. The arch is subjected to the following loading cases.

1. A vertical concentrated force
2. Vertical uniformly distributed forces

Example 2.   The reactions of a symmetric parabolic arch fixed at both ends are computed using the basic force method. The arch is subjected to a concentrated force.

Example 3.   The internal actions in the members of a statically indeterminate to the third degree symmetric frame are computed. The frame consists of a circular arch whose ends are rigidly connected to two columns. The frame is subjected to a uniformly distributed vertical force.

Example 1.   Consider a symmetric, hinged-at-both-ends, parabolic (with $I = I_c \sec \phi$) arch subjected to the following loading cases.

1. A vertical force P, as shown in Fig. a
2. A uniformly distributed load, as shown in Fig. b

Compute the horizontal reaction at the supports of the arch.

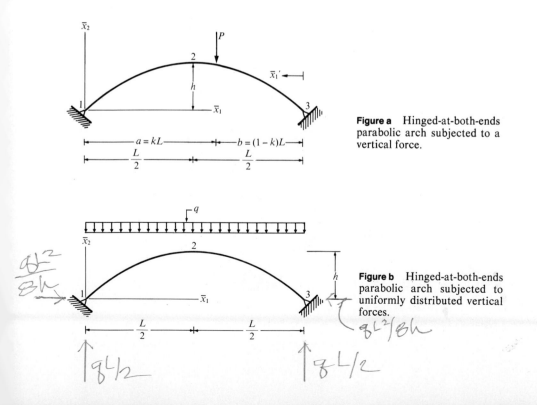

Figure a   Hinged-at-both-ends parabolic arch subjected to a vertical force.

Figure b   Hinged-at-both-ends parabolic arch subjected to uniformly distributed vertical forces.

solution   Referring to Fig. 7.11b, we have

$$ds = d\bar{x}_1 \sec \phi$$

Using the above relation and (7.41), we obtain

$$\frac{ds}{I} = \frac{d\bar{x}_1}{I_c} \tag{a}$$

Moreover, the equation of the arch is

$$\bar{x}_2 = 4h\left(\frac{\bar{x}_1}{L} - \frac{\bar{x}_1^2}{L^2}\right) \tag{b}$$

We choose as the redundant the horizontal reaction of the supports. Referring to Fig. c, we see that the moment in the primary structure subjected to $X_1 = 1$ is equal to

$$\tilde{M} = \bar{x}_2 \tag{c}$$

Substituting relation (c) into (5.66), disregarding the effect of the axial forces and using relation (a), we get

$$\Delta_1^{PL} = \int_0^L \frac{M^{PL}(\bar{x}_2)}{EI_c}\, d\bar{x}_1 \tag{d}$$

and

$$F_{11} = \int_0^L \frac{(\bar{x}_2)^2}{EI_c}\, d\bar{x}_1 \tag{e}$$

M^{PL} is the moment in the primary structure subjected to the given loads. Substituting relation (b) into (e) and integrating, we obtain

$$F_{11} = \frac{16h^2}{EI_c} \int_0^L \left(\frac{\bar{x}_1}{L} - \frac{\bar{x}_1^2}{L^2}\right)^2 d\bar{x}_1 = \frac{16h^2 L}{30EI_c} \tag{f}$$

Substituting relations (d) and (f) into the compatibility equations (7.20) and using relation (b), we obtain

$$X_1 = -\frac{\Delta_1^{PL}}{F_{11}} = -\frac{30\int_0^L M^{PL}\left(\frac{\bar{x}_1}{L} - \frac{\bar{x}_1^2}{L^2}\right) d\bar{x}_1}{4hL} \tag{g}$$

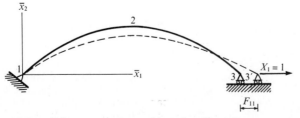

Figure c   Primary structure subjected to $X_1 = 1$.

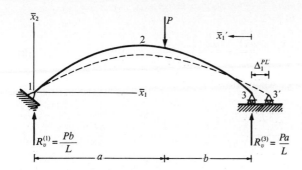

Figure d Free-body diagram of the primary structure subjected to the given force.

Loading Case of Fig. a.

Referring to Fig. d, the moment in the primary structure subjected to the given load is equal to

$$M^{PL} = \frac{Pb\bar{x}_1}{L} \qquad 0 \le \bar{x}_1 \le a$$

$$M^{PL} = \frac{Pa\bar{x}_1'}{L} \qquad 0 \le \bar{x}_1' \le b \tag{h}$$

Substituting relations (h) into (g), we get

$$X_1 = -\frac{30\left[\dfrac{Pb}{L}\displaystyle\int_0^a \bar{x}_1\left(\dfrac{\bar{x}_1}{L} - \dfrac{\bar{x}_1^2}{L^2}\right)dx_1 + \dfrac{Pa}{L}\displaystyle\int_0^b \bar{x}_1'\left[\dfrac{\bar{x}_1'}{L} - \dfrac{(\bar{x}_1')^2}{L^2}\right]dx_1'\right]}{4hL}$$

$$= -\frac{5PLk}{8h}(1-k)(1+k-k^2) \tag{i}$$

where $k = a/L$. When the load P is at the midspan ($k = \frac{1}{2}$), relation (i) reduces to

$$X_1 = -\frac{5PL}{16h}\left(\frac{1}{2}\right)\left(\frac{5}{4}\right) = -\frac{25}{128}\frac{PL}{h} \tag{j}$$

Loading Case of Fig. b

We choose the horizontal reaction at the supports of the arch as the redundant. The primary structure is shown in Fig. e subjected to the given loading. The moment in

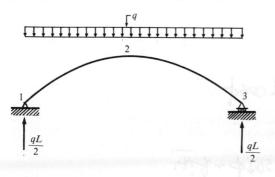

Figure e Free-body diagram of the primary structure subjected to the given loading.

the primary structure is equal to

$$M^{PL} = \frac{qL\bar{x}_1}{2} - \frac{q\bar{x}_1^2}{2} = \frac{q\bar{x}_1(L - \bar{x}_1)}{2} \tag{k}$$

Substituting relation (k) into (g), we obtain

$$X_1 = -\frac{15q \int_0^L \frac{\bar{x}_1}{2}(L - \bar{x}_1)\left(\frac{\bar{x}_1}{L} - \frac{\bar{x}_1^2}{L^2}\right)d\bar{x}}{4hL} = -\frac{qL^2}{8h} \tag{l}$$

From the equilibrium of the segment of the arch shown in Fig. f, we get

$$M = \frac{qL}{2}\bar{x}_1' - \frac{qL^2}{8h}\bar{x}_2 - \frac{q\bar{x}_1'^2}{2} = \frac{q}{2}\left[L\bar{x}_1' - L^2\left[\frac{\bar{x}_1'}{L} - \frac{(\bar{x}_1')^2}{L^2}\right] - (\bar{x}_1')^2\right]$$

$$= 0 \tag{m}$$

$$\Sigma\bar{F}_h = 0 \quad N\cos\phi - Q\sin\phi = \frac{qL^2}{8h} \tag{n}$$

$$\Sigma\bar{F}_v = 0 \quad N\sin\phi + Q\cos\phi = \frac{qL}{2} - q\bar{x}_1' \tag{o}$$

Multiplying relation (n) by $\sin\phi$ and relation (o) by $\cos\phi$ and subtracting the first of the resulting relations from the second we get

$$Q = \frac{qL\cos\phi}{2} - q\bar{x}_1'\cos\phi - \frac{qL^2\sin\phi}{8h} = q\cos\phi\left(\frac{L}{2} - \bar{x}_1' - \frac{L^2\tan\phi}{8h}\right) \tag{p}$$

The slope of the axis of the arch may be obtained by differentiating Eq. (b). Thus

$$\tan\phi = \frac{d\bar{x}_2}{d\bar{x}_1'} = \frac{8h}{L^2}\left(\frac{L}{2} - \bar{x}_1'\right) \tag{q}$$

Substituting relation (q) into (p) we obtain

$$Q = q\cos\phi\left[\frac{L}{2} - \bar{x}_1' - \frac{L^2}{8h}\left(\frac{8h}{L^2}\right)\left(\frac{L}{2} - \bar{x}_1'\right)\right] = 0$$

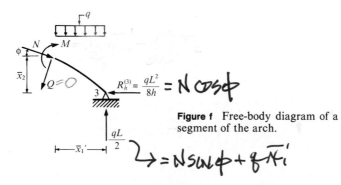

Figure f Free-body diagram of a segment of the arch.

Thus, when a symmetric, hinged-at-both-ends, parabolic arch with $I = I_c \sec \phi$ is subjected to a uniformly distributed vertical load, only an axial force acts on its cross section.

Example 2. Consider a symmetric, fixed-at-both-ends, parabolic arch with $I = I_c \sec \phi$, subjected to a concentrated force P, as shown in Fig. a. Compute the reactions at the supports of the arch.

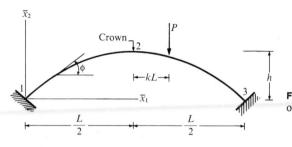

Figure a Geometry and loading of the arch.

solution Referring to Fig. 7.11b we have

$$ds = d\bar{x}_1 \sec \phi$$

Using the above relation and (7.41) we obtain

$$\frac{ds}{I} = \frac{d\bar{x}_1}{I_c} \tag{a}$$

Moreover, the equation of the arch is

$$\bar{x}_2 = 4h \left(\frac{\bar{x}_1}{L} - \frac{\bar{x}_1^2}{L^2} \right) \tag{b}$$

STEP 1. We choose the internal actions at the crown as the redundants.

STEP 2. Using the method of virtual work, we compute the dispacements of the primary structure corresponding to the redundants when the primary structure is subjected to the given force (see Fig. b). The corresponding moment is equal to

$$M^{PL} = -P(x_1 - kL) \qquad kL < x_1 < L \tag{b}$$

The primary structure subjected to the virtual loading for computing the displace-

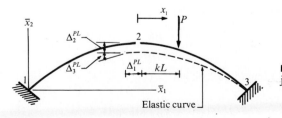

Figure b Primary structure subjected to the given loading.

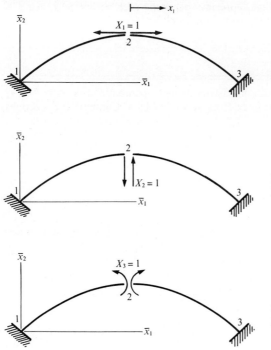

Figure c Primary structure subjected to $X_1 = 1$.

Figure d Primary structure subjected to $X_2 = 1$.

Figure e Primary structure subjected to $X_3 = 1$.

ments Δ_1^{PL}, Δ_2^{PL}, and Δ_3^{PL} is shown in Figs. c, d, and e, respectively. The corresponding moments are equal to

$$M^{X_1=1} = h - \bar{x}_2 = \frac{4hx_1^2}{L^2} \qquad 0 < x_1 < \frac{L}{2} \tag{c}$$

$$M^{X_2=1} = x_1 \qquad 0 < x_1 < \frac{L}{2} \tag{d}$$

$$M^{X_3=1} = 1 \qquad 0 < x_1 < \frac{L}{2} \tag{e}$$

Substituting relations (b) and (c) into (5.66) we get

$$\Delta_1^{PL} = -\frac{1}{EI_c} \int_{kL}^{L/2} P(x_1 - kL) \frac{4hx_1^2}{L^2} dx_1 = -\frac{PhL^2}{48EI_c} (3 - 8k + 16k^4) \tag{f}$$

Substituting relations (b) and (d) into (5.66) we obtain

$$\Delta_2^{PL} = -\frac{1}{EI_c} \int_{kL}^{L/2} P(x_1 - kL)x_1 \, dx_1 = -\frac{PL^3}{24EI_c} (1 - 3k + 4k^3) \tag{g}$$

Substituting relations (b) and (e) into (5.66), we get

$$\Delta_3^{PL} = -\frac{1}{EI_c} \int_{kL}^{L/2} P(x_1 - kL) \, dx_1 = -\frac{PL^2}{8EI_c} (1 - 4k + 4k^2) \tag{h}$$

STEP 3.   Using the method of virtual work, we compute the flexibility coefficients of the structure. Thus, referring to relations (c), (d), and (e), we obtain

$$F_{11} = \frac{2}{EI_c} \int_0^{L/2} (M^{X_1=1})^2 \, dx_1 = \frac{32h^2}{EI_cL^4} \int_0^{L/2} x_1^4 \, dx_1 = \frac{h^2L}{5EI_c} \tag{i}$$

$$F_{12} = F_{21} = 0 \tag{j}$$

$$F_{22} = \frac{2}{EI_c} \int_0^{L/2} (M^{X_2=1})^2 \, dx_1 = \frac{2}{EI_c} \int_0^{L/2} x_1 \, dx_1 = \frac{L^3}{12EI_c} \tag{k}$$

$$F_{13} = F_{31} = \frac{2}{EI_c} \int_0^{L/2} (M^{X_2=1})(M^{X_3=1}) \, dx_1 = \frac{8h}{EI_cL^2} \int_0^{L/2} x^2 \, dx_1 = \frac{hL}{3EI_c} \tag{l}$$

$$F_{33} = \frac{2}{EI_c} \int_0^{L/2} (M^{X_3=1})^2 \, dx_1 = \frac{2}{EI_c} \int_0^{L/2} dx_1 = \frac{L}{EI_c} \tag{m}$$

$$F_{23} = F_{32} = 0 \tag{n}$$

STEP 4.   We compute the redundants. Substituting relations (f) to (n) into the equations of compatibility (7.20), we have

$$-\frac{PhL^2}{48} (3 - 8k + 16k^4) + \frac{h^2LX_1}{5} + \frac{hLX_3}{3} = 0$$

$$-\frac{PL^3}{24} (1 - 3k + 4k^3) + \frac{L^3X_2}{12} = 0 \tag{o}$$

$$-\frac{PL^2}{8} (1 - 4k + 4k^2) + \frac{hLX_1}{3} + LX_3 = 0$$

Solving Eq. (o), we get

$$X_1 = \frac{15PL}{64h} (1 - 8k^2 + 16k^4)$$

$$X_2 = \frac{P}{2} (1 - 3k + 4k^3)$$

$$X_3 = \frac{PL}{64} (3 - 32k + 72k^2 - 80k^4)$$

The reactions at the supports of the arch are obtained by referring to the free-body diagrams shown in Fig. f. That is,

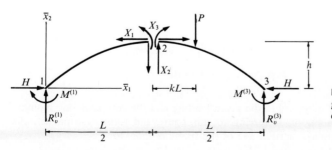

Figure f Free-body diagrams of the two halves of the arch.

$$H = X_1$$

$$R_v^{(1)} = X_2$$

$$M^{(1)} = X_3 + X_1 h - X_2 \frac{L}{2}$$

$$R_v^{(3)} = P - X_2$$

$$M^{(3)} = X_3 + X_1 h - X_2 \frac{L}{2} - P\left(\frac{L}{2} - kL\right)$$

Example 3.   Consider the frame subjected to a uniform load shown in Fig. a. Compute the internal actions in its members and plot its moment diagram. The members of the frame have the same constant cross section and are made from the same material.

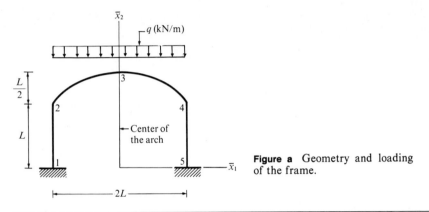

Figure a Geometry and loading of the frame.

solution   This frame is symmetric with respect to the axis $\bar{x}_2$. Moreover, it is subjected to a load which is symmetric with respect to the same axis. Consequently, the shearing force at point 3 vanishes. We choose as the redundants the internal moment and the axial force at point 3. The primary structure subjected to the given loading and to the chosen redundants is shown in Fig. b. Referring to this figure, from geometric considerations, we have

$$R^2 = L^2 + (R - L/2)^2 \tag{a}$$

$$\cos \psi = \frac{R - L/2}{R} \qquad \sin \psi = \frac{L}{R}$$

or

$$R = 1.25L \qquad \cos \psi = \tfrac{3}{5} \qquad \sin \psi = \tfrac{4}{5} \tag{b}$$

$$\psi = 53.13° = 0.93 \text{ rad}$$

The primary structure subjected to the given loading and the corresponding moment diagram are shown in Fig. c. Referring to this figure and using relation (a), the moment at a cross section of portion $\overset{\frown}{345}$ of the frame is equal to

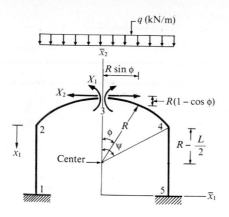

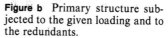

Figure b Primary structure subjected to the given loading and to the redundants.

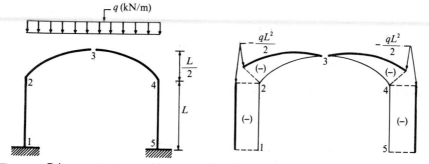

Figure c Primary structure subjected to the given loading and corresponding moment diagram.

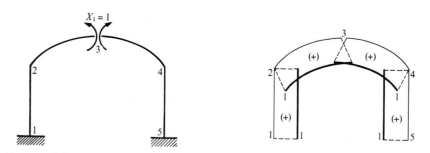

Figure d Primary structure subjected to $X_1 = 1$ and corresponding moment diagram.

$$M^{PL} = -\frac{q(R\sin\phi)^2}{2} = -0.78qL^2\sin^2\phi \qquad 0 \le \phi \le 53.13°$$

$$M^{PL} = -\frac{qL^2}{2} \qquad\qquad\qquad 0 \le x_1 \le L \qquad\qquad (c)$$

The primary structure subjected to the pair of unit moments $X_1 = 1$ is shown in Fig. d. The corresponding moment at the cross sections of the primary structure is equal to

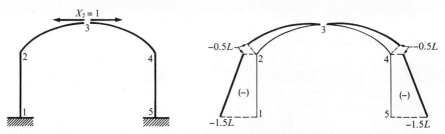

Figure e   Primary structure subjected to $X_2 = 1$ and corresponding moment diagram.

$$M^{(1)} = 1 \qquad\qquad\qquad\qquad \text{(d)}$$

The primary structure subjected to the pair of unit forces $X_2 = 1$ is shown in Fig. e. The corresponding moment at a cross section of the portion $\overset{\frown}{345}$ of the primary structure is equal to

$$M^{(2)} = -R(1 - \cos \phi) = -1.25L(1 - \cos \phi) \qquad 0 \le \phi \le 53.13° \qquad \text{(e)}$$

$$M^{(2)} = -\frac{L}{2} - x_1 \qquad\qquad\qquad\qquad 0 \le x_1 \le L$$

We compute the displacements Δ_1^{PL} and Δ_2^{PL} using the method of virtual work. Thus, substituting relations (c) and (d) into (5.66) and noting that for a circular arc $ds = R\,d\phi = 1.25L\,d\phi$, we obtain

$$\Delta_1^{PL} = \int \frac{M^{PL}M^{(1)}\,ds}{EI}$$

$$= -2\int_0^{0.93\ \text{rad}} \frac{0.78qL^2 \sin^2 \phi\,(1.25L)\,d\phi}{EI} - 2\int_0^L \frac{qL^2\,dx_1}{2EI} \qquad \text{(f)}$$

$$= -1.438\,\frac{qL^3}{EI}$$

Moreover, substituting relations (c) and (e) into (5.66), we get

$$\Delta_2^{PL} = \int \frac{M^{PL}M^{(2)}\,ds}{EI}$$

$$= 2\int_0^{0.93\ \text{rad}} \frac{0.78qL^2 \sin^2 \phi\,(1.25L)(1 - \cos \phi)(1.25L)\,d\phi}{EI}$$

$$+ 2\int_0^L \frac{qL^2(\tfrac{1}{2}L + x_1)\,dx_1}{2EI} \qquad \text{(g)}$$

$$= 2.44\frac{qL^4}{EI}\int_0^{0.93\ \text{rad}} (\sin^2 \phi - \sin^2 \phi \cos \phi)\,d\phi + \frac{qL^4}{EI} = 1.131\,\frac{qL^4}{EI}$$

We compute the flexibility coefficients of the frame by substituting relations (d) and (e) into (5.66). Thus

$$F_{11} = \int \frac{(M^{(1)})^2\,ds}{EI} = 2\int_0^{0.93\ \text{rad}} \frac{1.25L\,d\phi}{EI} + 2\int_0^L \frac{dx_1}{EI} = \frac{4.32L}{EI} \qquad \text{(h)}$$

$$F_{12} = \int \frac{M^{(1)} M^{(2)} \, ds}{EI} \tag{i}$$

$$= -2 \int_0^{0.93 \text{ rad}} \frac{1.25L(1 - \cos \phi)(1.25L) \, d\phi}{EI} - 2 \int_0^L \frac{(0.5L + x_1) \, dx_1}{EI}$$

$$= -2.40 \frac{L^2}{EI}$$

$$F_{22} = \int \frac{[M^{(2)}]^2 \, ds}{EI}$$

$$= \frac{2}{EI} \int_0^{0.93 \text{ rad}} [1.25L(1 - \cos \phi)]^2 (1.25L) \, d\phi + \frac{2}{EI} \int_0^L \left(\frac{L}{2} + x_1 \right)^2 dx_1$$

$$= \frac{3.90}{EI} \int_0^{0.93 \text{ rad}} (1 + \cos^2 \phi - 2 \cos \phi) \, d\phi + \frac{2.16}{EI} L^3 = 2.29 \frac{L^3}{EI} \tag{j}$$

Substituting relations (f), (g), (h), (i) and (j) into (7.20) we get

$$\begin{Bmatrix} 0 \\ 0 \end{Bmatrix} = \frac{1}{EI} \begin{Bmatrix} -1.438qL^3 \\ 1.131qL^4 \end{Bmatrix} + \frac{1}{EI} \begin{bmatrix} 4.32L & -2.40L^2 \\ -2.40L^2 & 2.29L^3 \end{bmatrix} \begin{Bmatrix} X_1 \\ X_2 \end{Bmatrix} \tag{k}$$

Thus

$$\begin{Bmatrix} X_1 \\ X_2 \end{Bmatrix} = \frac{1}{4.138L^4} \begin{bmatrix} 2.29L^3 & 2.40L^2 \\ 2.40L^2 & 4.32L \end{bmatrix} \begin{Bmatrix} -1.438qL^3 \\ 1.131qL^4 \end{Bmatrix} = \begin{Bmatrix} -0.138qL^2 \\ 0.350qL \end{Bmatrix}$$

The free-body diagram of the left half of the frame and the corresponding moment diagram are shown in Fig. f.

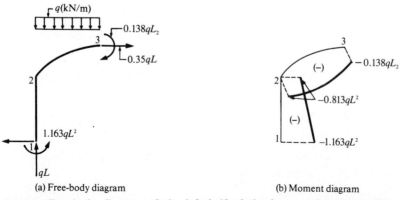

(a) Free-body diagram   (b) Moment diagram

Figure f Free-body diagram of the left half of the frame and corresponding moment diagram.

7.7   Derivation of the Compatibility Equations Using Castigliano's Second Theorem

In this section, we present a variation of the force method of structural analysis wherein the compatibility equations (7.20) are obtained by the use of Castigliano's second theorem (5.98) or (5.109) and (5.110). This variation of the force method can be used effectively in analyzing trusses, beams, and frames having a small degree of static indeterminacy. In applying this method to a statically indeterminate structure of the nth degree, the following steps are taken:

1. We choose the redundants X_i $(i = 1, \ldots, n)$.
2. If the structure is a truss, we express the internal force in each of its members in terms of the redundants X_i $(i = 1, 2, \ldots, n)$. If the structure is a beam or a frame we express the moment in every one of its members in terms of the axial coordinate of the member and the redundants X_i $(i = 1, 2, \ldots, n)$. If the effect of axial deformation of the members of the structure is not neglected, we also express the axial force in every member of the structure in terms of the redundants X_i $(i = 1, 2, \ldots, n)$.
3. On the basis of Castigliano's second theorem presented in Sec. 5.15, the displacement Δ_i^X corresponding to the redundant X_i is equal to the derivative of the total strain energy of the structure with respect to X_i. Castigliano's second theorem (5.109) or (5.110) may be rewritten as

$$\Delta_i^X = \frac{\partial (U_s)_T}{\partial X_i} = \sum_{k=1}^{NM} \left[\int_0^L \left(\frac{N}{EA} \frac{\partial N}{\partial X_i} + \frac{M}{EI} \frac{\partial M}{\partial X_i} \right) dx_1 \right]^{(k)} \quad (7.42)$$

We substitute the expressions for the moments and axial forces established in Step 2 into relation (7.42) and integrate to obtain the required equations of compatibility for the structure.

In what follows, we illustrate this method by solving the following three examples.

Example 1. We compute the redundant of a beam that is statically indeterminate to the first degree.

Example 2. We compute the redundants of a frame that is statically indeterminate to the second degree and plot its moment diagram.

Example 3. We establish the redundant internal force of a truss that is statically indeterminate to the first degree.

Example 1. Consider the beam of constant cross section fixed at one end and simply supported at the other end. The beam is loaded as shown in Fig. a. Compute the reaction at support 2 of this beam.

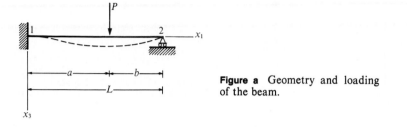

Figure a Geometry and loading of the beam.

solution The beam under consideration is statically indeterminate to the first degree. We choose the reaction at support 2 as the redundant. Referring to Fig. b, the moment in the beam may be expressed as

$$M = X_1 x_1' \qquad\qquad 0 \le x_1' \le b$$
$$M = X_1 x_1' - P(x_1' - b) \qquad b \le x_1' \le L \tag{a}$$

Substituting relations (a) into (7.42), we get

$$0 = \int_0^b \frac{(X_1 x_1') x_1'}{EI}\, dx_1' + \int_b^L [X_1 x_1' - P(x_1' - b)] x_1'\, dx_1' \tag{b}$$

Carrying out the integration in the above equation, we have

$$X_1 = \frac{P}{2L^3}(2L^3 + b^3 - 3bL^2) = \frac{Pa^2}{2L^3}(2L + b) \tag{c}$$

This is the compatibility equation for the beam. The value of the reaction X_1 which satisfies relation (c) is the one which renders the deflection at point 2 of the beam equal to zero. Notice, that if there were a given settlement of support 2, we would have to set the derivative of the total strain energy with respect to X_1 equal to the given settlement instead of equal to zero.

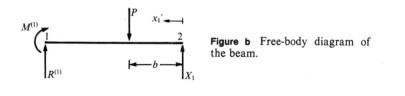

Figure b Free-body diagram of the beam.

Example 2. Plot the moment diagram for the frame loaded as shown in Fig. a. The members of the frame are made of the same material and have the same constant cross section.

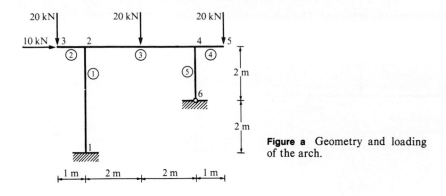

Figure a Geometry and loading of the arch.

solution This is a statically indeterminate structure to the second degree. Thus, we need two compatibility equations in addition to the equations of statics in order to establish the distribution of the internal actions. We obtain the compatibility equations using Castigliano's second theorem. We choose as the redundants the two reactions X_1 and X_2 acting at support 6.

The free-body diagrams of the members of the frame are shown in Fig. b. In these diagrams, the end actions are given in terms of the redundants X_1 and X_2. Referring to Fig. b, the moment distribution in the members of the frame is

Member 1: $M = (10 - X_1)x_1 - 160 + 2X_1 + 4X_2$ $\qquad 0 \le x_1 \le 4$

Member 3: $M = (40 - X_2)x_1 - 140 - 2X_1 + 4X_2$ $\qquad 0 \le x_1 \le 2$ $\qquad$ (a)

$\qquad\qquad\quad M = (X_2 - 20)x_1' - 20 - 2X_1$ $\qquad\qquad 0 \le x_1' \le 2$

Member 5: $M = X_1x_1 - 2X_1 = X_1(x_1 - 2)$ $\qquad\qquad 0 \le x_1 \le 2$

Notice that we do not require the expressions for the moment in members 2 and 4 because in these members the moment is not a function of the redundants. Thus, the

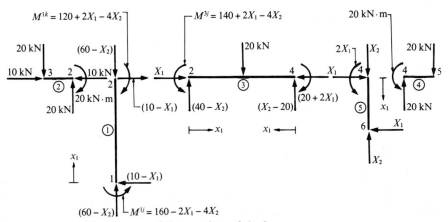

Figure b Free-body diagrams of the members of the frame.

derivative of the moment with respect to a redundant vanishes and, consequently, it does not contribute to Eq. (7.42).

Substituting relations (a) into Castigliano's theorem (7.42), we get

$$0 = \frac{\partial (U_s)_T}{\partial X_1}$$

$$= \int_0^4 \frac{1}{EI} (10x_1 - X_1x_1 - 160 + 2X_1 + 4X_2)(2 - x_1) \, dx_1$$

$$+ \int_0^2 \frac{1}{EI} (40x_1 - X_2x_1 - 140 - 2X_1 + 4X_2)(-2) \, dx_1$$

$$+ \int_0^2 \frac{1}{EI} (X_2x_1' - 20x_1' - 20 - 2X_1)(-2) \, dx_1'$$

$$+ \int_0^2 \frac{1}{EI} (X_1x_1 - 2X_1)(x_1 - 2) \, dx_1 \qquad \text{(b)}$$

$$0 = \frac{\partial (U_s)_T}{\partial X_2}$$

$$= \int_0^4 \frac{1}{EI} (10x_1 - X_1x_1 - 160 + 2X_1 + 4X_2)(4) \, dx_1$$

$$+ \int_0^2 \frac{1}{EI} (40x_1 - X_2x_1 - 140 - 2X_1 + 4X_2)(4 - x_1) \, dx_1$$

$$+ \int_0^2 \frac{1}{EI} (X_2x_1' - 20x_1' - 20 - 2X_1)x_1' \, dx_1' \qquad \text{(c)}$$

Integrating the above relations and simplifying, we obtain

$$9X_1 - 6X_2 + 190 = 0 \qquad 6X_1 - 32X_2 + 1110 = 0 \qquad \text{(d)}$$

These are the compatibility relations for the frame. The values of reactions X_1 and X_2 which satisfy relations (d) render the horizontal and vertical movement of point 6 of the frame equal to zero. Notice that if the components of translation of support 6 were

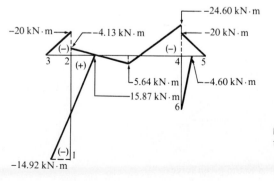

Figure c Moment diagram for the frame.

not zero, we would have to set the derivatives of the total strain energy with respect to X_1 and X_2 equal to the corresponding given components of translation instead of equal to zero. Solving relations (d) for X_1 and X_2, we obtain

$$X_1 = 2.30 \text{ kN} \qquad X_2 = 35.12 \text{ kN} \tag{e}$$

The end actions in the members of the frame may be established by referring to Fig. b and using results (e). The moment diagram for the frame is plotted in Fig. c.

Example 3. Compute the internal forces in the members of the truss shown in Fig. a. The members of the truss are made of the same material and have the same constant cross section. As discussed in Example 1 of Sec. 7.2.1, the internal forces in the members of this truss are the same whether or not members 5 and 6 are connected at their intersection.

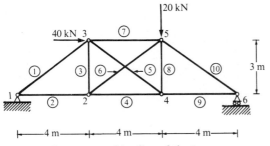

Figure a Geometry and loading of the truss.

solution This truss is internally statically indeterminate to the first degree. It has one member (the diagonal 5 or 6) in excess of those needed for the truss to be geometrically stable. We choose the internal force in member 5 as the redundant. Referring to Fig. b, from the equilibrium of the truss we have

$$\Sigma M^{(1)} = 0 = 12 R_v^{(6)} - (40)(3) - (20)(8)$$

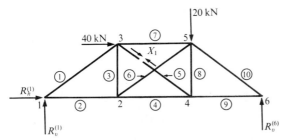

Figure b Primary structure subjected to the given forces and to the chosen redundant.

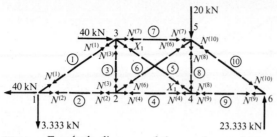

Figure c   Free-body diagrams of the members and joints of the truss.

or
$$R_v^{(6)} = 23.333 \text{ kN}$$

$$\Sigma \bar{F}_v = 0 \qquad R_v^{(1)} = -3.333 \text{ kN}$$

$$\Sigma \bar{F}_h = 0 \qquad R_h^{(1)} = -40 \text{ kN}$$

Referring to Fig. c, from the equilibrium of the joints of the truss we have

Joint 1

$$\Sigma \bar{F}_v = 0 = \tfrac{3}{5} N^{(1)} - 3.333 \qquad \Sigma \bar{F}_h = 0 = -40 + N^{(2)} + \tfrac{4}{5} N^{(1)}$$

or

$$N^{(1)} = 5.556 \text{ kN} \qquad N^{(2)} = 35.556 \text{ kN} \qquad (a)$$

Joint 3

$$\Sigma \bar{F}_h = 0 = 40 + N^{(7)} - \tfrac{4}{5}(5.556) = \tfrac{4}{5} X_1$$

$$\Sigma \bar{F}_v = 0 = -N^{(3)} - \tfrac{3}{5}(5.556) - \tfrac{3}{5} X_1$$

or

$$N^{(7)} = -35.556 - \tfrac{4}{5} X_1 \qquad N^{(3)} = -3.333 - \tfrac{3}{5} X_1 \qquad (b)$$

Joint 2

$$\Sigma \bar{F}_h = 0 = N^{(4)} + \tfrac{4}{5} N^{(6)} - N^{(2)} \qquad \Sigma \bar{F}_v = 0 = N^{(3)} + \tfrac{3}{5} N^{(6)}$$

or

$$N^{(6)} = 5.556 + X_1 \qquad N^{(4)} = 31.111 - \tfrac{4}{5} X_1 \qquad (c)$$

Joint 4

$$\Sigma \bar{F}_h = 0 = -N^{(4)} - \tfrac{4}{5} X_1 + N^{(9)} \qquad \Sigma \bar{F}_v = 0 = \tfrac{3}{5} X_1$$

or

$$N^{(9)} = 31.111 \qquad N^{(8)} = -\tfrac{3}{5} X_1 \qquad (d)$$

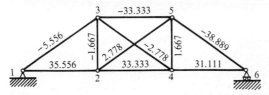

Figure d   Internal forces in the members of the truss.

Joint 6

$$\Sigma \overline{F}_h = 0 = -N^{(9)} - \tfrac{4}{5}N^{(10)}$$

$$\Sigma \overline{F}_v = 0 = 23.333 + \tfrac{3}{5}N^{(10)}$$

$$N^{(10)} = -38.889 \qquad N^{(9)} = 31.111 \qquad check \tag{e}$$

Substituting relations (a) to (e) in (7.42) and noting that for members 1, 2, 9, and 10 the term $\partial N/\partial X_1$ is equal to zero, we have

$$0 = \frac{\partial (U_s)_T}{\partial X_1} = \sum_{k=1}^{10} \left[\frac{N}{EA} \frac{\partial N4}{\partial X_1} L \right]^{(k)} = \frac{1}{EA} [\, (-3.333 - \tfrac{3}{5}X_1)(-\tfrac{3}{5})(3)$$

$$+ (31.111 - \tfrac{4}{5}X_1)(-\tfrac{4}{5})(4) + X_1(1)(5) + (5.556 + X_1)(1)(5)$$

$$+ (-35.556 - \tfrac{4}{5}X_1)(-\tfrac{4}{5})(4) + (-\tfrac{3}{5}X_1)(-\tfrac{3}{5})(3)] = \frac{1}{EA}(48 + 17.28X_1)$$

Thus

$$X_1 = -2.778 \text{ kN} \tag{f}$$

The internal forces in the members of the truss are found by substituting relation (f) into (b) to (d). The results are shown in Fig. d.

7.8 Derivation of the Compatibility Equations for Beams Using the Conjugate Beam Method

In this section, we present a variation of the force method of structural analysis wherein the compatibility equations (7.20) for beams are obtained using the conjugate beam method discussed in Chap. 6. This variation of the force method can be used effectively in analyzing beams that have a small degree (one or two) of static indeterminacy. In applying this method to a statically indeterminate beam of the nth degree, the following steps are taken.

1. We choose the redundants X_i ($i = 1, \ldots, n$).

2. We plot the moment diagram for the beam. The moments are expressed in terms of the redundants X_i.

3. We form the conjugate beam and load it with the M/EI diagram (elastic load). The conjugate beam forms a mechanism; that is, the number of its reactions is smaller than the number of independent equations of equilibrium. Their difference is equal to the number of the redundants X_i of the actual beam. However, the conjugate beam is in equilibrium under the influence of the elastic load (M/EI).

4. We write the equations of equilibrium for the conjugate beam that include the redundants X_i. These are the required equations of compatibility for the actual beam.

In what follows, we illustrate this method by an example.

Example 1.   Consider the beam of Fig. a fixed at one end and simply supported at the other end. The cross section of the beam is constant. Compute the reaction at support 2 of this beam.

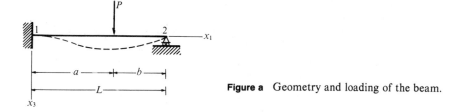

Figure a   Geometry and loading of the beam.

---

solution   The beam under consideration is statically indeterminate to the first degree. We choose the reaction at support 2 as the redundant (see Fig. b). The moment at the cross sections of the beam from $0 \le x_1' \le b$ is equal to

$$M_2 = X_1 x_1' \tag{a}$$

Moreover, the moment at the cross section of the beam from $b \le x_1' \le L$ is

$$M_2 = X_1 x_1' - P(x_1' - b) \tag{b}$$

In Fig. c, we plot each of the terms of Eq. (b) separately. The resulting moment diagram is called the moment diagram by parts. Referring to Fig. c, it is apparent that the area under the moment diagram for the beam and the moment of this area about any point can be readily established by plotting the moment diagram of the beam by parts (see Table B on the inside of the back cover). The conjugate beam corresponding to the beam under consideration is shown in Fig. d. The conjugate beam has only a reaction at point 2. Hence it is a mechanism which, however, is in equilibrium under the elastic loading M/EI. Thus, referring to Fig. d and setting equal to zero the sum

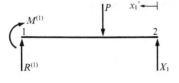

Figure b   Free-body diagram of the beam.

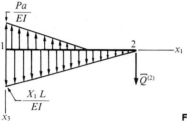

Figure c   Moment diagram for the beam by parts.

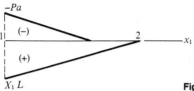

Figure d   Conjugate structure.

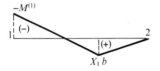

Figure e   Moment diagram for the beam.

of the moments about the axis parallel to the x_2 axis and passing through point 2, we get

$$\Sigma M^{(2)} = 0 = \frac{X_1 L}{EI_2}\left(\frac{L}{2}\right)\frac{2L}{3} - \frac{Pa^2}{2EI_2}\left(b + \frac{2a}{3}\right) \tag{a}$$

This is, in fact, a compatibility relation for the beam. For the value of X_1 that satisfies relation (a), the conjugate beam is in equilibrium without any force or moment at point 1, or a moment at point 2. This is equivalent to saying that the value of X_1 satisfying relation (a) is the one which, for the given loading, renders the deflection of point 2, and the slope and deflection at point 1 of the real beam equal to zero. Notice, that if there were a known settlement of support 2, we would have to put an external moment at the end 2 of the conjugate beam equal to this settlement. Solving relation (a) for X_1, we get

$$X_1 = \frac{Pa^2}{2L^3}(3b + 2a) = \frac{Pa^2}{2L^3}(2L + b)$$

Referring to Fig. b and taking moments about point 1, we obtain

$$M^{(1)} = X_1 L - Pa = -\frac{Pab}{2L^2}(L + b)$$

7.9 Diagonalization of the Flexibility Matrix—The Elastic Center Method

In this section, we present a version of the force method for analyzing statically indeterminate structures, referred to as the *elastic center method*. In this method, the analysis of the structure is referred to a set of axes with respect to which the flexibility matrix of the structure is diagonalized. The origin of this set of axes is referred to as the *elastic center* of the structure.

The elastic center method can be applied to planar structures which have both their ends fixed, such as beams, arches, one-bay frames, as well as closed box structures and rings, as shown in Fig. 7.13.

Consider the structure shown in Fig. 7.14a and the auxiliary structure shown in Fig. 7.14b. The two structures differ only in the detail of their right-hand support. The right-hand end of the auxiliary structure is not fixed directly to the supporting body but is rigidly connected to one end (point 2) of a stiff arm

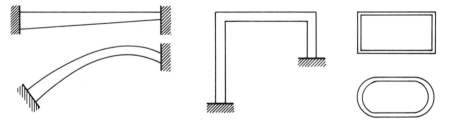

Figure 7.13 Structures which could be analyzed by the elastic center method.

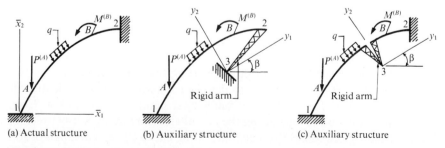

(a) Actual structure   (b) Auxiliary structure   (c) Auxiliary structure

Figure 7.14 Actual and auxiliary structures.

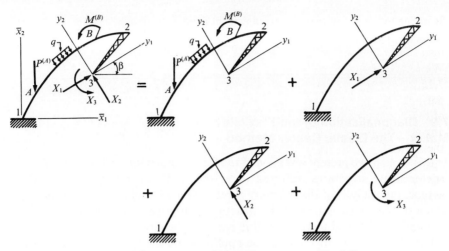

(a) Primary structure of the auxiliary structure of Fig. 7.14*a*

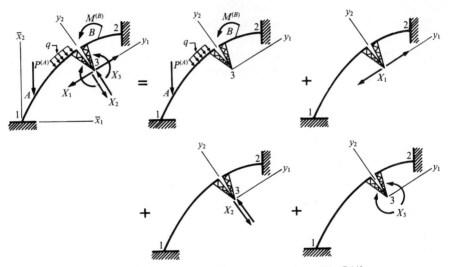

(b) Primary structure of the auxiliary structure of Fig. 7.14*b*

Figure 7.15 Superposition of the effects of the given actions and of the redundants on the primary structure.

(2,3) whose other end is fixed to the supporting body. It is apparent that the response of the auxiliary structure to the given external actions is identical to that of the actual structure. Another auxiliary structure is shown in Fig. 7.14*c*. This structure is formed by cutting the actual structure at a point between its supports and rigidly connecting each side of the cut to one end of a pair of stiff arms. The other ends of these arms are rigidly connected together at point 3. It is apparent, that the response of the auxiliary structure shown in Fig. 7.14*c* is identical to that of the actual structure of Fig. 7.14a. On the basis of the foregoing discussion, we can proceed to analyze one of the auxiliary structures instead of the actual one by employing the force method of structural analysis. We select the auxiliary structure of Fig. 7.14b. In this case, as shown in Fig. 7.15a, we choose the reactions at the support 3 as the redundants, and we refer our analysis to the set of axes y_1, and y_2 with origin at point 3. The direction of this set of axes with respect to the global axes $\bar{x}_1$ and $\bar{x}_2$ is specified by the angle β (see Fig. 7.14b). If we had chosen the auxiliary structure of Fig. 7.14c, then, as shown in Fig. 7.15b, we would have chosen the pairs of equal and opposite forces and moments acting on the end of the stiff arms at point 3 as the redundants. Referring to relation (7.20), the redundants X_1, X_2, and X_3 may be obtained from the following set of three linear algebraic equations.

$$\begin{Bmatrix} 0 \\ 0 \\ 0 \end{Bmatrix} = \begin{Bmatrix} \Delta_1^{PL} \\ \Delta_2^{PL} \\ \Delta_3^{PL} \end{Bmatrix} + \begin{bmatrix} F_{11} & F_{12} & F_{13} \\ F_{21} & F_{22} & F_{23} \\ F_{31} & F_{31} & F_{33} \end{bmatrix} \begin{Bmatrix} X_1 \\ X_2 \\ X_3 \end{Bmatrix} \qquad (7.43)$$

The components of displacement Δ_1^{PL}, Δ_2^{PL}, and Δ_3^{PL} of point 3 of the primary structure, due to the given external actions, are shown in Fig. 7.16. They may be computed using the method of virtual work (5.66) in conjunction with the virtual loading shown in Fig. 7.17a to c, respectively. Thus

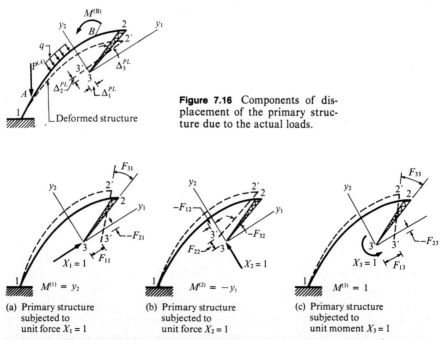

Figure 7.16 Components of displacement of the primary structure due to the actual loads.

(a) Primary structure subjected to unit force $X_1 = 1$

(b) Primary structure subjected to unit force $X_2 = 1$

(c) Primary structure subjected to unit moment $X_3 = 1$

Figure 7.17 Physical significance of the flexibility coefficients of the auxiliary structure.

$$\Delta_1^{PL} = \int_{pt.1}^{pt.2} \frac{M^{PL} y_2}{EI} \, ds \qquad \Delta_2^{PL} = - \int_{pt.1}^{pt.2} \frac{M^{PL} y_1}{Ei} \, ds$$

$$\Delta_3^{PL} = \int_{pt.1}^{pt.2} \frac{M^{PL}}{EI} \, ds \tag{7.44}$$

where M^{PL} is the moment in the primary structure subjected to the actual loads.

The flexibility coefficients F_{ij} may be computed by referring to Fig. 7.17a to c and using relation (5.66). Thus

$$F_{ij} = \int_{pt.1}^{pt.2} \frac{M^{(i)} M^{(j)}}{EI} \, ds \qquad i, j = 1, 2, 3 \tag{7.45}$$

or

$$F_{11} = \int_{pt.1}^{pt.2} \frac{y_2^2}{EI} \, ds \qquad F_{22} = \int_{pt.1}^{pt.2} \frac{y_1^2}{EI} \, ds \qquad F_{33} = \int_{pt.1}^{pt.2} \frac{ds}{EI}$$

$$F_{12} = F_{21} = - \int_{pt.1}^{pt.2} \frac{y_1 y_2}{EI} \, ds \qquad F_{13} = F_{31} = \int_{pt.1}^{pt.2} \frac{y_2}{EI} \, ds \tag{7.46}$$

$$F_{23} = F_{32} = - \int_{pt.1}^{pt.2} \frac{y_1}{EI} \, ds$$

We are interested in finding the coordinates of point 3 and the angle β which locates the y_1 and y_2 axes with respect to the global axes in order that the nondiagonal *flexibility* coefficients vanish ($F_{12} = F_{13} = F_{23} = 0$). This point is referred to as the *elastic center* of the structure, while the set of axes y_1 and y_2 are referred to as the *conjugate axes* of the structure. On the basis of the foregoing discussion, when relation (7.43) is referred to the conjugate axes, it reduces to

$$\begin{Bmatrix} 0 \\ 0 \\ 0 \end{Bmatrix} = \begin{Bmatrix} \tilde{\Delta}_1^{PL} \\ \tilde{\Delta}_2^{PL} \\ \tilde{\Delta}_3^{PL} \end{Bmatrix} + \begin{bmatrix} \tilde{F}_{11} & 0 & 0 \\ 0 & \tilde{F}_{22} & 0 \\ 0 & 0 & \tilde{F}_{33} \end{bmatrix} \begin{Bmatrix} \tilde{X}_1 \\ \tilde{X}_2 \\ \tilde{X}_3 \end{Bmatrix} \tag{7.47}$$

The process of converting a square matrix to its form with vanishing nondiagonal terms is called *diagonalization*. From Eq. (7.47), the unknown quantities $\tilde{X}_i$ can be obtained directly without having to solve three simultaneous equations. Physically, the diagonalization of the flexibility matrix denotes that if a force acts at point 3 of the stiff arm in the direction of a conjugate axis, it will displace the arm along its direction, without rotating it (see Fig. 7.18).

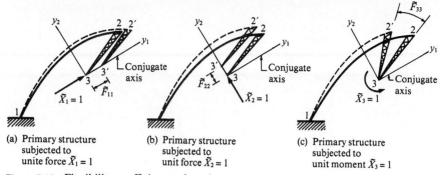

(a) Primary structure
subjected to
unite force $\tilde{X}_1 = 1$

(b) Primary structure
subjected to
unit force $\tilde{X}_2 = 1$

(c) Primary structure
subjected to
unit moment $\tilde{X}_3 = 1$

Figure 7.18 Flexibility coefficients referred to the conjugate axes.

Moreover, if a moment acts at point 3 of the stiff arm, it will only rotate the arm about the axis normal to the plane of the structure at point 3, without translating it (see Fig. 7.18c).

Referring to relations (7.44) and (7.46), Eq. (7.47) may be solved for $\tilde{X}_i$ to yield

$$\tilde{X}_1 = -\frac{\tilde{\Delta}_1^{PL}}{\tilde{F}_{11}} = -\int_{pt.1}^{pt.2} \frac{M^{PL}y_2}{EI}\,ds \bigg/ \int_{pt.1}^{pt.2} \frac{y_2^2}{EI}\,ds$$

$$\tilde{X}_2 = -\frac{\tilde{\Delta}_2^{PL}}{\tilde{F}_{22}} = \int_{pt.1}^{pt.2} \frac{M^{PL}y_1}{EI}\,ds \bigg/ \int_{pt.1}^{pt.2} \frac{y_1^2}{EI}\,ds$$

$$\tilde{X}_3 = -\frac{\tilde{\Delta}_3^{PL}}{\tilde{F}_{33}} = -\int_{pt.1}^{pt.2} \frac{M^{PL}}{EI}\,ds \bigg/ \int_{pt.1}^{pt.2} \frac{ds}{EI} \qquad (7.48)$$

From the redundants, the internal actions at any cross section (y_1, y_2) of the structure may be easily computed. For instance, the moment at a cross section (y_1, y_2) of the structure is equal to

$$M = M^{PL} + \tilde{X}_1 y_2 - \tilde{X}_2 y_1 + \tilde{X}_3 \qquad (7.49)$$

We now proceed to describe a procedure for locating the elastic center and the conjugate axes of a structure, that is, for locating point 3 and the direction of the y_1 and y_2 axes, with respect to which the flexibility matrix is diagonal. Consider the auxiliary plane surface shown in Fig. 7.19b, which has the axis of the actual structure as its center line and width equal to $1/EI$. The total area, the first moments, and the moments and product of inertia of this auxiliary surface with respect to the axes y_1 and y_2 are equal to

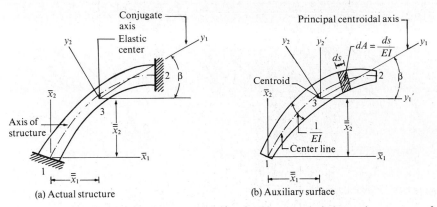

(a) Actual structure   (b) Auxiliary surface

Figure 7.19 Auxiliary surface for locating the elastic center and the conjugate axes of a structure.

$$A = \int_{\text{pt.1}}^{\text{pt.2}} \frac{ds}{EI} = F_{33} \qquad P_1 = \int_{\text{pt.1}}^{\text{pt.2}} \frac{y_2}{EI} ds = F_{13}$$

$$P_2 = \int_{\text{pt.1}}^{\text{pt.2}} \frac{y_1}{EI} ds = -F_{23} \qquad I_{11} = \int_{\text{pt.1}}^{\text{pt.2}} \frac{y_2^2}{EI} ds = F_{11} \qquad (7.50)$$

$$I_{22} = \int_{\text{pt.1}}^{\text{pt.2}} \frac{y_1^2}{EI} ds = F_{22} \qquad I_{12} = \int_{\text{pt.1}}^{\text{pt.2}} \frac{y_1 y_2}{EI} ds = -F_{12}$$

From relations (7.50), it is apparent that if we choose point 3 as the centroid of the auxiliary surface, and the y_1 and y_2 axes as its principal centroidal axes, the nondiagonal terms of the flexibility matrix of the structure vanish. Thus, we can locate the elastic center and the conjugate axis of a structure, as follows.

1. We construct the auxiliary plane surface having the axis of the structure as its center line and width equal to $1/EI$.

2. We locate the centroid of the auxiliary surface with respect to the global axes (point 3 in Fig. 7.19b). The coordinates $\bar{\bar{x}}_1$, $\bar{\bar{x}}_2$ of the centroid are

$$\bar{\bar{x}}_1 = \frac{1}{A} \int_{\text{pt.1}}^{\text{pt.2}} \frac{\bar{x}_1 \, ds}{EI} \qquad \bar{\bar{x}}_2 = \frac{1}{A} \int_{\text{pt.1}}^{\text{pt.2}} \frac{\bar{x}_2 \, ds}{EI} \qquad (7.51)$$

3. We locate the principal centroidal axes of the auxiliary surface (y_1 and y_2 in Fig. 7.19b). The angle β between the principal centroidal axis y_1 and the $\bar{x}_1$ axis (see Fig. 7.19b), is given by the relation

$$\tan 2\beta = \frac{2I'_{12}}{I'_{22} - I'_{11}} \qquad (7.52)$$

where I'_{11}, I'_{22}, and I'_{12} are the moments of inertia and the product of inertia of the auxiliary surface, with respect to the centroidal axes y_1' and y_2' which are parallel to the $\bar{x}_1$ and $\bar{x}_2$ axes, respectively (see Fig. 7.19b).

4. The elastic center of the structure has the same coordinates as the centroid of the auxiliary surface, while the conjugate axes of the structure are parallel to the principle axes of the auxiliary surface (see Fig. 7.19a and b).

Notice that if the auxiliary surface (Fig. 7.19b) has an axis of symmetry, then the centroid of this surface is located on the axis of symmetry. Moreover, this axis and the axis perpendicular to it through the centroid of the auxiliary surface are its principal centroidal axes. In general, the centroid of the auxiliary surface (Fig. 7.19b) may be easily located. However, if the surface does not have an axis of symmetry, the location of its principal centroidal axes and the computation of its moments of inertia with respect to these axes is laborious. For this reason, it is often more convenient to refer our analysis to a convenient set of centroidal axes y_1 and y_2 rather than to the principal centroidal axes. In this case, as can be seen from relations (7.50), the flexibility coefficients $F_{12} = F_{21}$ do not vanish and Eq. (7.43) reduces to

$$
\begin{Bmatrix} 0 \\ 0 \\ 0 \end{Bmatrix} = \begin{Bmatrix} \Delta_1^{PL} \\ \Delta_2^{PL} \\ \Delta_3^{PL} \end{Bmatrix} + \begin{bmatrix} I_{11} & -I_{12} & 0 \\ -I_{12} & I_{22} & 0 \\ 0 & 0 & A \end{bmatrix} \begin{Bmatrix} X_1 \\ X_2 \\ X_3 \end{Bmatrix} \tag{7.53}
$$

Solving for the redundants X_1, X_2, and X_3, we get

$$
X_1 = -\frac{\Delta_1^{PL} I_{22} + \Delta_2^{PL} I_{12}}{I_{11} I_{22} - I_{12}^2} \qquad X_2 = -\frac{\Delta_2^{PL} I_{11} + \Delta_1^{PL} I_{12}}{I_{11} I_{22} - I_{12}^2} \tag{7.54}
$$

$$
X_3 = -\frac{\Delta_3^{PL}}{A}
$$

Example 1.   Compute the reactions at support 5 of the frame loaded as shown in Fig. a, using the elastic center method. The members of the frame are made of the same material and have the same constant cross section.

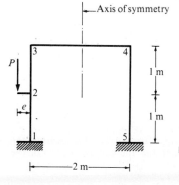

Figure a   Geometry and loading of the frame.

solution We analyze the auxiliary structure of Fig. b, whose elastic center is point 6. The internal actions in the members of this structure are identical to those in the corresponding members of the structure of Fig. a. We locate the elastic center of the structure by using the auxiliary surface shown in Fig. c. The elastic center of the structure corresponds to the centroid of this surface. Since this surface has an axis of symmetry (y_2), its centroid must be located on it. Moreover, the axis of symmetry is a principal axis for any of its points. Referring to Fig. c, the centroidal axis y_1 may be located as follows:

$$A\overline{\overline{x}}_2 = 2\frac{2}{EI} = \frac{4}{EI}$$

$$A = \frac{2}{EI} + \frac{2}{EI} + \frac{2}{EI} = \frac{6}{EI}$$

$$\overline{\overline{x}}_2 = \frac{2}{3} \text{ m}$$

Referring to Fig. c, the moments of inertia of the auxiliary surface with respect to the y_1 and y_2 axes are

$$I_{11} = \frac{2(2)^3}{12EI} + \frac{2(2)}{EI}\left(\frac{1}{3}\right)^2 + \frac{2}{EI}\left(\frac{2}{3}\right)^2 = \frac{8}{3EI}$$

$$I_{22} = 2\left(\frac{2}{EI}\right) + \frac{2^3}{12EI} = \frac{14}{3EI} \tag{a}$$

The primary structure of the auxiliary structure of Fig. b is shown in Fig. d, subjected to the given loading and to the redundants. The moment in this structure due to the given loading is equal to

$$M^{PL} = Pe, \qquad -\frac{4}{3} \le y_2 \le -\frac{1}{3}$$

The components of displacement $\tilde{\Delta}_1^{PL}$, $\tilde{\Delta}_2^{PL}$, and $\tilde{\Delta}_3^{PL}$ of point 6 of the primary structure

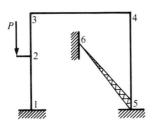

Figure b Auxiliary structure.

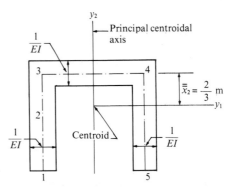

Figure c Auxiliary surface for locating the elastic center and the conjugate axes of the frame.

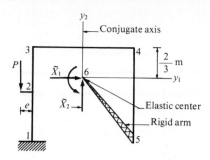

Figure d Primary structure for the auxiliary structure of Fig. b subjected to the given loading and to the redundants.

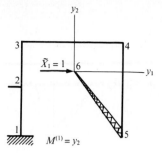

Figure e Primary structure for the auxiliary structure of Fig. b subjected to $\tilde{X}_1$ = 1.

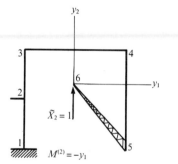

Figure f Primary structure for the auxiliary structure of Fig. b subjected to $\tilde{X}_2$ = 1.

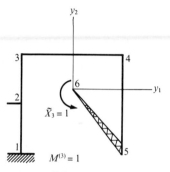

Figure g Primary structure for the auxiliary structure of Fig. b subjected to $\tilde{X}_3$ = 1

for the auxiliary structure of Fig. b are obtained using the virtual loading shown in Figs. e, f, and g, respectively. Referring to these figures, we have

$$\tilde{\Delta}_1^{PL} = \int_{-4/3}^{-1/3} \frac{M^{PL} y_2 \, dy_2}{EI} = \left[\frac{M^{PL}}{EI} \frac{y_2^2}{2} \right]_{-4/3}^{-1/3} = -\frac{5Pe}{6EI}$$

$$\tilde{\Delta}_2^{PL} = \int_{-4/3}^{-1/3} \frac{M^{PL} y_1 \, dy_2}{EI} = \left[\frac{M^{PL}}{EI} y_2 \right]_{-4/3}^{-1/3} = \frac{Pe}{EI} \qquad \text{(b)}$$

$$\tilde{\Delta}_3^{PL} = \int_{-4/3}^{-1/3} \frac{M^{PL} \, dy_2}{EI} = \left[\frac{M^{PL}}{EI} y_2 \right]_{-4/3}^{-1/3} = \frac{Pe}{EI}$$

Substituting relations (a) and (b) into (7.48), we obtain

$$\tilde{X}_1 = -\frac{\tilde{\Delta}_1^{PL}}{\tilde{F}_{11}} = -\frac{\tilde{\Delta}_1^{PL}}{I_{11}} = \frac{5Pe3(EI)}{6EI(8)} = \frac{5Pe}{16}$$

$$\tilde{X}_2 = -\frac{\tilde{\Delta}_2^{PL}}{\tilde{F}_{22}} = -\frac{\tilde{\Delta}_2^{PL}}{I_{22}} = -\frac{3Pe}{14}$$

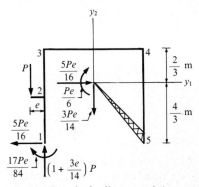

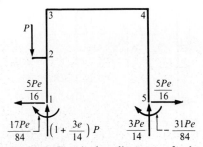

Figure h Free-body diagram of the auxiliary structure subjected to the given loading.

Figure i Free-body diagram of the actual frame subjected to the given loading.

$$\tilde{X}_3 = -\frac{\tilde{\Delta}_3^{PL}}{\tilde{F}_{33}} = -\frac{\tilde{\Delta}_3^{PL}}{A} = -\frac{Pe}{6}$$

Example 2. Consider a symmetric, fixed-at-both-ends, parabolic, arch with $I = I_c \sec \phi$ subjected to a concentrated force P as shown in Fig. a. Compute the reactions at the supports of the arch.

In Example 2 of Sec. 7.6 this arch was analyzed using the basic force method. In this example, it is analyzed very effectively using the elastic center method.

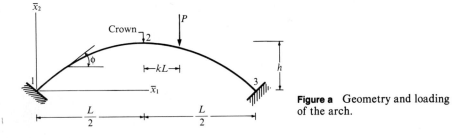

Figure a Geometry and loading of the arch.

solution The first step in the analysis of the arch by the elastic center method is to locate its elastic center. This has the same coordinates as the centroid of the auxiliary surface shown in Fig. b. The center line of this surface has the same geometry as the axis of the arch, while its thickness is equal to $1/EI$. The auxiliary surface for a symmetric arch has an axis of symmetry (the y_2 axis) and consequently, its centroid lies on this axis. The coordinate $\bar{x}_2$ of the centroid of the auxiliary surface is equal to

$$\overline{\overline{x}}_2 = \int_{pt.1}^{pt.3} \frac{\overline{x}_2 \, ds}{EI} \bigg/ \int_{pt.1}^{pt.3} \frac{ds}{EI} \tag{a}$$

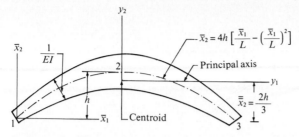

Figure b Auxiliary surface for locating the elastic center and the conjugate axes of the arch.

Referring to Fig. 7.11b we have

$$ds = d\bar{x}_1 \sec \phi \tag{b}$$

Using the above relation and $I = I_c \sec \phi$, we obtain

$$\frac{ds}{I} = \frac{d\bar{x}_1}{I_c} \tag{c}$$

Moreover, the equation of the arch is

$$\bar{x}_2 = 4h \left(\frac{\bar{x}_1}{L} - \frac{\bar{x}_1^2}{L} \right) \tag{d}$$

Substituting relations (c) and (d) into (a) we obtain

$$\bar{\bar{x}}_2 = 4h \int_0^L \left(\frac{\bar{x}_1}{L} - \frac{\bar{x}_1^2}{L^2} \right) d\bar{x}_1 \bigg/ \int_0^L d\bar{x}_1 = \frac{2h}{3} \tag{e}$$

We analyze the auxiliary structure of Fig. c. We choose as the redundants its reactions at support 4. The primary structure of the auxiliary structure subjected to the given load and to the redundants is shown in Fig. d. The primary structure of the auxiliary structure subjected to the given load is shown in Fig. e. The moment in this structure is

$$M^{PL} = -P(kL - y_1) \qquad -\frac{L}{2} < y_1 < kL \tag{f}$$

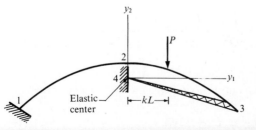

Figure c Auxiliary structure.

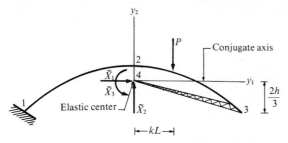

Figure d Primary structure of the auxiliary structure of Fig. c subjected to the given force and to the redundants.

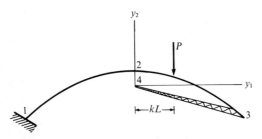

Figure e Primary structure subjected to the given force.

Noting that $ds = dy_1 \sec \phi$ and $I = I_c \sec \phi$, we get

$$\int_{pt.1}^{pt.3} \frac{M^{PL}\,ds}{EI} = -\frac{P}{EI_c}\int_{-L/2}^{kL}(kL - y_1)\,dy_1 = -\frac{PL^2}{8EI_c}(2k+1)^2 \tag{g}$$

$$\int_{pt.1}^{pt.3} \frac{M^{PL}y_1\,ds}{EI} = -\frac{P}{EI_c}\int_{-L/2}^{kL}(kL - y_1)y_1\,dy_1 = \frac{PL^3}{24EI_c}(1-k)(1+2k)^2 \tag{h}$$

$$\int_{pt.1}^{pt.3} \frac{M^{PL}y_2\,ds}{EI} = -\frac{P}{EI_c}\int_{-L/2}^{kL}\frac{h}{3L^2}(kL - y_1)(L^2 - 12y_1^2)\,dy_1 \tag{i}$$

$$= \frac{PhL^2}{48EI_c}(1 - 8k^2 + 16k^4)$$

Referring to Figs. a and b, it can be seen that the coordinates y_1, y_2 are related to the coordinates $\bar{x}_1$, $\bar{x}_2$ by the following relations.

$$\bar{x}_1 = y_1 + \frac{L}{2} \qquad \bar{x}_2 = y_2 + \frac{2h}{3} \tag{j}$$

Using the above relations, the equation of the axis of the arch (d) can be written as

$$y_2 = \frac{h}{3L^2}(L^2 - 12y_1^2) \tag{k}$$

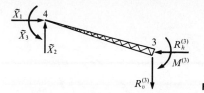

Figure f Free-body diagram of the stiff arm.

Referring to relations (7.50), using relations (c) and (k), and noting that $d\tilde{x}_1 = dy_1$, the area and the moments of inertia of the auxiliary surface with respect to the y_1 and y_2 axes are

$$A = \int_{\text{pt.1}}^{\text{pt.3}} \frac{ds}{EI} = \int_{\text{pt.1}}^{\text{pt.3}} \frac{dy_1}{EI_c} = \int_{-L/2}^{L/2} \frac{dy_1}{EI_c} = \frac{L}{EI_c}$$

$$I_{11} = \int_{\text{pt.1}}^{\text{pt.3}} \frac{y_2^2\, ds}{EI} = \int_{\text{pt.1}}^{\text{pt.3}} \frac{y_2^2\, dy_1}{EI_c} = \int_{-L/2}^{L/2} \frac{h^2}{qL^4 EI_c}\, (L^2 - 12y_1^2)^2\, dy_1 = \frac{4h^2 L}{45 EI_c} \qquad (1)$$

$$I_{22} = \int_{\text{pt.1}}^{\text{pt.3}} \frac{y_1^2\, ds}{EI} = \int_{\text{pt.1}}^{\text{pt.3}} \frac{y_1^2\, dy_1}{EI} = \int_{-L/2}^{L/2} \frac{y_1^2\, dy_1}{EI_c} = \frac{L^3}{12 EI_c}$$

Substituting relations (g), (h), (i), and (l) into relations (7.48), we obtain

$$\tilde{X}_1 = -\int_{\text{pt.1}}^{\text{pt.3}} \frac{M^{PL} y_2\, ds}{EI} \bigg/ I_{11} = -\frac{15PL}{64h}\, (1 - 2k)^2\, (1 + 2k)^2$$

$$\tilde{X}_2 = \int_{\text{pt.1}}^{\text{pt.3}} \frac{M^{PL} y_1\, ds}{EI} \bigg/ I_{22} = \frac{P}{2}\, (1 - k)(1 + 2k)^2 \qquad (m)$$

$$\tilde{X}_3 = -\int_{\text{pt.1}}^{\text{pt.3}} \frac{M^{PL}\, ds}{EI} \bigg/ A = \frac{PL}{8}\, (2k + 1)^2$$

The reactions at the support of the arch (point 3) may be obtained by considering the equilibrium of the stiff arm. Thus, referring to Fig. f, we have

$$R_h^{(3)} = -\frac{15PL}{64h}\, (1 - 2k)^2(1 + 2k)^2$$

$$R_v^{(3)} = \frac{P}{2}\, (1 - k)(1 + 2k)^2 \qquad (n)$$

$$M^{(3)} = \frac{PL}{32}\, (1 + 2k)^2(1 - 2k)(1 - 10k)$$

7.9.1 Applications of the elastic center method to the analysis of structures subjected to movement of a support

In this section, we extend the elastic center method to the analysis of structures subjected to a displacement of one of their supports. Let us consider the aux-

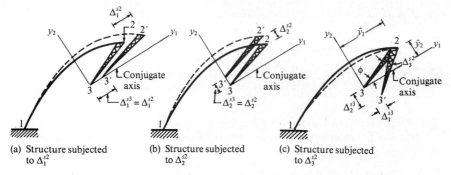

(a) Structure subjected to Δ_1^{s2}

(b) Structure subjected to Δ_2^{s2}

(c) Structure subjected to Δ_3^{s2}

Figure 7.20 Relations between the displacements of points 2 and 3 of the auxiliary structure of Fig. 7.14b.

iliary structure shown in Fig. 7.14b and suppose that support 2 moves by Δ_1^{s2} and Δ_2^{s2} in the y_1 and y_2 directions, respectively, and rotates by Δ_3^{s2} about the y_3 axis. Referring to Fig. 7.20, it is apparent that the components of displacement of point 3 of the auxiliary structure differ from those of point 2 by the rigid-body movement of the rigid arm. A translation Δ_1^{s2} ($\Delta_2^{s2} = \Delta_3^{s2} = 0$) of point 2 along the y_1 axis results in a translation $\Delta_1^{s3} = \Delta_1^{s2}$ of point 3 along the same axis. A translation Δ_2^{s2} ($\Delta_1^{s2} = \Delta_3^{s2} = 0$) along the y_2 axes results in a translation $\Delta_2^{s3} = \Delta_2^{s2}$ of point 3 along the same axes. A rotation Δ_3^{s2} ($\Delta_1^{s2} = \Delta_2^{s2} = 0$) about the y_3 axes at point 2 results in a rotation $\Delta_3^{s3} = \Delta_3^{s2}$ about the y_3 axes at point 3 and translations $\Delta_1^{s3} = \hat{y}_2\Delta_3^{s2}$ and $\Delta_2^{s3} = -\hat{y}_1\Delta_3^{s2}$ of point 3— where $\hat{y}_1$ and $\hat{y}_2$ are the coordinates of point 2. This may be proven by considering the displaced configuration of the rigid arm. Thus, referring to Fig. 7.20c and denoting by L_a the length of the rigid arm, we have

$$\sin\phi = \frac{\hat{y}_2}{L_a} \qquad \cos\phi = \frac{\hat{y}_1}{L_a} \tag{7.55}$$

$$\overline{3,3'} = (L_a)\Delta_3^{s2}$$
$$\Delta_1^{s3} = (\overline{3,3'})\sin\phi = \hat{y}_2\Delta_3^{s2} \tag{7.56}$$
$$\Delta_2^{s3} = -(\overline{3,3'})\cos\phi = -\hat{y}_1\Delta_3^{s2}$$

On the basis of the foregoing discussion, the components of displacement of support 2 of the actual structure Δ_1^{s2}, Δ_2^{s2}, and Δ_3^{s2} and the corresponding components of displacement of point 3 of the auxiliary structure are related by the following relation.

$$\begin{Bmatrix} \Delta_1^{s3} \\ \Delta_2^{s3} \\ \Delta_3^{s3} \end{Bmatrix} = \begin{bmatrix} 1 & 0 & \hat{y}_2 \\ 0 & 1 & -\hat{y}_1 \\ 0 & 0 & 1 \end{bmatrix} \begin{Bmatrix} \Delta_1^{s2} \\ \Delta_2^{s2} \\ \Delta_3^{s2} \end{Bmatrix} \tag{7.57}$$

$$\{\Delta^{s3}\} = [T]\{\Delta^{s2}\} \tag{7.58}$$

Thus, the redundants $\tilde{X}_1$, $\tilde{X}_2$, and $\tilde{X}_3$ due to the displacement of support 2 by Δ_1^{s2}, Δ_2^{s2}, and Δ_3^{s2} are obtained from relation (7.58) and the equation of compatibility (7.20) which, for the auxiliary structure of Fig. 7.14b, can be written as

$$\{\tilde{X}\} = [\tilde{F}]^{-1}\{\Delta^{s3}\} = [\tilde{F}]^{-1}[T]\{\Delta^{s2}\} \tag{7.59}$$

or

$$
\begin{Bmatrix} \hat{X}_1 \\[2mm] \hat{X}_2 \\[2mm] \hat{X}_3 \end{Bmatrix} = \begin{Bmatrix} (\Delta_1^{s2} + \hat{y}_2 \Delta_3^{s2}) \Big/ \displaystyle\int_{pt.1}^{pt.2} y_2^2 \frac{ds}{EI} \\[4mm] (\Delta_2^{s2} - \hat{y}_1 \Delta_3^{s2}) \Big/ \displaystyle\int_{pt.1}^{pt.2} y_1^2 \frac{ds}{EI} \\[4mm] \Delta_3^{s2} \Big/ \displaystyle\int_{pt.1}^{pt.2} \frac{ds}{EI} \end{Bmatrix} \tag{7.60}
$$

Substituting the values of the redundants $\tilde{X}_1$, $\tilde{X}_2$, and $\tilde{X}_3$ obtained from relations (7.59) or (7.60) into Eq. (7.49), we obtain the moment at any cross section of the structure due to the movement of support 2.

7.10 Three-Moment Equation for Continuous Beams

In the force method of structural analysis, the redundants are obtained by solving the compatibility equations (7.20). For continuous beams subjected to external actions and settlement of supports, these equations can yield a recurrence formula, referred to as *the three-moment equation,* relating the internal moments at three adjacent supports. This formula can be applied repeatedly to every three adjacent supports of a continuous beam with n supports, yielding $n - 2$ compatibility equations.

Consider a continuous beam with n supports subjected to given external actions and settlement of its supports. Each span of the beam has a constant cross section. In Fig. 7.21, the spans between supports $i - 1$, i, and $i + 1$ are shown. The length and the moment of inertia of a span have the number of its left support as a subscript or superscript, respectively. We analyze this beam using the basic force method, choosing the moments at its supports as the

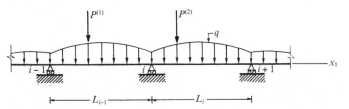

Figure 7.21 Segment of a continuous beam.

redundants. Thus, the primary structure is as shown in Fig. 7.22*a,* subjected to the given external actions and the given settlement of its supports. The corresponding moment diagram is shown in Fig. 7.22*b.* We employ the method of virtual work to compute Δ_i^{PL} which, in this case, represents the change of angle of the ends of the two beams resting on support *i.* The primary structure subjected to the virtual loading for computing Δ_i^{PL} is shown in Fig. 7.23*b.* Referring to this figure, we have

$$R^{(i-1)} = \frac{-1}{L_{i-1}} \qquad R^{(i+1)} = \frac{-1}{L_{i+1}} \tag{7.61}$$

Thus the moments in spans $(i - 1)$ and (i) of the primary structure are

$$\tilde{M} = \frac{x_1}{L_{i-1}} \quad 0 \le x_1 \le L_{i-1} \qquad \tilde{M} = \frac{x_1'}{L_i} \quad 0 \le x_1' \le L_1 \tag{7.62}$$

Substituting relations (7.61) and (7.62) into (5.66), we obtain

$$\Delta_i^{PL} = \frac{1}{EI^{(i-1)} L_{i-1}} \int_0^{L_{i-1}} M x_1 \, dx_1 + \frac{1}{EI^{(i)} L_i} \int_0^{L_i} M x_1' \, dx_1'$$

$$+ [(\Delta^{(i-1)} - \Delta^{(i)}) \frac{1}{L_{i-1}} + (\Delta^{(i+1)} - \Delta^{(i)}) \frac{1}{L_{i+1}}] \tag{7.63}$$

where $\Delta^{(i-1)}$, $\Delta^{(i)}$, and $\Delta^{(i+1)}$ are the settlements of support $(i - 1)$, (i), and $(i + 1)$, respectively. They are considered positive when they represent downward movement of the supports.

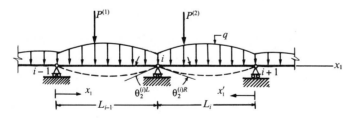

(a) Segment of the primary structure subjected to the actual loading

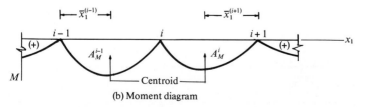

(b) Moment diagram

Figure 7.22 Segment of the primary structure subjected to the actual loading and corresponding moment diagram.

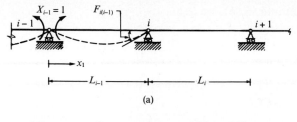

(a)

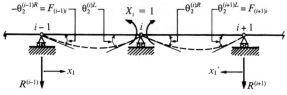

(b)

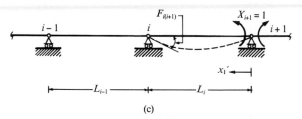

(c)

Fig. 7.23 Primary structure subjected to $X_{i-1} = 1$, $X_i = 1$ or $X_{i+1} = 1$.

Referring to Fig. 7.22b and denoting the areas of the moment diagram for the spans $i - 1$ and i by $A_M^{(i-1)}$ and $A_M^{(i)}$, respectively, it is apparent that

$$\int_0^{L_{i-1}} Mx_1 \, dx_1 = A_M^{(i-1)}\overline{x}^{(i-1)} \qquad \int_0^{L_i} Mx_1' \, dx_1' = A_M^{(i)}\overline{x}^{(i+1)} \quad (7.64)$$

Substituting the above relations into (7.63), we get

$$\Delta_i^{PL} = \frac{A_M^{(i-1)}\overline{x}^{(i-1)}}{EI^{(i-1)}L_{i-1}} + \frac{A_M^{(i)}\overline{x}^{(i+1)}}{EI^iL_i} + \frac{\Delta^{(i-1)} - \Delta^{(i)}}{L_{i-1}} + \frac{\Delta^{(i+1)} - \Delta^{(i)}}{L_i} \quad (7.65)$$

We employ the method of virtual work to compute the flexibility coefficients $F_{(i-1)i}$, F_{ii}, and $F_{(i+1)i}$ of the beam. The meaning of these flexibility coefficients is shown in Fig. 7.23. Notice that on the basis of Betti's principle (see Sec. 5.16), the rotation of the left end of the beam of span $i - 1$ of the primary structure when subjected to $X_i = 1$ is equal to the rotation of its right end when subjected to $X_{i-1} = 1$ (see Fig. 7.23). Thus

$$F_{(i-1)i} = F_{i(i-1)} \qquad F_{(i+1)i} = F_{i(i+1)} \quad (7.66)$$

Substituting relations (7.62) into (5.66), we get

$$
\begin{aligned}
F_{ii} &= \frac{1}{EI^{(i-1)}L_{i-1}^2} \int_0^{L_{i-1}} x_1^2 \, dx + \frac{1}{EI^{(i)}L_i^2} \int_0^{L_i} (x_1)^2 \, dx_1 \\
&= \frac{L_{i-1}}{3EI^{(i-1)}} + \frac{L_i}{3EI^{(i)}}
\end{aligned}
\tag{7.67}
$$

Moreover, referring to Fig. 7.23a, the moment in the beam of span $(i - 1)$ when the structure is subjected to $X_{i-1} = 1$ is equal to

$$
\tilde{M} = 1 - \frac{x_1}{L_{i-1}} \qquad 0 \le x_1 \le L_{i-1}
\tag{7.68}
$$

Substituting relations (7.62) and (7.68) into (5.66) we obtain

$$
F_{i(i-1)} = F_{(i-1)i} = \frac{1}{EI^{(i-1)}L_{i-1}} \int_0^{L_{i-1}} \left(1 - \frac{x_1}{L_{i-1}}\right) x_1 \, dx_1 = \frac{L_{i-1}}{6EI^{(i-1)}}
\tag{7.69}
$$

Furthermore, referring to Fig. 7.23c, we see that when the structure is subjected to $X_{i+1} = 1$, the moment in the beam of span (i) is equal to

$$
\tilde{M} = \frac{1 - x_1'}{L_i} \qquad 0 \le x_1' \le L_i
\tag{7.70}
$$

Substituting relations (7.62) and (7.70) into (5.66), we get

$$
F_{i(i+1)} = F_{(i+1)i} = \frac{1}{EI^{(i)}L_i} \int_0^{L_i} \left(1 - \frac{x_1'}{L_i}\right) x_1' \, dx_1' = \frac{L_i}{6EI^{(i)}}
\tag{7.71}
$$

The ith equation of compatibility (7.20) for the beam of Fig. 7.21 is

$$
0 = \Delta_i^{PL} + F_{i(i-1)}X_{(i-1)} + F_{ii}X_i + F_{i(i+1)}X_{i+1}
\tag{7.72}
$$

or $\quad 0 = \Delta_i^{PL} + F_{i(i-1)}M^{(i-1)} + F_{ii}M^{(i)} + F_{i(i+1)}M^{(i+1)}$

Substituting relations (7.65), (7.67), (7.69), and (7.71) into (7.72), we get

$$
\begin{aligned}
0 =\ & \frac{A_M^{(i-1)} \bar{x}_1^{(i-1)}}{EI^{(i-1)}L_{i-1}} + \frac{A_M^{(i)} \bar{x}_1^{(i+1)}}{EI^{(i)}L_i} + \frac{\Delta^{(i-1)} - \Delta^{(i)}}{L_{i-1}} + \frac{\Delta^{(i+1)}\Delta^{(i)}}{L_i} \\
& + \frac{L_{i-1}}{6EI^{(i-1)}} M^{(i-1)} + \left[\frac{L_{i-1}}{3EI^{(i-1)}} + \frac{L_i}{3EI^{(i)}}\right] M^{(i)} + \frac{L_i}{6EI^{(i)}} M^{(i+1)}
\end{aligned}
$$

or
$$\frac{L_{i-1}}{I^{(i-1)}} M^{(i-1)} + 2\left[\frac{L_{i-1}}{I^{(i-1)}} + \frac{L_i}{I^{(i)}}\right] M^{(i)} + \frac{L_i}{I^{(i)}} M^{(i+1)}$$

$$= -\frac{6A_M^{(i-1)}\overline{x}^{(i-1)}_1}{I^{(i-1)}L_{i-1}} - \frac{6A_M^{(i)}\overline{x}^{(i+1)}_1}{I^{(i)}L_i} - 6E\left[\frac{\Delta^{(i-1)} - \Delta^{(i)}}{L_{i-1}} + \frac{\Delta^{(i+1)} - \Delta^{(i)}}{L_i}\right]$$

$$(7.73)$$

This equation is referred to as the three-moment equation. It can be applied repeatedly to adjacent joints of any continuous beam to yield the required compatibility equations. This is illustrated by the following examples.

Example 1. For the continuous beam loaded as shown in Fig. a, plot the shear and moment diagrams.

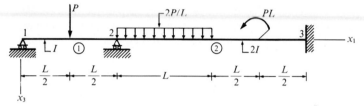

Figure a Geometry and loading of the beam.

---

solution This is a statically indeterminate structure to the second degree. We obtain the two required equations of compatibility by applying the three-moment equation (7.73) to joints 2 and 3. In applying Eq. (7.73) to joint 3, we can assume that, as shown in Fig. b. the beam extends to the right of the fixed support 3 by a span of finite length having a cross section with a sufficiently large moment of inertia so that the term $L_3/I^{(3)}$ in the three-moment equation (7.73) vanishes. The primary structure loaded with the given loads is shown in Fig. c. The moment in member 1 of the primary structure is

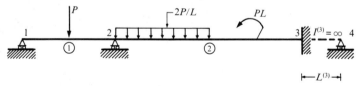

Figure b Beam with imaginary span of infinite stiffness.

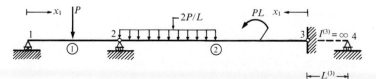

Figure c Primary structure subjected to the given actions.

$$M = \frac{P}{2}x_1 \qquad\qquad 0 \le x_1 \le \frac{L}{2}$$

$$M = \frac{P}{2}x_1 - P(L - x_1) \qquad \frac{L}{2} \le x_1 \le L \qquad\text{(a)}$$

Moreover, the moment in member 2 of the primary structure is

$$M = 0 \qquad\qquad 0 \le x_1' \le \frac{L}{2}$$

$$M = PL \qquad\qquad \frac{L}{2} \le x_1' \le L \qquad\text{(b)}$$

$$M = PL - \frac{2P}{L}\frac{(x'_1 - L)^2}{2} \qquad L \le x_1' \le 2L$$

The moment diagram is plotted in Fig. d. In this diagram each of the terms of Eqs. (a) and (b) are plotted separately. The resulting diagram is called moment diagram by parts. Referring to Fig. d, applying the three-moment equation (7.73) to joints 2 and 3, and noting that $M^{(1)} = 0$, we have

$$2\left(\frac{L}{I} + \frac{2L}{2I}\right)M^{(2)} + \frac{1}{2I}2LM^{(3)}$$

$$= -\frac{6}{IL}\left[\left(\frac{PL}{2}\right)\left(\frac{L}{2}\right)\left(\frac{2}{3}L\right) - \left(\frac{PL}{2}\right)\left(\frac{L}{4}\right)\left(\frac{2L}{6} + \frac{L}{2}\right)\right]$$

$$-\frac{6}{2L}\left(\frac{1}{2I}\right)\left[-(PL)\left(\frac{L}{3}\right)\left(\frac{3L}{4} + L\right) + (PL)\left(\frac{3L}{2}\right)\left(\frac{L}{2} + \frac{3L}{4}\right)\right]$$

$$2LM^{(2)} + 2(2L + 0)(M^{(3)})$$

$$= -\frac{6}{2L}\left[-(PL)\left(\frac{L}{4}\right)\left(\frac{L}{3}\right) + (PL)\left(\frac{3L}{2}\right)\left(\frac{3L}{4}\right)\right]$$

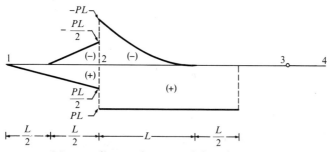

Figure d Moment diagram by parts of the primary structure subjected to the given actions.

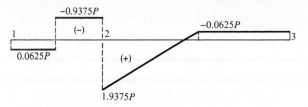

Figure e Shear diagram for the beam.

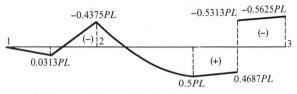

Figure f Moment diagram for the beam.

The above relations may be simplified to yield the following equations of compatibility.

$$4M^{(2)} + M^{(3)} = -PL + \frac{PL}{8}(5) + \frac{PL}{8}(7) - \frac{PL}{16}(45) = -\frac{37}{16}PL$$

$$M^{(2)} + 2M^{(3)} = \frac{PL}{8} - \frac{27PL}{16} = -\frac{25}{16}PL$$

Solving the above equations simultaneously, we get

$$M^{(2)} = -0.4375PL \qquad M^{(3)} = -0.5625PL$$

Using the above results, we plot the shear and moment diagrams for the beam in Figs. e and f.

Example 2. Compute the reactions of the continuous beam in Fig. a, resulting from a settlement of 20 mm of support 2. The beam is made of steel ($E = 210{,}000$ MPa). The moments of inertia of the cross section of members 1 and 2 of the beam are equal to

$$I^{(1)} = 200 \times 10^6 \text{ mm}^4 \qquad I^{(2)} = 400 \times 10^6 \text{ mm}^4$$

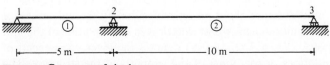

Figure a Geometry of the beam.

solution   We apply relation (7.73) to joint 2 of the beam. Noting that $M^{(1)} = M^{(3)} = 0$ and $A_M^{(1)} = A_M^{(2)} = 0$, we have

$$2\left(\frac{5}{200 \times 10^{-6}} + \frac{10}{400 \times 10^{-6}}\right)M^{(2)} = -6(210{,}000 \times 10^3)\left(-\frac{0.020}{5} - \frac{0.020}{10}\right)$$

$$\text{or } M^{(2)} = 75.6 \text{ kN} \cdot \text{m}$$

Referring to Fig. b, the reactions of the beam are

$$R_v^{(1)} = 15.12 \text{ kN} \qquad R_v^{(2)} = 22.68 \text{ kN} \qquad R_v^{(3)} = 7.56 \text{ kN}$$

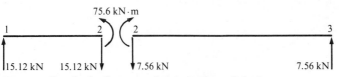

Figure b   Free-body diagrams of the members of the beam.

7.11 Computation of Components of Displacement of Points of Statically Indeterminate Structures

In this section we apply the method of virtual work to compute the components of displacement of points of statically indeterminate structures. That is, we employ the following relation.

$$d + \sum_{s=1}^{S} \tilde{R}^{(s)}\Delta^{(s)} = \sum_{k=1}^{NM}\left[\int_0^L \left[\frac{N\tilde{N}}{AE} + \frac{M_1\tilde{M}_1}{KG} + \frac{M_2\tilde{M}_2}{EI_2} + \frac{M_3\tilde{M}_3}{EI_3}\right]\right.$$

$$\left. + \int_0^L \left[\alpha\tilde{N}\,\Delta T_c + \alpha\tilde{M}_2\frac{\delta T_3}{h_3} + \alpha\tilde{M}_3\frac{\delta T_2}{h_2}\right]dx_1\right]^{(k)} \qquad (7.74)$$

where d is either a component of translation or a component of rotation of a point of the structure. The symbols $\tilde{N}$ and $\tilde{M}_i$ ($i = 1, 2, 3$) denote a set of statically admissible (not necessarily the actual) distributions of the internal axial force and the components of moment, respectively, in the members of the structure subjected to a virtual loading. This virtual loading consists of a unit load applied at the point of the structure where the displacement is desired. If we want to compute the component of translation of a point of a structure in the direction of the unit vector **n**, the virtual loading is a unit force acting in the direction of the unit vector **n**. If we want to compute the component of rotation of a point of a structure about an axis specified by the unit vector **n**, the virtual loading is a unit moment whose vector is in the direction of the unit vector **n**. The symbol $\Delta^{(s)}$ ($s = 1, 2, \ldots, S$) denotes the given components of displacements (translations or rotations) of the supports of the structure. The symbol $\tilde{R}^{(s)}$ ($s = 1, 2, \ldots, S$) denotes the statically admissible components of the reactions (forces and moments) at the supports of the structure corresponding to the given components of displacements $\Delta^{(s)}$, when the structure is subjected to the virtual loading. $R^{(s)}$ is considered positive when it acts in the direction of $\Delta^{(s)}$.

In what follows, we compute components of displacement of a statically indeterminate truss and a grid using the method of virtual work.

Example 1. Compute the horizontal and vertical components of translation of joint 3 of the statically indeterminate truss shown in Fig. a. The members of the truss have the same constant cross section and are made from the same material.

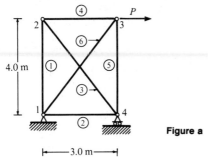

Figure a Geometry and loading of the truss.

---

solution The internal forces in the members of this truss have been established in Example 1 of Sec. 7.2.1. They are shown in Fig. b.

In order to compute the horizontal component of translation of joint 3, we consider the truss subjected at this joint to a horizontal unit force (see Fig. c). This truss is statically indeterminate to the first degree. That is, it has one member more than the minimum required for not forming a mechanism. Thus, we can arbitrarily select the value of the internal force in one member of the truss and compute a statically admissible set of internal forces in its other members. For example, we can set the force in member 3 equal to zero. By considering the equilibrium of joint 2, we can see (see

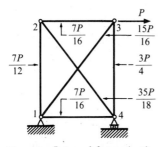

Figure b Internal forces in the members of the truss subjected to the given force.

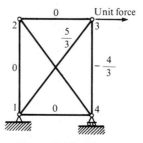

Figure c Statically admissible internal forces in the members of the truss subjected to a unit horizontal force.

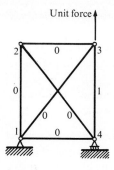

Unit force

Figure **d** Statically admissible internal forces in the members of the truss subjected to a unit vertical force.

Fig. c) that the forces in members 1 and 4 must vanish. Moreover, by considering the equilibrium of joint 3, we can establish the forces in members 6 and 5. The results are shown in Fig. c. Thus, using the internal forces shown in Figs. b and c, we find that Eq. (7.74) gives

$$u_n^{(3)} = \frac{1}{AE}\left[\frac{4}{3}\left(\frac{3P}{4}\right)4 + \frac{5}{3}\left(\frac{15}{16}\right)5\right] = \frac{189P}{16AE} \qquad \text{to the right}$$

In order to compute the vertical component of translation of joint 3, we consider the truss subjected at this joint to a vertical unit force (see Fig. d). We choose the convenient, statically admissible distribution of internal forces corresponding to this loading shown in Fig. d. Using the internal forces in the members of the truss shown in Figs. b and d, we find that Eq. (7.74) gives:

$$u_v^{(3)} = \left(-\frac{3P}{4}\right)\frac{4}{AE} = -\frac{3P}{AE} = \frac{3P}{AE} \qquad \text{downward}$$

The minus sign indicates that the displacement is in the direction opposite to that of the unit force applied to the truss in Fig. d.

Example 2. The members of the grid shown in Fig. a are tubular. The moment of inertia of member 2 is half that of the other members. The modulus of elasticity of

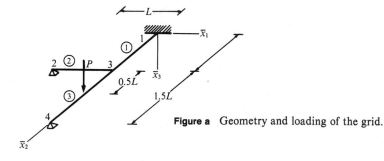

Figure a Geometry and loading of the grid.

the material from which the tubes are made is 2.6 times larger than its shear modulus. Compute the deflection of joint 3 of the grid.

solution The moment diagrams for the members of the grid have been plotted in Example 7 of Sec. 7.2.2. They are shown in Fig. b. In order to compute the deflection of joint 3, we consider the grid subjected to a vertical unit force at joint 3 and we choose a convenient set of statically admissible reactions (see Fig. c). The corresponding moment diagram is shown in Fig. d. We denote by I the moment of inertia of members 1 and 2. Using the method of virtual work (7.74) and referring to Figs. b and d and to Table A on the inside of the back cover of the book, we get

$$u_3^{(3)} = \int_2^3 \frac{M_2 \tilde{M}_2}{2EI} = \frac{1}{2EI} \left[-\frac{L}{6}(0.1372PL)L + \frac{L}{3}(0.4039PL)L \right]$$

$$= \frac{0.0559PL^3}{EI} \quad \text{downward}$$

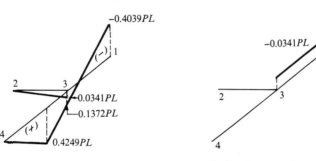

Bending moment diagram

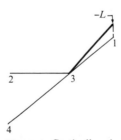

Torsional moment diagram

Figure b Moment diagrams for the grid subjected to the given actions.

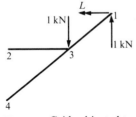

Figure c Grid subjected to a vertical unit force at joint 3 and a set of statically admissible reactions.

Figure d Statically admissible moment diagram for the grid of Fig. c.

7.12   Problems

Note: In the following problems disregard the effect of shear deformation of the members of beams and frames.

1 to 4.   Using the basic force method, compute the internal forces in the members of the truss resulting from the external forces shown in Fig. 7.P1. The members of the truss are made of the same material (E = 210,000 MPa). Repeat for Figs. 7.P2 to 7.P4.

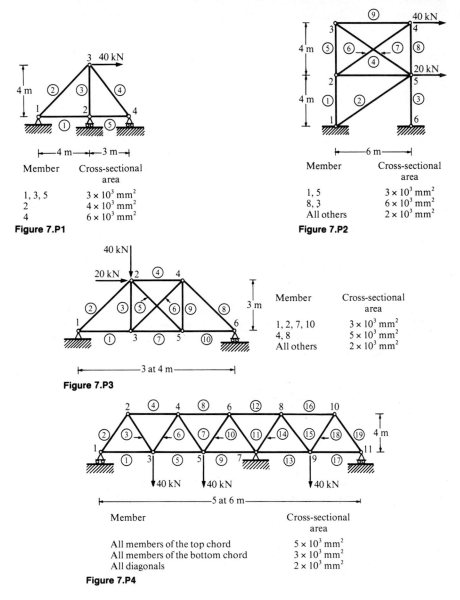

Member	Cross-sectional area
1, 3, 5	3×10^3 mm^2
2	4×10^3 mm^2
4	6×10^3 mm^2

Figure 7.P1

Member	Cross-sectional area
1, 5	3×10^3 mm^2
8, 3	6×10^3 mm^2
All others	2×10^3 mm^2

Figure 7.P2

Member	Cross-sectional area
1, 2, 7, 10	3×10^3 mm^2
4, 8	5×10^3 mm^2
All others	2×10^3 mm^2

Figure 7.P3

Member	Cross-sectional area
All members of the top chord	5×10^3 mm^2
All members of the bottom chord	3×10^3 mm^2
All diagonals	2×10^3 mm^2

Figure 7.P4

5 and 6.   Using the basic force method, establish the reactions of the beam resulting from the external actions shown in Fig. 7.P5. The members of the beam have the same constant cross section and are made of the same material. Choose the reaction at support 2 as the redundant. Plot the shear and moment diagrams for the beam. Repeat for Fig. 7.P6.

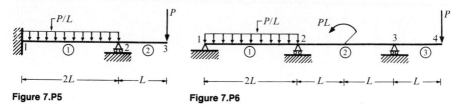

Figure 7.P5

Figure 7.P6

7.   Using the basic force method, establish the reactions of the structure resulting from the external forces shown in Fig. 7.P7. The members of the structure have the same constant cross section and are made from the same material. Choose the reaction at support 3 as the redundant. Plot the shear and moment diagrams for the structure. Disregard the effect of the axial deformation of the members of the structure.

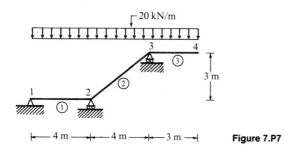

Figure 7.P7

8.   Using the basic force method, establish the reactions of the beam of Fig. 7.P6. Choose the internal moment at point 2 as the redundant. Plot the shear and moment diagrams for the beam.

9.   Using the basic force method, establish the reactions of the beam of Fig. 7.P5. Choose the internal moment of point 1 as the redundant. Plot the shear and moment diagrams for the beam.

10.   Using the basic force method, establish the reactions of the beam resulting from the external forces shown in Fig. 7.P10. The beam is made of a standard section with cover plates extending 2 m on each side of support 2. Plot the shear and moment diagrams for the beam.

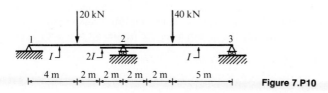

Figure 7.P10

11. Using the basic force method, compute the reactions of the structure resulting from the external forces shown in Fig. 7.P11. The members of the truss are made of steel and have the same constant cross section of area 3×10^3 mm^2. The columns are also made of steel and have a constant cross section $I = 369.7 \times 10^6$ mm^4. Disregard the effect of axial deformation of the columns.

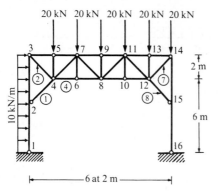

Figure 7.P11

12. Using the basic force method, compute the reactions of the steel structure subjected to the external force shown in Fig. 7.P12. The area of the cross section of the steel cable is $A_c = 800$ mm^2. The moment of inertia of the beam is $I = 369.7 \times 10^6$ mm^4. Disregard the effect of the axial deformation of beam 1,3,4. Plot the shear and moment diagrams for the beam.

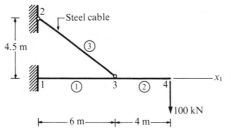

Figure 7.P12

13 to 15. Using the basic force method, plot the shear and moment diagrams for the frame subjected to the external actions shown in Fig. 7.P13. The members of the frame are made of the same material and have the same constant cross section. Disregard the effect of axial deformation of the members of the frame. Repeat with Figs. 7.P14 and 7.P15.

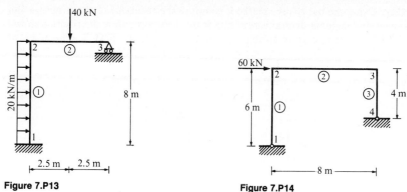

20 kN/m

40 kN

8 m

2.5 m   2.5 m

Figure 7.P13

60 kN

6 m

4 m

8 m

Figure 7.P14

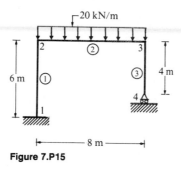

−20 kN/m

6 m

4 m

8 m

Figure 7.P15

16. Using the basic force method, compute the axial forces in members 3, 4, and 5 and draw the shear and moment diagram for beam 1,2,4 of the structure shown in Fig. 7.P16. The members of the structure are made of the same material and have constant cross sections. Compare the moment distribution in beam 1,2,4 with that in a simply supported beam of the same length.

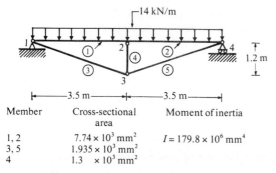

−14 kN/m

1.2 m

——3.5 m——   ——3.5 m——

Member	Cross-sectional area	Moment of inertia
1, 2	$7.74 \times 10^3 \text{ mm}^2$	$I = 179.8 \times 10^6 \text{ mm}^4$
3, 5	$1.935 \times 10^3 \text{ mm}^2$	
4	$1.3 \ \times 10^3 \text{ mm}^2$	

Figure 7.P16

17 to 19.   Using the basic force method, compute the reactions of the structure and plot its shear and moment diagrams resulting from the temperature distribution shown in Fig. 7.P17. The temperature during construction was $T_0 = 15\,°C$. The members of the structure are made of the same material ($E = 210,000$ MPa, $\alpha = 10^{-5}/°C$), and have the same constant cross section. Repeat for Figs. 7.P18 and 7.P19.

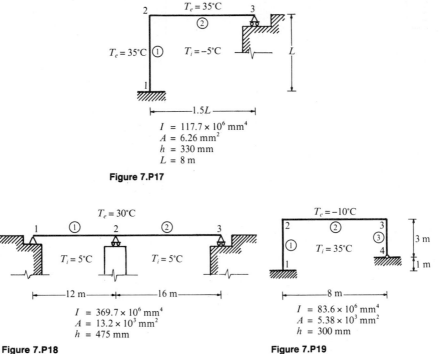

$I = 117.7 \times 10^6\ mm^4$
$A = 6.26\ mm^2$
$h = 330\ mm$
$L = 8\ m$

Figure 7.P17

$T_e = 30°C$

$T_i = 5°C$ $T_i = 5°C$

├──────12 m──────┼──────16 m──────┤

$I = 369.7 \times 10^6\ mm^4$
$A = 13.2 \times 10^3\ mm^2$
$h = 475\ mm$

Figure 7.P18

$T_e = -10°C$

$T_i = 35°C$

3 m

1 m

├──────8 m──────┤

$I = 83.6 \times 10^6\ mm^4$
$A = 5.38 \times 10^3\ mm^2$
$h = 300\ mm$

Figure 7.P19

20 to 22.   Using the basic force method, for the structure shown in Fig. 7.P17, establish the reactions and plot the shear and moment diagrams resulting from settlement of support 1 of 20 mm. The members of the structure are made of the same material ($E = 210,000$ MPa) and have the same constant cross section whose properties are given in the figure. Disregard the effect of axial deformation of the members of the structure. Repeat with Figs. 7.P18 and 7.P19.

23 to 28.   Using the basic force method, analyze the frame shown in Fig. 7.P23 for each of the following loading cases and plot its shear and moment diagrams:
  (*a*) The external actions shown in the figure.
  (*b*) A settlement of 20 mm of support 1.
  (*c*) A temperature of the external fibers $T_e = 35\,°C$ and of the internal fibers of $T_i = -5\,°C$. The temperature during construction was $15\,°C$.

   The members of the frame are made of the same material ($E = 210,000$ MPa, $\alpha = 10^{-5}/°C$) and have the same constant cross section ($I = 162.7 \times 10^6\ mm^4$, $h =$

360 mm). Disregard the effect of axial deformation of the members of the frame. Repeat for Figs. 7.P24 to 7.P28.

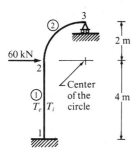

Figure 7.P23

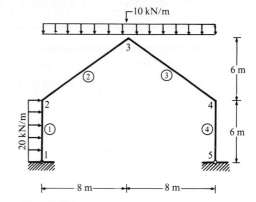

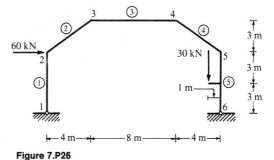

Figure 7.P24

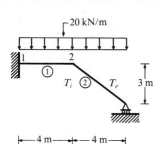

Figure 7.P25

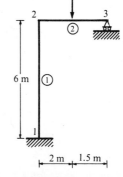

Figure 7.P26

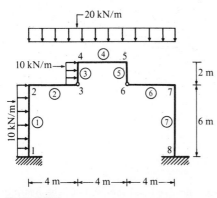

Figure 7.P27

Figure 7.P28

29 to 31. Using the basic force method, analyze the continuous beam subjected to the external action shown in Fig. 7.P29. The members of the beam have constant cross section whose moment of inertia is indicated in the figure. Plot the shear and moment diagrams for the beam. Repeat for Figs. 7.P30 and 7.P31.

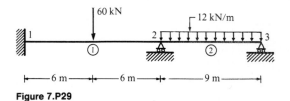

Figure 7.P29

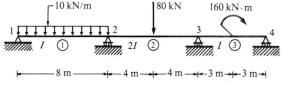

Figure 7.P30

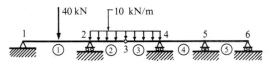

Figure 7.P31

32. Using the basic force method, analyze the beam loaded as shown in Fig. 7.P32. The beam is supported on an elastic spring at point 2 whose constant is equal to $k/EI = 0.01/m^3$. Plot the shear and moment diagrams for the beam.

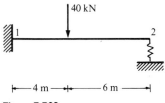

Figure 7.P32

33. Using the basic force method, compute the fixed-end moments of the beam of constant width and variable depth loaded as shown in Fig. 7.P33.

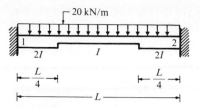

Figure 7.P33

34 and 35. Using the basic force method, compute the reactions of the beam of variable cross section, loaded as shown in Fig. 7.P34. The area of the cross section of each flange is constant and equal to $A_f = 6 \times 10^3$ mm². In order to simplify the analysis, neglect the effect of the web on the moment of inertia of the cross sections of the beam. Repeat for Fig. 7.P35.

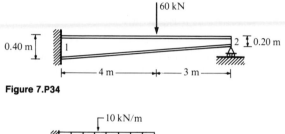

Figure 7.P34

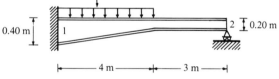

Figure 7.P35

36. Using the basic force method, compute the reactions of the beam shown in Fig. 7.P36 when it is subjected to the following loading cases:
  (*a*) The external actions shown in Fig. 7.P36.
  (*b*) A settlement of support 1 equal to 0.006 m.
The width of the beam is 0.20 m.

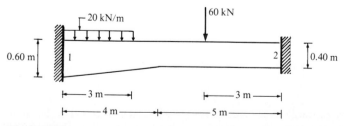

Figure 7.P36

37 and 38.   Using the basic force method, analyze the one-bay frame shown in Fig. 7.P37. The members of the frame are made of the same material and have the same constant cross section. Disregard the effect of axial deformation of the members of the frame. Repeat for Fig. 7.P38.

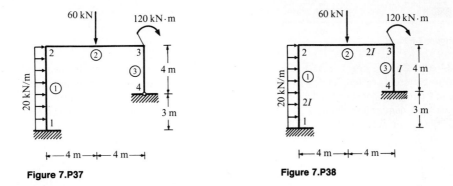

Figure 7.P37   **Figure 7.P38**

39 and 40   Using the basic force method, analyze the two-bay frame shown in Fig. 7.P39. The members of the frame are made of the same material and have the same constant cross section. Disregard the effect of axial deformation of the members of the frame. Repeat for Fig. 7.P40.

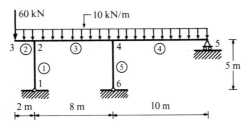

Figure 7.P39

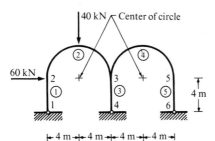

Figure 7.P40

41 and 42.   Using the basic force method, compute the internal forces in the members of the truss resulting from the external forces shown in Fig. 7.P41. The members of the truss are made of the same material. Repeat for Fig. 7.P42.

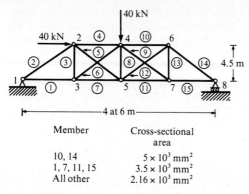

Figure 7.P41

Member	Cross-sectional area
10, 14	5×10^3 mm^2
1, 7, 11, 15	3.5×10^3 mm^2
All other	2.16×10^3 mm^2

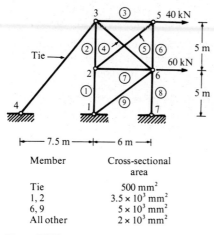

Figure 7.P42

Member	Cross-sectional area
Tie	500 mm^2
1, 2	3.5×10^3 mm^2
6, 9	5×10^3 mm^2
All other	2×10^3 mm^2

43. Using the basic force method, analyze the one-bay frame subjected to the external forces shown in Fig. 7.P43. The members of the frame are made of the same material. Disregard the effect of the axial deformation of the members of the frame.

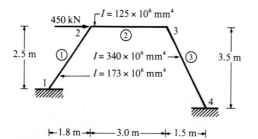

Figure 7.P43

44. Using the basic force method, compute the force in the cable and plot the shear and moment diagrams for the girder of the cable-stayed bridge shown in Fig. 7.P44. The girder is subjected to a uniform load of 20 kN/m. Assume that the steel cable slides without friction at the top of the tower. The girder is made of steel and has a constant cross section ($I = 576.80 \times 10^6$ mm^4) and is continuous at the tower where it is supported on rollers. The cross-sectional area of the cable is $A = 900$ mm^2. Disregard the effect of the axial deformation of the girder and the tower.

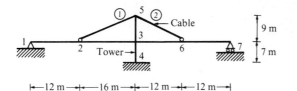

Figure 7.P44

45 to 56. Using the basic force method, analyze the symmetric structure subjected to the external actions shown in Fig. 7.P45. The members of the structure are made of the same material ($E = 210,000$ MPa) and have the same constant cross section ($I = 117.70 \times 10^6$ mm^4). Disregard the effect of axial deformation of the members of the structure. Repeat for Figs. 7.46 to 7.56, but include the effect of axial deformation at the tie in Fig. 7.47.

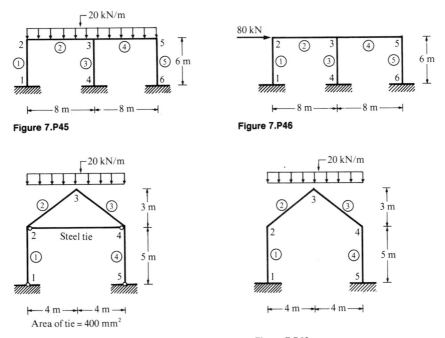

Figure 7.P45

Figure 7.P46

Figure 7.P47

Figure 7.P48

Area of tie = 400 mm^2

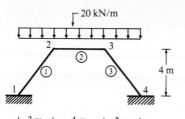

Figure 7.P49

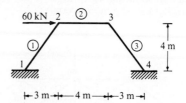

Figure 7.P50

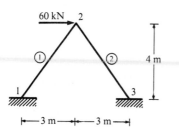

Figure 7.P51

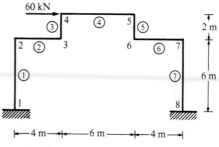

Figure 7.P52

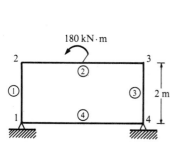

Figure 7.P53

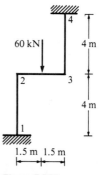

Figure 7.P54

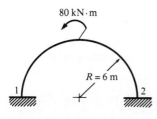

Figure 7.P55

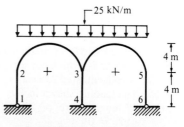

Figure 7.P56

57. Using the basic force method, analyze the symmetric truss subjected to the external force shown in Fig. 7.P57. The members of the truss are made of the same material ($E = 210,000$ MPa).

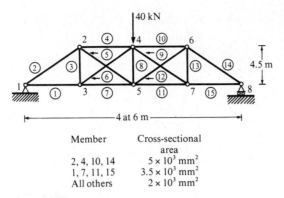

Member	Cross-sectional area
2, 4, 10, 14	5×10^3 mm^2
1, 7, 11, 15	3.5×10^3 mm^2
All others	2×10^3 mm^2

Figure 7.P57

58. Using the basic force method, analyze the concrete culvert shown in Fig. 7.P58. The specific weight of the soil and of the concrete are 17 kN/m^3 and 24 kN/m^3, respectively.

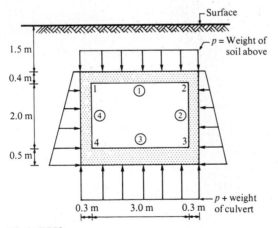

Figure 7.P58

59. Using the basic force method, compute the reactions of a beam of length L, that is fixed at both ends, which correspond to each of the following loading cases.
  (a) A concentrated force at $x_1 = a$
  (b) The loading shown in Fig. 7.P59
  (c) A temperature of its external fibers T_e and of its internal fibers T_i ($\Delta T_c = 0$)

The beam has a constant cross section ($EI = $ constant).

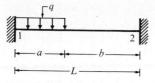

Figure 7.P59

60. The members of the planar frame of Fig. 7.P60 have a cross section of height h = 400 mm and $I = 133.33 \times 10^6$ mm^4. The modulus of elasticity and the coefficient of linear expansion of the material from which the members of the frame are made are E = 210,000 MPa and $\alpha = 10^{-5}/°$C, respectively. The frame is subjected to the following loading cases.

(a) The external actions shown in Fig. 7.P60. In this case, disregard the effect of axial deformation of the members of the frame.

(b) A temperature of its external fibers equal to $T_e = -5°$C and to its internal fibers equal to $T_i = 25°$C. The temperature during construction of the frame was $T_0 = 20°$.

For each case of loading, plot the shear and moment diagram of the frame.

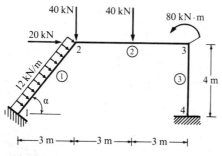

Figure 7.P60

61. Plot the shear and moment (bending and twisting) diagrams for the grid subjected to the loading shown in Fig. 7.P61. The members of the grid are circular steel tubes of constant cross section ($G = E/2.6$).

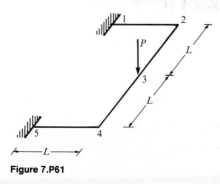

Figure 7.P61

62 to 68.   Using the basic force method, analyze the space frame subjected to the external forces shown in Fig. 7.P62. The members of the frame are made of the same material (E = 210,000 MPa, ν = 0.33) and have the same constant, tubular cross section (outside diameter 0.20 m, inside diameter 0.18 m). Repeat for Figs. 7.P63 to 7.P68.

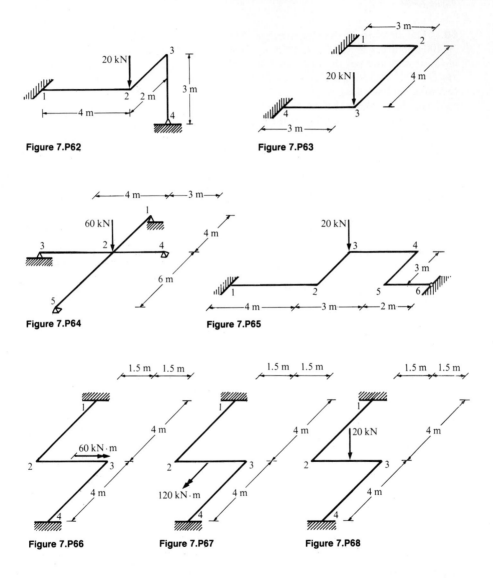

Figure 7.P62

Figure 7.P63

Figure 7.P64

Figure 7.P65

Figure 7.P66

Figure 7.P67

Figure 7.P68

69 to 71.   Using the conjugate beam method, analyze the beam of Fig. 7.P5. The members of the beam are made from the same material and have the same constant cross section. Repeat for Figs. 7.P6 and 7.P10.

72 and 73. Using Castigliano's theorem, analyze the truss loaded as shown in Fig. 7.P2. The members of the truss are made of the same material and have the cross sections indicated in the figure. Repeat for Fig. 7.P3.

74 to 78. Using Castigliano's theorem, analyze the structure loaded as shown in Fig. 7.P13. The members of the structure are made from the same material and have the same constant cross section. Disregard the effect of the axial deformation of the members of the structure. Repeat for Figs. 7.P14, 7.P16, 7.P23, and 7.P37.

79 to 81. Using the three-moment equation (7.73), analyze the beam loaded as shown in Fig. 7.P29. The members of the beam are made from the same material and have the same constant cross section. Repeat for Figs. 7.P30 and 7.P31.

82 to 84. Using the elastic center method, analyze the structure loaded as shown in Fig. 7.P82. Plot the shear and moment diagrams for the structure. The members of the structure are made of the same material and have the same constant cross section. Repeat for Figs. 7.P83 and 7.P84.

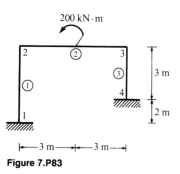

Figure 7.P82

Figure 7.P83

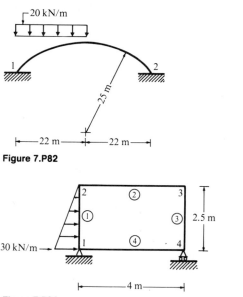

Figure 7.P84

85. Using the method of virtual work, compute the horizontal component of translation of joint 2 of the steel truss loaded as shown in Fig. 7.P1.

86. Using the method of virtual work, compute the horizontal component of translation of joint 4 of the steel truss loaded as shown in Fig. 7.P2.

87. Using the method of virtual work, compute the vertical component of translation (deflection) and the rotation of point 3 of the beam loaded as shown in Fig. 7.P5. The members of the beam are made of the same material and have the same constant cross section.

88 and 89. Using the method of virtual work, compute the vertical component of translation (deflection) and the rotation of point 4 of the beam loaded as shown in Fig. 7.P7 with $L = 3$ m and $P = 30$ kN. The members of the beam are made of the same material ($E = 210,000$ MPa) and have the same constant section ($I = 369.70 \times 10^6$ mm^4). Repeat for Fig. 7.P6.

90. Using the method of virtual work, compute the vertical component of translation (deflection) and the rotation of point 4 of the beam loaded as shown in Fig. 7.P12. The members of the beam are made of the same material ($E = 210,000$ MPa) and have the same constant cross section ($I = 369.70 \times 10^6$ mm^4). The area of the cross section of the cable is $A_c = 800$ mm^2. Disregard the effect of axial deformation of the members of the beam.

91 to 93. Using the method of virtual work, compute the horizontal component of translation of point 3 of the frame loaded as shown in Fig. 7.P14. The members of the frame are made of the same material ($E = 210,000$ MPa) and have the same constant cross section ($I = 369.70 \times 10^6$ mm^4). Disregard the effect of the axial deformation of the members of the frame. Repeat for Figs. 7.P15 and 7.P25.

94 and 95. Using the method of virtual work, compute the horizontal component of translation of point 3 of the frame, loaded as shown in Fig. 7.P17. The members of the frame are made of the same material ($E = 210,000$ MPa, $\alpha = 10^{-5}/°$C) and have the same constant cross section. Disregard the effect of the axial deformation of the members of the frame. Repeat for Fig. 7.P19.

96. The moment of inertia of the parabolic arch shown in Fig. 7.P96 varies as $I = I_c \sec \phi$, where ϕ is the angle that the tangent to the axis of the arch makes with the horizontal and I_c is the moment of inertia of the cross section at the crown of the arch. Establish the external moment $M^{(2)}$ which must be placed at the end of the arch adjacent to support 2 in order to produce a unit counterclockwise rotation of this end. This moment is called the rotational stiffness of end 2 of the arch. The ratio $M^{(2)}/M^{(1)}$ is called the carry-over factor at the end 2 of the arch.

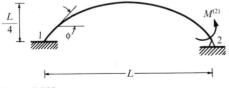

Figure 7.P96

97. The cable of the middle span of the suspension bridge shown in Fig. 7.P97 is a polyhedron inscribed in a parabola. Its sag is $s = 2L_p$, while the height of its towers above the deck is $h = 2s$. Denoting the moment of inertia of the girder of the middle span by I, we see that the area of the cable is $A^c = I/L_p^2$, while the area of the hangers is $A^{(h)} = A^c/25$. Compute the maximum tension in the cable. Plot the shear and

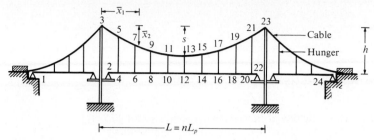

Figure 7.P97

moment diagrams for the girder of the middle span when a uniform load of q kN/m is acting on this span.

Long-span suspension bridges deflect considerably; thus, the effect of the change of their geometry on their internal actions is not negligible. Consequently, the internal actions in the members of such bridges, obtained on the basis of the theory of small deformation, can be considered only as an approximation. In order to further simplify the analysis of the bridge, the effect of the side spans will be eliminated by assuming that the cables are anchored to the top of the stiff towers. Notice that, since the cable is inscribed in a parabola and the hangers are equally spaced, the internal forces in the hangers will be equal.

REFERENCE

1. Alfred Freudenthal, "Deflection Theory for Arches," *Int. Assn. Bridge Struct. Engrs. (Zurich)*, vol. 3, 1935.

Photograph and Brief Description of a Winter Garden

The Winter Garden of the city of Niagara Falls is an elegant 175 ft × 155 ft (53.34 × 47.24 m) and 107 ft (32.61 m) high building with glass walls. It is filled with a variety of tropical plants. The building is designed to withstand high wind and snow loads as well as a variation in temperature of 100°F (37.8°C). The roof consists of a network of five transverse and 12 longitudinal trusses. Each transverse truss is supported by four reinforced concrete columns of which five are 58 ft (17.68 m) high. The trusses are continuous and cantilever 14 ft (4.27 m) on each end. The vertical trusses which support the exterior glass walls hang from the edges of the cantilever roof trusses. Thus, the columns are 14 ft away from the glass wall and do not interfere with the lightness and transparency of the garden.

Figure 7.24 The Winter Garden of the city of Niagara Falls, N.Y. *(Courtesy of De Simone and Chaplin and Associates, Consulting Engineers.)*

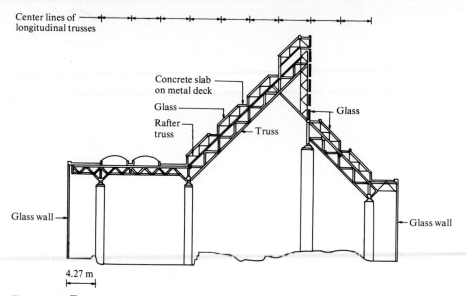

Figure 7.25 Transverse section of the Winter Garden of the city of Niagara Falls. *(Courtesy of De Simone and Chaplin and Associates, Consulting Engineers.)*

Chapter 8

Displacement Methods of Structural Analysis

8.1 Introduction

In this chapter, we present two classical displacement methods of structural analysis, *the slope deflection method* and the *displacement method with moment distribution.* These methods have been used extensively in analyzing planar beams and planar frames using hand calculation or the aid of a desk calculator, and disregarding the effects of axial and shear deformation of their members. These methods cannot be used directly in analyzing trusses or frames which have one or more members subjected only to an axial force. When the slope deflection method is employed in analyzing a beam or a frame with many joints (many unknown displacements), it results in a large number of simultaneous linear algebraic equations whose solution is difficult when a computer is not available. When the displacement method with moment distribution is employed in analyzing the same structure, it results in a considerably smaller number of simultaneous, linear, algebraic equations. For this reason, the use of the slope deflection method has declined considerably since the displacement method with moment distribution was introduced by Professor Hardy Cross in 1932. Today, however, simultaneous linear algebraic equations can be solved easily with the aid of a computer. Consequently, the slope deflection method can be used very effectively in analyzing beams and frames even when they have many joints.

In the displacement methods, it is convenient to consider as positive the local components of the internal actions (forces and moments) acting on the ends of a member if they are directed along the positive local axes (see Fig. 8.1). Moreover, the local components of displacement (translations and rotations) of the ends of a member are considered positive if they are directed along the positive local axes. We chose the x_3 axis normal to the plane of the structure. However after the reactions and the end actions in the members of a structure are established, the shear and moment diagrams are plotted using the sign convention for the internal actions specified in Fig. 7.1.

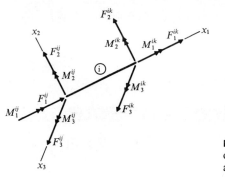

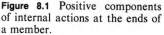

Figure 8.1 Positive components of internal actions at the ends of a member.

When analyzing a structure by the flexibility (force) method, we choose external redundant reactions or internal redundant actions as unknown quantities. The number of unknown quantities is equal to the degree of static indeterminacy of the structure. When analyzing a structure using a displacement or stiffness method, we choose components of displacements of the joints of the structure as unknown quantities.

8.2 Kinematic Indeterminacy of a Structure

When a structure is subjected to loads, some of its joints undergo displacements (translations and rotations) which are not known, while others may undergo displacements which are known. For instance, all the displacements of a fixed support are zero. The sum of the unknown components of displacements of the joints of a structure represent its *degree of freedom* or the *degree of its kinematic indeterminacy.*

In general, the movement of an unrestrained joint of a space frame can be specified by three components of translation and three components of rotation with respect to a rectangular system of axes. Moreover, the movement of an unrestrained joint of a planar frame can be specified by two components of translation with respect to a set of two rectangular axes lying in the plane of the frame and one component of rotation whose vector is normal to the plane of the frame. Furthermore, generally the joints of a truss do not rotate since the members of the truss are assumed connected to the joints by pins, and consequently the members cannot transfer a moment to the joints. Thus, the movement of an unrestrained joint of a space truss is specified by three components of translation, while the movement of an unrestrained joint or a planar truss is specified by two components of translation. For example, the movement of joints 2, 3, 4, and 5 of the simple planar truss shown in Fig. 8.2*a* is specified by their horizontal and vertical components of translation, while the movement of joint 6 is specified only by its horizontal component of translation because this joint cannot move in the vertical direction. Thus, the simple truss of Fig.

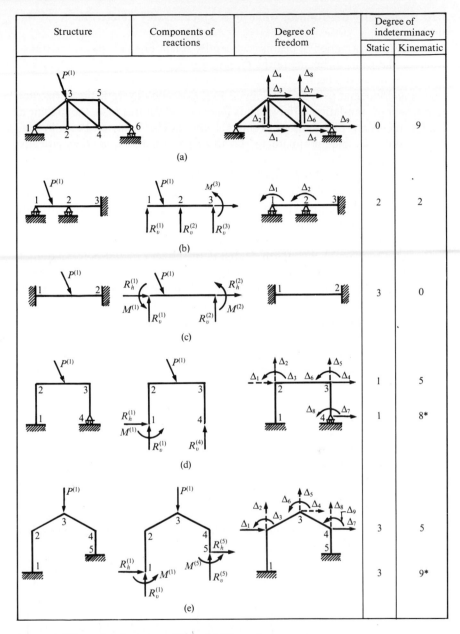

Structure	Components of reactions	Degree of freedom	Degree of indeterminacy	
			Static	Kinematic
(a)			0	9
(b)			2	2
(c)			3	0
(d)			1	5
			1	8*
(e)			3	5
			3	9*

* Degree of kinematic indeterminacy if the axial deformation
  of the members of the structure is not neglected

Figure 8.2   Static and kinematic indeterminacy.

8.2*a* is statically determined, while it is kinematically indeterminate to the ninth degree.

The axial deformation of beams is generally small. For this reason the translation of the joints of beams in the direction of their axis can be disregarded. Thus, the continuous beam of Fig. 8.2*b* is statically indeterminate to the second degree and kinematically indeterminate to the second degree if the horizontal component of translation of its joints is disregarded. In this case, the unknown displacements are the rotations of joints 1 and 2. The fixed-end beam in Fig. 8.2*c* is statically indeterminate to the third degree, but it is kinematically determinate.

The effect of the axial deformation of the members of frames on their internal actions is very small and is usually disregarded. For this reason, certain components of translation of some joints of frames can be disregarded and, consequently, their translational degree of freedom decreases. For instance, the frame of Fig. 8.2*d* is statically indeterminate to the first degree and kinematically indeterminate to the eighth degree if the effect of axial deformation of its members is taken into consideration. However, when the effect of axial deformation of the members of this frame is disregarded, the vertical components of translation of joints 2 and 3 are negligible, while the horizontal components of their translation differ negligibly. Thus, if the effect of axial deformation of its members is disregarded, the frame of Fig. 8.2*d* is kinematically indeterminate to the fifth degree.

When the effect of the axial deformation of the members of planar frames is neglected, the establishment of their translational degree of freedom often requires special attention. For instance, the sawtooth frame of Fig. 8.2*e* is statically indeterminate to the third degree and kinematically indeterminate to the ninth degree if the axial deformation of its members is considered. When the axial deformation of its members is neglected, joints 2 and 4 do not move in the vertical direction, while joint 3 can translate both in the vertical and horizontal directions. However, the components of translation of this joint can be established from the horizontal component of translations of joints 2 and 4 of the frame by considering the geometry of the deformed frame.

A systematic method is presented in Figs. 8.3 and 8.4 for establishing the

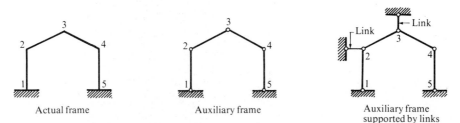

| Actual frame | Auxiliary frame | Auxiliary frame supported by links |

Figure 8.3 Procedure for establishing the translational degree of freedom of the joints of a planar frame.

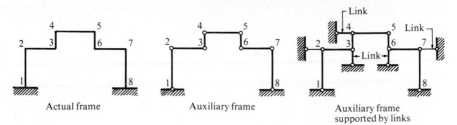

Actual frame   Auxiliary frame   Auxiliary frame
                                                 supported by links

Figure 8.4 Procedure for establishing the translational degree of freedom of the joints of a planar frame.

translational degrees of freedom of the joints of a planar frame when the effect of axial deformation of its members is neglected. An auxiliary frame is used which is obtained from the actual frame by inserting a hinge at the end of each of its members (see Figs. 8.3 and 8.4). The auxiliary frame is a mechanism. The translational degree of freedom of the actual frame is equal to the minimum number of link supports which must be connected to the joints of the auxiliary frame to prevent it from moving without deforming. For instance, the members of the auxiliary frame of Fig. 8.3 cannot move as rigid bodies when each of its joints 2 and 3 is supported by a link. Thus the joints of the frame of Fig. 8.3 have two translational degrees of freedom. Moreover, the members of the auxiliary frame of Fig. 8.4 cannot move as rigid bodies when each of its joints 2, 3, 4, 6, and 7 is supported by a link. Thus, the joints of the frame of Fig. 8.4 have five translational degrees of freedom.

8.3 Classification of the Members of Planar Structures

As mentioned previously, the methods presented in this chapter cannot be applied directly when analyzing trusses or frames having one or more members subjected only to axial force. That is, they can be applied directly in analyzing only beams and frames whose members are subjected to bending. The one end of the members of these structures is rigidly connected to another member or to a support, while their other end is either rigidly connected or pinned to another member or to a support (see Fig. 8.5).

In analyzing a structure by the slope deflection method, we first compute the components of displacements at its joints. The internal actions in the members of the structure are established from the displacements of its joints. The effort required to analyze a structure by the slope deflection method depends on the number of the unknown components of displacement (degree of freedom) of its joints; the smaller this number, the less effort required. The same is valid when the internal actions in the members of a structure are established using the displacement method with moment distribution. In order to have the minimum number of unknown components of displacements of the joints of a

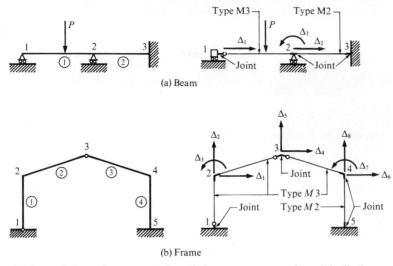

(a) Beam

(b) Frame

Figure 8.5 Schematic representation of planar structures and possible displacements of their joints.

planar beam or frame, we will classify its members as:

Type M1. Both ends are connected by pins to joints.

Type M2. Both ends are rigidly connected to joints.

Type M3. One end is rigidly connected while the other end is pinned to joints.

The beams and frames we are considering in this chapter have only type M2 and M3 members.

With the above classification of the members of a structure, its hinges are considered as parts of the adjacent members. That is, the hinge at joint 1 of the beam of Fig. 8.5*a* is regarded as part of member 1 and, consequently, it is assumed that this joint can translate only horizontally. Similarly, the hinge at joint 1 of the frame of Fig. 8.5*b* is regarded as part of member 1; consequently, this joint is considered fixed to the support. Moreover, the hinge at point 3 of the frame of Fig. 8.5*b* is regarded as part of members 2 and 3; consequently joint 3 does not rotate.

Notice that we can consider all the members of a planar beam or frame as type M2. With this classification the hinges are considered as parts of the joints. That is, the joint at point 1 of the frame of Fig. 8.6 is assumed connected to the ground by a pin; hence it is free to rotate. Thus this joint has one degree of freedom. Moreover, it is assumed that there is one joint on each side of an internal hinge. The components of translation of these two joints are equal while their rotations are not. Thus these two joints have four degrees of freedom. By comparing Figs. 8.5*b* and 8.6, it is apparent that we have introduced more unknown components of displacement with the second classification than

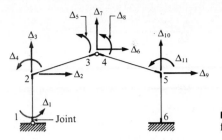
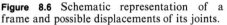

Figure 8.6 Schematic representation of a frame and possible displacements of its joints.

with the first. Therefore, it is anticipated that the analysis of a structure by a displacement method will be less laborious when the first classification of its members is used than the second. This is illustrated in Example 1 of Sec. 8.6 and Example 2 of Sec. 8.8.

8.4 Rotational Stiffnesses and Carry-Over Factors of a Member

8.4.1 Type M2 member

Consider a type M2 member (the ith) of a structure. Suppose that the ends of the member are clamped in the jaws of a testing machine which subjects its end p ($p = k$ or j) to a unit rotation ($\theta^{ip} = 1$), while its end q ($q = j$ or k, $q \neq p$) remains fixed ($\theta^{iq} = 0$). The bending moment applied by the machine at the end p of the member is called *the rotational stiffness of the end p of the member* and it is denoted by k^{ip} ($p = k$ or j). The ratio of the bending moment applied by the machine at the end q of the member to the moment applied to its end p is called *the carry-over factor of the end p of the member* and is denoted by C^{ip}. That is

$$C^{ip} = \frac{\text{moment at the end } q \text{ when } \theta^{iq} = 0, \theta^{ip} = 1}{k^{ip}} \qquad p, q = j \text{ or } k, q \neq p$$

$$(8.1)$$

On the basis of its definition the rotational stiffness k^{ij} and the carry-over factor C^{ij} of a type M2 member may be established by considering the member simply supported at both ends and subjected at its end j to a counterclockwise moment k^{ij} and at its end k to a counterclockwise moment $C^{ij}k^{ij}$ (see Fig. 8.7a). The moment at any cross section of this simply supported (statically determinate) member may be established in terms of its end moments k^{ij} and $C^{ij}k^{ij}$. Moreover, the rotation θ^{ij} and θ^{ik} of the ends of the member may be obtained in terms of the end moments k^{ij} and $C^{ij}k^{ij}$ using one of the methods presented in Chaps. 5 and 6 (i.e., the method of virtual work). The end moments k^{ij} and $C^{ij}k^{ij}$ are then established by requiring that the rotation θ^{ij} be equal to unity while the rotation θ^{ik} be equal to zero (see Example 1 at the end

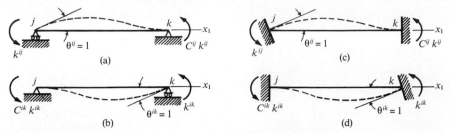

Figure 8.7 Physical significance of the rotational stiffnesses and the carry-over factors of a type M2 member.

of this section). Equivalently, the member can be considered fixed at both ends and subjected to a counterclockwise unit rotation of the support at its end j (see Fig. 8.7c). The reacting moments k^{ij} and $C^{ij}k^{ij}$ of this statically indeterminate member may be established using one of the methods presented in Chap. 7 (see Example 2 at the end of this section).

Notice that referring to Fig. 8.7a and b and using Betti's reciprocal theorem (see Sec. 5.16), we can show that

$$C^{ij}k^{ij} = C^{ik}k^{ik} \tag{8.2}$$

The end moments of a fixed-at-both-ends member of constant cross section subjected to a rotation of one of its ends are given in App. C. Referring to this appendix, for a member of constant cross section with moment of inertia $I^{(i)}$, we have

$$k^{ij} = k^{ik} = \frac{4E_iI^{(i)}}{L_i} \tag{8.3}$$

$$C^{ij}k^{ij} = C^{ik}k^{ik} = \frac{2E_iI^{(i)}}{L_i} \tag{8.4}$$

Thus, $$C^{ij} = C^{ik} = \tfrac{1}{2} \tag{8.5}$$

The same results are obtained in Example 1 at the end of this section.

Notice that if a type M2 member has a transverse axis of symmetry in its plane (see Fig. 8.8a and b), the rotational stiffness and the carry-over factor of its one end are equal to the rotational stiffness and the carry-over factor of its other end, respectively.

8.4.2 Type M3 member

Consider a type M3 member (the ith) and denote its rigidly connected end by p ($p = j$ or k). Suppose that the end q of the member is pinned at a nonmoving

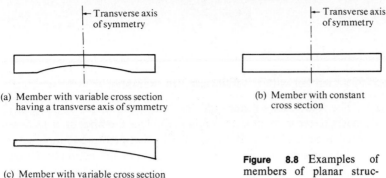

(a) Member with variable cross section
    having a transverse axis of symmetry

(b) Member with constant
    cross section

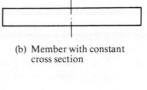

(c) Member with variable cross section
    not having an axis of symmetry

Figure 8.8 Examples of members of planar structures.

support while its end p is clamped in the jaws of a testing machine which subjects it to a unit rotation ($\theta^{ip} = 1$). The bending moment applied by the machine at the end p of the member is called the *rotational stiffness of the end p of the member*. It is denoted by k^{ip}. Notice that the other end of the member has not been subjected to any moment; it rotates freely.

On the basis of its definition, the rotational stiffness $\hat{k}^{ip}$ of a type M3 member may be established by considering the member simply supported at both ends and subjected at its end p ($p = j$ or k) to a counterclockwise moment of the magnitude required for this end to rotate by an amount equal to unity (see Fig. 8.9a). The moment at any cross section of this simply supported, statically determinate member may be established in terms of the end moment $\hat{k}^{ip}$. Moreover, the rotation θ^{ip} of the end p of the member may be obtained in terms of the end moment $\hat{k}^{ip}$ using one of the methods presented in Chaps. 5 and 6 (i.e., the method of virtual work). The end moment $\hat{k}^{ip}$ is then established by requiring that the rotation θ^{ip} be equal to unity. Equivalently, we can consider the member fixed at its end p ($p = j$ or k) while its other end is pinned and subjected to a counterclockwise unit rotation of the support at its end p (see Fig. 8.9b). It is apparent that the moment at the fixed end p of this statically indeterminate member is equal to its rotational stiffness k^{ip} of the end p of the member. This moment may be established using one of the methods presented in Chap. 7.

When a type M3 member of length L_i has a constant cross section of moment of inertia $I^{(i)}$, using the definition of the rotational stiffness and referring to App. C, we have

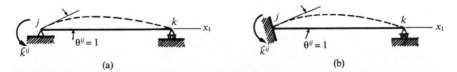

Figure 8.9 Physical significance of the rotational stiffness of a type M3 member.

$$\hat{k}^{ip} = \frac{3\bar{E}_i \bar{I}^{(i)}}{L_i} \qquad (8.6)$$

where p ($p = j$ or k) is the rigidly connected end of the member.

We now illustrate the computation of the rotational stiffnesses and the carry-over factors of a member with the following two examples.

Example 1. The rotational stiffness and the carry-over factor of a type M2 member of constant cross section are computed. The member is considered simply supported at both ends, and subjected at its end j to an end moment k^{ij} and at its end k to an end moment $C^{ij}k^{ij}$. The values of the moments k^{ik} and $C^{ij}k^{ij}$ are established by requiring that the rotation θ^{ij} be equal to unity while the rotation θ^{ik} be equal to zero.

Two procedures are employed in computing the end moments k^{ij} and $C^{ij}k^{ij}$ of the member.

Example 2. The rotational stiffness and the carry-over factor of a type M2 member of variable cross section having a transverse axis of symmetry are established. The member is considered fixed at both ends and its end j is subjected to a unit rotation. The end moments of the member are established using the force method. The flexibility coefficients of the primary structure are established using the method of virtual work. The values of the integrals are obtained employing Simpson's rule. The moment at the end j of the member is equal to the rotational stiffness $k^{ij} = k^{ik}$, while the moment at the end k is equal to $C^{ij}k^{ij}$.

Example 1. Compute the rotational stiffness $k^{ij} = k^{ik}$ and the carry-over factor $C^{ij} = C^{ik}$ of a type M2 member of constant cross section ($EI = $ constant).

---

solution Consider a type M2 member simply supported at its ends. Assume that, as shown in Fig. a, the member is subjected at its end j to a moment equal to its rotational stiffness k^{ij} and at its end k to a moment equal to $C^{ij}k^{ij}$. For this loading, the rotation of the end j of the member is equal to unity, while the rotation of the end k vanishes.

PROCEDURE 1.   The rotations θ^{ij} and θ^{ik} are computed in terms of k^{ij} and C^{ij} using the method of virtual work. The rotation θ^{ij} is set equal to unity while the rotation θ^{ik} is set equal to zero which results in a set of two linear algebraic equations from which k^{ij} and C^{ij} are computed.

Referring to Fig. b and using the sign convention for the moment specified in Fig. 7.1, the moment at any cross section of the member is equal to

$$M = -k^{ij} + \frac{k^{ij}(1 + C^{ij})}{L_i} x_1 \qquad (a)$$

We compute the rotations θ^{ij} and θ^{ik} of the ends of the member of Fig. a by employing the method of virtual work. The member subjected to the virtual loading for comput-

Figure a   Type M2 member fixed at its end k and simply supported at its end j.

Figure b   Free-body diagram of the member of Fig. a.

Figure c   Virtual loading for computing θ^{ij}.

Figure d   Virtual loading for computing θ^{ik}.

ing the rotation θ^{ij} is shown in Fig. c. Referring to this figure, the moment at any point of the member is given as

$$\tilde{M} = -1 + \frac{x_1}{L_i} \tag{b}$$

Substituting relations (a) and (b) into (5.66), we obtain

$$
\begin{aligned}
\theta^{ij} = 1 &= \int_0^{L_i} \left[-k^{ij} + \frac{k^{ij}(1 + C^{ij})x_1}{L_i} \right] \left(-1 + \frac{x_1}{L} \right) dx_1 \\
&= \frac{k^{ij}L_i}{2EI^{(i)}} - \frac{k^{ij}(1 + C^{ij})L_i}{6EI^{(i)}}
\end{aligned}
\tag{c}
$$

The member subjected to the virtual loading for computing the rotation θ^{ik} is shown in Fig. d. Referring to this figure, the moment at any point of the member is given as

$$\tilde{M} = \frac{x_1}{L_i} \tag{d}$$

Substituting relations (a) and (d) into (5.66), we obtain

$$\theta^{ik} = 0 = \frac{1}{EI^{(i)}} \int_0^{L_i} \left[-k^{ij} + \frac{k^{ij}(1 + C^{ij})x_1}{L_i} \right] \left(\frac{x_1}{L_i} \right) dx_1$$

$$= \frac{1}{EI^{(i)}} \left[-\frac{k^{ij}L_i}{2} + \frac{k^{ij}(1 + C^{ij})L_i}{3} \right] \tag{e}$$

Relation (e) gives

$$C^{ij} = \frac{1}{2} \tag{f}$$

Substituting the above result into relation (c), we get

$$k^{ij} = \frac{4EI^{(i)}}{L_i} \tag{g}$$

PROCEDURE 2.   The relation between the transverse component of external force and the transverse component of displacement established in Sec. A.8 [see relation (A.77b)] represents the condition for equilibrium of the actions acting on an element of length dx_1 of a member expressed in terms of the transverse component of displacement. For a member of constant cross section which is not subjected to external forces along its length, relation (A.77b) reduces to the following homogeneous equation

$$\frac{d^4u_3}{dx_1^4} = 0$$

The solution of this equation is

$$u_3 = a_0 + a_1x_1 + a_2x_1^2 + a_3x_1^3 \tag{h}$$

Thus referring to the first of relation (A.71), we obtain

$$\theta_2 = -\frac{du_3}{dx_1} = -a_1 - 2a_2x_1 - 3a_3x_1^2 \tag{i}$$

The constants a_0, a_1, a_2, and a_3 can be evaluated from the conditions at the ends of the member. Thus referring to Fig. a, we obtain

$$u_3(0) = 0 \qquad a_0 = 0$$

$$\theta_2(0) = 1 \qquad a_1 = -1$$

$$u_3(L) = 0 \qquad 0 = -L + a_2L^2 + a_3L^3$$

$$\theta_2(L) = 0 \qquad 0 = 1 - 2a_2L = 3a_3L^2$$

Consequently

$$a_0 = 0 \qquad a_1 = 1 \qquad a_2 = \frac{2}{L} \qquad a_3 = -\frac{1}{L^2} \tag{j}$$

Substituting the values of the constants into relation (h), we obtain:

$$u_3 = x_1 + \frac{2x_1^2}{L} - \frac{x_1^3}{L^2} \tag{k}$$

The moment at any point of the member may be obtained using the moment-displacement relation (A.74b). Thus

$$M(x_1) = -EI\frac{d^2u_3}{dx_1^2} = -EI\left(\frac{4}{L} - \frac{6x_1}{L^2}\right) \tag{1}$$

Notice that the moment $M(x_1)$ is positive if it compresses the top fibers of the member. Therefore the moment at $x_1 = 0$ is equal to $-k^{ij}$ while at $x_1 = L$ it is equal to $C^{ij}k^{ij}$. Hence

$$k^{ij} = -M(0) = \frac{4EI}{L} \qquad C^{ij}k^{ij} = M(L) = \frac{2EI}{L}$$

Example 2.   Compute the rotational stiffness $k^{ij} = k^{ik}$ and the carry-over factor $C^{ij} = C^{ik}$ of the type M2 member of constant width b whose geometry is shown in Fig. a.

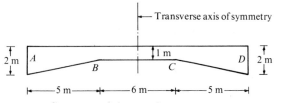

Figure a   Geometry of the member.

solution   We can establish the rotational stiffness and the carry-over factor of this member using the procedure employed in Example 1. However, the integrals in the expressions for the end rotations of the member must be evaluated numerically over the portions of the length of the member where the moment of inertia of its cross sections is not constant. Nevertheless, in order to illustrate a somewhat different approach we assume that, as shown in Fig. b, the ends of the member under consideration are fixed and that its end j is subjected to a unit rotation. The moment acting at the end j of the member is equal to k^{ij}, while the moment acting at its end k is equal to $C^{ij}k^{ij}$. We establish these moments employing the basic force method. We choose the moments at the ends of the member as the redundants. The primary structure subjected to the redundants is shown in Fig. c. Inasmuch as the member is not subjected to any external loads, the displacements Δ_1^{PL} and Δ_2^{PL} vanish. Thus, the equations of compatibility (7.20) reduce to

$$1 = F_{11}X_1 + F_{12}X_2 \qquad 0 = F_{21}X_1 + F_{22}X_2 \tag{a}$$

where $F_{12} = F_{21}$. Moreover, referring to Figs. d and e, since the member has a transverse axis of symmetry in its plane, we have

$$F_{11} = F_{22}$$

The primary structure subjected to $X_1 = 1$ is shown in Fig. d. Using the sign con-

Figure b Member fixed at its ends subjected to a unit rotation of $\theta^{ij} = 1$.

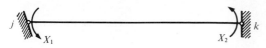

Figure c Primary structure subjected to the redundants.

Figure d Primary structure subjected to $X_1 = 1$.

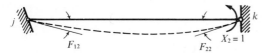

Figure e Primary structure subjected to $X_2 = 1$.

vention for the moment specified in Fig. 7.1, the corresponding moment distribution is

$$M^{(1)} = -1 + \frac{x_1}{L} \tag{b}$$

The primary structure subjected to $X_2 = 1$ is shown in Fig. e. The corresponding moment distribution is

$$M^{(2)} = \frac{x_1}{L} \tag{c}$$

Referring to Fig. a, the moments of inertia of the cross sections of the member are

$$I_{BC} = \frac{bd^3}{12} = I$$
$$I_{AB} = \frac{bd^3}{12}\left(2 - \frac{x_1}{5}\right)^3 = I\rho \tag{d}$$

where

$$\rho = \left(2 - \frac{x_1}{5}\right)^3$$

We establish the flexibility coefficients of the primary structure corresponding to the redundants using the method of virtual work (5.66). Thus,

$$F_{11} = F_{22} = \int_0^L \frac{[M^{(1)}]^2}{EI} = \frac{1}{EI}\left[\int_0^5 \frac{[M^{(1)}]^2}{\rho}\,dx_1 + \int_5^{11}\left(-1 + \frac{x_1}{L}\right)^2 dx_1\right.$$

$$\left. + \int_{11}^{16}\frac{[M^{(1)}]^2}{\rho}\,dx_1\right]$$

$$= \frac{1}{EI}\left[\int_0^5 \frac{[M^{(1)}]^2}{\rho}\,dx_1 + \int_{11}^{16}\frac{[M^{(1)}]^2}{\rho}\,dx_1 + 1.57\right] \qquad (e)$$

$$F_{12} = F_{21} = \frac{1}{EI}\left[\int_0^5 \frac{M^{(1)}M^{(2)}}{\rho}\,dx_1 + \int_5^{11}\left(\frac{x_1}{16}\right)\left(-1 + \frac{x_1}{16}\right)dx_1\right.$$

$$\left. + \int_{11}^{16}\frac{M^{(1)}M^{(2)}}{\rho}\,dx_1\right]$$

$$= \frac{1}{EI}\left[\int_0^5 \frac{M^{(1)}M^{(2)}}{\rho}\,dx_1 + \int_{11}^{16}\frac{M^{(1)}M^{(2)}}{\rho}\,dx_1 + 1.429\right] \qquad (f)$$

In obtaining the above result, we took into account that $M^{(1)}\,|_{x_1} = M^{(2)}\,|_{L-x_1}$.

We evaluate numerically the integrals in relations (e) and (f) by applying Simpson's rule. We subdivide each interval from $x_1 = 0$ to $x_1 = 5$ m, and from $x_1 = 11$ m to $x_1 = 16$ m into four equal segments (see Fig. f). That is, referring to relation (5.81), we can rewrite relations (e) and (f) as

$$F_{11} = F_{22} = \frac{1}{EI}\left[\frac{h}{3}\,(f_0 + 4f_1 + 2f_2 + 4f_3 + f_4)\right.$$

$$\left. + \frac{h}{3}\,(f_5 + 4f_6 + 2f_7 + 4f_8 + f_9) + 1.57\right]$$

$$F_{12} = F_{21} = \frac{1}{EI}\left[\frac{h}{3}\,(F_0 + 4F_1 + 2F_2 + 4F_3 + F_4)\right. \qquad (g)$$

$$\left. + \frac{h}{3}\,(F_5 + 4F_6 + 2F_7 + 4F_8 + F_9) + 1.429\right]$$

where f_i or F_i is the value of $f(x_1)$ or $F(x_1)$ evaluated at point i $(i = 0, 1, \ldots, 9)$ shown in Fig. f. The functions $f(x_1)$ and $F(x_1)$ are defined as

$$f(x_1) = \frac{[M^{(1)}]^2}{\rho} \qquad F(x_1) = \frac{M^{(1)}M^{(2)}}{\rho} \qquad (h)$$

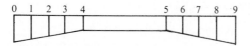

Figure f Member subdivided into segments.

TABLE a   Computation of the Integrals in Relations (e) and (f)

pt	ρ	$-M^{(1)}$	$M^{(2)}$	f_0, f_4	f_i	f_i	$-F_0, -F_4$	$-F_i$	$-F_i$
				Left Haunch					
0	8.0000	1.0000	0	0.1250	.....	.....	0		
1	5.3594	0.9219	0.0781	.....	0.1586	.....	.....	0.0134	
2	3.3750	0.8438	0.1563	.....	.....	0.2110	.....	.....	0.0391
3	1.9531	0.7656	0.2344	.....	0.3001	.....	.....	0.0919	
4	1.0000	0.6875	0.3125	0.4727	.....	.....	0.2148		
Total	.....	.....	.....	0.5977	0.4587	0.2110	0.2148	0.1053	0.0391
				Right Haunch					
5	1.0000	0.3125	0.6875	0.0977	.....	.....	0.2148		
6	1.9531	0.2344	0.7656	.....	0.0281	.....	.....	0.0919	
7	3.3750	0.1563	0.8438	.....	.....	0.0072	.....	.....	0.0391
8	5.3594	0.0781	0.9219	.....	0.0011	.....	.....	0.0134	
9	8.0000	0.0000	1.0000	0.0000	.....	.....	0.0000		
Total	.....	.....	.....	0.0977	0.0292	0.0072	0.2148	0.1053	0.0391

Substituting the values of f_i and F_i given in Table a into relations (g) and noting that the length h of each segment is equal to 1.25 m, we get

$$F_{11} = F_{22} = \frac{1}{EI}\left[\frac{1.25}{3}[0.5977 + 4(0.4587) + 2(0.2110)]\right.$$

$$\left.+ \frac{1.25}{3}[0.0977 + 4(0.0292) + 2(0.0072)] + 1.5703\right]$$

$$= \frac{2.8551}{EI}$$

and

$$F_{12} = F_{21} = -\frac{2}{EI}\left(\frac{1.25}{3}\right)[0.2148 + 4(0.1053) + 2(0.0391) + 1.4297]$$

$$= -\frac{2.0249}{EI}$$

Substituting the above results into the compatibility equations (a), we obtain

$$EI = 2.8551X_1 - 2.0249X_2$$

$$0 = -2.0249X_1 + 2.8551X_2$$

or

$$X_1 = 0.7047EI = k^{ij}$$

$$X_2 = 0.49998EI = C^{ij}k^{ij}$$

Thus

$$C^{ij} = C^{ik} = 0.7095$$

$$k^{ij} = k^{ik} = 0.7047EI$$

8.5 Relations between the End Moments and the End Displacements of a Member Subjected Only to End Actions

8.5.1 Type M2 member

Consider a type M2 member (the ith) of a structure, subjected only to end actions. Referring to Fig. 8.10a, we can see that the deformed configuration of this member may be obtained by subjecting it to a rigid-body translation by an amount specified by the vector $u_1^{ij}\mathbf{i}_1 + u_2^{ij}\mathbf{i}_2$, a rigid-body rotation about its end j equal to β^i, and a distortion. The latter is imposed by holding the ends of the member pinned and subjecting it to a moment and a shearing force at each of its ends having the magnitudes required in order that the member assumes its final deformed configuration. Referring to Fig. 8.10a, we obtain

$$\phi^{ij} = \theta^{ij} - \beta^i \qquad \phi^{ik} = \theta^{ik} - \beta^i \tag{8.7}$$

where for deformation within the range of validity of the theory of small deformation we have

$$\beta^i = \theta^{ij} - \phi^{ij} = \theta^{ik} - \phi^{ik} = \frac{u_2^{ik} - u_2^{ij}}{L_i} \tag{8.8}$$

In Fig. 8.10b the member is subject to the end moments required to produce a rotation ϕ^{ij} of its end j while the rotation ϕ^{ik} of its end k vanishes. On the basis of the definitions of the rotational stiffness k^{ij} and the carry-over factor C^{ij}, we have

$$M^{ij1} = \phi^{ij}k^{ij} \qquad M^{ik1} = \phi^{ij}C^{ij}k^{ij} \tag{8.9}$$

Moreover, in Fig. 8.10c the member is subjected to the end moments required to produce a rotation ϕ^{ik} of its end k while the rotation ϕ^{ij} of its end j vanishes.

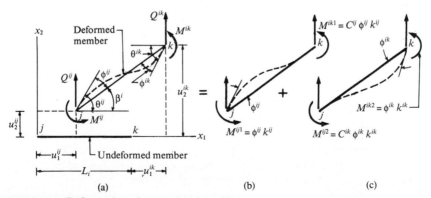

Figure 8.10 Deformation of a type M2 member.

On the basis of the definitions of the stiffness coefficient k^{ik} and the carry-over factor C^{ik}, we get

$$M^{ij2} = \phi^{ik} C^{ik} k^{ik} \qquad M^{ik2} = \phi^{ik} k^{ik} \tag{8.10}$$

It is apparent that the moment at any cross section of the member loaded as shown in Fig. 8.10a is equal to the sum of the moments at the same cross section of the member loaded as shown in Fig. 8.10b and c. Thus referring to Fig. 8.10 and using relations (8.9) and (8.10), we have

$$
\begin{aligned}
M^{ij} &= M^{ij1} + M^{ij2} = \phi^{ij} k^{ij} + \phi^{ik} C^{ik} k^{ik} \\
M^{ik} &= M^{ik1} + M^{ik2} = \phi^{ik} k^{ik} + \phi^{ij} C^{ij} k^{ij}
\end{aligned}
\tag{8.11}
$$

Substituting relations (8.7) into (8.11) and using relations (8.2), we obtain

$$
\begin{aligned}
M^{ij} &= k^{ij}(\theta^{ij} - \beta^i) + C^{ij} k^{ij}(\theta^{ik} - \beta^i) \\
M^{ik} &= k^{ik}(\theta^{ik} - \beta^i) + C^{ik} k^{ik}(\theta^{ij} - \beta^i)
\end{aligned}
\tag{8.12}
$$

The above relations give the moments acting at the ends of a type M2 member when it is subjected to the end rotations θ^{ij} and θ^{ik} as well as to the relative translation $u_2^{ik} - u_2^{ij} = L_i \beta^i$ of its ends.

8.5.2   Type M3 member

Consider a type M3 member (the ith) subjected only to end actions. This member is rigidly connected to other members or to a support at its end p ($p = i$ or j) and pinned to other members or to a support at its end q ($q = j$ or i, $p \neq q$). Referring to Fig. 8.11, we can see that the deformed configuration of this member may be obtained by first subjecting it to a rigid-body translation specified by the vector $u_1^{ip} \mathbf{i}_1 + u_2^{ip} \mathbf{i}_2$, a rigid-body rotation about its end j equal to $\beta^i = \theta^{ip} - \phi^{ip} = \theta^{iq} - \phi^{iq}$, and a distortion. The latter is imposed by holding the ends of the member pinned and subjecting it to a moment at its rigidly

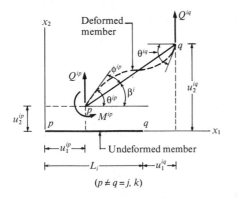

Figure 8.11 Deformation of a type M3 member.

connected end having the magnitude required in order that the member assume its final deformed configuration. For deformation within the range of validity of the theory of small deformation, the angle β^i is given by relation (8.8). Moreover, on the basis of the definition of the rotational stiffness $\hat{k}^{ip}$, it is apparent that the moment M^{ip} at the rigidly connected end of the member is equal to

$$M^{ip} = \hat{k}^{ip}(\theta^{ip} - \beta^i) \tag{8.13}$$

Suppose that the end j of a type M2 member is pinned ($M^{ij} = 0$), the first of relations (8.12) yields

$$\theta^{ij} = -C^{ij}\theta^{ik} + \beta^i(1 + C^{ij}) \tag{8.14}$$

This relation gives the rotation θ^{ij} of the pinned end of the type M3 member in terms of the rotation θ^{ik} of its rigidly connected end and the relative translation of its ends. Substituting relation (8.14) into the second of relations (8.13), we get

$$M^{ik} = k^{ik}(1 - C^{ik}C^{ij})(\theta^{ik} - \beta^i) \tag{8.15}$$

This relation gives the moment at the rigidly connected end of a type M3 member in terms of the stiffness and carry-over factor of a type M2 member of the same geometry and modulus of elasticity. Comparing relations (8.15) and (8.13), we obtain

$$\hat{k}^{ik} = k^{ik}(1 - C^{ik}C^{ij}) \tag{8.16}$$

This is the relation between the rotational stiffnesses of a type M2 and a type M3 member having the same geometry and modulus of elasticity.

8.6   The Slope Deflection Method

Consider the frame subjected to the external actions shown in Fig. 8.12a. The external actions acting on this frame are equal to the sum of the corresponding external actions acting on the frame loaded as shown in Fig. 8.12b and c. Consequently, since the principle of superposition is valid for the structures we are considering (see Sec. 1.9), the internal actions at any cross section of a member

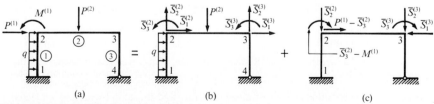

Figure 8.12   Principle of superposition.

and the components of displacement of any point of the frame, loaded as shown in Fig. 8.12a, are equal to the sum of the corresponding quantities of the frame loaded as shown in Fig. 8.12b and c. Notice that the values of the external actions $\overline{S}_1^{(2)}$, $\overline{S}_2^{(2)}$, $\overline{S}_3^{(2)}$, $\overline{S}_1^{(3)}$, $\overline{S}_2^{(3)}$, and $\overline{S}_3^{(3)}$ can be chosen so that the components of translation and the rotation of joints 2 and 3 of the frame loaded as shown in Fig. 8.12b vanish. If we regard member 3 as type M3, then the hinge at point 4 is considered part of member 3 and joint 4 is considered as fixed. Thus for the previously described choice of the external actions $\overline{S}_1^{(2)}$, $\overline{S}_2^{(2)}$, $\overline{S}_3^{(2)}$, $\overline{S}_1^{(3)}$, $\overline{S}_2^{(3)}$, and $\overline{S}_3^{(3)}$ the joints of the structure of Fig. 8.12b do not move. That is, the structure is kinematically determinate. In this case, the structure of Fig. 8.12b is called the *restrained structure* and the external actions $\overline{S}_1^{(2)}$, $\overline{S}_2^{(2)}$, $\overline{S}_3^{(2)}$, $\overline{S}_1^{(3)}$, $\overline{S}_2^{(3)}$, and $\overline{S}_3^{(3)}$ are called the *restraining actions*. Moreover, the external actions acting on the joints of the structure of Fig. 8.12c are called the *equivalent actions*.

Consider a member of a structure subjected to a general loading. Referring to Fig. 8.13, we can establish the relation between the moments acting at the ends of this member and the displacements of its ends by superimposing the end moments of the corresponding member of the following two structures.

1. An auxiliary structure, referred to as the *restrained structure* made from the actual structure by restraining the movement of its joints. In order to restrain the joints of the restrained structure, external actions, referred to as the *restraining actions* (see Fig. 8.13b), must be applied to them. The restrained structure is subjected to the given loading of the actual structure (external actions, temperature change, initial stress) except the concentrated actions acting at the joints of the structure and the given movement of the supports. The concentrated actions acting at the joints of the structure do not affect the internal actions in the members of the restrained structure. The given movement of the supports of the structure could be included in the loading of the restrained structure, as is done in Sec. 8.8. However, in the slope deflection

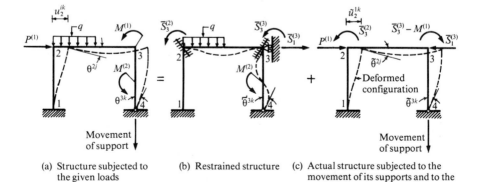

(a) Structure subjected to the given loads	(b) Restrained structure	(c) Actual structure subjected to the movement of its supports and to the equivalent actions at its joints

Figure 8.13 Superposition of the restrained structure and the actual structure subjected to the given movement of its supports and to the equivalent actions at its joints.

method, it is convenient to include it in the loading of the second structure (see Fig. 8.13c).

2. The actual structure subjected to the given movement of its supports and to external actions acting at its joints (see Fig. 8.13c). These actions are referred to as the *equivalent actions* and are equal to the sum of

   a. the given actions acting at the joints of the actual structure

   b. actions equal and opposite to the restraining actions of the restrained structure.

For instance joint 1 of the frame of Fig. 8.13a does not move. Moreover, if we consider element 3 as type M3, the hinge at point 4 is part of member 3 and consequently joint 4 is considered as fixed against translation and rotation. Furthermore inasmuch as we disregard the effect of axial deformation of the members of frames, the vertical components of translation and the difference of the horizontal components of translation of joints 2 and 3 are negligible. Thus the unknown components of displacement of the joints of the frame of Fig. 8.13a are the rotations of joints 2 and 3 and their horizontal translation. Hence, the restrained structure for this frame is obtained by restraining joints 2 and 3 from rotating and joint 3 from translating (see Fig. 8.13b). This is accomplished by applying to the joints of the frame the restraining actions $\overline{S}_3^{(2)}$, $\overline{S}_1^{(3)}$, and $\overline{S}_3^{(3)}$ shown in Fig. 8.13b.

The displacements of the joints of the frame loaded as shown in Fig. 8.13a are equal to those of the frame loaded as shown in Fig. 8.13c. This becomes apparent by noting that the displacements of the joints of the restrained structure vanish. Consequently, referring to Fig. 8.13 generally, we see that the components of displacements of the ends of a type M2 member and the components of displacement of the rigidly connected end of a type M3 member of a structure subjected to the given loading are equal to the corresponding components of displacement of the same member of the structure subjected to the equivalent actions at its nodes and to the given movement of its supports. However, the rotation of the pinned end of a type M3 member of the structure subjected to the given loading is equal to the sum of the rotations of this end of the corresponding member of the restrained structure and of the structure subjected to the equivalent actions and to the given movement of its supports. That is, referring to Fig. 8.13, we have

$$\theta^{3k} = \tilde{\theta}^{3k} + \vec{\theta}^{3k}$$

$$\theta^{2j} = \tilde{\theta}^{2j} \qquad u_2^{1k} = \tilde{u}_2^{1k} \tag{8.17}$$

The internal actions in the members of the frame, loaded as shown in Fig. 8.13a, are equal to the sum of the corresponding internal actions in the members of the restrained frame of Fig. 8.13b and of the frame of Fig. 8.13c subjected to the given movement of its supports and to the equivalent actions at its joints.

The analysis of the restrained structure is not a very difficult problem because the external loads affect only the members on which they act. That is, the restrained structure is made up of a number of members which are either fixed at both ends or fixed at one end and pinned at the other. If these members have a variable cross section, the internal actions at their ends can be established using one of the methods presented in Chap. 7. However, if these members have a uniform cross section, the internal actions at their ends can be obtained from handbooks or from App. C. The restraining actions which must be applied to a joint of the restrained structure are computed from the end actions of the members framing into this joint by considering the equilibrium of the joint.

On the basis of the foregoing, referring to Fig. 8.14, the moment M^{ip} at the end p ($p = j$ or k) of a type M2 member (the ith) of a structure subjected to a general loading is equal to

$$M^{ip} = \text{FEM}^{ip} + \tilde{M}^{ip} \tag{8.18}$$

FEM^{ip} is the moment (fixed-end moment, FEM) at the end p of a type M2 member the (ith) of the restrained structure subjected to the given loading (external actions, change of temperature, initial stress), except for the concentrated actions at the joints of the structure and the given movement of its supports. That is, in computing FEM^{ip}, the ends of the member are not permitted to rotate ($\theta^{ij} = \theta^{ik} = 0$) or translate. The fixed-end moments FEM^{ip} ($p = j$ or k) are obtained by using one of the methods presented in Chap. 7. The fixed-end moments of members with constant cross section subjected to some of the

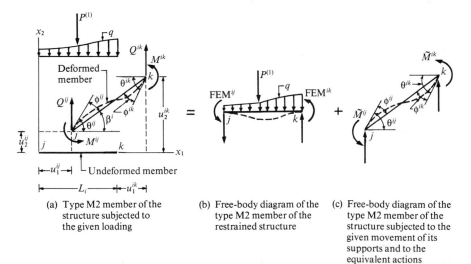

(a) Type M2 member of the structure subjected to the given loading

(b) Free-body diagram of the type M2 member of the restrained structure

(c) Free-body diagram of the type M2 member of the structure subjected to the given movement of its supports and to the equivalent actions at its joints

Figure 8.14 Superposition of the ith type M2 member of the restrained structure and of the structure subjected to the given movement of its supports and to the equivalent actions at its joints.

most frequently encountered loads along their length are given in App. C. $\tilde{M}^{ip}$ ($p = j$ or k) is the moment at the end p of the ith member of the structure subjected to given movement of its supports and to the equivalent actions at its joints. That is, this moment is due only to the relative movement of the ends of the member. Consequently, the relation between the moment $\tilde{M}^{ip}$ and the components of displacements of the ends of the member is given by relation (8.12). Notice, however, that the components of displacements of the ends of a type M2 member of the structure subjected to the equivalent actions at its joints and to the given movement of its supports are equal to the corresponding components of displacements of the ends of this member when the structure is subjected to the given loading. Substituting relation (8.12) into (8.18), we get

$$M^{ip} = \text{FEM}^{ip} + k^{ip}(\theta^{ip} - \beta^i) + C^{ip}k^{ip}(\theta^{iq} - \beta^i) \qquad p,q = j \text{ or } k, p \neq q$$

$$(8.19)$$

where $\beta^i = (u_2^{ik} - u_2^{ij})/L_i$. This equation is referred to as *the slope deflection equation* for a type M2 member.

Referring to Fig. 8.15, the moment M^{ip} at the rigidly connected end p ($p = j$ or k) of a type M3 member is

$$M^{ip} = \widehat{\text{FEM}}^{ip} + \hat{M}^{ip} \qquad (8.20)$$

$\widehat{\text{FEM}}^{ip}$ is the moment (fixed-end moment) at the rigidly connected end p ($p = j$ or k) of the type M3 member (the ith) of the restrained structure, subjected

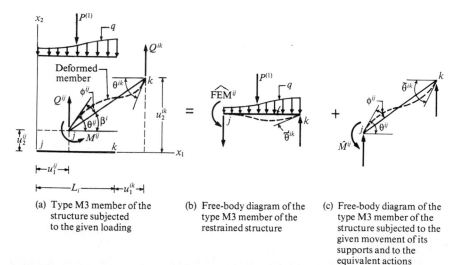

(a) Type M3 member of the structure subjected to the given loading

(b) Free-body diagram of the type M3 member of the restrained structure

(c) Free-body diagram of the type M3 member of the structure subjected to the given movement of its supports and to the equivalent actions

Figure 8.15 Superposition of the ith type M3 member of the restrained structure and of the structure subjected to the given movement of its supports and to the equivalent actions at its joints.

to the given loading (external actions, change of temperature, initial stress) except for the concentrated actions at the joints of the structure and the given movement of its supports. That is, in computing $\widehat{\text{FEM}}^{ip}$ the ends of the member are not permitted to translate, its end p is not permitted to rotate ($\theta^{ip} = 0$), but its other end is free to rotate. The fixed-end moment $\widehat{\text{FEM}}^{ip}$ is obtained using one of the methods presented in Chap. 7. The fixed-end moments of members with constant cross section subjected to some of the most frequently encountered loads along their length are given in App. C. $\hat{M}^{ip}$ is the moment at the rigidly connected end p of the ith member of the structure, subjected to the given movement of the supports and to the equivalent actions at its joints. That is, this is the moment due only to the relative movement of the ends of the member. Consequently, the relation between the moment $\hat{M}^{ip}$ and the components of displacements of the ends of the member is given by relation (8.13). Substituting this relation into (8.20), we get

$$M^{ip} = \widehat{\text{FEM}}^{ip} + \hat{k}^{ip}(\theta^{ip} - \beta^i) \tag{8.21}$$

This is *the slope deflection equation* for a type M3 member rigidly connected to its end p ($p = j$ or k).

The slope deflection equations (8.19) and (8.21) represent a convenient form of the relation between end moments and the end displacements of any member of a planar beam or a planar frame. They can be applied to the analysis of such structures using the following procedure.

Step 1. We note the types of the members of the structure, and we label the unknown displacements of its joints ($\Delta_1, \Delta_2, \ldots, \Delta_m$).

Step 2. Referring to the deformed configuration of the structure, we deduce all pertinent relations between the local components of displacements of the ends of its members and the unknown displacements of its joints. Moreover, from the known displacements of the supports of the structure we establish the displacements of the ends of its members which are connected to supports.

Step 3. We compute the rotational stiffnesses and the carry-over factors of each member of the structure, as well as the fixed-end moments of the members of the restrained structure subjected to following loads.

1. The given external actions not acting on the joints of the structure

2. The change of temperature

3. The initial stress

The specified settlements or rotations of the supports of the structure are incorporated in the boundary conditions.

The rotational stiffnesses, the carry-over factors, and the fixed-end moments of any member may be established as discussed in Sec. 8.4. Moreover, for members of certain geometry, these quantities are available in handbooks of structural analysis,† while for members with constant cross sections they are also available in App. C.

Step 4. We write the relations between the end moments of each member of the structure and the displacements of its joints. This is accomplished by substituting into Eqs. (8.19) or (8.21) the following.

1. The relations between the components of the displacements of the ends of each member of the structure and the unknown displacements of its joints deduced in step 2

2. The components of displacement of the ends of the members of the structure which are connected to its supports. They have been established in step 2

3. The stiffnesses, carry-over factors, and fixed-end moments of step 3

Step 5. For a structure with m unknown joint displacements, we establish m independent relations between the end moments of its members. In these relations, we substitute the relations among the end moments of the members of the structure and the displacements of its joints established in step 4, and we obtain a set of m linear algebraic equations for the m unknown displacements of the joints of the structure which we solve. This set of linear algebraic equations can be written in the following form.

$$\{P\} = [K]\,\{\Delta\}$$

or

$$\begin{Bmatrix} P_1 \\ P_2 \\ \cdots \\ P_n \end{Bmatrix} = \begin{bmatrix} K_{11} & K_{12} & \cdots K_{1m} \\ K_{21} & K_{22} & \cdots K_{2m} \\ \cdots & \cdots & \cdots \cdots \\ K_{n1} & K_{m2} & K_{nm} \end{bmatrix} \begin{Bmatrix} \Delta_1 \\ \Delta_2 \\ \cdots \\ \Delta_m \end{Bmatrix}$$

where $\{\Delta\}$ is called the *matrix of joint displacements of the structure*. Its terms represent the unknown components of displacements of the joints of the structure. $\{P\}$ is called *the matrix of external actions of the structure*. Its terms represent the components of the equivalent actions to be applied on the joints of the structure in order to account for all the loading (external actions, change of temperature, settlement of supports, initial stress) acting on the structure. The matrices $\{P\}$ and $\{\Delta\}$ are conjugate. That is, the term of the nth row of the matrix $\{P\}$ represents the component of equivalent action which performs work on the component of displacement represented by the term of the nth row of the matrix $\{\Delta\}$. For example, referring to Fig. 8.13a and c, the terms of the matrices $\{\Delta\}$ and $\{P\}$ for the structure of Fig. 8.13a are

$$\Delta_1 = u_2^{1k} = \tilde{u}_2^{1k} \qquad \Delta_2 = \theta^{2j} = \tilde{\theta}^{2j} \qquad \Delta_3 = \theta^{2k} = \tilde{\theta}^{2k}$$
$$P_1 = P^{(1)} - \overline{S}_1^{(3)} \qquad P_2 = \overline{S}_3^{(2)} \qquad P_3 = -\overline{S}_3^{(3)} + M^{(1)}$$

---

† See, for instance, *Handbook of Frame Constants*.[1]

The matrix $[K]$ is called the *stiffness matrix of the structure.* It is a square, symmetric matrix, and if the structure does not form a mechanism, it is non-singular. The terms of the matrix $[K]$ are called *the stiffness coefficients of the structure.*

If the structure has t unknown translations of its joints, we can find directly $(m - t)$ independent relations between the end moments of its members by setting the sum of the moments acting on those of its joint which are free to rotate equal to zero. Moreover, we can obtain t independent relations between the end moments of the members of the structure either by setting the sum of the forces acting on its joints equal to zero or by employing the principle of virtual work, as illustrated in Example 3 at the end of this section.

Step 6. We substitute the displacements computed in step 5 into the relations between the end moments of the members of the structure and the displacements of its joints established in step 4 to obtain the end moments in the members of the structure.

In what follows, we illustrate the use of the slope deflection method by solving the following examples.

Example 1. The internal actions in the members of a statically indeterminate beam of constant cross section are established. The beam is subjected to given external actions, to a given settlement of one of its supports, and to a given rotation of another support. The beam consists of two members; one is type M2, whereas the other may be considered either type M2 or type M3. The beam is analyzed in two ways: (a) by considering member 2 as type M3 and (b) by considering member 2 as type M2.

Example 2. The internal actions in the members of a statically indeterminate beam of variable cross section are established. The beam is subjected only to external actions. The computation of the rotational stiffnesses, the carry-over factors, and the fixed-end moments of the members of this beam involves considerable labor. However, in this example, part of these computations have been avoided by finding the rotational stiffnesses of two of the three members of the beam from their flexibility coefficients established in Example 4 of Sec. 7.2.1. The fixed-end moments of these members have also been established in the same example. Moreover, the rotational stiffnesses and the carry-over factors of the third member are obtained from Example 2 of Sec. 8.4.

Example 3. The internal actions in the members of a statically indeterminate, nonsymmetric frame are established when subjected to external actions, to a temperature difference between the external and internal fibers of its members, and to a settlement of one of its supports. There is only one unknown translation and one unknown rotation of the nodes of the frame. One of the required relations between the end moments of the members of the frame is established by considering the equilibrium of the moments acting on the rotating node of the frame. The second equation is established on the basis of two procedures.

Example 1. Analyze the continuous beam of constant cross section ($I = 75.90 \times 10^6$ mm^4) subjected to the external forces shown in Fig. a, and a 10-mm settlement of support 2 and a 0.001-rad counterclockwise rotation of support 1. Plot the shear and moment diagrams for this beam. The modulus of elasticity of the material from which this beam is made is equal to $E = 210{,}000$ MPa.

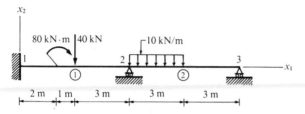

Figure a Geometry and loading of the beam.

---

solution

STEP 1. Referring to Fig. a, it can be seen that member 1 is type M2, while member 2 can be considered either as type M3 or as type M2. Moreover, joint 1 does not translate, but it rotates by the given rotation ($\theta^{ij} = 0.001$); joint 2 is free to rotate while it translates vertically by the given settlement (0.01 m). If member 2 is considered as type M3, the hinge at point 3 is considered as part of member 2 and consequently joint 3 does not rotate (see Fig. b*a*). Thus, in this case, we have only one unknown joint displacement, the rotation Δ_1 of joint 2. If member 2 is considered as type M2, the hinge at point 3 cannot be considered as part of member 2 and, consequently, joint 3 rotates (see Fig. b*b*). Thus, in this case, we have two unknown joint displacements, namely, the rotations Δ_1 and Δ_2 of joints 2 and 3, respectively.

Case 1: Member 2 Is Considered as Type M3

STEP 2. Referring to Fig. b*a*, from the deformed configuration of the beam, we can deduce the following boundary conditions and the following relations between the components of rotation of the ends of the members of the beam and the unknown rotation of joint 2:

$$\theta^{1j} = 0.001 \text{ rad} \qquad \theta^{1k} = \theta^{2j} = \Delta_1$$

$$u_2^{1j} = 0 \qquad u_2^{1k} = -0.01 \text{ m} \qquad u_2^{2j} = -0.01 \qquad u_2^{2k} = 0 \tag{a}$$

$$\beta^{(1)} = \frac{u_2^{1k} - u_2^{1j}}{L_1} = -\frac{0.01}{6} = -0.001667$$

$$\beta^{(2)} = \frac{u_2^{2k} - u_2^{2j}}{L_2} = 0.001667$$

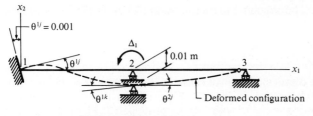

(a) Member 2 is considered type M3

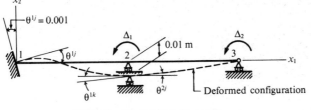

(b) Member 2 is considered type M2

Figure b   Deformed configuration of the beam.

STEP 3.   We compute the carry-over factors, the rotational stiffnesses, and the fixed-end moments of the members of the beam. Since the members of the beam have a constant cross section, by referring to relations (8.3) to (8.6), we find that their carry-over factors and their rotational stiffnesses are equal to:

$$C^{1j} = C^{1k} = \frac{1}{2}$$

$$k^{1j} = k^{1k} = \frac{4EI^{(1)}}{L_1} = \frac{2EI}{3} \qquad (b)$$

$$\hat{k}^{2j} = \frac{3EI^{(2)}}{L_2} = \frac{EI}{2}$$

Moreover, referring to App. C, the fixed-end moments of the members of the restrained beam subjected to the given external actions are:

$$\text{FEM}^{1j} = 30.0 \text{ kN} \cdot \text{m}$$

$$\text{FEM}^{1k} = -56.67 \text{ kN} \cdot \text{m} \qquad (c)$$

$$\widehat{\text{FEM}}^{2j} = 25.31 \text{ kN} \cdot \text{m}$$

Notice that $\widehat{\text{FEM}}^{2j}$ is the moment at the end j of member 2 when it is fixed at its end j and pinned at its end k.

STEP 4.   We substitute results (a), (b), and (c) into relations (8.19) and (8.21) to obtain the following relations between the end moments of the members of the beam and the displacements of its joints.

$$M^{1j} = \text{FEM}^{1j} + k^{1j}[\theta^{1j} - \beta^{(1)}] + C^{1j}k^{1j}[\theta^{1k} - \beta^{(1)}]$$

$$= 30.0 + \frac{2EI}{3}(0.001 + 0.001667) + \frac{EI}{3}(\Delta_1 + 0.001667)$$

$$M^{1k} = \text{FEM}^{1k} + k^{1k}[\theta^{1k} - \beta^{(1)}] + C^{1k}k^{1k}[\theta^{1j} - \beta^{(1)}] \qquad \text{(d)}$$

$$= -56.67 + \frac{2EI}{3}[\Delta_1 + 0.001667] + \frac{EI}{3}(0.001 + 0.001667)$$

$$M^{2j} = \widehat{\text{FEM}^{2j}} + \hat{k}^{2j}[\theta^{2j} - \beta^{(2)}]$$

$$= 25.31 + \frac{EI}{2}(\Delta_1 - 0.001667)$$

STEP 5.   We establish the required relation between the end moments of the members of the beam by considering the equilibrium of joint 2. Thus, referring to Fig. c, we get

$$M^{1k} + M^{2j} = 0 \qquad \text{(e)}$$

Substituting relations (d) into (e), we obtain the following equation for the unknown displacement Δ_1:

$$-56.67 + \frac{2EI}{3}(\Delta_1 + 0.001667) + \frac{EI}{3}(0.001 + 0.001667)$$

or

$$+ 25.31 + \frac{EI}{2}(\Delta_1 - 0.00161667) = 0$$

$$\Delta_1 = \frac{(31.36)6}{EI(7)} - 0.001 = \frac{31.36(6)}{7(210,000 \times 10^3)(7590 \times 10^{-8})} - 0.001 \qquad \text{(f)}$$

$$= 0.0006864 \text{ rad}$$

STEP 6.   We substitute the value of Δ_1 obtained in step 5 into relations (d) to obtain the values of the moments acting at the ends of the members of the beam. Thus

$$M^{1j} = 70.84 \text{ kN} \cdot \text{m}$$

$$M^{1k} = -17.49 \text{ kN} \cdot \text{m} \qquad \text{(g)}$$

$$M^{2j} = 17.49 \text{ kN} \cdot \text{m}$$

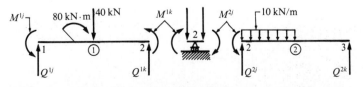

Figure c Free-body diagrams of the members and joint 2 of the beam.

Using the above results, the free-body diagrams of the members of the beam are plotted in Fig. d. The shear and moment diagrams are plotted in Figs. e and f, respectively, using the sign convention specified in Fig. 7.1.

Case 2: Member 2 is Considered Type M2

STEP 2.   Referring to Fig. b*b* and considering the deformed configuration of the beam in addition to relations (a), we have

$$\theta^{2k} = \Delta_2 \tag{h}$$

STEP 3.   We compute the carry-over factors, the rotational stiffnesses, and the fixed-end moments of the members of the beam. Since the members of the beam have a constant cross section, referring to relation (8.3) to (8.5), we see that their carry-over factors and their rotational stiffnesses are equal to

$$C^{1j} = C^{1k} = C^{2j} = C^{2k} = \frac{1}{2}$$

$$k^{1j} = k^{1k} = k^{2j} = k^{2k} = \frac{2EI}{L} \tag{i}$$

Moreover, referring to App. C, the moments at the ends (fixed-end moments) of the members of the restrained beam when subjected to the given external actions are:

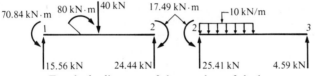

Figure d   Free-body diagrams of the members of the beam.

Figure e   Shear diagram for the beam.

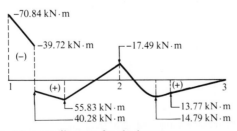

Figure f   Moment diagram for the beam.

$$\text{FEM}^{1j} = 30.00 \text{ kN} \cdot \text{m} \qquad \text{FEM}^{1k} = -56.67 \text{ kN} \cdot \text{m}$$
$$\text{FEM}^{2j} = 39.38 \text{ kN} \cdot \text{m} \qquad \text{FEM}^{2k} = -9.375 \text{ kN} \cdot \text{m} \tag{j}$$

STEP 4.   We substitute relations (a), (h), (i), and (j) into (8.19) to obtain the following relations between the end moments of the members of the beam and the displacement of its joints.

$$M^{1j} = \text{FEM}^{1j} + k^{1j}[\theta^{1j} - \beta^{(1)}] + C^{1j}k^{1j}[\theta^{1k} - \beta^{(1)}]$$

$$= 30.0 + \frac{2EI}{3}(0.001 + 0.001667) + \frac{EI}{3}(\Delta_1 + 0.001667)$$

$$M^{1k} = \text{FEM}^{1k} + k^{1k}[\theta^{1k} - \beta^{(1)}] + C^{1k}k^{1k}[\theta^{1j} - \beta^{(1)}]$$

$$= -56.67 + \frac{2EI}{3}[\Delta_1 + 0.001667] + \frac{EI}{3}(0.001 + 0.001667) \tag{k}$$

$$M^{2j} = 20.63 + \frac{2EI}{3}(\Delta_1 - 0.001667) + \frac{EI}{3}(\Delta_2 - 0.001667)$$

$$M^{2k} = -9.375 + \frac{2EI}{3}(\Delta_2 - 0.001667) + \frac{EI}{3}(\Delta_1 - 0.001667)$$

STEP 5.   We establish the required relations between the end moments of the members of the beam by considering the equilibrium of joints 2 and 3. That is

$$M^{1k} + M^{2j} = 0 \qquad M^{2k} = 0 \tag{l}$$

Substituting relations (k) into (l), we obtain a set of two equations involving the two unknown rotations Δ_1 and Δ_2 which may be solved to give

$$\Delta_1 = 0.0006864 \text{ rad} \qquad \Delta_2 = 0.0030391 \text{ rad}$$

These results are substituted in relations (k) to give the values of the internal moments M^{1j}, $M^{1k} = -M^{2j}$, and $M^{2k} = 0$.

On the basis of the foregoing, it is apparent that the analysis of the beam is simplified when member 2 is considered as type M3.

Example 2.   One of the continuous concrete beams of a bridge is loaded as shown in Fig. a. The cross sections of the beam are rectangular of constant width b. Plot the shear and moment diagrams for the beam.

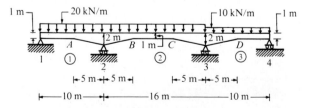

Figure a   Geometry and loading of the beam.

---

solution

STEP 1.   Members 1 and 3 of the beam are considered type M3, while member 2 is type M2. Thus, as shown in Fig. b, we have two unknown displacements: the rotations Δ_1 and Δ_2 of joints 2 and 3, respectively.

STEP 2.   Referring to Fig. b, from the deformed configuration of the beam, we can deduce the following boundary conditions and the following relations between the components of displacements of the ends of the members of the beam and the unknown rotations Δ_1 and Δ_2 of its joints.

$$\theta^{1k} = \theta^{2j} = \Delta_1 \qquad \theta^{2k} = \theta^{3j} = \Delta_2$$

$$u_2^{1j} = u_2^{1k} = u_2^{2j} = u_2^{2k} = u_2^{3j} = u_2^{3k} = 0 \qquad \text{(a)}$$

Thus
$$\beta^{(1)} = \beta^{(2)} = \beta^{(3)} = 0$$

STEP 3.   Members 1 and 3 have the same geometry and are subjected to the same loading as the beam of Example 4 of Sec. 7.2.1. This beam is fixed at its deep end and is simply supported at its other end. Hence, referring to Example 4 in Sec. 7.2.1, we have

$$\widehat{\text{FEM}}^{1k} = -392 \text{ kN} \cdot \text{m} \qquad \widehat{\text{FEM}}^{3j} = 196 \text{ kN} \cdot \text{m} \qquad \text{(b)}$$

In order to obtain the rotational stiffness of the deep end of members 1 or 3, the member is considered simply supported. The rotational stiffness of its deep end is the moment applied to this end in order to rotate it by a unit rotation. We can compute the rotational stiffness of members 1 and 3 using the procedure described in Sec. 8.4. However, in Example 4 of Sec. 7.2.1, we have computed the flexibility coefficient corresponding to an external moment applied at the deep end of members 1 or 3 simply supported at both ends. This flexibiity coefficient is equal to the angle of rotation of the deep end of the member when a unit moment is applied to this end. Thus, referring to Example 4 in Sec. 7.2.1, we have

$$\hat{k}^{1k} = \hat{k}^{3j} = \frac{1}{F_{11}} = \frac{EI}{1.281} = 0.7806EI \qquad \text{(c)}$$

Member 2 has the same geometry as the member of Example #2 in Sec. 8.4. Thus referring to this example we obtain

$$C^{2j} = C^{2k} = 0.7095$$
$$k^{2j} = k^{2k} = 0.7047EI \qquad \text{(d)}$$

Member 2 of the restrained structure is shown in Fig. c. We compute its fixed end moments using the basic force method. Notice that due to its symmetry and to the symmetry of the loading, the moments at the two ends of member 2 of the restrained structure must be equal. Thus we have only one unknown moment, which we choose

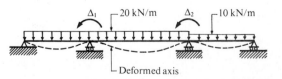

Figure b Deformed configuration of the beam and unknown joint displacements.

as the redundant. The primary structure of member 2 of the restrained structure subjected to the given loading and to the chosen redundant is shown in Fig. d. The primary structure, subjected to the given loading, is shown in Fig. e. Using the sign convention for the moment defined in Fig. 7.1, the moment in the primary structure of Fig. e is equal to

$$M = 8qx_1 - \frac{qx_1^2}{2} \tag{e}$$

The primary structure of member 2 of the restrained structure subjected to $X_1 = 1$ is shown in Fig. f. The corresponding moment is equal to:

$$\tilde{M} = -1 \tag{f}$$

Referring to Fig. a, the moments of inertia of the cross sections of member 2 are

$$I_{BC} = \frac{bd^3}{12} = I \qquad I_{2B} = \frac{bd^3}{12}\left(2 - \frac{x_1}{5}\right)^3 = I\rho \tag{g}$$

where

$$\rho = \left(2 - \frac{x_1}{5}\right)^3$$

We establish the displacement Δ_1^{PL} (see Fig. e) and the stiffness coefficient F_{11} (see Fig. f) using the method of virtual work (5.66). Thus,

$$
\begin{aligned}
\Delta_1^{PL} &= \frac{1}{EI}\left[\int_0^5 \frac{M\tilde{M}}{\rho}\,dx_1 - \int_5^{11} q\left(8x_1 - \frac{x_1^2}{2}\right)dx_1 + \int_{11}^{16} \frac{M\tilde{M}}{\rho}\,dx_1 \right] \\
&= \frac{1}{EI}\left[2\int_0^5 \frac{M\tilde{M}}{\rho}\,dx_1 - q\left[4x_1^2 - \frac{x_1^3}{6}\right]_5^{11} \right] \\
&= \frac{1}{EI}\left[2\int_0^5 \frac{M\tilde{M}}{\rho}\,dx_1 - 183q \right]
\end{aligned}
\tag{h}
$$

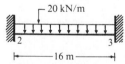

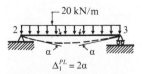

Figure c Member 2 of the restrained structure subjected to the given loading.

Figure e Primary structure of member 2 of the restrained structure subjected to the given loading.

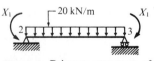

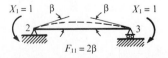

Figure d Primary structure of member 2 of the restrained structure subjected to the given loading and to the redundant.

Figure f Primary structure of member 2 of the restrained structure subjected to $X_1 = 1$.

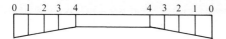

Figure g Member subdivided into segments.

$$F_{11} = \frac{1}{EI} \left[\int_0^5 \frac{\tilde{M}^2 \, dx_1}{\rho} + \int_5^{11} dx_1 + \int_{11}^{16} \frac{\tilde{M}^2 \, dx_1}{\rho} \right]$$

$$= \frac{1}{EI} \left[2 \int_0^5 \frac{\tilde{M}^2 \, dx_1}{\rho} + 6 \right] \tag{i}$$

We evaluate the integrals in relations (h) and (i) numerically by applying Simson's rule. We subdivide the intervals from $x_1 = 0$ to $x_1 = 5$ m and from $x_1 = 11$ m to $x_1 = 16$ m into four equal segments each by points 0 to 4 (see Fig. g). That is, referring to relation (5.81), we may write relations (h) and (i) as

$$\Delta_1^{PL} = \frac{1}{EI} \left[-\frac{2qh}{3} (f_0 + 4f_1 + 2f_2 + 4f_3 + f_4) - 183q \right]$$

$$F_{11} = \frac{1}{EI} \left[\frac{2h}{3} (F_0 + 4F_1 + 2F_2 + 4F_3 + F_4) + 6 \right] \tag{j}$$

where

$$f(x_1) = \frac{M\tilde{M}}{q\rho} \qquad F(x_1) = \frac{\tilde{M}^2}{\rho} \tag{k}$$

Substituting the values of f_i and F_i given in Table a into relations (j) and noting that the length h of each segment is equal to 1.25 m, we get[†]

$$\Delta_1^{PL} = \frac{1}{EI} \left\{ -2q \left(\frac{1.25}{3} \right) [27.5 + 4(13.4802) + 2(5.0)] - 183q \right\}$$

$$= -\frac{259.184q}{EI}$$

[†] The flexibility coefficient F_{11} may also be obtained by referring to the figures below and using the rotational stiffness and carry-over factor of the member [see relations (d)]. Thus

$$F_{11} = 2\alpha = \frac{2}{k^{2j}(1 - C^{2j})} = \frac{2}{0.7047EI(1 - 0.7095)} = \frac{9.760}{EI}$$

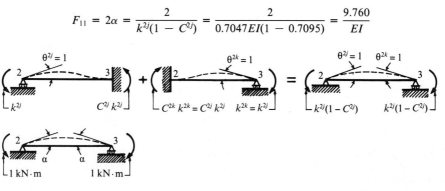

TABLE a   Computation of the Integrals in Relations (h) and (i)

Point	ρ	$\tilde{M}$	M/q	f_0, f_4	f_i	f_i	F_0, F_4	F_i	F_i
				Left Hunch					
0	8.0000	−1	0	0			0.125		
1	5.3594	−1	9.2187		−1.7201			0.1866	
2	3.3750	−1	16.8750			−5.0			0.2963
3	1.9531	−1	22.9687		−11.7601			0.5120	
4	1.0000	−1	27.500	−27.50			1.000		
Total	.....	...	......	−27.50	−13.4802	−5.0	1.125	0.6986	0.2963

$$F_{11} = \frac{1}{EI}\left\{2\left(\frac{1.25}{3}\right)[1.125 + 4(0.698) + 2(0.2963)] + 6\right\} = \frac{9.758}{EI}$$

Substituting the above results into the compatibility equations (7.20), we obtain

$$X_1 = -\frac{\Delta_1^{PL}}{F_{11}} = \frac{259.184q}{9.758} = 26.561q = 531.22 \text{ kN·m}$$

Thus $\text{FEM}^{2j} = 531.22 \text{ kN·m}$ $\text{FEM}^{2k} = -531.22 \text{ kN·m}$ (l)

STEP 4.   We substitute results (a) to (d) and (l) into relations (8.19) and (8.21) to obtain

$$M^{1k} = \widehat{\text{FEM}}^{1k} + \hat{k}^{1k}[\theta^{1k} - \beta^{(1)}] = -392.0 + 0.7806EI\Delta_1$$

$$M^{2j} = \text{FEM}^{2j} + k^{2j}[\theta^{2j} - \beta^{(2)}] + C^{2j}k^{2j}[\theta^{2k} - \beta^{(2)}]$$

$$= 531.22 + (0.7047)EI\Delta_1 + (0.7095)(0.7047)EI\Delta_2 \qquad (m)$$

$$M^{2k} = -531.22 + (0.7095)(0.7047)EI\Delta_1 + (0.7047)EI\Delta_2$$

$$M^{3j} = \widehat{\text{FEM}}^{3j} + \hat{k}^{3j}[\theta^{3j} - \beta^{(3)}] = 196.0 + 0.7806EI\Delta_2$$

STEP 5.   We establish the required relations between the end moments of the members of the beam by considering the equilibrium of joints 2 and 3. That is,

$$M^{1k} + M^{2j} = 0 \qquad M^{2k} + M^{3j} = 0 \qquad (n)$$

Substituting relations (m) into (n), we obtain

$$1.4853\Delta_1 + 0.49998\Delta_2 = -\frac{139.22}{EI}$$

$$0.49998\Delta_1 + 1.4853\Delta_2 = \frac{335.22}{EI} \qquad (o)$$

The set of linear algebraic equations (o) can be written in the following form.

$$\{P\} = [K]\{\Delta\} \qquad (p)$$

where
$$\{P\} = \begin{Bmatrix} -139.22 \\ 335.22 \end{Bmatrix} \qquad \{\Delta\} = \begin{Bmatrix} \Delta_1 \\ \Delta_2 \end{Bmatrix} \tag{q}$$

$$[K] = EI \begin{bmatrix} 1.4853 & 0.49998 \\ 0.49998 & 1.4853 \end{bmatrix} \tag{r}$$

$\{\Delta\}$ is called the *matrix of joint displacements*. Its terms represent the unknown components of displacement of the joints of the beam. $\{P\}$ is called the *matrix of external actions*. Its terms represent the equivalent actions which must be placed on the joints of the beam in order to account for the uniform load acting along the length of its members. That is, the rotations Δ_1 and Δ_2 of joints 2 and 3 of the beam, loaded as shown in Fig. a, are equal to those of the beam subjected to a clockwise moment of 139.22 kN·m at joint 2 and to a counterclockwise moment of 335.22 kN·m at joint 3. $[K]$ is called *the stiffness matrix of the structure*.

Solving equations (o), we obtain

$$\Delta_1 = -\frac{191.39}{EI} \qquad \Delta_2 = \frac{290.22}{EI}$$

STEP 6.   We substitute the values of Δ_1 and Δ_2 obtained in step 5 into relations (m) to get the end moments of the members of the beam. Thus

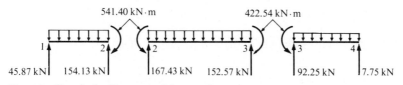

Figure h Free-body diagrams of the members of the beam.

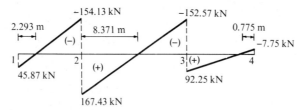

Figure i Shear diagram for the beam.

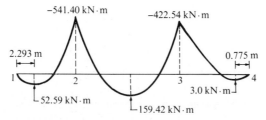

Figure j Moment diagram for the beam.

$$M^{1k} = -541.40 \text{ kN} \cdot \text{m} \qquad M^{2j} = 541.40 \text{ kN} \cdot \text{m}$$

$$M^{2k} = -422.54 \text{ kN} \cdot \text{m} \qquad M^{3j} = 422.54 \text{ kN} \cdot \text{m}$$

Using the above results, the free-body diagrams of the members of the beam are drawn in Fig. h. Referring to this figure, the shear and moment diagrams for the beam are plotted in Figs. i and j, respectively, using the sign convention specified in Fig. 7.1.

Example 3. The planar frame shown in Fig. a is subjected to the following loads.

1. To external actions shown in Fig. a.
2. A settlement of 10 mm of support 1.
3. A difference in temperature. The outside and inside temperatures are, respectively, $T_e = 25°C$ and $T_i = -5°C$.

The members of the frame have the same constant cross section ($I = 240 \times 10^6$ mm⁴, $h = 420$ mm) and are made of the same material ($E = 210,000$ MPa). Compute the end actions in the members of this frame. Disregard the effect of axial deformation of its members.

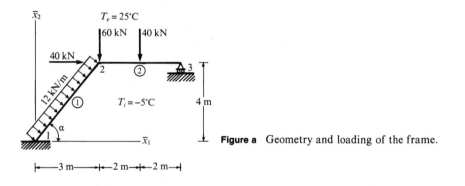

Figure a Geometry and loading of the frame.

---

solution This structure is statically indeterminate to the first degree and kinematically indeterminate to the second degree. Consequently, it is preferable to analyze it using the force method. However we will use the slope defection method in order to illustrate its application.

STEP 1. When the axial deformation of the members of the frame is disregarded, as shown in Fig. b, there are two unknown independent displacements of the joints of the frame denoted in Fig. b as Δ_1 and Δ_2. Notice that although joint 2 translates both vertically and horizontally (see Fig. b), the components of its translation may be established in terms of Δ_1. This may be accomplished by considering the orthogonal triangles 2,2′,2″ and 2,2‴,2″. For deformation within the range of validity of the theory of small deformation, side 2″,2‴ can be considered normal to member 1 and side 2′,2″ can be considered normal to member 2; consequently, the angle 2′,2″,2‴ is equal to the angle α. Thus,

$$\sin \beta = \frac{0.008}{2,2″} \qquad \cos \beta = \frac{\overline{2‴,2″}}{2,2″} \tag{a}$$

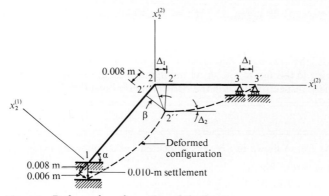

Figure b   Deformed configuration of the frame.

$$\sin (\alpha - \beta) = \frac{\Delta_1}{\overline{2,2''}} \tag{b}$$

Using relation (a), we may rewrite relation (b) as

$$\sin (\alpha - \beta) = \sin \alpha \cos \beta - \sin \beta \cos \alpha = \frac{4}{5}\left(\frac{\overline{2''',2''}}{\overline{2,2''}}\right) - \left(\frac{0.008}{\overline{2,2''}}\right)\left(\frac{3}{5}\right) = \frac{\Delta_1}{\overline{2,2''}}$$

or $$\overline{2'',2'''} = \frac{5}{4}\Delta_1 + 0.006 \tag{c}$$

From the triangles $2,2',2''$ and $2,2'',2'''$, we obtain

$$\overline{2,2''} = (0.008)^2 + (\overline{2'',2'''})^2 = \Delta_1^2 + (\overline{2',2''})^2 \tag{d}$$

Substituting relation (c) into (d) and simplifying, we get

$$\overline{2',2''} = \frac{3}{4}\Delta_1 + 0.010 \tag{e}$$

STEP 2.   We choose the local x_3 axes of the members of the structure normal to its plane. Referring to Fig. b, from the deformed configuration of the structure, we can deduce the following displacement boundary conditions as well as the following relations between the local components of the displacements of the ends of the members of the structure and the unknown displacements of its joints.

$$\theta^{1j} = 0 \qquad\qquad \theta^{2j} = \Delta_2$$

$$u_2^{1j} = -0.006 \qquad\qquad u_2^{2j} = -\tfrac{3}{4}\Delta_1 - 0.010$$

$$\theta^{1k} = \Delta_2 \qquad\qquad \theta^{2k} = \Delta_3 \tag{f}$$

$$u_2^{1k} = -\tfrac{5}{4}\Delta_1 - 0.006 \qquad u_2^{2k} = 0$$

Thus $$\beta^{(1)} = -\tfrac{1}{4}\Delta_1 \qquad \beta^{(2)} = \tfrac{3}{16}\Delta_1 + 0.0025 \tag{g}$$

STEP 3.   Since the members of the frame have a constant cross section, referring to relations (8.3) and (8.5), their carry-over factors and their rotational stiffnesses are equal to

$$C^{1j} = C^{1k} = C^{2j} = C^{2k} = \tfrac{1}{2}$$

$$k^{1j} = k^{1k} = \frac{4EI^{(1)}}{L_1} = \frac{4EI}{5} \tag{h}$$

$$k^{2j} = k^{2k} = \frac{3EI^{(2)}}{L_2} = \frac{3EI}{4}$$

Moreover, referring to App. C, the fixed-end moments of the members of the structure are as follows.

Due to the external actions

$$\text{FEM}^{1j} = 25.0 \text{ kN} \cdot \text{m}$$

$$\text{FEM}^{1k} = -25.0 \text{ kN} \cdot \text{m} \tag{i}$$

$$\text{FEM}^{2j} = 30.0 \text{ kN} \cdot \text{m}$$

Due to the temperature difference

$$\delta T_2 = 25 + 5 = 30°C$$

$$\text{FEM}^{1j} = -\frac{\alpha EI \, \delta T_2}{h} = -\frac{10^{-5}(210)(240)(30)}{420} = -36 \text{ kN} \cdot \text{m}$$

$$\text{FEM}^{1k} = 36 \text{ kN} \cdot \text{m}$$

$$\text{FEM}^{2j} = -\frac{3\alpha EI \, \delta T_2}{2h} = -54 \text{ kN} \cdot \text{m}$$

Total

$$\text{FEM}^{1j} = 25 - 36 = -11 \text{ kN} \cdot \text{m}$$

$$\text{FEM}^{1k} = -25 + 36 = 11 \text{ kN} \cdot \text{m} \tag{j}$$

$$\text{FEM}^{2j} = 30 - 54 = -24 \text{ kN} \cdot \text{m}$$

STEP 4.   Substituting relations (f), (g), (h), and (j) into (8.19) and (8.21), we get

$$M^{1j} = -11.0 + \tfrac{4}{5}EI(0 + \tfrac{1}{4}\Delta_1) + \tfrac{2}{5}EI(\Delta_2 + \tfrac{1}{4}\Delta_1)$$

$$M^{1k} = 11.0 + \tfrac{4}{5}EI(\Delta_2 + \tfrac{1}{4}\Delta_1) + \tfrac{2}{5}EI(0 + \tfrac{1}{4}\Delta_1) \tag{k}$$

$$M^{2j} = -24 + \tfrac{3}{4}EI(\Delta_2 - \tfrac{3}{16}\Delta_1 - 0.0025)$$

STEP 5.   We obtain one equation relating the end moments of the members of the frame by setting the sum of the moments acting on joint 2 equal to zero. That is,

$$\Sigma \overline{M} = 0 \qquad M^{1k} + M^{2j} = 0 \tag{l}$$

Moreover, we obtain a second independent equation by referring to Fig. c, setting the sum of the forces acting on joint 2 equal to zero and eliminating the axial force from the resulting equations. That is,

$$\Sigma \overline{F}_v = 0 \qquad 60 + 20 + \tfrac{1}{4}M^{2j} + 18 - \tfrac{3}{25}(M^{1j} + M^{1k}) + \tfrac{4}{5}F_1^{1k} = 0 \tag{m}$$

$$\Sigma \overline{F}_h = 0 \qquad 40 + 24 - \tfrac{4}{25}(M^{1j} + M^{1k}) - \tfrac{3}{5}F_1^{1k} = 0 \tag{n}$$

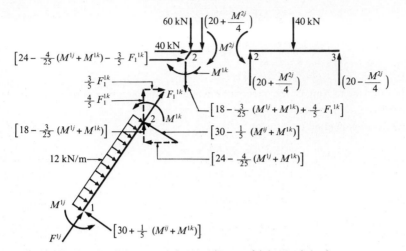

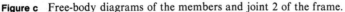

Figure c   Free-body diagrams of the members and joint 2 of the frame.

Eliminating F_1^{1k} from Eqs. (m) and (n), we get

$$M^{1j} + M^{1k} - \tfrac{3}{4}M^{2j} = 550 \tag{o}$$

Relations (l) and (o) are a set of two independent equilibrium equations. Notice that in Fig. c, the shearing forces acting on the ends of each member of the frame have been expressed in terms of the unknown end moments of the member by considering its equilibrium. Substituting relation (k) into (l) and (o) we obtain:

$$0.15938\Delta_1 + 1.55\Delta_2 = \frac{107.5}{EI}$$
$$0.54609\Delta_1 - 0.9125\Delta_2 = \frac{353.625}{EI} \tag{p}$$

Equations (p) are two linear algebraic equations for the two unknown displacements Δ_1 and Δ_2. They can be solved to give

$$\Delta_1 = \frac{651.50}{EI} \qquad \Delta_2 = \frac{2.36}{EI} \tag{q}$$

The above results may be substituted into relations (k) to yield

$$M^{1j} = 185.40 \text{ kN} \cdot \text{m}$$
$$M^{1k} = 208.34 \text{ kN} \cdot \text{m} \tag{r}$$
$$M^{2j} = -208.34 \text{ kN} \cdot \text{m}$$

Substituting the results (r) into relation (n) and into the expressions for the shearing forces given in Fig. c, we get

$$F_1^{1k} = 1.65 \text{ kN} \qquad F_2^{1j} = 108.75 \text{ kN} \qquad F_2^{1k} = 48.75 \text{ kN}$$
$$F_2^{2j} = -32.08 \text{ kN} \qquad F_2^{2k} = 72.08 \text{ kN}$$

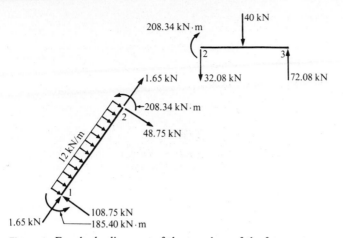

Figure d   Free-body diagrams of the members of the frame.

Using the results (r) and (s), we draw the free-body diagrams of the members of the frame in Fig. d.

Alternative Procedure for Obtaining Relation (o)

The structure of Fig. a has one unknown rotation of its joints (that of joint 2) and one unknown translation. The independent equilibrium equation (l) has been established directly by referring to Fig. c and setting the sum of the moments acting on joints 2 equal to zero. The second independent equilibrium equation (o) has been obtained by referring to Fig. c, setting the sum of the forces acting on joint 2 equal to zero, and then eliminating the unknown force from the resulting relations.

Relation (o) can also be established by considering the auxiliary frame shown in Fig. e, which has the same geometry as the actual frame except that it has hinges at the ends of its members (points 1 and 2). The auxiliary frame is subjected to the given actions of the actual frame and to external moments at the end of its members as shown in Fig. e. These moments are equal to the corresponding end moments in the members of the actual structure. It is apparent that the distribution of internal actions in the auxiliary frame is identical to that in the actual frame. Suppose that the afore-mentioned auxiliary frame is subjected to the virtual displacement shown in Fig. f. During this displacement, the joints of the frame do not rotate while its members are displaced as rigid bodies. Consequently the external moments acting on the joints of the frame and the internal forces in its members do not perform work. In Fig. f, lines $\overline{2,2''}$ and $\overline{2'',2''}$ are perpendicular to members 1 and 2, respectively, and thus angle $2,2'',2'$ is equal to α. Hence,

$$\overline{2',2''} = \frac{\Delta_1}{\tan \alpha} = \frac{3\Delta_1}{4}$$

$$\overline{2,2''} = \frac{\Delta_1}{\sin \alpha} = \frac{5\Delta_1}{4}$$

(s)

On the basis of the principle of virtual work, the work of the external actions acting on the frame of Fig. e during the virtual displacement shown in Fig. f must vanish. Notice, that since the joints of the frame do not rotate, the external moments acting

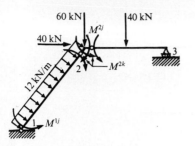

Figure e Auxiliary frame.

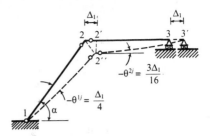

Figure f Virtual displacement of the auxiliary frame.

on them do not perform work during the virtual displacement. Hence, referring to Figs. e and f, we obtain

$$-M^{1j}(\tfrac{1}{4}\Delta_1) - M^{1k}(\tfrac{1}{4}\Delta_1) + M^{2j}(\tfrac{3}{16}\Delta_1)$$

$$+ 12(5)(\tfrac{5}{8}\Delta_1) + 40\Delta_1 + 60(\tfrac{3}{4}\Delta_1) + 40(\tfrac{3}{8}\Delta_1) = 0 \quad (t)$$

Simplifying the above relation, we obtain relation (o).

Comments

Consider a structure having t translational (Δ_1, Δ_2, Δ_3, $\cdots$ Δ_t) and r rotational degrees of freedom of its joints. In general, we can directly establish r independent relations between the end moments of its members by setting to zero the sum of the moments acting at each of its r joints which are free to rotate. Moreover, we can establish t independent relations between the unknown end moments of the members of this structure by adhering to one of the following two procedures.

1. Set the sum of the forces acting on each of its translating joints equal to zero and eliminate the unknown forces from the resulting relations.
2. Consider an auxiliary frame which has the same geometry as the actual frame, except that it has hinges at the ends of its members. The auxiliary frame is subjected to the given external actions of the actual frame and to external moments acting at the ends of its members as shown in Fig. e. These moments are equal to the corresponding internal end moments in the members of the actual structure. The principle of virtual work is then applied t times to this auxiliary structure each time using one of the following "virtual" displacements.

$$1. \quad \Delta_1 \ne 0 \qquad \Delta_2 = \Delta_3 = \cdots = \Delta_t = 0$$

$$2. \quad \Delta_2 \ne 0 \qquad \Delta_1 = \Delta_3 = \cdots = \Delta_t = 0$$

$$\cdots\cdots\cdots\cdots\cdots\cdots\cdots\cdots\cdots\cdots\cdots\cdots$$

$$t. \quad \Delta_t \ne 0 \qquad \Delta_1 = \Delta_2 = \Delta_3 = \cdots = \Delta_{t-1} = 0$$

For instance, the joints of the frame of Fig. g have two translational degrees of freedom (Δ_1, Δ_2). To obtain the two independent equations of equilibrium corresponding

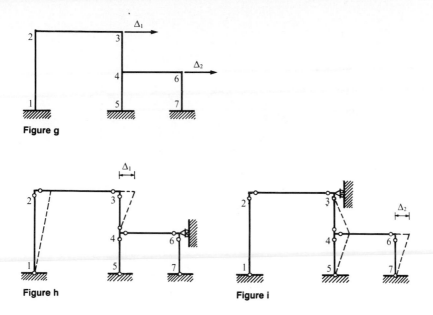

Figure g

Figure h   **Figure i**

to the translational degrees of freedom, we form an auxiliary structure which has the geometry of the actual structure, except that its members are connected to its joints by hinges. We then twice apply the principle of virtual work to this auxiliary structure using the virtual displacements shown in Figs. h and i.

8.7   Comments on the Slope Deflection Method

In the force or flexibility method, the unknown quantities are internal actions or reactions of the structure under consideration. The number of unknowns is equal to the degree of static indeterminacy of the structure. In the slope deflection method, the unknowns are the displacements of the joints of the structure. Consequently, the number of unknowns is equal to the degree of kinematic indeterminacy of the structure. Inasmuch as the most suitable method for the analysis of a particular structure is the one which involves the smallest number of unknowns, the flexibility method should be preferred in analyzing structures having a larger degree of kinematic than static indeterminacy. On the other hand, the slope deflection method should be preferred in analyzing structures having a smaller degree of kinematic than static indeterminacy. For example, the structure of Fig. 8.16a is statically indeterminate to the third degree and, if the effect of axial deformation of its members is disregarded, it is kinematically indeterminate to the fifth degree (the rotations of joints 2, 3, and 4 and the translations of joints 2 and 4). Thus it is easier to analyze this structure using the force method. The structure of Fig. 8.16b is statically indeterminate

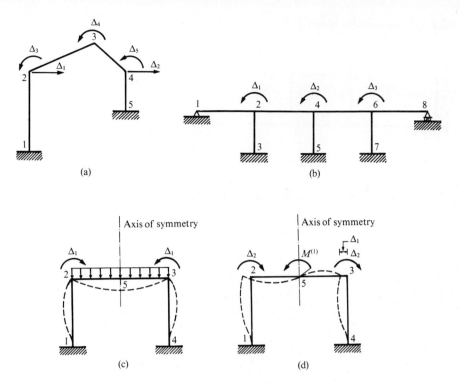

Figure 8.16   Degree of freedom of the joints of structures.

to the ninth degree, while it is kinematically indeterminate only to the third degree. Thus it is easier to analyze this structure using the slope deflection method.

The structure of Fig. 8.16c has an axis of symmetry. When it is subjected to a symmetric loading its joints do not translate. Moreover, the rotations of joints 2 and 3 are equal and opposite. Thus, it is kinematically indeterminate to the first degree. Furthermore, because of the symmetry of the structure and the loading, the shearing force at point 5 vanishes. Thus, the structure is statically indeterminate to the second degree. When the same structure is subjected to an antisymmetric loading (see Fig. 8.16d), joints 2 and 3 translate and rotate by the same amount. Thus, the structure is kinematically indeterminate to the second degree. Moreover, because of the symmetry of the structure and the antisymmetry of the loading, the axial force and the bending moment vanish at point 5. Hence, the structure is statically indeterminate to the first degree. Consequently, when the structure under consideration is subjected to a symmetric loading it can be analyzed more easily by the slope deflection method, while if it is subjected to an antisymmetric loading, it can be analyzed more easily by the force method.

8.8 Moment Distribution

In the previous section, we have seen that the slope deflection method leads to a set of simultaneous, linear algebraic equations for the unknown displacements (translations and rotations) of the joints of the structure. For structures with many joints, the number of simultaneous equations is large; consequently, their solution using hand calculations is difficult. An iterative method, referred to as *moment distribution,* which eliminates the need for solving simultaneous equations for structures whose joints do not translate was presented by Hardy Cross[2] in 1932. This iterative method is also applicable to the analysis of structures that have joints that are free to translate. In this case, it is referred to as the *displacement method with moment distribution,* and it reduces the number of simultaneous equations which must be solved in analyzing a structure to that of its unknown joint translations.

As becomes evident subsequently, in order to apply the moment distribution method, we must establish the end moments in the members of an auxiliary structure subjected to a given moment M^u at one of its joints which is free to rotate. The auxiliary structure is obtained from the actual structure by restraining the movement of its joints, except the rotation of the joint on which the moment M^u is applied. We refer to this auxiliary structure as the *partially restrained structure.* The partially restrained structure of Fig. 8.17b has the same geometry as the actual structure, except that its joint 2 is restrained from rotating. Thus, only its joint 3 is free to rotate. When a moment M^u is applied to joint 3 of this structure, a restraining moment $\overline{S}^{(23)}$ is required at joint 2 in order to prevent it from rotating. Similarly, only joint 2 of the partially restrained structure of Fig. 8.17c is free to rotate. When a moment M^u is applied to this joint, a restraining moment $\overline{S}^{(32)}$ is required at joint 3 in order to prevent it from rotating.

In Fig. 8.18, the free-to-rotate joint of a partially restrained structure is shown subjected to a given moment M^u. We assume that r members are connected to this joint with their end p ($p = j$ or k). From these members, r_1 are type M2, r_2 are type M3 with the hinge at their end q, while the remaining r_3 members are type M3 with the hinge at their end p. Because of application of the given moment M^u, the ends of the $(r_1 + r_2)$ members framing into the

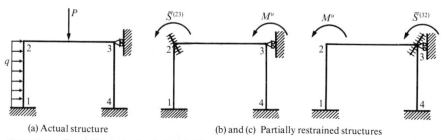

(a) Actual structure   (b) and (c) Partially restrained structures

Figure 8.17 Partially restrained structures.

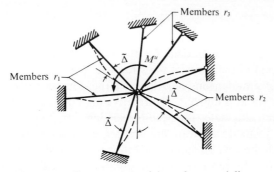

Figure 8.18 Free-to-rotate joint of a partially restrained structure.

joint under consideration will rotate by the same angle which we denote by $\tilde{\Delta}$. The moment to which the end p ($p = j$ or k) of the ith member is subjected, as a result of the application of the moment M^u to the joint under consideration, is equal to

$$M^{ip} = k^{ip}\tilde{\Delta} \quad \text{for each of the } r_1 \text{ type M2 members} \qquad (8.22)$$

$$\hat{M}^{ip} = \hat{k}^{ip}\tilde{\Delta} \quad \text{for each of the } r_2 \text{ type M3 members} \qquad (8.23)$$

In relations (8.22) or (8.23), k^{ip} or $\hat{k}^{ip}$ represents the rotational stiffness of the end p of the ith type M2 or M3 member, respectively, framing into the joint under consideration. From the equilibrium of the joint, it is apparent that the moment M^u applied to this joint is equal to the algebraic sum of the moments M^{ip} ($i = 1, 2, \ldots, r_1$) or $\hat{M}^{ip}$ ($i = 1, 2, \ldots, r_2$), acting at the ends of the members of the structure which frame into this joint. Consequently, using relations (8.22) and (8.23) we get

$$M^u = \sum_{i=1}^{r_1} M^{ip} + \sum_{i=1}^{r_2} \hat{M}^{ip} = \tilde{\Delta}\left[\sum_{i=1}^{r_1} k^{ip} + \sum_{i=1}^{r_2} k^{ip} \right] \qquad (8.24)$$

or

$$\tilde{\Delta} = \frac{M^u}{\displaystyle\sum_{i=1}^{r_1} k^{ip} + \sum_{i=1}^{r_2} \hat{k}^{ip}} \qquad (8.25)$$

Substituting this result into relation (8.22) or (8.23), we obtain

$$M^{jp} = M^u\text{DF}^j \quad \text{or} \quad \hat{M}^{ip} = M^u\widehat{\text{DF}}^j \qquad (8.26)$$

where $\widehat{\text{DF}}^j$ or DF^j is referred to as the *distribution factor* for the jth member at the joint under consideration and is equal to

$$DF^j = \frac{k^{jp}}{\displaystyle\sum_{i=1}^{r_1} k^{ip} + \sum_{i=1}^{r_2} \hat{k}^{ip}} \qquad (8.27)$$

or

$$\widehat{DF}^j = \frac{\hat{k}^{jp}}{\displaystyle\sum_{i=1}^{r_1} k^{ip} + \sum_{i=1}^{r_2} \hat{k}^{ip}} \qquad (8.28)$$

It is apparent that the distribution factors for the members which frame into a joint of a structure can be computed from the rotational stiffnesses of the ends of the members of the structure which frame into this joint. Thus, when the free-to-rotate joint of a partially restrained structure is subjected to a moment M^u, the moment M^{jp} or $\hat{M}^{jp}$ at the end of the jth member framing into this joint with its end p ($p = j$ or k) may be computed on the basis of relations (8.26). Moreover, if the jth member of the partially restrained structure is type M2, its end q ($q = k$ or $j, q \neq p$) is subjected to a moment as a result of the application of the moment M^u. This moment is equal to

$$M^{jq} = C^{jp} M^{ip} \qquad q, p = j \text{ or } k, q \neq p \qquad (8.29)$$

The moment M^{jq} is referred to as the *carry-over moment*. C^{jp} is the carry-over factor from the end p to the end q of the jth member of the structure [see relation (8.1)]. If the member under consideration is type M3, its end q is not subjected to any moment.

In applying the moment distribution method, we adhere to the following procedure.

Step 1. We compute the rotational stiffnesses and the carry-over factors for the members of the structure. Moreover we obtain the distribution factors for the members of the structure framing into each of its rotating joints by substituting their rotational stiffnesses in relations (8.27) or (8.28).

Step 2. We form the restrained structure by restraining the joints of the actual structure from rotating. This is accomplished by the application of external moments to the joints of the actual structure, referred to as the *restraining moments*. We refer to the restrained structure as the RS$^{(1)}$.

We analyze the restrained structure subjected to all the loads acting on the actual structure, including the external actions, the change of temperature, the translation or rotation of its supports, and the initial stress of its members. The analysis of the restrained structure can be readily performed because the external loads (except the movement of the supports) affect only the members on which they act. That is, the restrained structure is made up of a number of members fixed at both ends or fixed at one end and pinned at the other. If these members have a variable cross section, the internal moments at their ends can be established using one of the methods presented in Chap. 7. However, if these members have a constant cross section, the internal moments acting at their ends can be obtained from handbooks or from App. C. The internal moments at the ends of the members of the restrained structure are referred to as *their fixed-end moments*. The fixed-end moment at the end j of the ith member is denoted as FEM^{ij} or $\widehat{\text{FEM}}^{ij}$, depending upon whether the member is type M2 or M3, respectively.

We compute the restraining moments acting at the joints of the restrained structure from the fixed-end moments of its members by considering the equilibrium of its joints. The restraining moment at a joint is equal to the algebraic sum of the fixed-end moments in the members framing into this joint, minus the given external moment acting on this joint.

Step 3. We perform the moment distribution. We start by releasing the rotational restraint of one joint (say the mth) of the restrained structure $\text{RS}^{(1)}$ generally, the one subjected to the largest restraining moment. We refer to the resulting structure as the $\text{RS}^{(2)}$. It is apparent that there is no external moment acting on the mth joint of the $\text{RS}^{(2)}$. We compute the end moments in the members of the $\text{RS}^{(2)}$. These moments are equal to the sum of the corresponding moments in the members of the $\text{RS}^{(1)}$ and in the members of a partially restrained structure, formed from the actual structure by restraining all its joints from rotating except the mth. We refer to the partially restrained structure as the $\text{PRS}^{(1)}$. The free-to-rotate joint (the mth) of the $\text{PRS}^{(1)}$ is subjected to an external moment which is equal and opposite to the restraining moment acting on the mth joint of the $\text{RS}^{(1)}$. We restrain the mth joint of the $\text{RS}^{(2)}$ against further rotation before proceeding to the second iteration of the moment distribution process.

In each iteration, say the jth, we start with the $\text{RS}^{(j)}$ resulting from the previous iteration and we superimpose on the end moments of its members the corresponding moments in the members of the $\text{PRS}^{(j)}$, subjected to an external moment on its free-to-rotate joint. This moment is equal and opposite to the restraining moment acting on the same joint of the $\text{RS}^{(j)}$ and it is referred to as the *unbalanced moment*. The moments at the ends of the members of the $\text{PRS}^{(j)}$ can be obtained from the "unbalanced" moment using relations (8.26).

The moment distribution process continues for as many iterations as necessary to reduce to zero, to within the desired degree of accuracy, the restraining moments in the resulting RS. It is evident that in this case the loading of the resulting structure approaches that of the actual structure. Consequently, the end moments in the members of the resulting structure approach the corresponding moments in the members of the actual structure to the desired degree of accuracy. This iteration process converges very rapidly, particularly if, in every iteration, the joint of the $\text{PRS}^{(j)}$ which is free to rotate corresponds to that of the $\text{RS}^{(j)}$ having the maximum restraining moment.

Step 4. From the end moments of the members of the structure we compute their end actions.

In what follows, we illustrate the moment distribution method by the following two examples involving structures whose joints do not translate.

Example 1. The end moments of the members of a statically indeterminate frame are established when it is subjected only to given external actions. The physical interpretation of the successive iterations during the process of moment distribution is illustrated in Fig. e of this example. Notice that we would have adhered to the same procedure if the frame were subjected to a given settlement or rotation of one of its supports or to a given difference of temperature of its internal and external fibers, or if it were fabricated using an initially stressed member. The effect of any loading is included in the fixed-end moments of the members of the frame.

Example 2. The internal actions in the members of a statically indeterminate beam of constant cross section are established. The beam is subjected to given external actions; a given settlement of one of its supports and a given rotation of another support. The beam consists of two members; the one is type M2, whereas the other may be considered either type M2 or type M3. The beam is analyzed in two ways: (a) by considering member 2 as type M3; (b) by considering member 2 as type M2.

Example 1. Using the moment distribution method, establish the internal actions at the ends of the members of the frame of Fig. a, and plot its shear and moment diagrams. The members of the frame have the same constant cross section and are made of the same material.

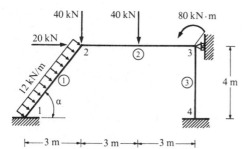

Figure a   Geometry and loading of the frame.

---

solution

STEP 1.   Inasmuch as the members of the frame have constant cross section and referring to relation (8.3), we see that their rotational stiffnesses are equal to

$$k^{1j} = k^{1k} = \frac{4EI^{(1)}}{L_1} = \frac{4EI}{5}$$

$$k^{2j} = k^{2k} = \frac{4EI^{(2)}}{L_2} = \frac{4EI}{6} \tag{a}$$

$$k^{3j} = k^{3k} = \frac{4EI^{(3)}}{L_3} = EI$$

Thus, referring to relation (8.27), we see that the distribution factors at joint 2 are

$$DF^{(1)} = \frac{4EI/5}{4EI/5 + 4EI/6} = 0.545$$

$$DF^{(2)} = \frac{4EI/6}{4EI/5 + 4\,EI/6} = 0.455$$

(b)

while the distribution factors at joint 3 are

$$DF^{(2)} = \frac{4EI/6}{4EI/6 + EI} = 0.40$$

$$DF^{(3)} = \frac{EI}{4EI/6 + EI} = 0.60$$

(c)

The distribution factors are placed as shown in Fig. b.

STEP 2.   The restrained structure subjected to the given loads is shown in Fig. c. The end moments [fixed-end moments (FEM)] of the members of the restrained structure are obtained from App. C and are shown in Fig. d. From those moments, the restraining moments at joints 2 and 3 are computed by considering the equilibrium of these joints. It is apparent that the restraining moment required at a joint is equal to the algebraic sum of the fixed-end moments in the members framing into this joint minus the external moment acting on this joint.

STEP 3.   We perform the moment distribution. Referring to Fig. ea, the largest

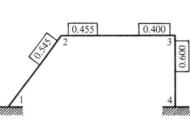

Figure b   Distribution factors.

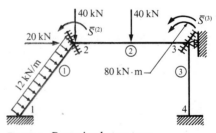

Figure c   Restrained structure.

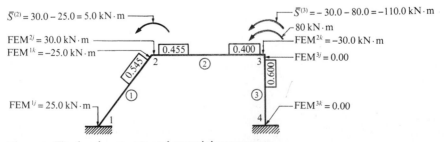

Figure d   Fixed-end moments and restraining moments.

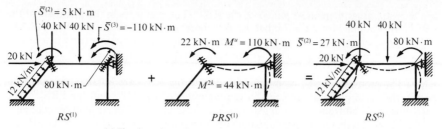

(a) First iteration – the rotation of joint 3 is released

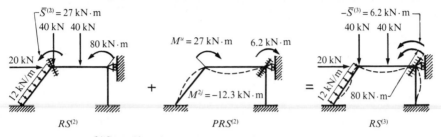

(b) Second iteration – the rotation of joint 2 is released

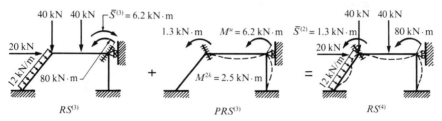

(c) Third iteration – the rotation of joint 3 is released

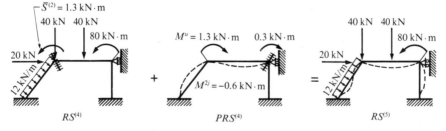

(d) Fourth iteration – the rotation of joint 2 is released

Figure e Process of moment distribution.

restraining moment in the $RS^{(1)}$ acts at joint 3. Thus, for the first iteration, we release the rotational constraint of joint 3 of the restrained structure. The moments in the members of the resulting structure ($RS^{(2)}$) are obtained by superimposing the corresponding moments in the members of the $RS^{(1)}$ and the $PRS^{(1)}$. The $PRS^{(1)}$ is obtained from the actual structure by restraining all its joints from rotating except joint 3. This joint is subjected to an "unbalanced" moment ($M^u = 110$ kN·m) equal and opposite to the restraining moment at joint 3 of the $RS^{(1)}$. The moments in the members of the $PRS^{(1)}$ are obtained from the "unbalanced" moment using relations (8.26) and (8.29). Before proceeding to the second iteration, we restrain joint 3 against further rotation.

As can be seen from Fig. *eb* the largest restraining moment of the $RS^{(2)}$ acts at joint 2. Thus in the second iteration we release the rotational constraint of joint 2 of the $RS^{(2)}$. The moments in the members of the resulting $RS^{(3)}$ are obtained by superimposing the corresponding moments in the members of the $RS^{(2)}$ and the $PRS^{(2)}$. The $PRS^{(2)}$ is obtained from the actual structure by restraining all of its joints except joint 2. This joint is subjected to an unbalanced moment ($M^u = -27$ kN·m) equal and opposite to the restraining moment at joint 2 of the $RS^{(2)}$. The moments in the members of the $PRS^{(2)}$ are obtained from the unbalanced moment using relations (8.26) and (8.29).

In Fig. f, we present a convenient way to perform the moment distribution. In this figure, we first list the distribution factors and the fixed-end moments FEM^{ip} ($p = j$ or k) in the members of the restrained structure. The restraining moment at each joint of the structure, although not listed, may be readily obtained inasmuch as it is equal to the algebraic sum of the fixed-end moments in the members of the structure framing into the joint minus the external moment acting on the joint. The maximum restraining moment acts at joint 3 and is equal to -110 kN·m. Thus, we first release joint 3 of the restrained structure. To accomplish this, we must distribute the unbalanced moment of $M^u = 110$ kN·m to members 2 and 3 which frame into joint 3. Each member will take a portion of the unbalanced moment proportional to its distribution factor. That is, member 2 takes $M^{2k} = (110)(0.4) = 44$ kN·m, while member 3 takes $M^{3j} = (110)(0.6) = 66$ kN·m. We refer to these moments as the *distributed moments*. In Fig. f, the distributed moments are written below the fixed-end moments acting in the ends of members 2 and 3 before joint 3 was released. The moments resulting from the summation of the distributed and the fixed-end moments satisfy the equilibrium of joint 3. We say that the moments acting on joint 3 have been balanced. To indicate this, in Fig. f we draw a line underneath the distributed moments of the members framing into joint 3.

The change of the moment at one end of members 2 and 3 affects the moment at their other end. Inasmuch as the carry-over factors of these members are equal to one-half, the moment at end j of member 2 changes by an amount equal to $44(\tfrac{1}{2}) = 22$

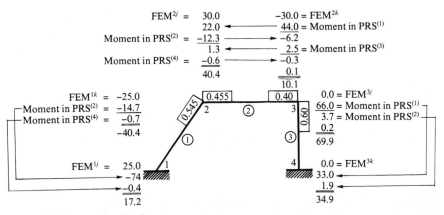

Figure f Moment distribution.

kN·m, while the moment at end k of member 3 changes by an amount equal to $66(\frac{1}{2})$ = 33 kN·m. These changes are written in Fig. f, and the first iteration of the moment distribution process has been completed. What we have done up to now is equivalent to superimposing the end moments in the members of the PRS[1] to the corresponding end moments in the members of the PS[1]. If at this stage we add the columns in Fig. f, the resulting moments are those acting at the ends of the corresponding members of the RS[2] shown in Fig. e. Before proceeding to the second iteration we fix joint 3 against further rotation.

In the second iteration we release joint 2 of the RS[2]. The restraining moment on this joint is equal to $30 - 25 + 22 = 27$ kN·m. Consequently, an unbalanced moment equal to -27 kN·m must be distributed to members 1 and 2 in accordance with their distribution factors. That is, as shown in Fig. f, member 1 takes a moment equal to $(-27)(0.545) = -14.7$ kN·m while member 3 takes a moment equal to $(-27)(0.455) = -12.3$ kN·m. These moments are then added to the moments acting on the ends of members 1 and 3 before joint 2 was released; thus, the moments acting on joint 2 have been balanced. Again, we draw a line underneath these moments to indicate that the moments acting on joint 2 have been balanced. Moreover, as shown in Fig. f half of the change of the moment at the end of each member framing into joint 2 is carried over to its other end. Thus, the second iteration of the moment distribution process has been completed. As shown in Fig. e, this iteration is equivalent to superimposing the end moments in the members of the PRS[2] to the end moments in the members of the RS[2]. The resulting moments are those in the ends of the corresponding members of the RS[3] shown in Fig. e. We then release joints 3 and 2, alternately and proceed as above until the moments at these joints are balanced to the desired degree of accuracy, that is, until the restraining moments acting on the joints of the resulting structure vanish to the desired degree of accuracy. Subsequently, we add the numbers in each column to obtain the end moments in the members of the structure.

The moment distribution may also be carried out as illustrated in Table a below.

TABLE a   Moment Distribution

Joint	1	2		3		4
Member	①	①	②	②	③	④
DF	—	0.545	0.455	0.40	0.60	
FEM	25.0	−25.0	30.0	−30.0	00.0	00.0
Released joint — 3	....	.....	22.0	−80 44.0	66.0	33.0
Released joint — 2	−7.4	−14.7	−12.3	−6.2		
Released joint — 3	....	.....	1.3	2.5	3.7	1.9
Released joint — 2	−0.4	−0.7	−0.6	−0.3	0.2	
				0.1		
Final moments	17.2	−40.4	40.4	10.1	69.9	34.9

STEP 4.   From the end moments of the members of the structure, we compute their end actions. The free-body diagrams of the members and the joints of the structure are shown in Fig. g. Referring to this figure, from the equilibrium of joint 2, we get

$$\Sigma \bar{F}_h = 0 = \tfrac{3}{5} F_1^{1k} - 27.7 + F_1^{2j} - 20$$

$$\Sigma \bar{F}_v = 0 = \tfrac{4}{5} F_1^{1k} + 20.8 + 28.42 + 40$$

or

$$F_1^{1k} = (-89.22)(\tfrac{5}{4}) = -111.52 \text{ kN}$$ (d)

$$F_1^{2j} = 114.6 \text{ kN} = -F_1^{2k}$$

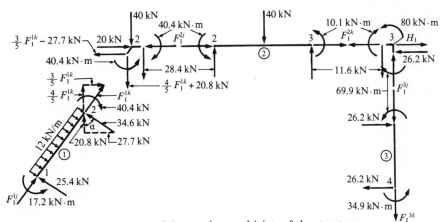

Figure g   Free-body diagrams of the members and joints of the structure.

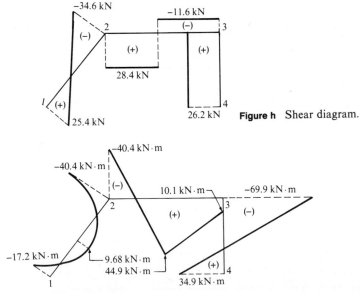

Figure h   Shear diagram.

Figure i   Moment diagram.

Moreover, from the equilibrium of joint 3 we obtain

$$\Sigma \bar{F}_h = 0 = -H_1 + 114.61 - 26.2$$

or
$$H_1 = 88.41 \text{ kN} \tag{e}$$

Referring to Fig. g, we draw the shear and moment diagrams in Figs. h and i, respectively, using the sign convention indicated in Fig. 7.1.

Example 2. Analyze the continuous beam of constant cross section ($I = 759 \times 10^6$ mm⁴) subjected to the external actions shown in Fig. a, as well as to a 10-mm settlement of support 2 and a 0.001-rad counterclockwise rotation of support 1. The modulus of elasticity of the material from which the beam is made is $E = 210,000$ MPa.

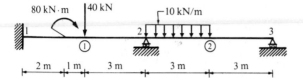

Figure a Geometry and loading of the beam.

solution

Case 1: Member 2 Is Considered as Type M3

STEP 1.   This structure consists of two members. Member 1 is type M2 while member 2 is considered type M3. Inasmuch as the members of the beam have a constant cross section, referring to relations (8.3) and (8.6), we find that their rotational stiffnesses are equal to

$$k^{1j} = k^{1k} = \frac{4EI^{(1)}}{L_1} = \frac{2EI}{3}$$

$$\hat{k}^{2j} = \frac{3EI^{(2)}}{L_2} = \frac{EI}{2} \tag{a}$$

Thus referring to relations (8.27) and (8.28), we find that the distribution factors at joint 2 are

$$DF^{(1)} = \frac{2EI/3}{2EI/3 + EI/2} = 0.571$$

$$\widehat{DF}^{(2)} = \frac{EI/2}{2EI/3 + EI/2} \tag{b}$$

The distribution factors are tabulated as shown in Fig. b.

STEP 2.   The restrained structure is made from the actual structure by restraining joint 2 from rotating. Thus member 1 of the restrained structure is fixed at both ends, while member 2 is fixed at the one end and pinned at the other end. The fixed-end

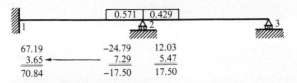

Figure b Moment distribution.

moments of the members of the restrained structure subjected to all the given loads are obtained from App. C. They are:

1. Fixed-end moments due to the external actions

$$\text{FEM}^{1j} = 30 \text{ kN} \cdot \text{m}$$

$$\text{FEM}^{1k} = -56.67 \text{ kN} \cdot \text{m} \qquad (c)$$

$$\widehat{\text{FEM}}^{2j} = 25.31 \text{ kN} \cdot \text{m}$$

2. Fixed-end moments due to the settlement of support 2

$$\text{FEM}^{1j} = \frac{6EI\Delta^s}{L_1^2} = \frac{6(210,000 \times 10^3)(7590 \times 10^{-8})(0.01)}{6^2}$$

$$= 26.57 \text{ kN} \cdot \text{m} \qquad (d)$$

$$\text{FEM}^{1k} = \frac{6EI\Delta^s}{L_1^2} = 26.57 \text{ kN} \cdot \text{m}$$

$$\widehat{\text{FEM}}^{2j} = -\frac{3EI\Delta^s}{L_2^2} = -13.28 \text{ kN} \cdot \text{m}$$

3. Fixed-end moments due to the rotation of support 1

$$\text{FEM}^{1j} = \frac{4EI(0.001)}{L_1} = \frac{4(210.000 \times 10^3)(7590 \times 10^{-8})(0.001)}{6}$$

$$= 10.62 \text{ kN} \cdot \text{m} \qquad (e)$$

$$\text{FEM}^{1k} = \frac{2EI(0.001)}{L_1} = 5.31 \text{ kN} \cdot \text{m}$$

$$\widehat{\text{FEM}}^{2j} = 0.0$$

Thus, the total fixed-end moments in the members of the restrained structure due to all the disturbances acting on the structure are

$$\text{FEM}^{1j} = 67.19 \text{ kN} \cdot \text{m}$$

$$\text{FEM}^{1k} = -24.79 \text{ kN} \cdot \text{m} \qquad (f)$$

$$\widehat{\text{FEM}}^{2j} = 12.03 \text{ kN} \cdot \text{m}$$

STEP 3.   The moment distribution is performed in Fig. b. The resulting moments are the end moments in the members of the beam. They are in agreement with those obtained in Example 1 of Sec. 8.6.

Case 2: Member 2 Is Considered as Type M2

STEP 1.   If we consider member 2 as type M2 its rotational stiffnesses are equal to

$$k^{2j} = k^{2k} = \frac{4EI^{(2)}}{L_2} = \frac{2EI}{3} \tag{g}$$

Consequently, referring to relation (8.27), we find that the distribution factors at joint 2 are equal to

$$DF^{(1)} = DF^{(2)} = \frac{1}{2} \tag{h}$$

Notice that since only one member of the beam is connected to joint 3, when this joint is released, all of the unbalanced moment is taken by this member. That is, the distribution factor at joint 3 is

$$DF^{(2)} = 1$$

STEP 2.   The restrained structure is made from the actual beam by restraining its joints 2 and 3 from rotating. Thus, both members of the beam are fixed at both ends. The fixed-end moments in member 1 of the restrained structure subjected to all the given loads are given by relations (f). Moreover, referring to App. C, we find that the fixed-end moments of member 2 of the restrained structure are

1. Fixed-end moments due to external actions

$$FEM^{2j} = 20.63 \text{ kN} \cdot \text{m} \qquad FEM^{2k} = -9.37 \text{ kN} \cdot \text{m} \tag{i}$$

2. Fixed-end moments due to the settlement of support 2

$$FEM^{2j} = -\frac{6EI\Delta^s}{L_2^2} = -26.57 \text{ kN} \cdot \text{m}$$
$$FEM^{2k} = -\frac{6EI\Delta^s}{L_2^2} = -26.57 \text{ kN} \cdot \text{m} \tag{j}$$

3. Fixed-end moments due to the rotation of support 1

$$FEM^{2j} = 0 \qquad FEM^{2k} = 0 \tag{k}$$

The total fixed-end moments in member 2 are

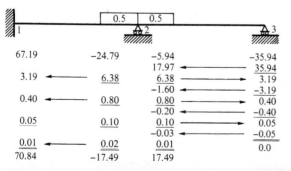

Figure c   Moment distribution.

$$FEM^{2j} = -5.94 \text{ kN} \cdot \text{m} \qquad FEM^{2k} = -35.94 \text{ kN} \cdot \text{m} \qquad \text{(f)}$$

STEP 3. The moment distribution is performed in Fig. c. Notice that joint 3 of the restrained structure is released first since it is subjected to the largest restraining moment.

On the basis of the foregoing, it is apparent that the moment distribution process is simplified when member 2 of the beam is considered as type M3 rather than when it is considered as type M2.

8.9 Moment Distribution for Symmetric Structures Subjected to Symmetric Loading

The joints of structures having an axis of symmetry and subjected to symmetric loads do not translate if the effect of the axial deformation of their members is disregarded (see Fig. 8.19b). Thus the moment distribution method can be applied directly to the analysis of these structures. As illustrated in the examples at the end of this section, in order to accelerate the convergence of the moment distribution method when it is applied to symmetric structures subjected to symmetric loads, we establish the end moments in the members of a partially restrained structure, which has two symmetric joints free to rotate, when this structure is subjected to symmetric moments at its free-to-rotate joints.

Consider the symmetric structure of Fig. 8.19a subjected to symmetric loads and the partially restrained structure of Fig. 8.19c, obtained from the actual structure by restraining its joints from rotating except for joints 3 and 4. Assume that joints 3 and 4 of this structure are simultaneously subjected to a symmetric pair of moments of magnitude M^u. We will establish the internal moments at the ends of members 2 and 3 framing into joint 3. These moments are symmetric to the internal moments at the ends of members 4 and 3 framing into joint 4. Referring to Fig. 8.19c and using the slope deflection equation (8.27), we have

$$M^{3j} = k^{3j}(\tilde{\Delta}_1 - C^{3j}\tilde{\Delta}_1) \qquad M^{2k} = k^{2k}(\tilde{\Delta}_1) \qquad (8.30)$$

From the equilibrium of joint 3 and using the above relations, we obtain

$$M^u = M^{2k} + M^{3j} = [k^{3j}(1 - C^{3j}) + k^{2k}]\tilde{\Delta}_1$$

Solving for $\tilde{\Delta}_1$, we get

$$\tilde{\Delta}_1 = \frac{M^u}{k^{3j}(1 - C^{3j}) + k^{2k}} \qquad (8.31)$$

Substituting this result in relations (8.30), we obtain

$$M^{3j} = \frac{k^{3j}(1 - C^{3j})M^u}{k^{3j}(1 - C^{3j}) + k^{2k}} = \widehat{DF}^{(3)} M^u \qquad (8.32)$$

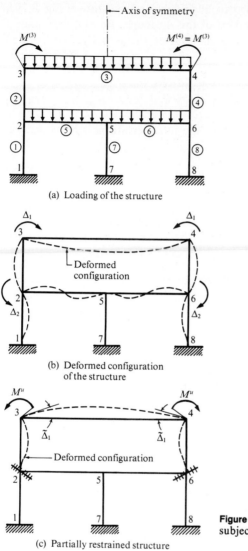

(a) Loading of the structure

(b) Deformed configuration of the structure

(c) Partially restrained structure

Figure 8.19 Symmetric structure subjected to a symmetric loading.

$$M^{2k} = \frac{k^{3k} M^u}{k^{3j}(1 - C^{3j}) + k^{2k}} = \widehat{\text{DF}}^{(2)} M^u$$

The internal moments M^{3k} and M^{4j} are symmetric to M^{3j} and M^{2k}, respectively. That is

$$M^{3k} = -M^{3j} \qquad M^{4j} = -M^{2k}$$

Thus it is apparent that when a symmetric pair of moments of magnitude

M^u is applied to joints 3 and 4 of the partially restrained structure of Fig. 8.19c, it is distributed in the ends of members of this structure, framing into joint 3 in accordance to distribution factors obtained from the actual stiffness of end k of member 2 and a modified stiffness of end j of member 3. The latter is equal to

$$\hat{k}^{3j} = k^{3j}(1 - C^{3j}) \tag{8.33}$$

If member 3 has a constant cross section, the modified stiffness of its end j is equal to

$$\hat{k}^{3j} = \frac{1}{2}k^{3j} \tag{8.34}$$

In the sequel, we employ the moment distribution method in obtaining the internal moments in the members of two symmetric structures. The one is subjected to symmetric external actions while the other is subjected to different temperatures at the external and internal fibers of its members.

Example 1.   Using the moment distribution method, analyze the symmetric structure loaded with the symmetric external actions shown in Fig. a. The members of the structure are made of the same material and have constant cross sections. The moments of inertia of their cross sections are indicated in Fig. a. Plot the shear and moment diagrams for this structure.

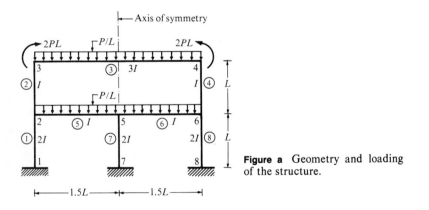

Figure a Geometry and loading of the structure.

---

solution   When a symmetric structure is subjected to a symmetric loading its joints do not translate. Moreover, referring to Fig. b, we can easily see that member 7 is not subjected to bending while the translation and rotation of joint 5 are zero; thus this joint may be considered fixed. In order to accelerate the convergence of the moment distribution method, we use the modified stiffness of member 3 [see relation (8.33)] in computing the distribution factors at joint 3.

STEP 1.   The rotational stiffness of members 1, 2, and 5 and the modified stiffness of

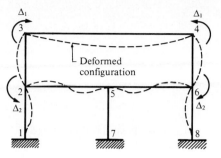

Figure b Deformed configuration of the frame.

member 3 [see relation (8.34)] of the structure are

$$k^{1k} = \frac{4EI^{(1)}}{L_1} = \frac{4E(2I)}{L} = \frac{8EI}{L}$$

$$k^{2j} = k^{2k} = \frac{4EI^{(2)}}{L_2} = \frac{4E(3I)}{3L} = \frac{4EI}{L}$$

$$\hat{k}^{3j} = \frac{1}{2}k^{3j} = \frac{2EI^{(3)}}{L_3} = \frac{2E(3I)}{3L} = \frac{2EI}{L}$$

$$k^{5j} = \frac{4EI^{(5)}}{L_5} = \frac{4EI}{1.5L} = \frac{2.667EI}{L}$$

Thus referring to relation (8.27) the distribution factors at joint 2 are

$$DF^{(1)} = \frac{8}{8 + 4 + 2.667} = 0.545$$

$$DF^{(2)} = \frac{4}{8 + 4 + 2.667} = 0.273$$

$$DF^{(5)} = \frac{2.667}{8 + 4 + 2.667} = 0.182$$

While the distribution factors at joint 3 are

$$\widehat{DF}^{(2)} = \frac{4}{4 + 2} = 0.667 \qquad \widehat{DF}^{(3)} = \frac{2}{4 + 2} = 0.333$$

STEP 2. The restrained structure subjected to the given loads is shown in Fig. c. The fixed-end moments of the members of the restrained structure are obtained by referring to App. C. They are

$$FEM^{1j} = (F.E.M.)^{1k} = 0$$

$$FEM^{2j} = (F.E.M.)^{2k} = 0$$

$$FEM^{3j} = \frac{3P(3L)}{12} = 0.75PL$$

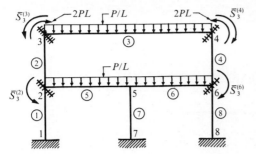

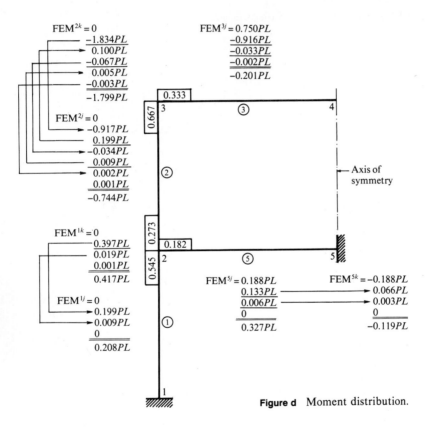

Figure c Restrained structure subjected to the given loads.

FEM$^{2k} = 0$
-1.834PL
0.100PL
-0.067PL
0.005PL
-0.003PL
-1.799PL

FEM$^{3j} = 0.750PL$
-0.916PL
-0.033PL
-0.002PL
-0.201PL

FEM$^{2j} = 0$
-0.917PL
0.199PL
-0.034PL
0.009PL
0.002PL
0.001PL
-0.744PL

FEM$^{1k} = 0$
0.397PL
0.019PL
0.001PL
0.417PL

FEM$^{1j} = 0$
0.199PL
0.009PL
0
0.208PL

FEM$^{5j} = 0.188PL$
0.133PL
0.006PL
0
0.327PL

FEM$^{5k} = -0.188PL$
0.066PL
0.003PL
0
-0.119PL

Figure d Moment distribution.

$$\text{FEM}^{5j} = \frac{1.5P(1.5L)}{12} = 0.1875PL$$

$$\text{FEM}^{5k} = -0.1875PL$$

STEP 3.   The moment distribution is carried out in Fig. d and also in Table a.

STEP 4.   From the end moments of the members of the structure, we compute their end actions. The free-body diagrams of the members and joints of the structure are

TABLE a Moment Distribution

Joint	1		2			3		5
Member	①	①	②	⑤	②	③	⑤
DF→	. . . .	0.545	0.273	0.182	0.667	0.333	
FEM*	0	0	0	0.188	0	0.750	−0.188
Released joint 3*	. . . .	. . . . . .	−0.917	. . . . . .	−1.834	−0.916	
2*	0.199	0.397	0.199	0.133	0.100		0.066
3*	. . . .	. . . . . .	−0.034	. . . . . .	−0.067	−0.033	
2*	0.009	+0.019	0.009	−0.006	0.005	. . . . . .	−0.003
3*	. . . .	. . . . . .	−0.002	. . . . . .	−0.003	−0.002	
2*	~0	0.001	0.001	~0	~0		
Final moment*	0.208	0.417	−0.744	0.327	−1.799	−0.201	−0.119

Above the released-joint rows, spanning the joint 3 columns: 2.00

* In each of these lines, moment = numerical value × *PL*.

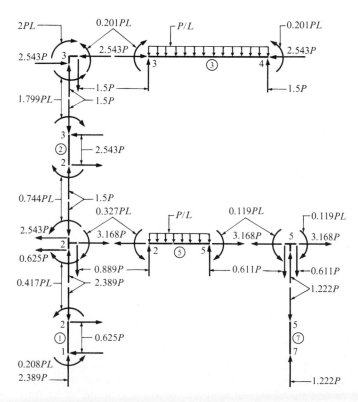

Figure e Free-body diagrams of the members and joint 2 of the structure.

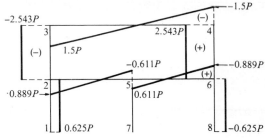
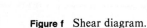

Figure f Shear diagram.

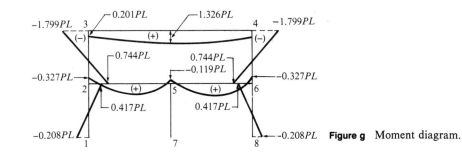

Figure g Moment diagram.

shown in Fig. e. The shear and moment diagrams are shown in Figs. f and g, respectively, using the sign convention indicated in Fig. 7.1.

Example 2. Using the moment distribution method, analyze the symmetric structure shown in Fig. a when the external fibers of its members are subjected to a temperature of $-10°C$ while their internal fibers are subjected to a temperature of $30°C$. The temperature during construction was $10°C$. The members of the frame are made of the same material ($E = 210,000$ MPa, $\alpha = 10^{-5}/°C$) and have the same constant cross section ($I = 133.33 \times 10^6$ mm^4, depth of cross section $h = 400$ mm).

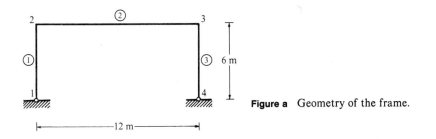

Figure a Geometry of the frame.

solution The difference in temperature δT at the external and internal fibers of the members of the frame and the change of temperature at the centroids of its cross sections are:

$$\delta T = 30 - (-10) = 40°C \tag{a}$$

$$\Delta T_c = \frac{30 + (-10)}{2} - 10 = 0$$

STEP 1.   The frame of Fig. a has an axis of symmetry and moreover the effect of the temperature difference is symmetric with respect to that axis. Thus in order to accelerate the convergence of the moment distribution method, we use the modified stiffness of member 2 and release joints 2 and 3 simultaneously. Hence referring to relations (8.6) and (8.34), we have

$$k^{1k} = k^{3j} = \frac{3EI}{L/2} = \frac{6EI}{L}$$

$$\hat{k}^{2j} = \frac{k^{2j}}{2} = \frac{2EI}{L} \tag{b}$$

Referring to relation (8.27) the distribution factors at joint 2 are:

$$DF^1 = 0.75 \qquad DF^2 = 0.25$$

STEP 2.   The fixed-end moments of the members of the restrained structure are obtained by referring to App. C. Thus

$$FEM^{1k} = -FEM^{1j} = \frac{3EI\alpha\,\delta T}{2h} = \frac{3(210)(133.33 \times 10^6)(10^{-5})(30)}{2(400 \times 10^3)} = 31.5 \text{ kN·m}$$

$$FEM^{2k} = -FEM^{2j} = \frac{EI\alpha\,\delta T}{h} = \frac{(210)(133.33 \times 10^6)(10^{-5})30}{(400 \times 103)} = 21 \text{ kN·m}$$

STEP 3.   The moment distribution is carried out in Fig. b.

STEP 4.   From the end moments of the members of the frame, we compute their end actions. The free-body diagrams of the members of the frame are shown in Fig. c.

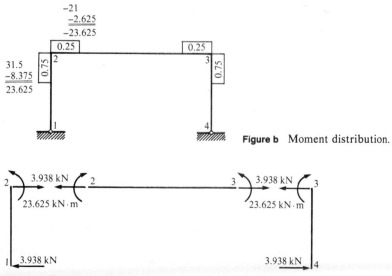

Figure b Moment distribution.

Figure c Free-body diagrams of the members of the frame.

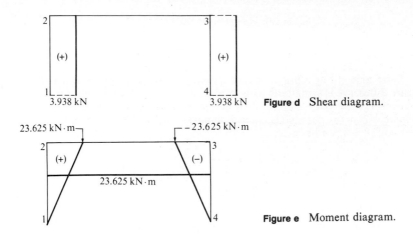

Figure d Shear diagram.

Figure e Moment diagram.

The shear and moment diagrams are shown in Figs. d and e, respectively, using the sign convention indicated in Fig. 7.1.

8.10 Displacement Method with Moment Distribution

In the displacement method with moment distribution, the internal actions in the members of a statically indeterminate structure are obtained by superimposing the corresponding internal actions of the following two structures (see Fig. 8.20):

1. An auxiliary structure, referred to as *the structure restrained against translation*. This structure is made from the actual structure by preventing from translation those joints which are free to do so by applying external forces to them. These forces are referred to as *holding forces* (see Fig. 8.20*b*). The structure restrained against translation is subjected to all the loads acting on the actual structure (external actions, change of temperature, movement of

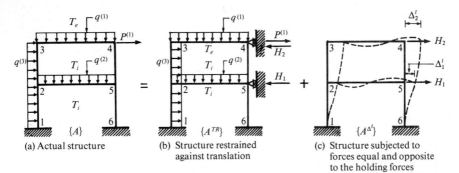

Figure 8.20 Superposition of the structure restrained against translation and the structure subjected to forces equal and opposite to the holding forces.

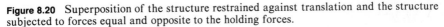

supports, initial stress). This structure can be analyzed using the moment distribution method discussed in Sec. 8.8. The internal actions at the ends of its members are stored in a matrix denoted by $\{A^{TR}\}$.

2. The actual structures subjected to external forces at its free-to-translate joints which are equal and opposite to the holding forces (see Fig. 8.20c). The internal actions at the end of the members of this structure are stored in a matrix denoted by $\{A^{\Delta'}\}$.

As mentioned previously, the effect of axial deformation of the members of frames is not taken into account. Thus, the vertical movement of joints 2, 3, 4, and 5 of the structure of Fig. 8.20a is disregarded. Likewise, the relative horizontal movements of joints 2 and 5 and of joints 3 and 4 of this structure are disregarded. Consequently, as shown in Fig. 8.20b, the structure is restrained against translation by preventing the horizontal movement of joints 3 or 4 and 2 or 5.

Consider a structure whose joints have t degrees of freedom in translation and r degrees of freedom in rotation, and assume that it is subjected to forces equal and opposite to the holding forces H_i ($i = 1, 2, \ldots, t$) which have been computed by analyzing the structure restrained against translation. As discussed in Sec. 1.9, the relations between the actions which are applied to the joints of a structure and the resulting displacements of its joints are linear. Thus, denoting by Δ_i^t ($i = 1, 2, \ldots, t$) the component of translation of a joint of a structure corresponding to the holding forces H_i, we have

$$\left\{ \begin{array}{c} H_1 \\ H_2 \\ \cdots \\ H_t \end{array} \right\} = \left[\begin{array}{cccc} K_{11}^t & K_{12}^t & \cdots & K_{1t}^t \\ K_{21}^t & K_{22}^t & \cdots & K_{2t}^t \\ \cdots & \cdots & \cdots & \cdots \\ K_{t1}^t & K_{t2}^t & \cdots & K_{tt}^t \end{array} \right] \left\{ \begin{array}{c} \Delta_1^t \\ \Delta_2^t \\ \cdots \\ \Delta_t^t \end{array} \right\} \tag{8.35}$$

or
$$\{H\} = [K^t]\{\Delta^t\} \tag{8.36}$$

The matrix $[K^t]$ is referred to as *the translational stiffness matrix of the structure*. Its elements are referred to as the *translational stiffness coefficients*. Notice that the component of translation Δ_i^t is considered positive when it is in the direction of the force H_i.

Suppose that the structure under consideration is subjected to a deformation pattern involving a unit value of the component of translation Δ_j^t, while the other components of translation of its joints vanish. Referring to relation (8.35), we get

$$\left\{ \begin{array}{c} H_1 \\ H_2 \\ \cdots \\ H_i \\ \cdots \\ H_t \end{array} \right\} = \left\{ \begin{array}{c} K_{1j}^t \\ K_{2j}^t \\ \cdots \\ K_{ij}^t \\ \cdots \\ K_{tj}^t \end{array} \right\} \tag{8.37}$$

Consequently, the stiffness coefficient K_{ij}^t represents the force H_i which must act on the structure when its deformation pattern involves a unit value of the component of translation Δ_j^t, while the other components of translation of the joints of the structure vanish (see Fig. 8.21).

On the basis of the foregoing, referring to Figs. 8.20 and 8.21, we may subdivide the analysis of beams or frames using the displacement method with moment distribution into the following three parts.

Part 1. The internal actions $\{A^{TR}\}$ are established in the members of the structure subjected to the given loading when its joints are restrained against translation. Moreover, the forces which must be applied to the joints of the structure in order to restrain them from translating (holding forces) are computed. This is accomplished by employing the moment distribution method described in Sec. 8.8.

Part 2. The internal actions $\{A^{\Delta^t}\}$ are established in the members of the structure subjected to forces equal and opposite to the holding forces. This is accomplished by adhering to the following steps.

Step 1. The translational stiffness matrix of the structure is computed. The structure is considered subjected to a deformation pattern involving a unit value of the component of translation $\Delta_j^t = 1$ of one of its joints, while all other components of displacements (translations and rotations) of the joints of the structure vanish. The members of this structure are either fixed at both ends or fixed at one end and pinned at the other end. Moreover the one end of some members of the structure has translated transversely relative to their other end by a unit translation. The structure is held in that position by restraining forces and moments applied to its joints. The end actions in the members of the structure can be established using one of the methods described in Chap. 7. However, the end actions in members having a constant cross section may also be obtained by referring to App. C. The restraining forces and moments can be computed by considering the equilibrium of the joints of the structure.

The restraining moments acting on the joints of the structure, deformed as described above, are distributed. The resulting moments are the end moments of the members of the structure when subjected to a deformation pattern

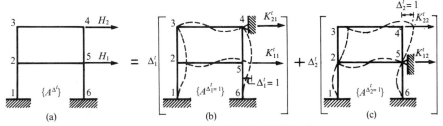

Figure 8.21 Analysis of the structure subjected to the holding forces.

involving a unit value of the component of translation $\Delta_j^t = 1$, while all other components of translation vanish (see Fig. 8.21b and c). The end actions of the members of the structure are established by considering their equilibrium and are stored in a matrix denoted by $\{A^{\Delta_j=1}\}$. From these actions, the external forces K_{ij}^t ($i = 1, 2, \ldots, t$) acting on the joints of the structure are established by considering their equilibrium. The forces K_{ij}^t ($i = 1, 2, \ldots, t$) are translational stiffness coefficients of the structure.

The process is repeated for each translational degree of freedom of the structure ($j = 1, 2, \ldots, t$) and the matrices $[K^t]$ and $[A^{\Delta^t=1}]$ are formed. The latter is defined as

$$[A^{\Delta^t=1}] = [\{A^{\Delta_1^t=1}\}, \{A^{\Delta_2^t=1}\}, \ldots, \{A^{\Delta_j^t=1}\}]$$ (8.38)

Step 2. The components of translation of the joints of the structure are computed when subjected to forces equal and opposite to the holding forces. Referring to relation (8.36), we obtain

$$\{\Delta^t\} = [K^t]^{-1}\{H\}$$ (8.39)

Moreover, the internal actions in the members of the structure are computed when subjected to forces equal and opposite to the holding forces. Referring to Fig. 8.21, we have

$$\{A^{\Delta^t}\} = [A^{\Delta^t=1}]\{\Delta^t\}$$ (8.40)

Part 3. The results of parts 1 and 2 are superimposed to obtain the internal actions in the members of the actual structure. Thus, referring to Fig. 8.20, we obtain

$$\{A\} = \{A^{TR}\} + \{A^{\Delta^t}\} = \{A^{TR}\} + [A^{\Delta^t=1}]\{\Delta^t\}$$ (8.41)

The number of simultaneous equations which must be solved in analyzing a structure by the displacement method with moment distribution is equal to the number of unknown translations, while the number of simultaneous equations which must be solved in analyzing a structure by the slope deflection method is equal to the total number of unknown displacements (translations and rotations). Thus, it is apparent that the number of simultaneous equations which must be solved in analyzing a structure by the displacement method with moment distribution is less than the number of equations which must be solved in analyzing the same structure by the slope displacement method.

The displacement method with moment distribution is applied to the following three examples.

Example 1. The end moments of the members of a statically indeterminate frame are established when it is subjected only to external actions. The joints of the frame have one translational and two rotational degrees of freedom.

Example 2. The end moments in the members of a statically indeterminate beam are established when it is subjected only to external actions. One support of the beam is elastic. That is, its displacement is proportional to its reaction. The beam has one translational and one rotational degree of freedom. The settlement of the elastic support is the translational degree of freedom.

Example 3. The end actions in the members of a statically indeterminate frame are computed. The joints of the frame have two degrees of freedom: one translational and one rotational. The frame is subjected to external actions: different temperatures at its exterior and interior fibers and a settlement of one of its supports.

Example 1. Using the displacement method with moment distribution, compute the internal actions at the ends of the members of the structure of Fig. a. The members of the structure are made from the same material and have the same constant cross section. Plot the shear and moment diagrams.

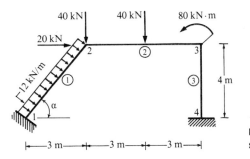

Figure a Geometry and loading of the structure.

solution We obtain the internal actions in the members of this structure by superimposing the corresponding internal actions in the members of the structure restrained against translation (see Fig. b*b*) and the structure subjected to a force equal and opposite to the holding force (see Fig. b*c*).

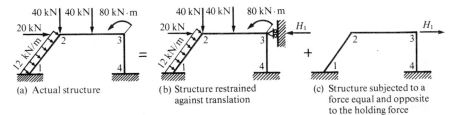

(a) Actual structure

(b) Structure restrained against translation

(c) Structure subjected to a force equal and opposite to the holding force

Figure b Superposition of the structure restrained against translation and the structure subjected to a force equal and opposite to the holding force.

Part 1

The internal actions at the ends of the members of the structure loaded as shown in Fig. b*b* and the holding force H_1 have been established in Example 1 of Sec. 8.8. Referring to Fig. g and to relations (d) and (e) of that example, we have

$$[A^{TR}]^T = [111.5, 25.4, 17.2, -111.5, 34.6, -40.4, -114.6, 28.4, 40.4,$$

$$-114.6, 11.6, 10.1, 11.6, 26.2, 69.9, -11.6, -26.2, 34.9] \quad (a)$$

and

$$H_1 = 88.41 \text{ kN} \quad (b)$$

Part 2

In this part, we establish the internal actions in the members of the structure subjected to a force equal and opposite to the holding force H_1. To accomplish this, we adhere to the following steps.

STEP 1.   We establish the translational stiffness coefficient K_{11}^t of the structure. This is equal to the horizontal force which must be applied to joint 3 of the structure in order to translate it horizontally by an amount equal to unity ($\Delta_1 = 1$) (see Fig. c).

We start by considering the structure subjected to the required restraining actions at its joints in order to produce a deformation pattern specified by a unit value of the component of translation Δ_1, while all rotations of its joints vanish ($\Delta_2 = \Delta_3 = 0$). Referring to Fig. d and App. C, the resulting fixed-end moments FEM of the members of the structure are

$$\text{FEM}^{1j} = \text{FEM}^{1k} = \frac{6EI}{L_1^2} u_2^{1k} = \frac{6EI}{25}\left(\frac{5}{4}\right) = 0.3EI$$

$$\text{FEM}^{2j} = \text{FEM}^{2k} = -\frac{6EI}{L_2^2} u_2^{2j} = -\frac{6EI}{6^2}\left(\frac{3}{4}\right) = -0.125EI \quad (c)$$

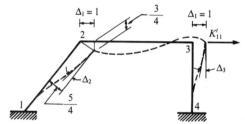

Figure c Translational stiffness coefficient K_{11}^t.

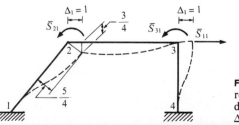

Figure d Structure subjected to restraining actions inducing a deformation pattern specified by $\Delta_1 = 1, \Delta_2 = \Delta_3 = 0$.

$$\text{FEM}^{3j} = \text{FEM}^{3k} = \frac{6EI}{L_3^2} u_2^{3j} = \frac{6EI}{4^2} = 0.375EI$$

The unbalanced moments at the joints of the structure are distributed as shown in Fig. e. The resulting moments are the end moments in the members of the structure when its joint 3 translates by a unit due to the application of the force K_{11}^t (see Fig. c). These moments are used in drawing the free-body diagrams of the members and the joints of the structure shown in Fig. f. Referring to these diagrams, the end actions $\{A^{\Delta t=1}\}$ in the members of the structure as well as the force K_{11}^t are computed. Thus,

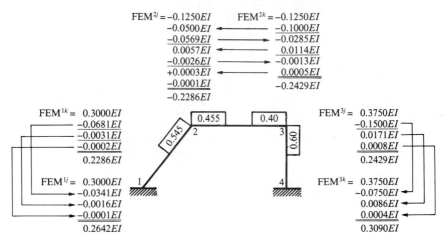

Figure e Moment distribution.

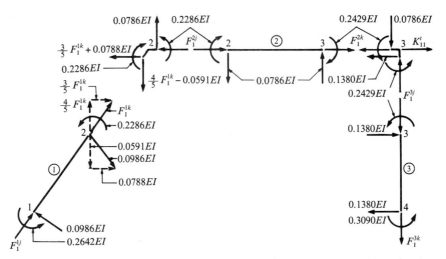

Figure f Free-body diagrams of the members and joints of the structure subjected to $\Delta_1 = 1$.

from the equilibrium of joint 2, we have

$$\Sigma \bar{F}_v = 0 = \tfrac{4}{3}F_1^{1k} - 0.0591EI - 0.0786EI$$

$$\Sigma \bar{F}_h = 0 = F_1^{2j} + \tfrac{3}{5}F_1^{1k} + 0.0788EI$$

or
$$F_1^{1k} = (0.1377EI)(\tfrac{5}{4}) = 0.1721EI = -F_1^{1j}$$
$$F_1^{2j} = -0.1821EI = -F_1^{2k}$$

(d)

Moreover, from the equilibrium of joint 3, we get

$$K_{11}^t = 0.1380EI + 0.1821EI = 0.3201EI \tag{e}$$

Referring to Fig. f and using the results (d), the end actions $\{A^{\Delta^t=1}\}$ in the members of the structure are

$$\{A^{\Delta^t=1}\}^T = EI[-0.1721, 0.0986, 0.2642, 0.1721, -0.0986, 0.2286, -0.1821,$$

$$-0.0786, -0.2286, 0.1821, 0.0786, -0.2429, 0.0786, 0.1380,$$

$$0.2429, -0.0786, -0.1380, 0.3090] \tag{f}$$

STEP 2.   Using relation (8.39), we compute the component of translation Δ_1 of joint 3 of the frame subjected to a force equal and opposite to the holding force $H_1 = 88.41$ kN. Thus

$$\Delta_1 = \frac{H_1}{K_{11}^t} = -\frac{88.41}{0.3201EI} = -\frac{276.2}{EI} \tag{g}$$

Part 3

The internal actions in the members of the actual structure are obtained by substituting relations (a), (f), and (g) into (8.41). Thus,

$$
\{A\} = \left\{
\begin{array}{r}
111.5 \\
25.4 \\
17.2 \\
-111.5 \\
34.6 \\
-40.6 \\
\text{-----} \\
114.6 \\
28.4 \\
40.4 \\
-114.6 \\
11.6 \\
10.1 \\
\text{-----} \\
11.6 \\
26.7 \\
69.9 \\
-115.8 \\
26.2 \\
34.9
\end{array}
\right\}
+
\left\{
\begin{array}{r}
-0.1721 \\
0.0986 \\
0.2642 \\
0.1721 \\
-0.0986 \\
0.2286 \\
\text{-------} \\
-0.1821 \\
-0.0786 \\
-0.2286 \\
0.1821 \\
0.0786 \\
-0.2429 \\
\text{-------} \\
0.0786 \\
0.1380 \\
0.2429 \\
-0.0786 \\
-0.1380 \\
0.3090
\end{array}
\right\}
(276.2) =
\left\{
\begin{array}{r}
64.0 \\
52.6 \\
90.3 \\
-64.0 \\
7.4 \\
22.7 \\
\text{-----} \\
64.3 \\
6.7 \\
-22.7 \\
-64.3 \\
33.3 \\
-56.9 \\
\text{-----} \\
33.3 \\
64.3 \\
136.9 \\
-33.3 \\
-64.3 \\
120.3
\end{array}
\right\}
$$

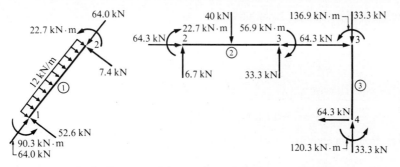

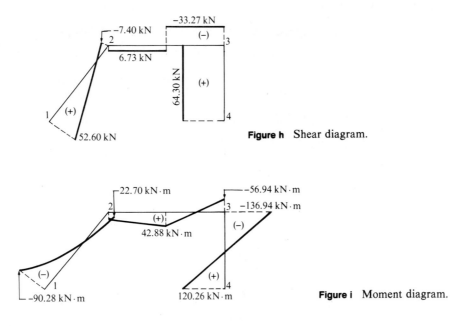

Figure g Free-body diagrams of the members of the frame subjected to the given loading.

Figure h Shear diagram.

Figure i Moment diagram.

Using the above results, the free-body diagrams of the members of the frame are shown in Fig. g. Referring to this figure, the shear and moment diagrams for the frame of Fig. a are plotted in Figs. h and i, using the sign convention specified in Fig. 7.1.

Example 2. Plot the shear and moment diagrams for the continuous steel ($E = $ 210,000 MPa) beam loaded as shown in Fig. a. The members of the beam have the same constant cross section with moment of inertia $I = 74.8 \times 10^6$ mm^4. The support of the beam at point 2 is elastic with constant $k/EI = 0.01$ per m^3.

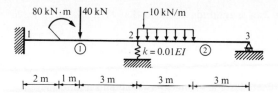

Figure a Geometry and loading of the beam.

---

solution Member 2 is considered type M3. Hence, the hinge at point 3 is considered a part of member 2. Thus as shown in Fig. b, the joints of the beam under consideration have one degree of freedom of rotational motion and one degree of freedom of translational motion.

It is apparent that the internal actions in the members of the beam are equal to the sum of the corresponding internal actions in the members of the two auxiliary beams shown in Fig. c. The one auxiliary beam is made from the beam of Fig. a with joint 2 restrained against translation, and it is subjected to the given external actions and to the force exerted by the spring support (see Fig. cb). The other auxiliary beam is made from the beam of Fig. a with the spring support removed, and it is subjected to a

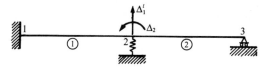

Figure b Degree of kinematic indeterminacy.

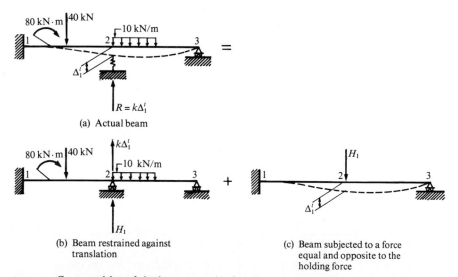

(a) Actual beam

(b) Beam restrained against translation

(c) Beam subjected to a force equal and opposite to the holding force

Figure c Superposition of the beam restrained against translation and the beam subjected to a force equal and opposite to the holding force.

downward vertical force at point 2 of magnitude equal to that of the holding force (see Fig. c*c*).

Part 1

In this part, we use the moment distribution method to analyze the structure restrained against translation, subjected to the given external actions and to the force exerted by the spring support (see Fig. c*b*).

STEP 1. This structure consists of two members. Member 1 is type M2 while member 2 is considered type M3. Inasmuch as the members of the beam have a constant cross section, referring to relations (8.3) and (8.6), their rotational stiffnesses are equal to

$$k^{1j} = k^{1k} = \frac{4EI^{(1)}}{L_1} = \frac{2EI}{3} \qquad \hat{k}^{2j} = \frac{3EI^{(2)}}{L_2} = \frac{EI}{2}$$

Thus referring to relations (8.27) and (8.28), we find that the distributon factors at joint 2 are

$$DF^{(1)} = \frac{2EI/3}{2EI/3 + EI/2} = 0.571 \qquad \widehat{DF^{(2)}} = \frac{EI/2}{2EI/3 + EI/2} = 0.429$$

The distribution factors are tabulated as shown in Fig. d.

STEP 2. The restrained structure is made from the structure of Fig. c*b* by restraining joint 2 from rotating. The fixed-end moments of the members of the restrained structure subjected to the given loads are obtained by referring to App. C. They are

$$FEM^{1j} = 30.0 \text{ kN} \cdot \text{m}$$

$$FEM^{1k} = -56.67 \text{ kN} \cdot \text{m} \qquad \text{(a)}$$

$$FEM^{2j} = 25.31 \text{ kN} \cdot \text{m}$$

STEP 3. We perform the moment distribution in Fig. d. Using the results of Fig. d, we draw in Fig. e the free-body diagrams of the members and joint 2 of the beam

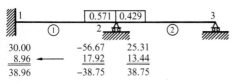

Figure d Moment distribution.

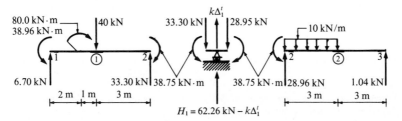

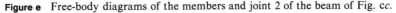

Figure e Free-body diagrams of the members and joint 2 of the beam of Fig. c*c*.

restrained against translation. Referring to these diagrams, we find that the end actions in the members of the beam and the required vertical holding force H_1 are

$$\{A^{TR}\}^T = [6.70, 38.96, 33.30, -38.75, 28.96, 38.75, 1.04, 0] \tag{b}$$

$$H_1 = 62.26 \text{ kN} - k\Delta_1^t = 62.26 \text{ kN} - 0.01EI\Delta_1^t \tag{c}$$

Part 2

In this part, we use the moment distribution method to establish the internal actions in the members of the beam subjected to a downward vertical force H_1 (see Fig. cc). STEP 1.   We compute the translational stiffness matrix of the beam. That is, referring to Fig. f, we compute the force K_{11}^t required to move joint 2 of the beam by a unit vertical translation ($\Delta_1^t = 1$). Referring to App. C, we establish the end moments in the members of the beam when it is subjected to a translation $\Delta_1^t = 1$, while its joint 2 is fixed against rotation. They are

$$\text{FEM}^{1j} = \text{FEM}^{1k} = 0.1667EI$$

$$\widehat{\text{FEM}}^{2j} = -0.0833EI$$

We compute the end moments in the members of the beam of Fig. f using moment distribution (see Fig. g). From the end moments in the members of the beam, their free-body diagrams as well as that of joint 2, are drawn in Fig. h. Referring to these

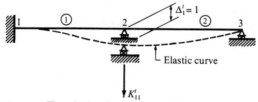

Figure f Translational stiffness coefficient K_{11}^t of the beam.

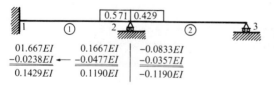

Figure g Moment distribution.

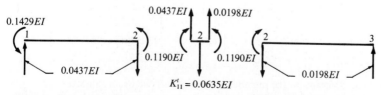

Figure h Free-body diagrams of the members and joint 2 of the beam subjected to $\Delta_1^t = 1$.

diagrams, the end actions in the members of the beam and the stiffness coefficient K_{11}^t are

$$\{A^{\Delta^t=1}\}^T = EI[0.0437, 0.1429, -0.0437, 0.1190, -0.0198, \tag{d}$$

$$-0.1190, 0.0198, 0]$$

$$K_{11}^t = 0.0635EI \tag{e}$$

STEP 2.   Using relation (8.39), we compute the translation of joint 2 of the beam. Thus,

$$H_1 = K_{11}^t \Delta_1^t \tag{f}$$

Substituting relations (e) and (c) into (f), we obtain

$$62.26 = (0.01\,EI + 0.0635EI)\Delta_1^t$$

or

$$\Delta_1^t = \frac{847.13}{EI} \tag{g}$$

Part 3

In this part, we superimpose the results of parts 1 and 2 to obtain the internal actions acting at the ends of the members of the beam. Substituting results (b), (e), and (g) into relation (8.41) we get

$$\{A\} = \begin{Bmatrix} 6.7 \\ 38.96 \\ 33.30 \\ -38.75 \\ ----- \\ 28.96 \\ 38.75 \\ 1.04 \\ 0 \end{Bmatrix} + \begin{Bmatrix} 0.0437 \\ 0.1429 \\ -0.0437 \\ 0.1190 \\ ------ \\ -0.0198 \\ -0.1190 \\ 0.0198 \\ 0 \end{Bmatrix} \left(\frac{847.13}{EI} \right) = \begin{Bmatrix} 43.68 \\ 159.98 \\ -3.68 \\ 62.10 \\ ----- \\ 12.15 \\ -62.10 \\ 17.85 \\ 0 \end{Bmatrix} \tag{h}$$

The free-body diagrams of the members of the beam are shown in Fig. i. The moment and shear diagrams are plotted in Figs. j and k, respectively, using the sign convention indicated in Fig. 7.1.

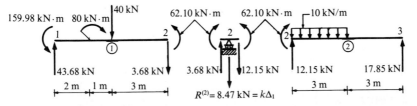

Figure i   Free-body diagrams of the members and joint 2 of the beam subjected to the given loading.

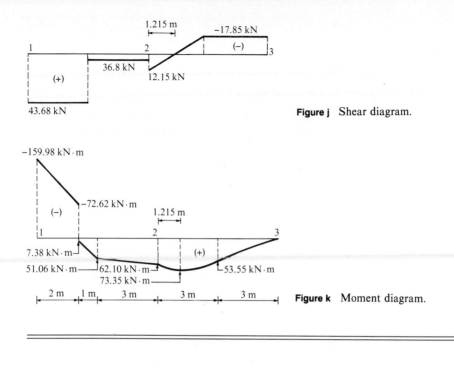

Figure j Shear diagram.

Figure k Moment diagram.

Example 3: The planar frame shown in Fig. a is subjected to the following loads.

1. The external actions shown in Fig. a
2. A settlement of 10 mm of support 1
3. A difference in temperature. The outside and inside temperatures are, respectively, $T_e = 25°C$ and $T_i = -5°C$.

The members of the frame have the same constant cross section ($I = 240 \times 10^6$ mm^4, $h = 420$ mm] and are made of the same material ($E = 210,000$ MPa). Compute the end actions in the members of this frame. Disregard the effect of axial deformation of its members.

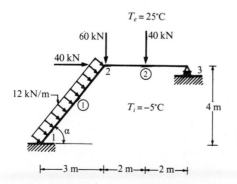

Figure a Geometry and loading of the frame.

solution

Part 1

We compute the internal actions in the members of the frame subjected to the given loading with its joints restrained against translation. Moreover, we compute the holding force.

STEP 1.   Inasmuch as the members of the frame have a constant cross section, referring to relations (8.3) and (8.5), we see that their rotational stiffnesses are equal to

$$k^{1j} = k^{1k} = \frac{4EI^{(1)}}{L_1} = \frac{4EI}{5} \qquad k^{2j} = k^{2k} = \frac{3EI^{(2)}}{L_2} = \frac{3EI}{4} \qquad \text{(a)}$$

Thus, referring to relations (8.27) and (8.29), the distribution factors at joint 2 are

$$\text{DF}^{(1)} = \frac{4EI/5}{4EI/5 + 3\,EI/4} = 0.516 \qquad \widehat{\text{DF}^{(2)}} = \frac{3EI/4}{4EI/5 + 3\,EI/4} = 0.484 \quad \text{(b)}$$

STEP 2.   The end moments [fixed-end moments (FEM)] of the members of the restrained structure subjected to the given loads are obtained from App. C. Thus,

Due to the external actions

$$\text{FEM}^{1j} = 25.0 \text{ kN} \cdot \text{m}$$

$$\text{FEM}^{1k} = -25.0 \text{ kN} \cdot \text{m}$$

$$\widehat{\text{FEM}^{2j}} = 30.0 \text{ kN} \cdot \text{m}$$

Due to the temperature difference

$$\delta T_2 = 25 + 5 = 30°\text{C}$$

$$\text{FEM}^{1j} = -\frac{\alpha EI\,\delta T_2}{h} = -\frac{10^{-5}(210)(240)(30)}{420} = -36 \text{ kN} \cdot \text{m}$$

$$\text{FEM}^{1k} = 36 \text{ kN} \cdot \text{m}$$

$$\widehat{\text{FEM}^{2j}} = -\frac{3\alpha EI\,\delta T_2}{2h} = -54 \text{ kN} \cdot \text{m}$$

Due to the settlement of support 1.   Referring to Fig. b and App. C, we have

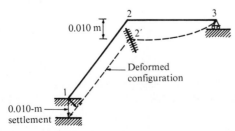

Figure b Deformed configuration of the restrained structure due to the settlement of support 1.

$$\widehat{\text{FEM}}^{2j} = -\frac{3EI\Delta^s}{L^2} = -\frac{3(210)(240)(0.01)}{4^2} = -94.5 \text{ kN·m} \qquad \text{(c)}$$

Total end moments

$$\text{FEM}^{ij} = 25 - 36 = -11 \text{ kN·m}$$

$$\text{FEM}^{1k} = -25 + 36 = 11 \text{ kN·m}$$

$$\widehat{\text{FEM}}^{2j} = 30 - 54 - 94.5 = -118.5 \text{ kN·m}$$

STEP 3. We perform the moment distribution in Fig. c.

STEP 4. From the end moments of the members of the structure we compute their end actions. The free-body diagrams of the members and joints of the structure are shown in Fig. d. Referring to these diagrams, from the equilibrium of joint 2, we get

$$\Sigma \bar{F}_h = 0 = \tfrac{3}{5}F_1^{1k} - 10.68 + F_1^{2j} - 40$$

$$\Sigma \bar{F}_v = 0 = \tfrac{4}{5}F_1^{1k} + 8.02 + 3.39 + 60 \qquad \text{(d)}$$

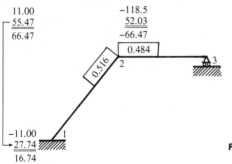

Figure c Moment distribution.

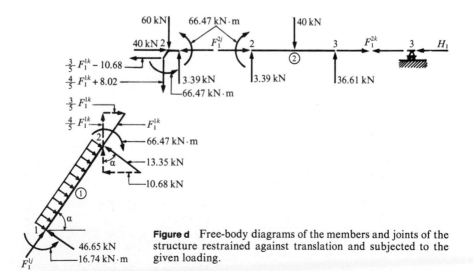

Figure d Free-body diagrams of the members and joints of the structure restrained against translation and subjected to the given loading.

or

$$F_1^{1k} = (-7.41)\tfrac{5}{4} = -89.26 \text{ kN} = -F_1^{1j}$$
$$F_1^{2j} = 104.23 \text{ kN} = -F_1^{2k}$$

Moreover, from the equilibrium of joint 3, we obtain

$$\Sigma \overline{F}_1 = 0 \qquad H_1 = 104.23 \text{ kN} \tag{e}$$

Referring to Fig. d and using results (d), we see that the end actions $\{A^{TR}\}$ in the members of the structure restrained against translation are

$$\{A^{TR}\}^T = [89.26, 46.65, 16.74, -89.26, 13.35, 66.47,$$

$$104.23, 3.39, -66.47, -104.43, 36.61, 0] \tag{f}$$

Part 2

In this part, we establish the internal actions in the members of the structure subjected to a force equal and opposite to the holding force H_1. To accomplish this, we adhere to the following steps.

STEP 1.   We establish the translational stiffness coefficient K_{11}^t of the structure. This is equal to the horizontal force which must be applied to joint 3 of the structure in order to translate it horizontally by an amount equal to unity ($\Delta_1 = 1$) (see Fig. e).

We start by considering the structure subjected to the required restraining actions at its joints in order to produce a deformation pattern specified by a unit value of the component of translation Δ_1, while the rotation of joint 2 vanishes ($\Delta_2 = 0$). Referring to App. C, the resulting fixed-end moments (FEM) of the members of the structure are

$$\text{FEM}^{1j} = \text{FEM}^{1k} = \frac{6EI}{L_1^2} u_2^{1k} = \frac{6EI}{25}\left(\frac{5}{4}\right) = 0.3EI \tag{g}$$

$$\widehat{\text{FEM}}^{2j} = -\frac{3EI}{L_2^2} u_2^{2k} = -\frac{3EI}{4^2}\left(\frac{3}{4}\right) = -0.140625EI$$

The unbalanced moments of the joints of the structure are distributed as shown in Fig. f. The resulting moments are the end moments in the members of the structure when its joint 3 translates by a unit due to the application of the force K_{11}^t. These moments are used in drawing the free-body diagrams of the members and the joints of the structure shown in Fig. g. Referring to these diagrams, we compute end actions $\{A^{\Delta t=1}\}$ in the members of the structure as well as the force K_{11}^t. Thus, from the equilibrium of joint 2, we have

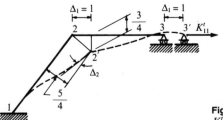

Figure e Translational stiffness coefficient K_{11}^t of the frame.

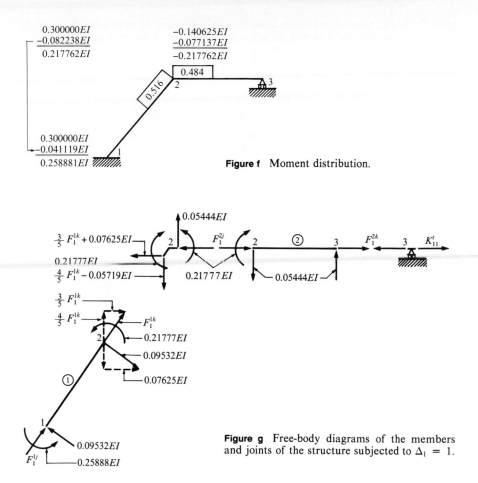

Figure f Moment distribution.

Figure g Free-body diagrams of the members and joints of the structure subjected to $\Delta_1 = 1$.

$$\Sigma \bar{F}_v = 0 = \tfrac{4}{5}F_1^{1k} - 0.5719EI - 0.05444EI$$

$$\Sigma \bar{F}_h = 0 = F_1^{2j} + \tfrac{3}{5}F_1^{1k} + 0.07625EI \tag{h}$$

or

$$F_1^{1k} = 0.13954EI = -F_1^{1j}$$

$$F_1^{2j} = -0.15998EI = -F_1^{2k} \tag{i}$$

Moreover, from the equilibrium of joint 3 we get

$$K_{11}^t = 0.15998 \, EI \tag{j}$$

Referring to Fig. g, and using the results (i), the end actions $\{A^{\Delta^t=1}\}$ in the members of the structure are

$$\{A^{\Delta^t=1}\}^T = EI[-0.139.54, \, 0.09532, \, 0.25888, \, 0.13954, \, -0.09532,$$

$$0.21777, \, -0.15998, \, -0.05444, \, -0.21776, \, 0.15998, \, 0.05444, \, 0] \tag{k}$$

STEP 2. Using relation (8.39), we compute the component of translation Δ_1 of joint 3 of the frame subjected to a force equal and opposite to the holding force $H_1 = 104.23$ kN. Thus

$$\Delta_1 = \frac{H_1}{K_{11}^t} = -\frac{104.23}{0.15998EI} = \frac{651.5}{EI} \tag{1}$$

Part 3

The internal actions in the members of the actual structure are obtained by substituting relations (f), (k), and (l) into (8.41). Thus,

$$
\left\{
\begin{array}{c}
89.26 \\
46.65 \\
16.74 \\
-89.26 \\
13.35 \\
66.47 \\
\hline
104.23 \\
3.39 \\
-66.47 \\
-104.23 \\
36.61 \\
0
\end{array}
\right\}
+
\left\{
\begin{array}{c}
-0.13954 \\
-0.09532 \\
0.25888 \\
0.13954 \\
-0.09532 \\
0.21777 \\
\hline
-0.15998 \\
-0.05444 \\
-0.21776 \\
0.15998 \\
0.05444 \\
0
\end{array}
\right\}
(651.5) =
\left\{
\begin{array}{c}
-1.65 \\
108.75 \\
185.40 \\
1.65 \\
-48.75 \\
208.34 \\
\hline
0 \\
-32.07 \\
-208.34 \\
0 \\
72.07 \\
0
\end{array}
\right\}
$$

These results are identical to those obtained in Example 4 of Sec. 8.6.

8.11 Problems

1. Examine the structures shown in Fig. 8.P1 and state the number of unknowns involved in analyzing each one of them using (a) the basic force method and (b) the slope deflection method. Explain.

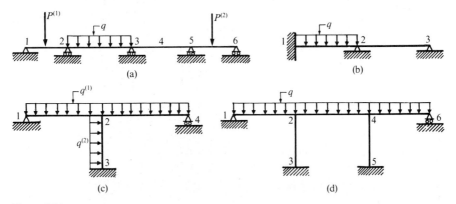

Figure 8.P1

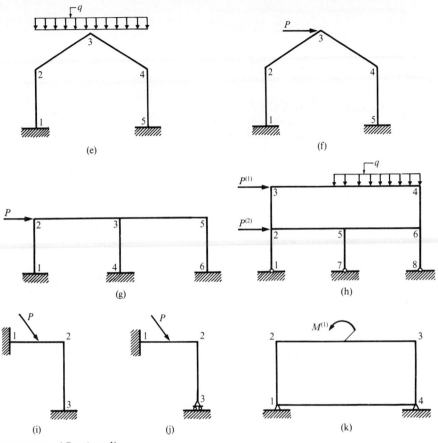

(e)

(f)

(g)

(h)

(i)

(j)

(k)

Figure 8.P1 (*Continued*)

2. Determine the stiffnesses and carry-over factors of the type M2 member shown in the Fig. 8.P2.

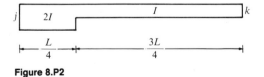

Figure 8.P2

3 and 4. Determine the stiffnesses and carry-over factors of the type M2 member shown in Fig. 8.P3. The area of the cross section of each flange of the member is equal to $A_f = 4 \times 10^3$ mm². Neglect the effect of the web on the moment of inertia of the cross sections of the member. Moreover, assume that the thickness of the flange is very small compared to the depth of the cross section. Repeat for Fig. 8.P4.

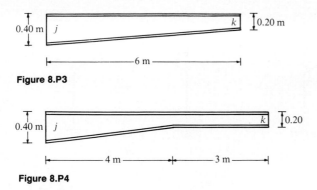

Figure 8.P3

Figure 8.P4

5 and 6. Determine the rotational stiffnesses and the carry-over factors of the type M2 member shown in Fig. 8.P5. The cross sections of the member have a constant width of 0.80 m. Repeat for Fig. 8.P6.

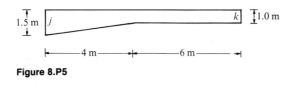

Figure 8.P5

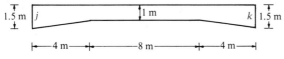

Figure 8.P6

7. Determine the rotational stiffness and carry-over factor of the circular type M2 member shown in Fig. 8.P7. Disregard the effect of axial deformation.

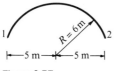

Figure 8.P7

8 to 11. Using the slope deflection method, compute the reactions of the beam subjected to the external actions shown in Fig. 8.P8. The members of the beam are made of the same material and have the same constant cross section. Plot the shear and moment diagrams for the beam. Repeat for Figs. 8.P9 to 8.P11.

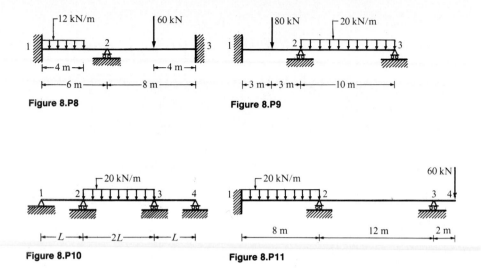

Figure 8.P8

Figure 8.P9

Figure 8.P10

Figure 8.P11

12 and 13. Using the slope deflection method, compute the reactions of the beam subjected to the external actions shown in Fig. 8.P9 and to a settlement of 10 mm of support 2. The members of the beam are made of the same material (E = 210,000 MPa) and have the same constant cross section (I = 576.80 × 10^6 mm⁴). Repeat for Fig. 8.P11.

14 to 18. Using the slope deflection method, compute the reactions of the frame subjected to the external actions shown in Fig. 8.P14. The members of the frame are made of the same material and have the same constant cross section. Plot the shear and moment diagrams. Repeat for Figs. 8.15 to 8.P18.

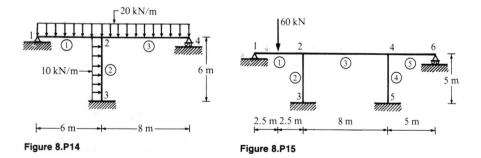

Figure 8.P14

Figure 8.P15

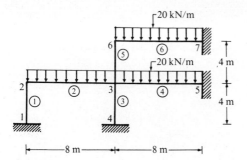

Figure 8.P16

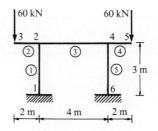

Figure 8.P17

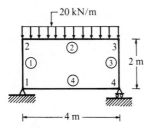

Figure 8.P18

19 and 20.   Using the slope deflection method, compute the reactions of the beam subjected to the external actions shown in Figs. 8.P19. The members of the beam are made of the same material and have the same constant cross section. Repeat for Fig. 8.P20.

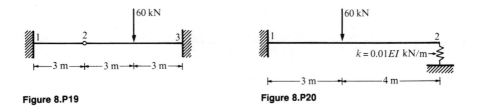

Figure 8.P19

Figure 8.P20

21 to 26.   Using the slope deflection method, analyze the frame shown in Fig. 8.P21 for the following loading and plot its shear and moment diagrams.
   (*a*) The external actions shown in the figure.
   (*b*) Settlement of support 1 by 10 mm.
   (*c*) Temperature of the external fibers of the members of the frame equal to T_e = 40°C and of the internal fibers equal to T_i = 20°C. The temperature during construction of the frame was 30°C.
     The members of the frame are made of the same material (E = 210,000 MPa = 10^{-5}/°C) and have the same constant cross section (I = 369.70 × 10^6 mm^4, A = 19.8 × 10^3 mm^2, h = 400 mm). Repeat for Figs. 8.P22 to 8.P26.

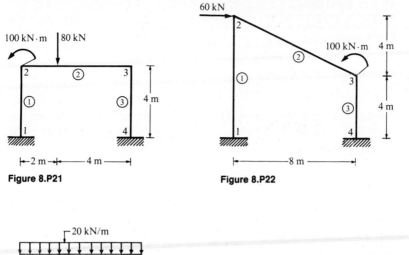

Figure 8.P21

Figure 8.P22

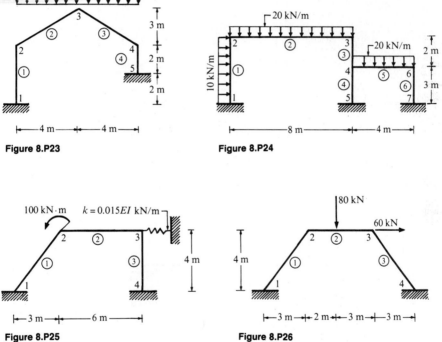

Figure 8.P23

Figure 8.P24

Figure 8.P25

Figure 8.P26

27. Using the slope deflection method, analyze the continuous girder subjected to the loading shown in Fig. 8.P27. The area of the cross section of each flange of the girder is constant and equal to $A_f = 8 \times 10^3$ mm². Neglect the effect of the web on the moment of inertia of the cross section of the girder. Moreover, assume that the thickness of the flange is very small compared to the depth of the cross section.

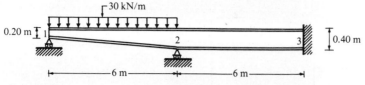

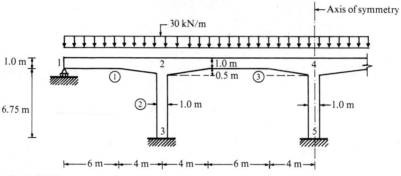

Figure 8.P27

28. Using the slope deflection method, analyze the frame subjected to the uniform loading shown in Fig. 8.P28. The beams and columns have the same constant width and are made of the same material.

Figure 8.P28

29. Using the slope deflection method, analyze the frame subjected to the forces shown in Fig. 8.P29 and to a temperature difference between the top and bottom fibers of the beams equal to $\delta T = T_{bot} - T_{top} = 15°C$. The cross section of the columns is a square of dimensions 0.50 m × 0.50 m. The cross section of the beams is a rectangle of 0.50 m width and 1.0 m depth. All the members of frame are made of the same material ($E = 24,000$ MPa, $\alpha = 10^{-5}/°C$).

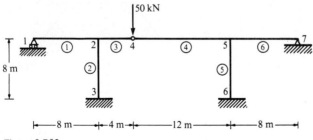

Figure 8.P29

30 to 33. Using moment distribution, analyze the continuous beam subjected to the loading shown in the Fig. 8.P8. The beam has a constant cross section. Repeat for Figs. 8.P9 to 8.P11.

34 to 38. Using moment distribution, compute the reactions of the frame subjected to the external actions shown in Fig. 8.P14. Plot the shear and moment diagrams. The members of the frame are made of the same material and have the same constant cross section. Repeat for Figs. 8.P15 to 8.P18.

39. The earth on the sides of the rectangular ditch shown in Fig. 8.P39 is supported by horizontal planks, which in turn are supported by vertical beams located 1.5 m apart. The vertical beams are supported by horizontal beams as shown in the figure. The cross section of the beams is 0.12×0.12 m. Assume that the vertical beams are fixed at their lower support and that the earth pressure is equivalent to that of a fluid having specific weight 8 kN/m³. Using moment distribution, analyze the beams and plot their shear and moment diagrams making one of the following assumptions.

(*a*) The horizontal beams are pinned to the vertical beams.

(*b*) The horizontal beams are rigidly connected to the vertical beams.

Disregard the effect of axial deformation of the members of the structure.

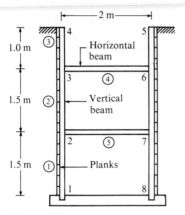

Figure 8.P39

40 to 43. Using moment distribution, analyze the frame loaded as shown in Fig. 8.P40. The members of the frame are made of the same material, and if the moment of inertia of their cross sections is not given in the figure, they have the same constant cross section. Repeat for Figs. 8.P41 to 8.P43.

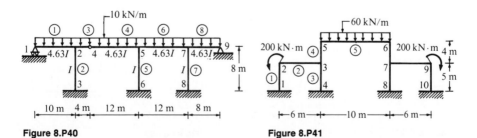

Figure 8.P40

Figure 8.P41

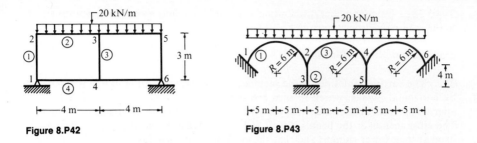

Figure 8.P42

Figure 8.P43

44 to 49. Using the displacement method with moment distribution, analyze the frame shown in Fig. 8.P21 when it is subjected to the following loading, and plot its shear and moment diagrams:
  (a) The external actions shown in the figure
  (b) A settlement of 10 mm of support 1
  (c) A temperature of the external fibers $T_e = 40°C$ and of the internal fibers of $T_i = 20°C$. The temperature during construction was $30°C$
The members of the frame are made of the same material ($E = 210,000$ MPa, $a = 10^{-5}/°C$) and have the same constant cross section ($I = 576.80 \times 10^6$ mm^2, $A = 19.8 \times 10^3$ mm^2, $h = 500$ mm). Repeat for Figs. 8.P22 to 8.P26.

50. Using the moment distribution method, analyze the structure subjected to the loading shown in Fig. 8.P27. The area of the cross section of each flange of the girder is constant and equal to $A_f = 8 \times 10^3$ mm^2. Neglect the effect of the web on the moment of inertia of the cross section of the girder. Moreover, assume that the thickness of the flange is very small compared to the depth of the cross section.

51. Using the moment distribution method, analyze the structure subjected to the loading shown in Fig. 8.P28. The beams and columns have the same constant width and are made of the same material.

52 to 57. Using the displacement method with moment distribution, analyze the frame loaded as shown in Fig. 8.P52. The members of the frame are made of the same material and, if their moment of inertia is not given in the figure, they have the same constant cross section. Repeat for Figs. 8.P53 to 8.P57.

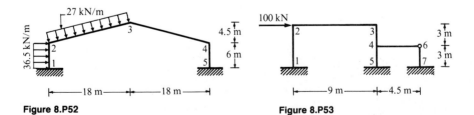

Figure 8.P52

Figure 8.P53

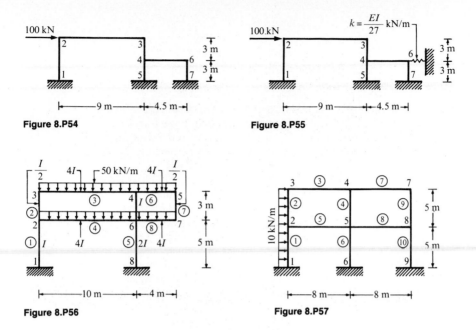

Figure 8.P54

Figure 8.P55

Figure 8.P56

Figure 8.P57

REFERENCES

1. *Handbook of Frame Constants,* Portland Cement Association, 5420 Old Orchard Road, Skokie, Illinois 60077.
2. H. Cross "Analysis of Continuous Frames by Distributing Fixed End Moments," *Transactions ASCE,* Paper No. 1793, **96,** 1932.

Photographs and Brief Description of a Water Pollution Control Plant

The North River Water Pollution Control Plant of New York City serves a resident population of 550,000 people and a large transient population of daily commuters. The plant is located on the west side of Manhattan Island. It is constructed on a reinforced concrete 2300- × 750-ft (710- × 228-m) platform which extends approximately 712 ft (217 m) into the Hudson River at an elevation of 5 ft (1.52 m) above mean high-water level. That is, almost 95 percent of the platform is over the Hudson River. The platform is supported by drilled-in caissons of either 3 ft (0.91 m) or 4 ft (1.22 m) in diameter with an average length of 200 ft (60.96 m). The platform is protected by icebreakers consisting of a cluster of piles anchored into the river bedrock. At the top of the piles, there is a rail which breaks the ice as it floats downstream. Two docking areas for mooring the sludge vessels have been provided as part of the platform. The plant consists of a main building, primary settling tanks, aeration tanks, secondary settling tanks, and sludge facilities, including a sludge storage tank and eight reinforced concrete, circular cylindrical digester tanks (see Fig. 8.22b). The plant will treat 200 million gallons per day of flow, and it will remove 90% or better of the B.O.D. The plant will be covered by a concrete slab roof supported on concrete columns. On this slab, provisions will be made for park and recreational facilities which will camouflage the plant (see artist's rendition, Fig. 8.22a).

(a)

(b)

Figure 8.22 The North River Water Pollution Control Plant of the City of New York. *(Courtesy of U.S. Army Corp of Engineers)*

Influence Lines for Statically Indeterminate Structures

9.1 Introduction

As in the case of statically determinate structures (see Chap. 4), when we analyze an indeterminate structure for live loads, it is convenient to prepare influence lines for some of its reactions and internal actions. We can then employ the influence line for a quantity (reaction or internal action) in order to establish the position of the live loads on the structure for which this quantity assumes its maximum value and to compute this maximum value.

In Chap. 4, it is shown that the influence lines for statically determinate structures consist of straight line segments; consequently, in plotting them it is necessary to establish only a few key ordinates. The influence lines for statically indeterminate structures are either curved lines or a series of broken lines. The latter occur when a statically indeterminate structure is loaded by floor beams at certain points along its length.

In order to establish the influence lines for reactions or internal actions of statically indeterminate structures, we first select the redundants and establish their influence lines. From the ordinates of the influence lines for the redundants, we can establish the corresponding ordinates of the influence line for any other reaction or internal action by considering the equilibrium of appropriate portions of the structure.

In this chapter, for the internal actions acting in the members of a structure, we use the sign convention described in Fig. 7.1.

In Sec. 9.2, we present the principle of Müller-Breslau. This principle is a direct application of Betti's reciprocal theorem (see Sec. 5.16) and is used in sketching and obtaining the coordinates of influence lines for reactions and/or internal actions of structures (statically determinate or indeterminate).

In Sec. 9.3 we apply the principle of Müller-Breslau in sketching the influence lines for beams and in computing their ordinates.

9.2 Principle of Müller-Breslau

The principle of Müller-Breslau states that the ordinates of the influence line for any reaction or internal action of a structure are obtained as follows.

1. An auxiliary structure is formed by eliminating the external constraint of the actual structure which causes the reaction whose influence line we want to plot or by introducing a mechanism in the structure which renders it unable to transfer the internal action whose influence line we want to plot. For instance, the auxiliary beam for computing the influence line for the reaction at support 2 of the beam of Fig. 9.1a is obtained by removing support 2 of the actual beam (see Fig. 9.1b). Moreover, the auxiliary beam for computing the influ-

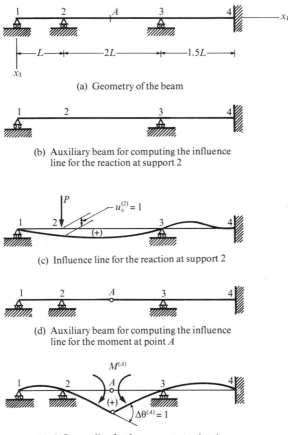

(a) Geometry of the beam

(b) Auxiliary beam for computing the influence line for the reaction at support 2

(c) Influence line for the reaction at support 2

(d) Auxiliary beam for computing the influence line for the moment at point A

(e) Influence line for the moment at point A

Figure 9.1 Illustration of the sketching of the influence lines for reactions and internal actions of a beam using the Müller-Breslau principle.

ence line for the moment at point A of the beam of Fig. 9.1a is obtained by introducing a hinge at point A of the beam (see Fig. 9.1d). Furthermore, the auxiliary beam for computing the influence line for the shearing force at point A of the beam of Fig. 9.1a is obtained by introducing a mechanism at point A of the actual beam which does not transmit shearing force (see Fig. 9.1f).

2. The auxiliary structure is subjected to an external action corresponding to the reaction whose influence line we want to plot or to a pair of equal and opposite external actions corresponding to the internal action whose influence line we want to plot. The magnitude of these external actions is such that the corresponding displacement is equal to unity, while their sense is that of the negative of the reaction or the internal action whose influence line we want to plot. Thus, in order to compute the influence line for the reaction at support 2 of the beam of Fig. 9.1a, the auxiliary beam of Fig. 9.1b is loaded by a downward force P, i.e., having the sense of the negative of the reaction at support 2. The magnitude of the force P is the required to produce a downward dis-

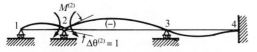

(f) Auxiliary beam for computing the influence line for the shearing force at point A

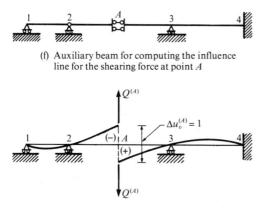

(g) Influence line for the shearing force at point A

(h) Auxiliary beam for computing the influence line for the moment at point 2

$M^{(2)}$

(−)

$\Delta\theta^{(2)} = 1$

(i) Influence line for the moment at support 2

Figure 9.1 (*Continued*)

placement of point 2 equal to unity. Moreover, in order to compute the influence line for the moment at point A of the beam of Fig. 9.1a, the auxiliary beam of Fig. 9.1d is loaded by a pair of equal and opposite external moments having the magnitude required to produce a unit relative rotation of the ends of the beam segments framing into the hinge at A (see Fig. 9.1e). The sense of this pair of moments is that of the negative internal moment (a negative moment subjects the top fibers of the beam to tension). Furthermore, in order to compute the influence line for the shearing force at point A of the beam of Fig. 9.1a, the auxiliary beam of Fig. 9.1f is loaded by a pair of equal and

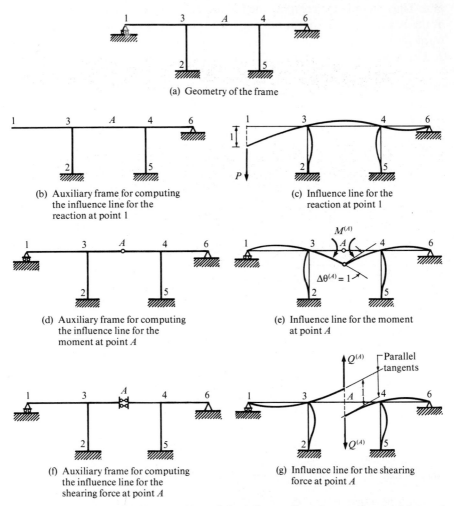

(a) Geometry of the frame

(b) Auxiliary frame for computing the influence line for the reaction at point 1

(c) Influence line for the reaction at point 1

(d) Auxiliary frame for computing the influence line for the moment at point A

(e) Influence line for the moment at point A

(f) Auxiliary frame for computing the influence line for the shearing force at point A

(g) Influence line for the shearing force at point A

Figure 9.2 Illustration of the sketching of the influence lines for reactions and internal actions of a frame using the Müller-Breslau principle.

opposite external forces having the magnitude required to produce a unit relative translation and no relative rotation of the ends of the beam segments framing into the mechanism at point A (see Fig. 9.1g). These external forces have
the sense of the negative internal shearing force.

3. The ordinates of the elastic curve of the auxiliary structure subjected to
the load described in 2 are computed. They are the desired ordinates of the
influence line for the reaction or internal action under consideration.

The sketching of influence lines for reactions and internal actions of a frame
using the Müller-Breslau principle is illustrated in Fig. 9.2.

Proof of the principle of Müller-Breslau.   Consider the beam shown in Fig. 9.3a.
By definition, the ordinate $\hat{M}^{(AB)}$ at point B of the influence line for the moment
at point A is equal to the value of the moment at point A when a unit load is

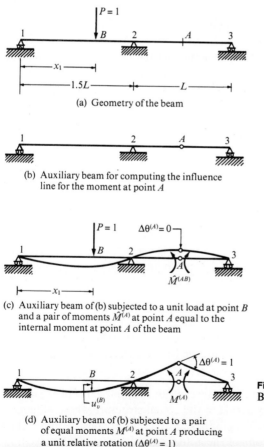

(a) Geometry of the beam

(b) Auxiliary beam for computing the influence
line for the moment at point A

(c) Auxiliary beam of (b) subjected to a unit load at point B
and a pair of moments $\hat{M}^{(A)}$ at point A equal to the
internal moment at point A of the beam

(d) Auxiliary beam of (b) subjected to a pair
of equal moments $M^{(A)}$ at point A producing
a unit relative rotation ($\Delta\theta^{(A)} = 1$)

Figure 9.3 Proof of the Müller-
Breslau principle.

applied at point B. We consider the auxiliary beam shown in Fig. 9.3b which has the same geometry as the actual beam, except that it has a hinge at point A. In Fig. 9.3c, we subject the auxiliary beam to a unit force at point B and a pair of external moments $\hat{M}^{(AB)}$ at point A. The moments $\hat{M}^{(AB)}$ are equal to the moment at point A of the actual beam when subjected to a unit force at point B. It is apparent that, the distributions of moments and displacements in the auxiliary beam loaded as shown in Fig. 9.3c are identical to those in the actual beam loaded as shown in Fig. 9.3a. Therefore, the elastic curve of the auxiliary beam of Fig. 9.3c at point A is smooth (there is no change of slope). In Fig. 9.3d, we subject the auxiliary beam to a pair of moments at point A whose magnitude $M^{(A)}$ is such that the relative rotation of the ends of the members of the auxiliary beam which are connected to the hinge at point A is equal to unity. We denote the deflection of point B of the auxiliary beam loaded as shown in Fig. 9.3d by $u_v^{(B)}$. Application of Betti's law to the auxiliary beam, loaded as shown in Fig. 9.3c and d, yields the following relation.

$$1xu_v^{(B)} + \hat{M}^{(AB)}(1) = M^{(A)}(0)$$

or
$$\hat{M}^{(AB)} = -u_v^{(B)}$$

Thus, the ordinate at point B of the influence line for the moment at point A of the beam of Fig. 9.3a is equal to minus the deflection $u_v^{(B)}$ of point B of the auxiliary beam, loaded as shown in Fig. 9.3d. Consequently, the ordinate at point B of the influence line for the moment $M^{(A)}$ at point A of the beam of Fig. 9.3a is equal to the deflection $u_v^{(B)}$ of the auxiliary beam when it is loaded with a pair of moments $M^{(A)}$ opposite to the one shown in Fig. 9.3d, that is, having the sense of the negative internal moment.

9.3 Application of the Principle of Müller-Breslau to Beams

In this section, we sketch the influence lines of beams (statically determinate and indeterminate) and compute their ordinates using the principle of Müller-Breslau. In computing the ordinates of influence lines of statically indeterminate structures, we adhere to the following steps:

Step 1. We form an auxiliary structure from the actual structure, either by eliminating the external constraint which causes the reaction whose influence line we want to establish, or by introducing a mechanism in the actual structure which renders it unable to transfer the internal action whose influence line we want to establish.

Step 2. We subject the auxiliary structure either to a unit value of the external action corresponding to the reaction of the actual structure whose influence line we want to plot, or to a pair of opposite external actions of unit value corresponding to the internal action of the actual structure whose influence line

we want to plot. The sense of these external actions is that of the negativ́
reaction or internal action whose influence line we want to plot.

Step 3. We establish the end actions of the members of the auxiliary struc
ture loaded as described in step 2, using a force method (see Chap. 7) or $
displacement method (see Chap. 8).

Step 4. We compute the value of the displacement of the auxiliary structur̥
corresponding to the reaction or internal action whose influence line we wan̄
to plot. For instance, in order to establish the ordinates of the influence line fơ
the moment at point A of the beam of Fig. 9.1a, we compute the relative rot̄
tion $\Delta\theta^{(A)}$ of the ends of the members of the auxiliary structure connected t̥
the hinge to point A (see Fig. 9.1e). In general, this rotation is not equal t̥
unity.

Step 5. We establish the ordinates of the elastic curve of the members ớ
the auxiliary structure loaded as described in step 2.

Step 6. We divide the ordinates of the elastic curve computed in step 5 b́
the value of the displacement computed in step 4 to obtain the ordinates of th̥
desired influence line.

Notice that when the influence lines for n reactions and/or internal actiős
of a statically indeterminate structure to the nth degree have been established̩,
the influence line for any other reaction or internal action can easily be computed from them by considering the equilibrium of parts of the structure. For
instance, if the ordinates $\hat{R}^{(1)}$, $\hat{R}^{(2)}$, and $\hat{R}^{(3)}$ of the influence lines for the reactions at supports 1, 2, and 3 of the beam of Fig. 9.4 are known, the ordinates
$\hat{R}^{(5)}$ of the influence line for the moment at point 4 of this beam may be established by setting the sum of the moments about point 4 equal to zero. Thus,
referring to Fig. 9.4, we have

$$\hat{R}^{(5)} = L\hat{R}^{(3)} + 3L\hat{R}^{(2)} + 4L\hat{R}^{(1)} - (4L - x_1)$$

In the following examples, we plot the influence lines for reactions or internal
actions of beams using the principle of Müller-Breslau. In the first example,
the beam is statically determinate; in the second example, the beam is statically
indeterminate to the second degree (fixed at both ends); in the third example,
the beam is statically indeterminate to the second degree.

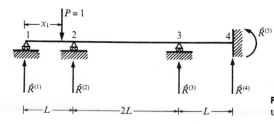

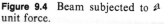

Figure 9.4 Beam subjected to a
unit force.

Example 1. Establish the influence lines for the reaction at support 2, the shearing force and moment at point A, the moment at point 2, and the shearing force at point B of the beam shown in Fig. a.

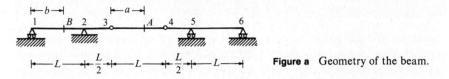

Figure a Geometry of the beam.

solution This is a statically determinate structure. The auxiliary beam for computing the influence line for the reaction $R^{(2)}$ is obtained by removing the support of the actual beam at point 2 (see Fig. b). Notice that inasmuch as the actual beam is statically determinate, the auxiliary beam is a mechanism. When it is subjected to a force P at point 2, it will collapse. The auxiliary beam is displaced so that the displacement of point 2 is equal to unity. The ordinates of the displaced axis of the auxiliary beam are equal to the ordinates of the influence line for the reaction at point 2 of the actual beam (see Fig. c).

The auxiliary beam for computing the influence line for the moment at point A of the beam of Fig. a is obtained by introducing a hinge at point A of the actual beam (see Fig. d). The auxiliary beam is then displaced so that the relative rotation of its segments $A,4$ and $3,A$ is equal to unity (see Fig. e). The ordinates of the displaced axis of the auxiliary beam are the ordinates of the influence line for the moment at point A of the actual beam (see Fig. e).

The auxiliary beam for computing the influence line for the moment at point 2 is obtained by introducing a hinge at point 2 of the actual beam (see Fig. f). The auxiliary beam is then displaced so that the rotation of segment 2,3 is equal to unity (see

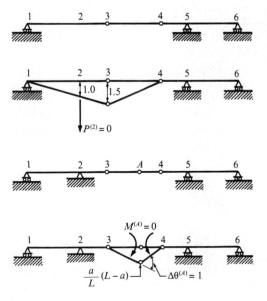

Figure b Auxiliary beam for computing the influence line for the reaction at support 2.

Figure c Influence line for the reaction at support 2.

Figure d Auxiliary beam for computing the influence line for the moment at point A.

Figure e Influence line for the moment at point A.

Fig. g). The ordinates of the displaced axis of the auxiliary beam are the ordinates of the influence line for the moment at support 2 of the beam of Fig. a (see Fig. g).

The auxiliary beam for computing the influence line for the shearing force at point A or point B is obtained by introducing a mechanism which does not transmit a shearing force at point A or B of the actual beam (see Figs. h and j). The auxiliary beam is then displaced so that the relative translation of the ends of its two segments which are adjacent to the mechanism is equal to unity, while their relative rotation is zero (see Figs. i and k). The ordinates of the displaced axis of the auxiliary beam are the ordinates of the influence line for the shearing force at point A or point B of the beam of Fig. a. These ordinates may be obtained from geometric considerations, i.e., referring to Figs. i or k.

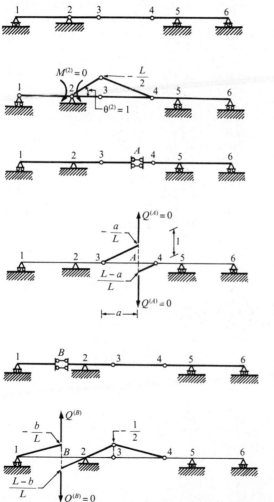

Figure f Auxiliary beam for computing the influence line for the moment at support 2.

Figure g Influence line for the moment at support 2.

Figure h Auxiliary beam for computing the influence line for the shearing force at point A.

Figure i Influence line for the shearing force at point A.

Figure j Auxiliary beam for computing the influence line for the shearing force at point B.

Figure k Influence line for the shearing at point B.

Example 2. Establish the ordinates of the influence lines for the reaction and the moment at support 1 and the shearing force and moment at point A $(x_1 = L/4)$ of the fixed-at-both-ends beam of constant cross section shown in Fig. a.

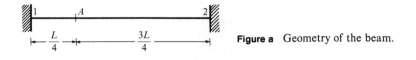

Figure a Geometry of the beam.

solution   When this beam is subjected only to vertical loads, it is a statically indeterminate structure to the second degree. However, because of its symmetry, the influence line of any reaction or internal action of the beam may be established from the influence line of only one quantity, say the end moment $M^{(1)}$. Thus we will establish the ordinates of the influence line for $M^{(1)}$, and from these we will compute the ordinates of the influence lines of the other quantities.

The auxiliary beam for computing the influence line for $M^{(1)}$ is obtained by introducing a hinge at point 1 of the actual beam (see Fig. b). In this example, we know the value of the moments $M^{(1)}$ and $M^{(2)}$ which must be applied to the ends of the auxiliary beam in order to produce a unit counterclockwise rotation at point 1 and zero rotation at point 2. The first is equal to minus the rotational stiffness of end 1 of the beam, while the second is equal to the product of the carry-over factor and the rotational stiffness of this end. Thus for a beam of constant cross section and referring to relations (8.3) and (8.4), we have

$$M^{(1)} = -\frac{4EI}{L} \qquad M^{(2)} = \frac{2EI}{L} \tag{a}$$

Consider a member of constant cross section of a planar structure subjected only to end actions, as shown in Fig. e, and assume that there is no relative movement of its ends. We will establish a formula for computing the ordinates of the elastic curve of this member in terms of its end moments, using the method of "virtual" work. The moment at any point x_1' of the member subjected to end actions, as shown in Fig. e,

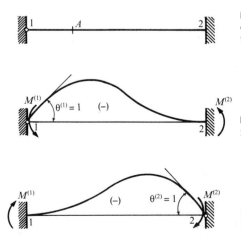

Figure b Auxiliary beam for computing the influence line for the moment at support 1.

Figure c Influence line for the moment at support 1.

Figure d Influence line for the moment at support 2.

is equal to

$$M = M^{ij} + \left(\frac{M^{ik} - M^{ij}}{L} \right) x_1' \tag{b}$$

The moment at any point x_1' of the member subjected to the "virtual" loading (see Fig. f) for computing the deflection $u_v(x_1)$ is equal to

$$\tilde{M} = \left(\frac{L - x_1}{L} \right) x_1' \qquad\qquad 0 \le x_1' \le x_1$$

$$\tilde{M} = \left(\frac{L - x_1}{L} \right) x_1' - (x_1' - x_1) \qquad x_1 \le x_1' \le L \tag{c}$$

Substituting relations (b) and (c) into (5.66), we get

$$u_v(x_1) = \frac{1}{EI} \int_0^{x_1} \left(\frac{L - x_1}{L} \right) x_1' \left[M^{ij} + \left(\frac{M^{ik} - M^{ij}}{L} \right) x_1' \right] dx_1'$$

$$+ \frac{1}{EI} \int_{x_1}^{L} \left[\left(\frac{L - x_1}{L} \right) x_1' - (x_1' - x_1) \right] \left[M^{ij} + \left(\frac{M^{ik} - M_{ij}}{L} \right) x_1' \right] dx_1'$$

$$= \frac{L^2}{6EI} \left\{ M^{ij} \left[\frac{2x_1}{L} - 3 \left(\frac{x_1}{L} \right)^2 + \left(\frac{x_1}{L} \right)^3 \right] + M^{ik} \left[\frac{x_1}{L} - \left(\frac{x_1}{L} \right)^3 \right] \right\} \tag{d}$$

It is apparent that the ordinates $\hat{M}^{(1)}$ of the influence line for the moment at support 1 of the beam of Fig. a are obtained by substituting the values of $M^{ij} = M^{(1)}$ and $M^{ik} = M^{(2)}$ from relations (a) into relation (d). Thus

$$\hat{M}^{(1)}(x_1) = u_v(x_1) = -x_1 + \frac{2x_1^2}{L} - \frac{x_1^3}{L^2} \tag{e}$$

The influence line for $M^{(1)}$ is sketched in Fig. c. Because of the symmetry of the beam, the ordinates of the influence line for the moment at support 2 of the beam are

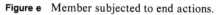

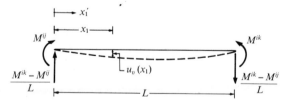

Figure e Member subjected to end actions.

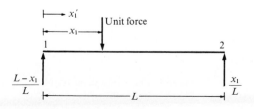

Figure f Free-body diagram of the member subjected to the virtual loading for computing the deflection $u_v(x_1)$.

obtained by substituting $(L - x_1)$ for x_1 in relation (e). Thus,

$$\hat{M}^{(2)}(x_1) = -(L - x_1) + \frac{2(L - x_1)^2}{L} - \frac{(L - x_1)^3}{L^2} \qquad \text{(f)}$$

The influence line for $M^{(2)}$ is plotted in Fig. d.

The influence line for the reaction at point 1 is obtained by considering the equilibrium of the beam. Referring to Fig. g, we get

$$\hat{R}^{(1)} = \frac{(L - x_1)}{L} - \frac{\hat{M}^{(1)}}{L} + \frac{\hat{M}^{(2)}}{L} \qquad \text{(g)}$$

Substitution of relations (e) and (f) into (g) gives the ordinates of the influence line for $R^{(1)}$ as

$$\hat{R}^{(1)} = 1 - \frac{3x_1^2}{L^2} + \frac{2x_1^3}{L^3} \qquad \text{(h)}$$

The influence line for $R^{(1)}$ is shown in Fig. h.

The influence line for the shearing force and the moment at point A is obtained by considering the equilibrium of the segment of the beam shown in Fig. i. That is,

$$\left. \begin{aligned} \hat{Q}^{(A)} &= \hat{R}^{(1)} - 1 \\ \hat{M}^{(A)} &= \hat{M}^{(1)} + a\hat{R}^{(1)} - (a - x_1) \end{aligned} \right\} \qquad 0 \le x_1 \le a \qquad \text{(i)}$$

$$\left. \begin{aligned} \hat{Q}^{(A)} &= \hat{R}^{(1)} \\ \hat{M}^{(A)} &= \hat{M}^{(1)} + a\hat{R}^{(1)} \end{aligned} \right\} \qquad a \le x_1 \le L \qquad \text{(j)}$$

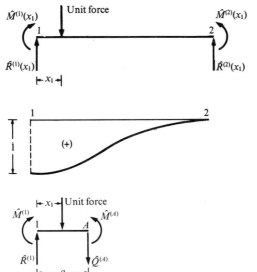

Figure g Free-body diagram of the fixed-end beam subjected to a unit force.

Figure h Influence line for the reaction at support 1.

Figure i Free-body diagram of a segment of a beam.

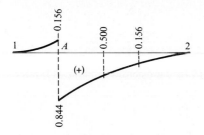

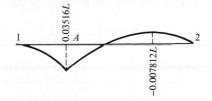

Figure j Influence line for the shearing force at point A.

Figure k Influence line for the moment at point A.

Substitution of relations (a) and (h) into (i) and (j) gives the ordinates of the influence lines for $Q^{(A)}$ and $M^{(A)}$ as functions of x_1. The influence lines for $Q^{(A)}$ and $M^{(A)}$ are sketched in Figs. j and k.

Example 3. Establish the ordinates of the influence line for the moments at points 2 and 3, and the shearing force and moment at point A of the beam of constant cross section shown in Fig. a.

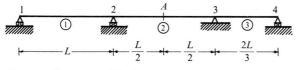

Figure a Geometry of the beam.

solution

Part a. Computation of the Ordinates of the Influence Line for the Moment at Point 2

STEP 1.   We form an auxiliary beam by introducing a hinge at point 2 of the actual beam (see Fig. b).

STEP 2.   We subject the auxiliary beam to a pair of unit moments. The one moment acts on the right end of member 1, while the other acts on the left end of member 2 (see Fig. c). The sense of this pair of moments is that of the negative, internal moment at point 2.

STEP 3.   Member 1 of the auxiliary beam of Fig. c is statically determinate. However, members 2 and 3 of the auxiliary beam form a continuous beam which is statically indeterminate to the first degree. We use the moment distribution method to compute the end moments in members 2 and 3 of the auxiliary beam. We consider these members as type M3. Hence, referring to relation (8.6), their rotational stiffnesses are

$$\hat{k}^{2k} = \frac{3EI}{L} \qquad \hat{k}^{3j} = \frac{3EI}{2L/3} = \frac{4.5EI}{L} \tag{a}$$

Thus the distribution factors at point 3 are

$$\widehat{DF}^{(2)} = \frac{3}{3 + 4.5} = 0.40 \qquad \widehat{DF}^{(3)} = \frac{4.5}{3 + 4.5} = 0.60 \qquad (b)$$

Since members 2 and 3 of the auxiliary beam of Fig. c are considered type M3, referring to App. C, their fixed-end moments are

$$\widehat{FEM}^{2k} = 0.5 \text{ kN·m} \qquad \widehat{FEM}^{3j} = 0 \qquad (c)$$

The moment distribution is carried out in Fig. d. The free-body diagrams of the members of the auxiliary beam of Fig. c are shown in Fig. e. Referring to this figure, we have

$$M = -\frac{x_1'}{L} \qquad\qquad 0 \le x_1' \le L$$

Figure b Auxiliary beam for computing the ordinates of the influence line for the moment at point 2.

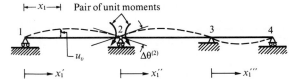

Figure c Auxiliary beam subjected to a pair of external unit moments at point 2.

0.40	0.60
0.5	
0.2	−0.3
0.3	−0.3

Figure d Moment distribution at joint 3.

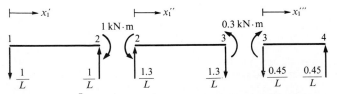

Figure e Free-body diagrams of the members of the auxiliary beam of Fig. c.

$$M = -1 + \frac{1.3x_1''}{L} \qquad 0 \leq x_1'' \leq L \tag{d}$$

$$M = 0.3 - \frac{0.45x_1'''}{L} \qquad 0 \leq x_1''' \leq \frac{2L}{3}$$

STEP 4.   We use the method of virtual work in order to compute

1. The value of the relative rotation of the right end of a member 1 and the left end of member 2 of the auxiliary beam loaded as shown in Fig. c.

2. The deflection u_v of points $x_1 = 0.25cL$ ($c = 1, 2, \ldots, 8$) and $x_1 = 2L + \dfrac{kL}{6}$

   ($k = 1, 2, 3$) of the auxiliary beam loaded as shown in Fig. c.

The auxiliary beam of Fig. b subjected to the virtual loading for computing the angle $\Delta\theta^{(2)}$ and to statically admissible reactions is shown in Fig. f. The corresponding statically admissible moment distribution in this beam is

$$\tilde{M} = \frac{x_1'}{L} \qquad 0 \leq x_1' \leq L$$

$$\tilde{M} = -1 + \frac{x_1''}{L} \qquad 0 \leq x_1'' \leq L \tag{e}$$

$$\tilde{M} = 0 \qquad 0 \leq x_1''' \leq \frac{2L}{3}$$

Substituting relations (d) and (e) into the principle of virtual work (5.66) and integrating, we obtain

$$\Delta\theta^{(2)} = \int \frac{M\tilde{M}}{EI}\, dx_1$$

$$= \frac{1}{EI}\left[\int_0^L \left(-\frac{x_1'}{L}\right)^2 dx_1' + \int_0^L \left(-1 + \frac{1.3x_1''}{L}\right)\left(-1 + \frac{x_1''}{L}\right) dx_1'' \right] = \frac{3.7L}{6EI} \tag{f}$$

The auxiliary beam of Fig. c subjected to the virtual loading for computing the deflection $u_v(x_1)$ of member 1 ($0 \leq x_1 \leq L$) and to statically admissible reactions is

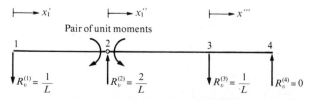

Figure f   Auxiliary beam of Fig. b subjected to the virtual loading for computing $\Delta\theta^{(2)}$ and to statically admissible reactions.

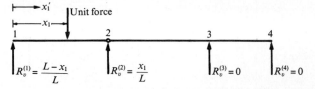

Figure g Auxiliary beam of Fig. b subjected to the virtual loading for computing $u_v(x_1)$ $(0 \le x_1 \le L)$ and to statically admissible reactions.

shown in Fig. g. The corresponding statically admissible moment distribution is equal to

$$\tilde{M} = \left(\frac{L - x_1}{L}\right) x_1' \qquad\qquad 0 \le x_1' \le x_1$$

$$\tilde{M} = \left(\frac{L - x_1}{L}\right) x_1' - (x_1' - x_1) \qquad x_1 \le x_1' \le L \tag{g}$$

Substituting relations (d) and (g) into (5.66), we find that the equation for the deflection of member 1 $(0 \le x_1 \le L)$ of the beam of Fig. c is

$$u_v(x_1) = \frac{1}{EI}\left[\int_0^{x_1} -\frac{x_1'}{L}\left(\frac{L - x_1}{L}\right) x_1'\, dx_1'\right.$$

$$\left. + \int_{x_1}^{L} -\frac{x_1'}{L}\left[\frac{L - x_1}{L} x_1' - (x_1' - x_1)\right] dx_1'\right]$$

$$= \frac{x_1 L}{6EI}\left[-1 + \left(\frac{x_1}{L}\right)^2\right] \qquad 0 \le x_1 \le L \tag{h}$$

The auxiliary beam of Fig. c subjected to the virtual loading for computing the deflection $u_v(x_1)$ of member 2 $(L \le x_1 \le 2L)$ and to statically admissible reactions is shown in Fig. h. The corresponding statically admissible moment distribution is equal to

$$\tilde{M} = \frac{2L - x_1}{L} x_1'' \qquad\qquad 0 \le x_1'' \le x_1 - L$$

$$\tilde{M} = \frac{2L - x_1}{L} x_1'' - (x_1'' + L - x_1) \qquad x_1 - L \le x_1'' \le L \tag{i}$$

Substituting relations (d) and (i) into (5.66), we find that the equation for the deflection of member 2 $(L \le x_1 \le 2L)$ of the beam of Fig. c is equal to

$$u_v(x_1) = \int \frac{M\tilde{M}}{EI}\, dx_1'' = \frac{1}{EI}\left\{\int_0^{x_1 - L}\left(-1 + \frac{1.3 x_1''}{L}\right)\frac{2L - x_1}{L} x_1''\, dx_1''\right.$$

$$\left. + \int_{x_1 - L}^{L}\left(-1 + \frac{1.3 x_1''}{L}\right)\left[\frac{2L - x_1}{L} x_1'' - (x_1'' + L - x_1)\right] dx_1''\right\}$$

$$= \frac{(x_1 - L)L}{6EI}\left[1.7 - \frac{3(x_1 - L)}{L} + 1.3\left(\frac{x_1 - L}{L}\right)^2\right] \qquad L \le x_1 \le 2L \tag{j}$$

The auxiliary beam of Fig. c subjected to the virtual loading for computing the deflection $u_v(x_1)$ of member 3 $(2L \le x_1 \le 8L/3)$ and to statically admissible reactions is shown in Fig. i. The corresponding statically admissible moment distribution is equal to

$$\tilde{M} = \left(4 - \frac{3x_1}{2L}\right) x_1''' \qquad\qquad 0 \le x_1''' \le x_1 - 2L$$

$$\tilde{M} = \left(4 - \frac{3x_1}{2L}\right) x_1''' - (x_1''' + 2L - x_1) \qquad x_1 - 2L \le x_1''' \le \frac{2L}{3} \tag{k}$$

Substituting relations (d) and (k) into (5.66) and integrating, we find that the equation for the deflection of member 3 of the beam of Fig. c is equal to

$$u_v(x_1) = \int \frac{M\tilde{M}}{EI}\, dx_1'''$$

$$= \frac{1}{EI}\left[\int_0^{x_1-2L}\left(0.3 - \frac{0.45x_1'''}{L}\right)\left(4 - \frac{3x_1}{2L}\right) x_1''' \, dx_1'''\right.$$

$$\left. + \int_{x_1-2L}^{2L/3}\left(0.3 - \frac{0.45x_1'''}{L}\right)\left[\left(4 - \frac{3x_1}{2L}\right) x_1''' - (x_1''' + 2L - x_1)\right] dx_1'''\right]$$

$$= \frac{(x_1 - 2L)L}{6EI}\left[0.4 + 0.45\left(\frac{x_1 - 2L}{L}\right)\left(\frac{x_1 - 4L}{L}\right)\right]$$

$$2L \le x_1 \le 8L/3 \tag{l}$$

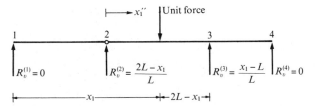

Figure h   Auxiliary beam of Fig. b subjected to the virtual loading for computing $u_v(x_1)$ $(L \le x_1 \le 2L)$ and to statically admissible reactions.

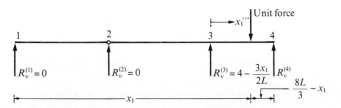

Figure i   Auxiliary beam of Fig. b subjected to the virtual loading for computing $u_v(x_1)$ $(2L \le x_1 \le 8L/3)$ and to statically admissible reactions.

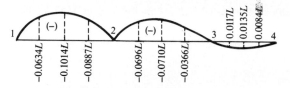

Figure j   Influence line for the moment of point 2.

The ordinates of the influence line for the moment at point 2 are equal to the ordinates of the elastic curve of the auxiliary beam, loaded as shown in Fig. c, divided by the absolute value of the relative rotation $\Delta\theta^{(2)}$ (see Fig. c). Thus using relation (f) we get

$$\tilde{M} = \frac{u_v}{|\Delta\theta^{(2)}|} = \frac{6EI}{3.7L}\, u_v \tag{m}$$

The ordinates of the influence line for the moment at point 2 are obtained by substituting relation (h), (j), or (l) into (m) and evaluating the resulting expressions at the points where the values of the ordinates are desired. The results are shown in Fig. j.

Part b. Computation of the Ordinates of the Influence Line for the Moment at Point 3

STEP 1.   We form an auxiliary beam by introducing a hinge at point 3 of the actual beam (see Fig. k).

STEP 2.   We subject the auxiliary beam to a pair of unit moments. The one moment acts on the right end of member 2, while the other acts on the left end of member 3 (see Fig. l). The sense of this pair of moments is that of the negative internal moment at point 3 of the actual beam.

STEP 3.   Member 3 of the auxiliary beam of Fig. l is statically determinate. However, members 1 and 2 of the auxiliary beam form a statically indeterminate to the first degree continuous beam. We use the moment distribution method to compute the end moments in members 1 and 2 of the auxiliary beam. We consider these members as type M3. Hence their rotational stiffnesses are

$$k^{1k} = \frac{3EI}{L} \qquad k^{2j} = \left(\frac{3EI}{2L/3}\right) = \frac{4.5EI}{L}$$

Thus the distribution factors at joint 2 are

$$\widehat{DF}^{(1)} = \widehat{DF}^{(2)} = 0.50$$

Since members 1 and 2 of the auxiliary beam of Fig. l are considered as type M3, referring to Fig. j, their fixed-end moments are

$$\widehat{FEM}^{1k} = 0 \qquad \widehat{FEM}^{2j} = 0.5 \text{ kN·m}$$

The moment distribution is carried out in Fig. m. The resulting moment diagram is plotted in Fig. n.

STEP 4.   We use the conjugate beam method in order to compute the following.

1. The value of the relative rotation of the right end of member 2 and the left end of member 3 of the auxiliary beam loaded as shown in Fig. l.

2. The deflection u_v of points $x_1 = 0.25cL$ $(c = 1, 2, 3, \ldots, 8)$ and $x_1 = 2L + \dfrac{kL}{6}$

$(k = 1, 2, 3)$ of the auxiliary beam loaded as shown in Fig. 1.

The conjugate beam for the auxiliary beam of Fig. 1 is shown in Fig. o. Its reactions are obtained by considering its equilibrium. Thus

$$\Sigma \overline{M}^{(2)} = 0 = L\overline{R}^{(1)} - \frac{0.25}{EI} \left(\frac{L}{2}\right)\left(\frac{L}{3}\right)$$

Figure k Auxiliary beam for computing the ordinates of the influence line for the moment at point 3.

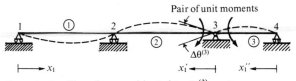

Figure l Auxiliary beam subjected to $M^{(3)} = 1$.

Figure m Moment distribution at joint 2.

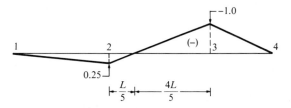

Figure n Moment diagram for the auxiliary beam loaded as shown in Fig. 1.

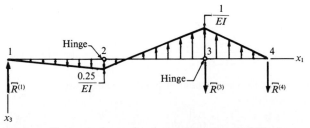

Figure o Conjugate beam for the auxiliary beam of Fig. 1.

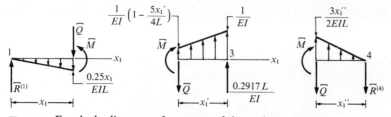

Figure p   Free-body diagrams of segments of the conjugate beam.

$$\Sigma \overline{M}^{(3)} = 0 = \frac{2L}{3}\overline{R}^{(4)} + \frac{1}{EI}\left(\frac{L}{3}\right)\left(\frac{2L}{9}\right)$$

$$\Sigma \overline{F}_v = 0 = -\overline{R}^{(1)} + \overline{R}^{(4)} + \overline{R}^{(3)}$$

$$- \frac{1}{EI}\left(\frac{L}{3}\right) - \frac{1}{EI}\left(\frac{2L}{5}\right) + \frac{0.25}{EI}\left(\frac{L}{10}\right) + \frac{0.25}{EI}\left(\frac{L}{2}\right)$$

or
$$\overline{R}^{(1)} = \frac{L}{24EI}$$

$$\overline{R}^{(4)} = -\frac{L}{9EI} \tag{n}$$

$$\overline{R}^{(3)} = 0.5139\frac{L}{EI} = -\Delta\theta^{(3)}$$

The moment at any point of the conjugate beam of Fig. o may be established by considering the equilibrium of its segments shown in Fig. p. Thus

For member 1 $0 \le x_1 \le L$

$$-u_v(x_1) = \overline{M}(x_1) = \overline{R}^{(1)}x_1 + \frac{0.25x_1}{EIL}\left(\frac{x_1}{2}\right)\left(\frac{x_1}{3}\right) = -\frac{Lx_1}{24EI}\left[1 - \left(\frac{x_1}{L}\right)^2\right] \tag{o}$$

For member 2 we measure x_1' ($0 \le x_1' \le L$) from point 3.

$$-u_v(x_1') = \overline{M}(x_1') = 0.2917\frac{Lx_1'}{EI} - \frac{1}{EI}\left(1 - \frac{5x_1'}{4L}\right)\frac{(x_1')^2}{2} - \frac{5x_1'}{4EIL}\left(\frac{x_1'}{2}\right)\frac{2x_1'}{3} \tag{p}$$

$$= \frac{Lx_1'}{EI}\left[0.2917 - \frac{0.5x_1'}{L} + \frac{5}{24}\left(\frac{x_1'}{L}\right)^2\right]$$

For member 3 we measure x_1'' ($0 \le x_1'' \le \frac{2}{3}L$) from point 4.

$$-u_v(x_1'') = \overline{M}(x_1'') = \overline{R}^{(4)}x_1'' - \frac{3x_1''}{2EIL}\left(\frac{x_1''}{2}\right)\frac{x_1''}{3} = \frac{Lx_1''}{9EI}\left[1 - \frac{9}{4}\left(\frac{x_1''}{L}\right)^2\right] \tag{q}$$

The ordinates of the influence line for the moment at point 3 are equal to

$$\hat{M} = \frac{u_v}{\overline{R}^{(3)}} = \frac{EIu_v}{0.5139L} \tag{r}$$

The ordinates of the influence line for the moment at point 3 are obtained by substituting relation (o), (p), or (q) into (r) and evaluating the resulting expressions at the points where the values of the ordinates are desired. The results are shown in Fig. q.

Part c. Computation of the Ordinates of the Influence Lines for the Shearing Force and the Moment at Point A

The beam of Fig. a is statically indeterminate to the second degree. Thus the influence line for any quantity may be established from the influence lines for the moments at points 2 and 3. For instance, at the points of members 1 and 3 the ordinates of the influence line for the shearing force or bending moment at point A are obtained by referring to Fig. r. That is

$$\hat{Q}^{(A)} = \frac{\hat{M}^{(3)} - \hat{M}^{(2)}}{L} \tag{s}$$

$$\hat{M}^{(A)} = \frac{\hat{M}^{(2)} + \hat{M}^{(3)}}{2} \tag{t}$$

At the points of member 2, the ordinates of the influence lines for the shearing force and moment at point A are obtained by referring to Fig. s. That is,

For $0 \leq x_1' < \dfrac{L}{2}$

$$\hat{Q}^{(A)} = \frac{\hat{M}^{(3)} - \hat{M}^{(2)}}{L} - \frac{x_1'}{L} \tag{u}$$

$$\hat{M}^{(A)} = \hat{M}^{(3)} - \frac{\hat{M}^{(3)} - \hat{M}^{(2)}}{2} + \frac{x_1'}{2} = \frac{\hat{M}^{(2)} + \hat{M}^{(3)}}{2} + \frac{x_1'}{2} \tag{v}$$

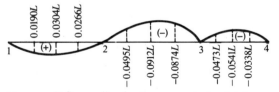

Figure q   Influence line for the moment of point 3.

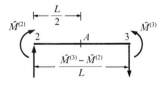

Figure r Free-body diagram of member 2 of the beam when the unit load is on member 1 or 3.

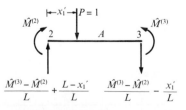

Figure s Free-body diagram of member 2 of the beam when the unit load is on this member.

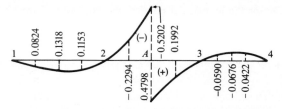

Figure t   Influence line for the shearing force at point A.

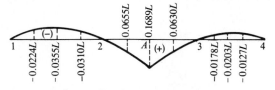

Figure u   Influence line for the moment at point A.

For $\dfrac{L}{2} < x_1' \le L$

$$\hat{Q}^{(A)} = \frac{\hat{M}^{(3)} - \hat{M}^{(2)}}{L} + \frac{L - x_1'}{L} \qquad (w)$$

$$\hat{M}^{(A)} = \frac{\hat{M}^{(2)} + \hat{M}^{(3)}}{2} + \frac{L - x_1'}{2} \qquad (x)$$

The ordinates of the influence lines for the shearing force and moment at point A are established by substituting in relations (s) to (x) the values of the ordinates of the influence lines for the moments at points 2 and 3, given in Figs. j and q, respectively. The influence lines for the shearing force and moment at point A are plotted in Figs. t and u, respectively.

9.4   Problems

1 to 3.   Using the principle of Müller-Breslau, sketch the influence lines for the reacting force at support 1, the shearing force and moment at point A, and for the moment at point 2 of the beam of Fig. 9.P1. Repeat for Figs. 9.P2 and 9.P3.

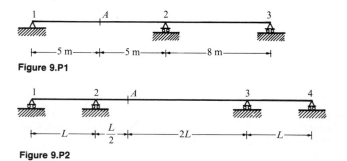

Figure 9.P1

Figure 9.P2

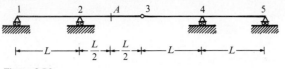

Figure 9.P3

4. Using the principle of Müller-Breslau, sketch the influence lines for the reacting force and moment at support 5 and the shearing force and moment at point A of the beam of Fig. 9.P4.

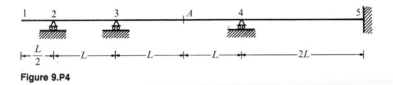

Figure 9.P4

5. Using the principle of Müller-Breslau, sketch the influence lines for the following quantities of the frame shown in Fig. 9.P5.
    (*a*) For the reacting force at support 1
    (*b*) For the reacting forces and moment at support 3
    (*c*) For the shearing force and moment in the beam to the right of point 2
    (*d*) For the shearing force and moment at point A
The unit force is moving only on the horizontal members.

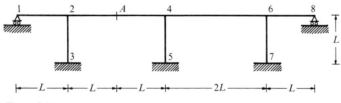

Figure 9.P5

6. Using the principle of Müller-Breslau, plot the influence lines for the reaction at support 1, the moment at point 2, and the shearing force and moment at point A of the beam shown in Fig. 9.P6. Give the ordinates of the influence lines every meter.

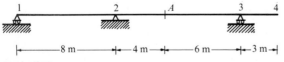

Figure 9.P6

7. Using the principle of Müller-Breslau plot the influence lines for all the reactions, the moment at point 2, and the shearing force and moment at point A of the beam shown in Fig. 9.P7. Give the ordinates of the influence lines every meter.

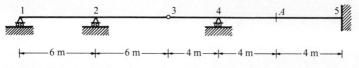

Figure 9.P7

8. Using the principle of Müller-Breslau plot the influence lines for all the reactions, the moment at point 2, and the shearing force and moment at point A of the beam shown in Fig. 9.P8. Give the ordinates of the influence lines every meter.

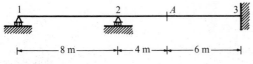

Figure 9.P8

9. Using the principle of Müller-Breslau, plot the influence lines for the reactions at points 1 and 3 and the shearing force and moment at the cross section adjacent and to the right of point 2 of the frame of Fig. 9.P9. Assume that the unit force is moving only on the beams. Give the ordinates of the influence lines every 2 m.

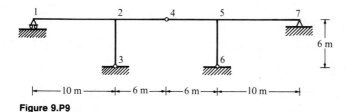

Figure 9.P9

Photographs and Brief Description of a Nuclear Power Plant

The St. Lucia Nuclear Power Plant consists of two units (890 MWe each). The one unit went into commercial operation in December 1976 and the other in August 1983.

Subsurface exploration borings of the soil where the plant was to be placed indicated poorly consolidated sand with thin layers of clay to a depth of 65 ft (19.85 m) below existing grade. To meet seismic criteria, the unsuitable soil was excavated and the hole was filled with well-graded sand and then compacted to the required specifications. An 80-ft (54.86-m) diameter circular sheet pile cofferdam was constructed by driving 72-ft (21.95-m) long sheet piling through the compacted sand with electric vibratory hammers. The cofferdam was braced with internal compression beams (see Fig. 9.6). The reactor containment consists of a free-standing, cylindrical steel vessel and an external 3-ft (0.91-m) thick cylindrical, reinforced concrete shield structure of 74-ft (22.56-m) inside diameter (see Figs. 9.5 and 9.7). The annulus between the steel vessel and the concrete shield structure provides an interspace whose atmosphere can be monitored and controlled. The concrete shield structure is supported on a 9-ft (2.47-m) thick ring wall which rests on the base concrete mat.

Figure 9.5 The St. Lucia nuclear power plant of Florida Power and Light Company on Hutchinson Island, Fla. *(Courtesy of Ebasco Services, Inc.)*

Figure 9.6   View of the circular cofferdam of the St. Lucia power plant during construction. *(Courtesy of Ebasco Services, Inc.)*

Figure 9.7   View of the cylindrical concrete shield structure of the St. Lucia power plant during construction. *(Courtesy of Ebasco Services, Inc.)*

Fundamental Relations of Mechanics of Materials

A.1 Introduction

In this appendix, we derive and discuss the relations between the quantities which characterize the state of stress and the state of deformation of a member of a structure. More precisely, we include the following.

1. In Secs. A.2 to A.5, we define the quantities (displacement, strain, stress, strain energy density, etc.) which characterize the state of deformation and the state of stress of a body and we present the relations between them (strain-displacement relations, equilibrium equations, stress-strain relations).

2. In Secs. A.6 to A.12, we focus our attention on straight line members. That is, we do the following:

   a. In Sec. A.6, we establish the relations between the external and internal actions acting on straight members (equations of equilibrium).

   b. In Sec. A.7, we establish the relations between the components of displacement of a cross section of a straight member and the components of internal actions acting on this cross section.

   c. In Sec. A.8, we compute the components of displacement of straight members by direct integration of the relations established in item b.

   d. In Secs. A.9 to A.12, we establish the components of stress and strain at a particle of a cross section of a straight member in terms of the components of the internal actions acting on this cross section.

A.2   Displacement Vector and Components of Strain of a Particle

Consider an infinitesimal portion (particle) of a body, located at point $P_0(x_1, x_2, x_3)$ prior to deformation, whose position vector, referred to a fixed origin O, is designated by $\mathbf{r}_0 = x_1\mathbf{i}_1 + x_2\mathbf{i}_2 + x_3\mathbf{i}_3$. After deformation, this particle moves to a point $P(\xi_1, \xi_2, \xi_3)$ whose position vector, referred to the same fixed origin O, is designated by $\mathbf{r} = \xi_1\mathbf{i}_1 + \xi_2\mathbf{i}_2 + \xi_3\mathbf{i}_3$. It is apparent that the coordinates ξ_i ($i = 1, 2, 3$) are functions of x_1, x_2, and x_3. We assume that the functions $\xi_i(x_1, x_2, x_3)$ have continuous partial derivatives of any order required. Referring to Fig. A.1, we define the *displacement vector* $\hat{\mathbf{u}} = \hat{u}_1\mathbf{i}_1 + \hat{u}_2\mathbf{i}_2 + \hat{u}_3\mathbf{i}_3$ of the particle of the body located at point P_0, prior to deformation, by the following vector equation.

$$\mathbf{r}(x_1, x_2, x_3) = \mathbf{r}_0 + \hat{\mathbf{u}}(x_1, x_2, x_3) \tag{A.1}$$

or $$\xi_i(x_1, x_2, x_3) = x_i + \hat{u}_i(x_1, x_2, x_3) \qquad i = 1, 2, 3 \tag{A.2}$$

Consider a material line segment, which is straight prior to deformation, extending from point P_0 to Q_0. After deformation, this straight material line will generally be a curve extending from point P to point Q. If point Q_0 is very close to point P_0, then, as shown in Fig. A.2, the curved segment PQ approaches a straight line. We denote the unit vectors in the directions $\overrightarrow{P_0Q_0}$ and $\overrightarrow{PQ}$ by $\mathbf{n}_0$ and $\mathbf{n}$. Thus, an infinitesimal material straight line segment extending, prior to deformation, between two neighboring points P_0 and Q_0 $(\overrightarrow{P_0Q_0} = d\mathbf{r}_0 = dx_1\,\mathbf{i}_1 + dx_2\,\mathbf{i}_2 + dx_3\,\mathbf{i}_3 = |d\mathbf{r}_0|\mathbf{n}_0)$ deforms into an infinitesimal straight line segment extending between two neighboring points P and Q $(\overrightarrow{PQ} = d\mathbf{r} = d\xi_1\,\mathbf{i}_1 + d\xi_2\,\mathbf{i}_2 + d\xi_3\,\mathbf{i}_3 = |d\mathbf{r}|\mathbf{n})$. The angle between the

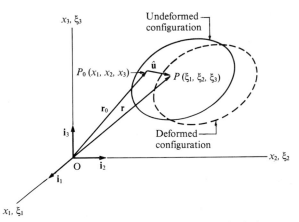

Figure A.1 Displacement vector of a particle of a body.

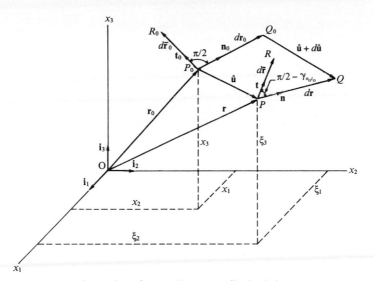

Figure A.2 Deformation of mutually perpendicular infinitesimal material line segments.

infinitesimal material line segment $\overrightarrow{P_0Q_0}$ and $\overrightarrow{PQ}$ represents the total rotation of the material line segment $\overrightarrow{P_0Q_0}$. We denote by $E_{n_0n_0}$ *the unit elongation* in the direction $\mathbf{n}_0$ of the particle located at point P_0 prior to deformation, and we define it as

$$E_{n_0n_0} = \frac{|d\mathbf{r}| - |d\mathbf{r}_0|}{|d\mathbf{r}_0|} \tag{A.3}$$

Referring to Fig. A.2, consider another infinitesimal material line segment extending prior to deformation from point P_0 to a neighboring point R_0 ($\overrightarrow{P_0R_0}$ $= d\tilde{\mathbf{r}}_0 = d\tilde{x}_1\,\mathbf{i}_1 + d\tilde{x}_2\,\mathbf{i}_2 + d\tilde{x}_3\,\mathbf{i}_3 = |d\tilde{\mathbf{r}}_0|\mathbf{t}_0$) such that the angle $\widehat{Q_0P_0R_0}$ is a right angle. Subsequent to deformation, this material line segment extends from point P to a neighboring point R ($\overrightarrow{PR} = d\tilde{\mathbf{r}} = d\tilde{\xi}_1\,\mathbf{i}_1 + d\tilde{\xi}_2\,\mathbf{i}_2 + d\tilde{\xi}_3\,\mathbf{i}_3$ $= |d\tilde{\mathbf{r}}|\mathbf{t}$). Generally, the angle $\widehat{QPR}$ is different than 90°. We define the *unit shear* $\gamma_{n_0t_0}$ in the directions $\mathbf{n}_0$ and $\mathbf{t}_0$ of the particle located prior to deformation at point P_0 as the change of the angle, in radians, between the two infinitesimal material lines P_0Q_0 and P_0R_0 due to deformation. Thus, the before-deformation right angle $\widehat{Q_0P_0R_0}$ becomes $\widehat{QPR} = (\pi/2) - \gamma_{n_0t_0}$ subsequent to deformation. With this definition, we have established the convention that positive unit shear $\widehat{\gamma_{n_0t_0}}$ indicates reduction of the original right angle. For example, consider two material, infinitesimal line segments $\overrightarrow{P_0Q_0} = dx_1\,\mathbf{i}_1$ and $\overrightarrow{P_0R_0} =$ $dx_2\,\mathbf{i}_2$ directed along the x_1 and x_2 axes, respectively, prior to deformation. After deformation, these line segments will translate, rotate, and elongate. In

Fig. A.3, the deformed configuration of these line segments is $\overrightarrow{PR}$ and $\overrightarrow{PQ}$, respectively. Referring to Fig. A.3, we have

$$
\overrightarrow{PQ} = \left(dx_1 + \frac{\partial \hat{u}_1}{\partial x_1} dx_1 \right) \mathbf{i}_1 + \left(\frac{\partial \hat{u}_2}{\partial x_1} dx_1 \right) \mathbf{i}_2 + \left(\frac{\partial \hat{u}_3}{\partial x_1} dx_1 \right) \mathbf{i}_3
$$

$$
\overrightarrow{PR} = \left(\frac{\partial \hat{u}_1}{\partial x_2} dx_2 \right) \mathbf{i}_1 + \left(dx_2 + \frac{\partial \hat{u}_2}{\partial x_2} dx_2 \right) \mathbf{i}_2 + \left(\frac{\partial \hat{u}_3}{\partial x_2} dx_2 \right) \mathbf{i}_3
$$

(A.4)

The unit elongation E_{11} is equal to

$$
E_{11} = \frac{|\overrightarrow{PQ}| - |\overrightarrow{P_0Q_0}|}{|\overrightarrow{P_0Q_0}|} = \frac{|\overrightarrow{PQ}| - dx_1}{dx_1}
$$

(A.5)

where, referring to relation (A.4), we see that the magnitude of the line segment $\overrightarrow{PQ}$ is equal to

$$
|\overrightarrow{PQ}| = \sqrt{\left(dx_1 + \frac{\partial \hat{u}_1}{\partial x_1} dx_1 \right)^2 + \left(\frac{\partial \hat{u}_2}{\partial x_1} dx_1 \right)^2 + \left(\frac{\partial \hat{u}_3}{\partial x_1} dx_1 \right)^2 }
$$

(A.6)

Substituting relation (A.6) into (A.5), we obtain

$$
E_{11} = \sqrt{ 1 + 2\left(\frac{\partial \hat{u}_1}{\partial x_1} \right) + \left(\frac{\partial \hat{u}_1}{\partial x_1} \right)^2 + \left(\frac{\partial \hat{u}_2}{\partial x_1} \right)^2 + \left(\frac{\partial \hat{u}_3}{\partial x_1} \right)^2 } - 1
$$

(A.7)

Moreover, referring to Fig. A.3, we see that the unit shear γ_{12} is equal to

$$
\gamma_{12} = \frac{\pi}{2} - \widehat{RPQ}
$$

(A.8)

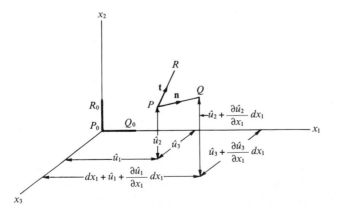

Figure A.3 Unit elongations and unit shears.

Thus, using relations (A.4) and noting by referring to relation (A.5) that $|\overrightarrow{PQ}|$ $= (1 + E_{11})\,dx_1$ and similarly that $|\overrightarrow{PR}| = (1 + E_{22})\,dx_2$, we get

$$\sin \gamma_{12} = \cos \widehat{RPQ} = \frac{(\overrightarrow{PQ}) \cdot (\overrightarrow{PR})}{|\overrightarrow{PQ}||\overrightarrow{RQ}|}$$

$$= \frac{\left(1 + \dfrac{\partial \hat{u}_1}{\partial x_1}\right)\dfrac{\partial \hat{u}_1}{\partial x_2} + \left(1 + \dfrac{\partial \hat{u}_2}{\partial x_2}\right)\dfrac{\partial \hat{u}_2}{\partial x_1} + \dfrac{\partial \hat{u}_3}{\partial x_1}\dfrac{\partial \hat{u}_3}{\partial x_2}}{(1 + E_{11})(1 + E_{22})} \qquad (A.9)$$

Consider a particle of a body which prior to deformation is an orthogonal parallelepiped having dimensions dx_1, dx_2, and dx_3 (see Fig. A.4). In general, because of deformation, this particle translates, rotates, elongates, or shrinks and distorts. The translation and rotation of the particle are rigid-body motions and are the result of the deformation of all the other particles comprising the total volume of the body. Notice that within an infinitesimal region of a deforming body, straight lines, planes, and parallelism of straight lines and planes are preserved. Therefore, in general, the deformed particle under consideration is a nonorthogonal parallelepiped with edges of length $(1 + E_{11})$ dx_1, $(1 + E_{22})\,dx_2$, $(1 + E_{33})\,dx_3$ and angles $(\pi/2 - \gamma_{12})$, $(\pi/2 - \gamma_{13})$, $(\pi/2 - \gamma_{23})$, where E_{ii} $(i = 1, 2, 3)$ are the unit elongations in the $\mathbf{i}_i$ direction and γ_{ij} $(i, j = 1, 2, 3, i \neq j)$ are the shears in the $\mathbf{i}_i$, $\mathbf{i}_j$ directions. The geometry of the deformed particle is completely specified if the length of its edges and the angles between its edges are known. Thus, referring to Fig. A.4, the unit

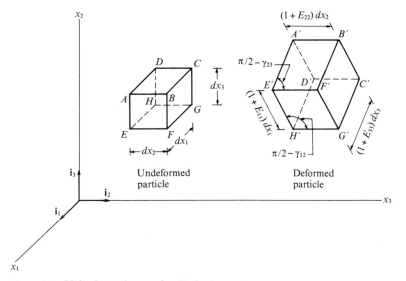

Figure A.4 Unit elongations and unit shears.

elongations and the unit shears of a particle of a body along any three mutually perpendicular directions i_1, i_2, and i_3 completely specify the deformation of this particle.

In this text, we limit our attention to structures whose deformation is such that the unit elongations, unit shears, and rotations of their particles are very small compared to unity, and, moreover, the rotations are not of a higher order of magnitude than the unit elongations and the unit shears. In this case, the undeformed and the deformed cross-sectional areas of the members of a structure differ negligibly. Consequently, when computing the components of stress in the members of a structure from the corresponding internal actions, the effect of the change of their cross sectional areas due to the deformation is disregarded. Moreover, in this case, the rate of change (gradient) of the components of displacement ($\partial \hat{u}_i / \partial x_j$ $i, j = 1, 2, 3$) is small compared to unity and the difference between the unit shear γ_{12} and $\sin \gamma_{12}$ is negligible. Consequently, in relation (A.7) the terms $(\partial \hat{u}_1 / \partial x_1)^2$, $(\partial \hat{u}_2 / \partial x_1)^2$, and $(\partial \hat{u}_3 / \partial x_1)^2$ can be disregarded as compared to unity, while in relation (A.9) the products $(\partial \hat{u}_1 / \partial x_1)(\partial \hat{u}_1 / \partial x_2)$, $(\partial \hat{u}_2 / \partial x_1)(\partial \hat{u}_2 / \partial x_2)$, and $(\partial \hat{u}_3 / \partial x_1)(\partial \hat{u}_3 / \partial x_2)$ can be disregarded as compared to the terms $\partial \hat{u}_1 / \partial x_2$ and $\partial \hat{u}_2 / \partial x_1$. Thus, using the binomial expansion, we reduce† relations (A.7) and (A.9), to

$$E_{11} \approx \sqrt{1 + 2\frac{\partial \hat{u}_1}{\partial x_1}} - 1 = 1 + \frac{\partial \hat{u}_1}{\partial x_1} - \frac{1}{3}\left(\frac{\partial u_1}{\partial x_1}\right)^3 + \cdots - 1$$

$$\approx \frac{\partial \hat{u}_1}{\partial x_1} = e_{11} \tag{A.10}$$

$$\gamma_{12} \approx \sin \gamma_{12} = \frac{\partial u_1}{\partial x_2} + \frac{\partial u_2}{\partial x_1} = 2e_{12}$$

or in general

$$\begin{array}{ll} E_{ii} = e_{ii} & i = 1, 2, 3 \\ \gamma_{ij} = 2e_{ij} & i \neq j\ i, j = 1, 2, 3 \end{array} \tag{A.11}$$

where
$$e_{11} = \frac{\partial \hat{u}_1}{\partial x_1} \qquad e_{22} = \frac{\partial \hat{u}_2}{\partial x_2} \qquad e_{33} = \frac{\partial \hat{u}_3}{\partial x_3}$$

$$e_{21} = e_{12} = \frac{1}{2}\left(\frac{\partial \hat{u}_2}{\partial x_1} + \frac{\partial \hat{u}_1}{\partial x_2}\right)$$

---

† For a more detailed discussion, see Novozhilov.[1]

$$e_{31} = e_{13} = \frac{1}{2}\left(\frac{\partial \hat{u}_3}{\partial x_1} + \frac{\partial \hat{u}_1}{\partial x_3}\right) \tag{A.12}$$

$$e_{32} = e_{23} = \frac{1}{2}\left(\frac{\partial \hat{u}_2}{\partial x_3} + \frac{\partial \hat{u}_3}{\partial x_2}\right)$$

are called *the components of strain of the particle*. Referring to relations (A.12), it is apparent that

$$e_{ij} = e_{ji} \tag{A.13}$$

It can be shown that e_{ij} $(i, j = 1, 2, 3)$ are components of a symmetric tensor of the second rank, referred to as the *strain tensor*. A tensor of the second rank $[e]$ is an entity which has the following properties.

1. At any point of the region wherein it is defined, it is specified by an array of nine numbers e_{ij} $(i, j = 1, 2, 3)$ referred to a rectangular system of axes x_1, x_2, x_3. These nine numbers are called the cartesian components of the tensor; they are expressed in matrix form as

$$[e] = \begin{bmatrix} e_{11} & e_{12} & e_{13} \\ e_{21} & e_{22} & e_{23} \\ e_{31} & e_{32} & e_{33} \end{bmatrix} \tag{A.14}$$

2. Its components at a point, referred to any two right-handed rectangular system of axes, are related by the following relation.

$$e'_{ij} = \sum_{k=1}^{3} \sum_{m=1}^{3} \lambda_{ik}\lambda_{jm}e_{km} \qquad i, j = 1, 2, 3 \tag{A.15}$$

where e_{km} $(k, m = 1, 2, 3)$ are the components of the tensor referred to the rectangular system of axes x_1, x_2, and x_3 while e'_{ij} $(i, j = 1, 2, 3)$ are the components of the tensor referred to the rectangular system of axes x'_1, x'_2, and x'_3; λ_{ik} and λ_{jm} are the direction cosines of the x'_1, x'_2, and x'_3 axes with respect to the x_1, x_2, and x_3 axes; that is, λ_{12} is the cosine of the angle between the axes x'_1 and x_2. Relation (A.15) can be written in matrix form as

$$[e'] = [\Lambda_S][e][\Lambda_S]^T \tag{A.16}$$

where $[\Lambda_S]^T$ is the transpose of the matrix $[\Lambda_S]$. The latter is defined as

$$[\Lambda_S] = \begin{bmatrix} \lambda_{11} & \lambda_{12} & \lambda_{13} \\ \lambda_{21} & \lambda_{22} & \lambda_{23} \\ \lambda_{31} & \lambda_{32} & \lambda_{33} \end{bmatrix} \tag{A.17}$$

As discussed previously, the three-unit elongations E_{11}, E_{22}, and E_{33} and the three unit shears γ_{12}, γ_{13}, and γ_{23} of a particle of a structure referred to a rectangular system of axes x_1, x_2, and x_3 specify completely the deformation of this particle. In the case in which the unit elongations, the unit shears, and the rotations are very small compared to unity and the rotations are not of a higher order of magnitude than the unit elongations and unit shears, referring to relations (A.11), we can readily see that the nine components of strain (A.12) of a particle referred to a rectangular system of axes x_1, x_2, and x_3 specify completely the deformation of this particle; that is, the deformation of a particle of a structure is specified by a symmetric tensor of the second rank whose components are linearly related to the components of the displacement vector $\hat{\mathbf{u}}$ [see relations (A.12)].

A.3 Components of Stress at a Particle

Consider a particle located at point B inside a body which has deformed as a result of the application of external actions, and imagine that a plane nn through point B severs the body into portions I and II (see Fig. A.5). The unit vector normal to this plane is designated as $\mathbf{n}$. Generally, in order to maintain the equilibrium of a part of a body, a force $\mathbf{F}$ and a moment $\mathbf{M}$ must be applied at the cut. The force $\mathbf{F}$ is the resultant of all the forces $\Delta\mathbf{F}$ acting on each element ΔA of the area of the surface of the cut (see Fig. A.5). The moment $\mathbf{M}$ is the resultant of the moments of all the forces $\Delta\mathbf{F}$ about the point of application of the resultant $\mathbf{F}$. We assume that in the limit, as $\Delta A \to 0$, the moment $\Delta\mathbf{M}$ acting on the element ΔA vanishes. It is apparent that if the external actions and the reactions acting on a body are known, the resultant force $\mathbf{F}$ and the resultant moment $\mathbf{M}$ can be established by considering the equilibrium of the one portion of the body.

We denote by $\Delta F_n \mathbf{i}_n$ and $\Delta F_t \mathbf{i}_t$ the normal and tangential components of

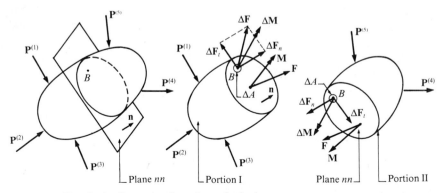

Figure A.5 Free-body diagrams of portions of a body.

the force $\Delta\mathbf{F}$, respectively; that is, $\Delta\mathbf{F} = \Delta F_n\mathbf{i}_n + \Delta F_t\mathbf{i}_t$. The normal component of stress at particle B acting on the plane nn is defined as

$$\tau_{nn} = \lim_{\Delta A \to 0} \frac{\Delta F_n}{\Delta A} \quad (FL^{-2}) \tag{A.18}$$

Moreover, the shearing component of stress at particle B acting on the plane nn is defined as

$$\tau_{nt} = \lim_{\Delta A \to 0} \frac{\Delta F_t}{\Delta A} \quad (FL^{-2}) \tag{A.19}$$

In relations (A.18) and (A.19), it is implied that ΔA includes B as it approaches zero. Notice that ΔA is a small area taken in the deformed config-uration. However, on the basis of the assumption that the unit elongations, unit shears, and rotations are small compared to unity and the rotations are not of a higher order than the unit elongations and the unit shears, the difference between the dimensions before and after deformation of any element of length, area, or volume of a body is small and can be neglected when compared to the original dimensions of the element.

Notice that the normal and the tangential component of stress at particle B acting on a plane nn is generally unequal to the normal and the tangential component of stress at particle B respectively, acting on another plane mm.

If the plane nn is chosen normal to the x_1 axis of an orthogonal system of axes (see Fig. A.6), then the normal component of stress acting on the particle at point B on the surface perpendicular to the x_1 axis is designated as τ_{11}. Moreover, the tangential component $\Delta\mathbf{F}_t$ of the force acting on an area ΔA containing point B may be decomposed into its components in the x_2 and x_3 directions, $(\Delta\mathbf{F}_t = \Delta F_2\mathbf{i}_2 + \Delta F_3\mathbf{i}_3)$. The shearing components of stress τ_{12} and

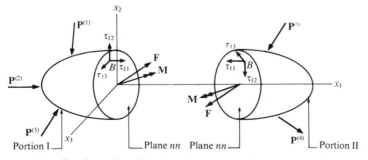

Figure A.6 Reorientation of the cut plane so that its normal is along the x_1 axis.

τ_{13} acting on the particle at point B on the plane perpendicular to the x_1 axis may be defined as

$$\tau_{12} = \lim_{\Delta A \to 0} \frac{\Delta F_2}{\Delta A} \qquad \tau_{13} = \lim_{\Delta A \to 0} \frac{\Delta F_3}{\Delta A} \qquad (A.20)$$

Generally, on the basis of the foregoing discussion, on any plane through a particle there are three components of stress—one normal and two tangential. We represent schematically three mutually perpendicular planes passing through a particle by the faces of a parallelepiped. For instance, if the parallelepiped shown in Fig. A.7 were to be considered as a schematic representation of three mutually perpendicular planes through particle B of the body shown in Fig. A.6, we may observe that face $A'E'D'C'$ of the parallelepiped represents plane nn as viewed on portion I of the body (see Fig. A.6), whereas face $AEDC$ represents the same plane nn as viewed on portion II of the body. Similarly, faces $AA'EE'$ and $DD'CC'$ represent the plane through particle B which is perpendicular to the x_2 axis.

The components of stress at a particle are assumed positive, as shown in Fig. A.7. Notice that a positive normal component of stress tends to elongate the element along the line of its action; it is referred to as *tension*. A negative normal component of stress tends to compress the element; it is referred to as *compression*. The nine components of stress at a particle acting on three

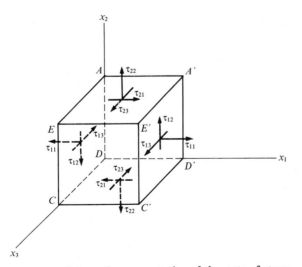

Figure A.7 Schematic representation of the state of stress at a particle. Definition of positive components of stress.

mutually perpendicular planes form the stress matrix of the particle. That is,

$$[\tau] = \begin{bmatrix} \tau_{11} & \tau_{12} & \tau_{13} \\ \tau_{21} & \tau_{22} & \tau_{23} \\ \tau_{31} & \tau_{32} & \tau_{33} \end{bmatrix} \tag{A.21}$$

The first subscript indicates the direction of the normal to the plane on which the stress acts, while the second subscript indicates the direction of the stress component. In Sec. A.4, it is shown that the stress matrix is symmetric. Moreover, it can be shown that the components of stress τ_{ij} ($i, j = 1, 2, 3$) are components of a symmetric tensor of the second rank. That is, the nine components of stress τ_{ij} at a particle acting on the planes normal to the rectangular system of axes x_1', x_2', x_3' are related to the nine components of stress at the same particle acting on the planes normal to another rectangular system of axes x_1, x_2, x_3 by a relation analogous to (A.15) or (A.16). That is,

$$[\tau'] = [\Lambda_S][\tau][\Lambda_S]^T \tag{A.22}$$

where $[\Lambda_S]$ is defined by relation (A.17).

The components of the resultant internal force $\mathbf{F} = N\mathbf{i}_1 + Q_2\mathbf{i}_2 + Q_3\mathbf{i}_3$ acting on the plane normal to the x_1 axis are given by the following relations:

$$N = \iint_A \tau_{11} \, dA \tag{A.23a}$$

$$Q_2 = \iint_A \tau_{12} \, dA \tag{A.23b}$$

$$Q_3 = \iint_A \tau_{13} \, dA \tag{A.23c}$$

Moreover, the components of the resultant internal moment $\mathbf{M}$ acting on the plane normal to the x_1 axis about the origin of the axes of reference are given by the following relations:

$$M_1 = \iint_A (-\tau_{12}x_3 + \tau_{13}x_2) \, dA \tag{A.24a}$$

$$M_2 = \iint_A \tau_{11}x_3 \, dA \tag{A.24b}$$

$$M_3 = -\iint_A \tau_{11}x_2 \, dA \tag{A.24c}$$

A.4 Equilibrium Equations—Symmetry of the Components of Stress at a Particle

Consider an infinitesimal element of a body at the deformed configuration having dimensions dx_1, dx_2, and dx_3. Inasmuch as the stress varies from point to point of the body, the stress components acting on any face of the element will, in general, vary throughout this face. Thus, referring to Fig. A.8, we may consider $\tau_{11}(x_1, x_2, x_3)$, $\tau_{12}(x_1, x_2, x_3)$, and $\tau_{13}(x_1, x_2, x_3)$ to be the average components of stress acting on face *OGFD* and $\tau_{11}(x_1 + dx_1, x_2, x_3)$, $\tau_{12}(x_1 + dx_1, x_2, x_3)$, and $\tau_{13}(x_1 + dx_1, x_2, x_3)$ to be the average components of stress acting on face *ABCE*. Inasmuch as a well-behaved function $\tau_{11}(x_1 + dx_1, x_2, x_3)$ can be expressed in terms of its values at a neighboring point x_1, x_2, x_3 by a Taylor series expansion, we can write

$$\tau_{11}(x_1 + dx_1, x_2, x_3) = \tau_{11}(x_1, x_2, x_3) + \frac{\partial \tau_{11}}{\partial x_1} dx_1 \qquad \text{(A.25)}$$

where the partial derivative of τ_{11} is evaluated at point (x_1, x_2, x_3), and the terms involving second or higher powers of dx_1 have been disregarded.

As shown in Fig. A.8, in addition to the components of stress acting on the faces of the element, the body force $\mathbf{B} = B_1\mathbf{i}_1 + B_2\mathbf{i}_2 + B_3\mathbf{i}_3$ per unit volume acts at the center of mass of the element. The body force could be the result of the presence of the body in a gravitational field. Since the element is in equilibrium, the sum of the forces acting on it must vanish. Therefore, we may write

$$\Sigma F_1 = 0 = -\tau_{11}\, dx_2\, dx_3 + \left(\tau_{11} + \frac{\partial \tau_{11}}{\partial x_1} dx_1 \right) dx_2\, dx_3$$

$$-\tau_{21}\, dx_1\, dx_3 + \left(\tau_{21} + \frac{\partial \tau_{21}}{\partial x_2} dx_2 \right) dx_1\, dx_3 - \tau_{31}\, dx_1\, dx_2$$

$$+ \left(\tau_{31} + \frac{\partial \tau_{31}}{\partial x_3} dx_3 \right) dx_1\, dx_2 + B_1\, dx_1\, dx_2\, dx_3 \qquad \text{(A.26)}$$

Collecting terms, we have

$$\frac{\partial \tau_{11}}{\partial x_1} + \frac{\partial \tau_{21}}{\partial x_2} + \frac{\partial \tau_{31}}{\partial x_3} + B_1 = 0 \qquad \text{(A.27a)}$$

Similarly,

$$\Sigma F_2 = 0 \qquad \frac{\partial \tau_{12}}{\partial x_1} + \frac{\partial \tau_{22}}{\partial x_2} + \frac{\partial \tau_{32}}{\partial x_3} + B_2 = 0 \qquad \text{(A.27b)}$$

$$\Sigma F_3 = 0 \qquad \frac{\partial \tau_{13}}{\partial x_1} + \frac{\partial \tau_{23}}{\partial x_2} + \frac{\partial \tau_{33}}{\partial x_3} + B_3 = 0 \qquad \text{(A.27c)}$$

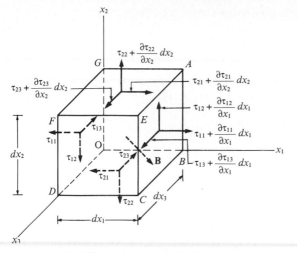

Figure A.8 Equilibrium of a material element.

Equations (A.27) are referred to as the *equations of equilibrium*. They must be satisifed by the components of stress acting on any three mutually perpendicular planes at a particle of a body.

Finally, the equilibrium of the element requires that the sum of the moments of all the forces acting on it about any point must vanish. We may, therefore, write

$$\Sigma M_3 = -\tau_{11}\, dx_2\, dx_3\, \frac{dx_2}{2} + \left(\tau_{11} + \frac{\partial \tau_{11}}{\partial x_1}\, dx_1\right) dx_2\, dx_3\, \frac{dx_2}{2}$$

$$- \left(\tau_{12} + \frac{\partial \tau_{12}}{\partial x_1}\, dx_1\right) dx_2\, dx_3\, dx_1 + \tau_{22}\, dx_1\, dx_3\, \frac{dx_1}{2}$$

$$- \left(\tau_{22} + \frac{\partial \tau_{22}}{\partial x_2}\, dx_2\right) dx_1\, dx_3\, \frac{dx_1}{2} + \left(\tau_{21} + \frac{\partial \tau_{21}}{\partial x_2}\, dx_2\right) dx_1\, dx_3\, dx_2$$

$$- \tau_{31}\, dx_1\, dx_2\, \frac{dx_2}{2} + \left(\tau_{31} + \frac{\partial \tau_{31}}{\partial x_3}\, dx_3\right) dx_1\, dx_2\, \frac{dx_2}{2}$$

$$+ \tau_{32}\, dx_1\, dx_2\, \frac{dx_1}{2} - \left(\tau_{32} + \frac{\partial \tau_{32}}{\partial x_3}\, dx_3\right) dx_1\, dx_2\, \frac{dx_1}{2}$$

$$+ B_1\, dx_1\, dx_2\, dx_3\, \frac{dx_2}{2} + B_2\, dx_1\, dx_2\, dx_3\, \frac{dx_1}{2} = 0 \qquad \text{(A.28)}$$

Simplifying and neglecting higher order terms, we obtain

$$\tau_{12} = \tau_{21} \tag{A.29a}$$

Similarly, we have

$$\Sigma M_2 = 0 \qquad \tau_{13} = \tau_{31} \tag{A.29b}$$

$$\Sigma M_1 = 0 \qquad \tau_{23} = \tau_{32} \tag{A.29c}$$

Therefore, we may conclude that the matrix of the components of stress at a particle is a symmetric matrix.

A.5 Stress-Strain Relations—Strain Energy Density and Complementary Energy Density

It has been established from uniaxial tension tests, performed in an environment of constant temperature, that for many materials of engineering importance and for values of the stress component below a certain limit, referred to as the *elastic limit,* the relation between the component of stress and strain is unique. It can be considered independent not only of time, but also of the loading history. That is, a value of the component of stress corresponds to the same value of the component of strain, irrespective of whether this value has been reached by loading or unloading, subsequent to one or more cycles. These materials are called *elastic.*

When a body made of an elastic material is subjected to a general three-dimensional state of stress in an environment of varying temperature, the components of stress at a particle are single-valued functions of the components of strain and the temperature of this particle. Moreover the components of strain of a particle are single-valued functions of the components of stress and the temperature of this particle. That is

$$\left.\begin{array}{l} \tau_{ij} = \tau_{ij}(e_{11}, e_{12}, e_{13}, e_{21}, e_{22}, e_{23}, e_{13}, e_{23}, e_{33}, T) \\[2mm] e_{ij} = e_{ij}(\tau_{11}, \tau_{12}, \tau_{13}, \tau_{21}, \tau_{22}, \tau_{23}, \tau_{13}, \tau_{23}, \tau_{33}, T) \end{array}\right\} \quad i, j = 1, 2, 3 \quad (A.30)$$

Elastic materials are subdivided into *linearly elastic* and *nonlinearly elastic,* depending on whether the relation between the components of stress, strain and temperature of their particles are linear or nonlinear, respectively (see Fig. A.9).

A.5.1 Strain energy density and complementary energy density

Consider a body made of an elastic material subjected to a state of uniaxial tension or compression ($\tau_{11} \neq 0$, $\tau_{12} = \tau_{13} = \tau_{23} = \tau_{22} = \tau_{33} = 0$) in an

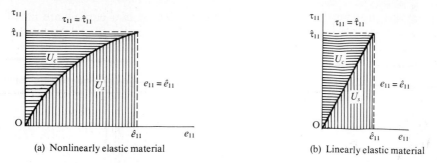

(a) Nonlinearly elastic material   (b) Linearly elastic material

Figure A.9 Stress-strain curve of elastic materials in uniaxial tension or compression.

environment of constant temperature. The *strain energy density* U_s and the *complementary energy density* U_c of a particle of this body are defined as

$$U_s = \int_0^{e_{11}} \tau_{11}\, de'_{11} \qquad U_c = \int_0^{\tau_{11}} e_{11}\, d\tau'_{11} \qquad (A.31)$$

Referring to Fig. A.9, we see that the strain energy density U_s, corresponding to any value of the strain $\hat{e}_{11}$, is the area between the stress-strain curve, the e_{11} axis, and the line $e_{11} = \hat{e}_{11}$, while the complementary energy density U_c for any value of the stress $\hat{\tau}_{11}$ is the area between the stress-strain curve, the τ_{11} axis, and the line $\tau_{11} = \hat{\tau}_{11}$. It is apparent that for a given value of the component of strain or stress, there is only one corresponding value of the strain energy density U_s or of the complementary energy density U_c.

The *strain energy density* U_s and the *complementary energy density* U_c at a particle of a body made of an elastic material, subjected to a general three-dimensional state of stress in an environment of constant temperature, are defined as

$$U_s = \int_0^{e_{mn}} \sum_{i=1}^{3} \sum_{j=1}^{3} \tau_{ij}(e'_{pq})\, de'_{ij} \qquad (A.32a)$$

$$m, n, p, q = 1, 2, 3$$

$$U_c = \int_0^{\tau_{mn}} \sum_{i=1}^{3} \sum_{j=1}^{3} e_{ij}(\tau'_{pq})\, d\tau'_{ij} \qquad (A.32b)$$

In relations (A.32a), integration is carried out over a curve in the nine-dimensional space of the nine components of strain e_{pq} connecting the origin (undeformed state) with the point $P(e_{mn})$ (deformed state). In relation (A.32b), integration is carried out over a curve in the nine-dimensional space of the nine components of stress τ_{pq} connecting the origin (unstressed state) with the point $P(\tau_{mn})$ (stressed state).

We assume that the relations between the components of stress and strain at the particles of a body made of an elastic material are such that the strain energy density and the complementary energy density of a particle are independent of the history of loading. They depend only on the final value of the components of stress or strain. Thus, the integrals in relations (A.32) are independent of the path of integration; consequently, their integrands must be perfect differentials. It is known from calculus that for $\tau_{ij}\, de_{ij}$ to be a perfect differential it is necessary and sufficient that

$$\frac{\partial \tau_{ij}}{\partial e_{mn}} = \frac{\partial \tau_{mn}}{\partial e_{ij}} \tag{A.33a}$$

Moreover for $e_{ij}\, d\tau_{ij}$ to be a perfect differential, it is necessary and sufficient that

$$\frac{\partial e_{ij}}{\partial \tau_{mn}} = \frac{\partial e_{mn}}{\partial \tau_{ij}} \tag{A.33b}$$

These relations impose certain restrictions on the stress-strain relations for an elastic body (A.30).

Inasmuch as the integrands in relations (A.32) are perfect differentials we have

$$dU_s = \sum_{i=1}^{3} \sum_{j=1}^{3} \frac{\partial U_s}{\partial e_{ij}}\, de_{ij} = \sum_{i=1}^{3} \sum_{j=1}^{3} \tau_{ij}\, de_{ij} \tag{A.34}$$

and

$$dU_c = \sum_{i=1}^{3} \sum_{j=1}^{3} \frac{\partial U_c}{\partial \tau_{ij}}\, d\tau_{ij} = \sum_{i=1}^{3} \sum_{j=1}^{3} e_{ij}\, d\tau_{ij} \tag{A.35}$$

Consequently

$$\tau_{ij} = \frac{\partial U_s}{\partial e_{ij}} \tag{A.36}$$

and

$$e_{ij} = \frac{\partial U_c}{\partial \tau_{ij}} \tag{A.37}$$

In the above relations, U_s is considered a function of the nine components of strain, while U_c is considered a function of the nine components of stress. Therefore, their partial derivative with respect to any component of strain or stress, respectively, implies that the other eight components are considered constant during differentiation.

From relation (A.36) it is apparent that the relation between the components of stress and strain at a particle of an elastic body subjected to deformation in an environment of constant temperature may be established if the strain energy density U_s is a known function of the components of strain. If U_s is a second-degree polynomial of the components of strain, the relations between the components of stress and strain will be linear. If U_s is higher than a second-degree polynomial or any nonlinear function of the components of strain, then the relation between the components of stress and strain will be nonlinear.

The total strain energy and the total complementary energy of a body of volume V made of an elastic material are equal to

$$(U_s)_T = \int\int\int_V U_s \, dV \qquad (U_c)_T = \int\int\int_V U_c \, dV \qquad \text{(A.38)}$$

Referring to Fig. A.9b, we can readily see that for linearly elastic materials, relations (A.31) reduce to

$$U_s = U_c = \tfrac{1}{2}\tau_{11}e_{11}$$

This result may be extended to a body made from a linearly elastic material subjected to a general state of stress in an environment of constant temperature. In this case, the strain energy density and the complementary energy density are equal to

$$U_s = U_c = \tfrac{1}{2}\sum_{i=1}^{3}\sum_{j=1}^{3}\tau_{ij}e_{ij} \qquad \text{(A.39)}$$

A.5.2 Physical significance of the total strain energy of a body

Consider an infinitesimal element of dimensions dx_1, dx_2, and dx_3 of a body that is made of an elastic material subjected to external actions and is in an environment of constant temperature. The forces acting on the faces of this element normal to the x_1 and x_2 axes and the components of the body force **B** along those axes are shown in Fig. A.11. Suppose that the external actions applied to the body induce a state of plane strain [$\hat{u}_3 = 0$, $\hat{u}_1 = \hat{u}_1(x_1, x_3)$, $\hat{u}_2 = \hat{u}_2(x_1, x_2)$, $e_{33} = e_{13} = e_{23} = 0$]. In this case, the components of stress are functions only of the coordinates x_1 and x_2.

Assume that the external actions change by an infinitesimal amount in a way that the state of plane strain is preserved. The corresponding changes of the components of displacement of the vertices of the element are shown in Fig.

A.10. The work performed by the forces acting on this element (see Fig. A.11), as a result of the additional deformation described in Fig. A.10, is equal to

$$d(dW) = -(\tau_{11} \, dx_2 \, dx_3) \left(d\hat{u}_1 + \frac{1}{2} \frac{\partial d\hat{u}_1}{\partial x_2} \, dx_2 \right)$$

$$+ \left(\tau_{11} + \frac{\partial \tau_{11}}{\partial x_1} \, dx_1 \right) dx_2 \, dx_3 \left(d\hat{u}_1 + \frac{\partial d\hat{u}_1}{\partial x_1} \, dx_1 + \frac{1}{2} \frac{\partial d\hat{u}_1}{\partial x_2} \, dx_2 \right)$$

$$+ B_1 \, dV \left(d\hat{u}_1 + \frac{\partial d\hat{u}_1}{\partial x_1} \, dx_1 + \frac{\partial d\hat{u}_1}{\partial x_2} \, dx_2 \right)$$

$$- (\tau_{22} \, dx_1 \, dx_3) \left(d\hat{u}_2 + \frac{1}{2} \frac{\partial d\hat{u}_2}{\partial x_1} \, dx_1 \right)$$

$$+ \left(\tau_{22} + \frac{\partial \tau_{22}}{\partial x_2} \, dx_2 \right) dx_1 \, dx_3 \left(d\hat{u}_2 + \frac{\partial d\hat{u}_2}{\partial x_2} \, dx_2 + \frac{1}{2} \frac{\partial d\hat{u}_2}{\partial x_1} \, dx_1 \right)$$

$$+ B_2 \, dV \left(d\hat{u}_2 + \frac{\partial d\hat{u}_2}{\partial x_2} \, dx_2 + \frac{1}{2} \frac{\partial d\hat{u}_2}{\partial x_1} \, dx_1 \right)$$

$$- (\tau_{21} \, dx_1 \, dx_3) \left(d\hat{u}_1 + \frac{1}{2} \frac{\partial d\hat{u}_1}{\partial x_1} \, dx_1 \right)$$

$$+ \left(\tau_{21} + \frac{\partial \tau_{21}}{\partial x_2} \, dx_2 \right) dx_1 \, dx_3 \left(d\hat{u}_1 + \frac{\partial d\hat{u}_1}{\partial x_2} \, dx_2 + \frac{1}{2} \frac{\partial du_1}{\partial x_1} \, dx_1 \right)$$

$$- (\tau_{12} \, dx_2 \, dx_3) \left(d\hat{u}_2 + \frac{1}{2} \frac{\partial d\hat{u}_2}{\partial x_2} \, dx_2 \right)$$

$$+ \left(\tau_{12} + \frac{\partial \tau_{12}}{\partial x_1} \, dx_1 \right) dx_2 \, dx_3 \left(d\hat{u}_2 + \frac{\partial d\hat{u}_2}{\partial x_1} \, dx_1 + \frac{1}{2} \frac{\partial d\hat{u}_2}{\partial x_2} \, dx_2 \right)$$

Disregarding infinitesimals of higher order, using relations (A.27) and (A.12), and taking into account that the components of stress are functions only of the coordinates x_1 and x_2, the above relation reduces to

$$d(dW) = \left(\tau_{11} \frac{\partial d\hat{u}_1}{\partial x_1} + \tau_{22} \frac{\partial d\hat{u}_2}{\partial x_2} + \tau_{21} \frac{\partial d\hat{u}_1}{\partial x_2} + \tau_{12} \frac{\partial d\hat{u}_2}{\partial x_1} \right) dV$$

$$= (\tau_{11} \, de_{11} + \tau_{22} \, de_{22} + \tau_{12} \, de_{12} + \tau_{21} \, de_{21}) \, dV$$

(A.40)

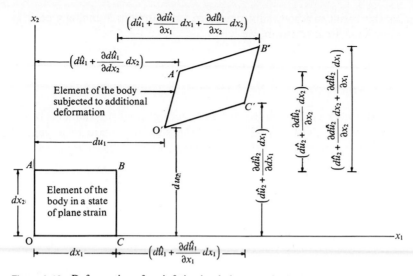

Figure A.10   Deformation of an infinitesimal element of a body subjected to a state of plane strain.

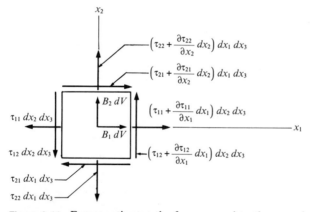

Figure A.11   Forces acting on the faces normal to the x_1 and x_2 axis of an infinitesimal element of a body.

The sum of the work performed by the forces acting on all the elements of the body, which is due to the change of its deformation as a result of the increment of the external actions, is equal to

$$dW = \int\int\int_V d(dW)$$

$$= \int\int\int_V (\tau_{11}\, de_{11} + \tau_{22}\, de_{22} + \tau_{12}\, de_{12} + \tau_{21}\, de_{21})\, dV \quad \text{(A.41)}$$

Extending this result to a body subjected to external actions inducing a general three-dimensional state of strain, we have

$$dW = \int\int\int_V \sum_{i=1}^{3} \sum_{j=1}^{3} \tau_{ij} \, de_{ij} \, dV \tag{A.42}$$

The sum of the work performed by the forces acting on all the elements of the body, as a result of its deformation from the stess-free state to a general deformed state characterized by the components of strain e_{mm} is equal to

$$W = \int\int\int_V \left(\int_0^{e_{mn}} \sum_{i=1}^{3} \sum_{j=1}^{3} \tau_{ij} \, de_{ij} \right) dV \tag{A.43}$$

Referring to relation (A.32a), we get from the above relation

$$W = \int\int\int_V U_s \, dV = (U_s)_T \tag{A.44}$$

where $(U_s)_T$ represents the total strain energy of the body. Since the common boundary of two adjacent elements is subjected to equal and opposite components of stress and is displaced by the same amount, the sum of the work performed by these equal and opposite components of stress vanishes. Consequently, the total work performed by the forces acting on every element of the body consists only of the work of the components of stress acting on the faces of the elements which are part of the surface of the body and the work of the body forces. Thus, W in Eq. (A.44) is equal to the work of the known external actions (body forces and surface actions) acting on the body and of the reactions of its supports. If the supports of the body do not move, the work of the reactions must vanish. On the basis of the foregoing, Eq. (A.44) becomes

$$W_{\text{ext act}} + W_{\text{reac}} = (U_s)_T \tag{A.45}$$

Thus, we may conclude that *the total strain energy of an elastic body subjected to surface actions and body forces in an environment of constant temperature is equal to the work performed by these actions in bringing it from its stress-free strain-free reference state to its deformed state.*

A.5.3   Stress-strain relations

To the order of accuracy of most engineering calculations, the behavior of many engineering materials can be considered independent of the direction of the stress field. That is, the constants involved in the relations between the components of stress and strain of a particle of a body made of one of these materials do not change with the system of axes to which the components of stress and strain are referred. These materials are called *isotropic*. In this text, we limit our attention to isotropic, linearly elastic materials.

When a body made from an isotropic, linearly elastic material is subjected to external actions in an environment of constant temperature, using the sign convention for the components of stress shown in Fig. A.7, we see that the relations between the components of stress and strain of a particle of this body are

$$\tau_{11} = \frac{E}{(1 + \nu)(1 - 2\nu)} [(1 - \nu)e_{11} + \nu(e_{22} + e_{33})]$$

$$\tau_{22} = \frac{E}{(1 + \nu)(1 - 2\nu)} [(1 - \nu)e_{22} + \nu(e_{11} + e_{33})] \qquad \text{(A.46)}$$

$$\tau_{33} = \frac{E}{(1 + \nu)(1 - 2\nu)} [(1 - \nu)e_{33} + \nu(e_{22} + e_{11})]$$

$$\tau_{12} = 2Ge_{12} \qquad \tau_{13} = 2Ge_{13} \qquad \tau_{23} = 2Ge_{23}$$

Relations (A.46) may be inverted to yield

$$e_{11} = \frac{1}{E} [\tau_{11} - \nu(\tau_{22} + \tau_{33})]$$

$$e_{22} = \frac{1}{E} [\tau_{22} - \nu(\tau_{11} + \tau_{33})] \qquad \text{(A.47)}$$

$$e_{33} = \frac{1}{E} [\tau_{33} - \nu(\tau_{11} + \tau_{22})]$$

$$e_{12} = \frac{\tau_{12}}{2G} \qquad e_{13} = \frac{\tau_{13}}{2G} \qquad e_{23} = \frac{\tau_{23}}{2G}$$

For a state of uniaxial tension ($\tau_{22} = \tau_{33} = \tau_{12} = \tau_{13} = \tau_{23} = 0, \tau_{11} \neq 0$), Eqs. (A.47) reduce to

$$e_{11} = \frac{\tau_{11}}{E} \qquad e_{22} = e_{33} = -\frac{\nu\tau_{11}}{E} = -\nu e_{11}$$

The material (elastic) constants E and ν are referred to as the *modulus of elasticity* and the *Poisson's ratio,* respectively. They may be evaluated from a uniaxial tension test. The material constant G is referred to as the *shear modulus.* It can be evaluated from a torsion test.

It can be shown[2] that the elastic constants E, ν, and G of an isotropic linearly elastic material are related by the following relation:

$$G = \frac{E}{2(1 + \nu)} \tag{A.48}$$

This implies that the response to an isotropic, linearly elastic material is characterized by two material constants. Physically, this makes sense for it indicates that the response of these materials to shearing stress is not independent of their response to normal stress.

The elastic constants of materials depend on the temperature at which the deformation process takes place. However, the changes of the elastic constants of most engineering materials, within a temperature range in the vicinity of room temperature, are rather small (see Fig. A.12).

In relations (A.47) the components of strain of a particle are only functions of its components of stress. Moreover the strain-free state corresponds to the stress-free state. However, in general the normal components of strain of a particle are also functions of the change of temperature from the reference temperature T_0 at the stress-free state. Moreover, when analyzing framed structures it may be necessary to consider some of their members as being in a state of initial strain (see Sec. 1.3.4). That is, it is assumed that the strain-free state of the particles of these members does not correspond to their stress-free state at the reference temperature T_0. Thus, in general, the components of strain of a particle may be expressed as

$$e_{ii} = e_{ii}^S + e_{ii}^T + e_{ii}^I \qquad i, j = 1, 2, 3$$

$$e_{ij} = e_{ij}^S \qquad\qquad i \neq j$$

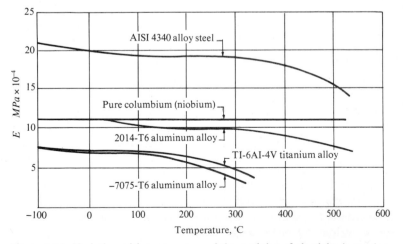

Figure A.12 Variation with temperature of the modulus of elasticity in tension.

where e_{ij}^S is the part of the component of strain e_{ij} $(i, j = 1, 2, 3)$ of the particle generated by the components of stress. It is related to the components of stress by relations (A.47). e_{ii}^T is the part of the normal component of strain e_{ii} $(i = 1, 2, 3)$ of the particle generated by the change of temperature ΔT $(\Delta T = T - T_0)$ from the reference temperature T_0 at the stress-free state $(e_{ii}^T = 0$ when $T = T_0)$. For linearly elastic materials it is usually assumed that e_{ii}^T is a linear function of the change of temperature. e_{ii}^I is the part of the normal component of strain e_{ii} $(i = 1, 2, 3)$ of the particle which exists at its stress-free state at the reference temperature T_0.

In certain instances a member of a framed structure is manufactured with its length or curvature slightly different than that required to fit in the geometry of the structure. In order that such a member be connected to the structure, it must be subjected to the external actions required to change its length or curvature and make it fit. When these actions are subsequently removed, the structure assumes a deformed configuration. The members of statically indeterminate structures restrain their nonfitting members from assuming their undeformed geometry, while the members of statically determinate structures do not. For this reason the members of statically indeterminate structures deform and are subjected to internal actions while the members of statically determinate structures neither deform nor are subjected to internal actions; they only move as rigid bodies.

In order to establish the effect of nonfitting members of a structure on the components of displacement of its points and on the internal actions in its members, we may regard its nonfitting members as being subjected to a state of initial strain. Thus a member which is manufactured with a length which is either shorter or longer than that required in order to fit may be regarded as being subjected to a uniform axial component of initial strain. That is

$$e_{11}^I = \frac{\text{manufactured length} - \text{required length}}{\text{required length}}$$

Moreover, a member which is manufactured with a different curvature than that required in order to fit may be regarded as being subjected to an initial axial component of strain. For example, if a straight member of a structure has an initial curvature, $k_2^I(x_1)$ in the $x_1 x_2$ plane may be regarded as being subjected to a nonuniform axial component of initial strain. Referring to relations (A.63a) and to Fig. A.13, we see that this initial strain is equal to

$$e_{11}^I = -x_2 \frac{d\theta_3^I}{dx_1} = -x_2 k_2^I = -\frac{x_2}{\rho_2^I}$$

where ρ_2^I is the initial radius of curvature.

As discussed in Sec. 1.3.4, the steel bars of a prestressed concrete beam are subjected to tensile forces until they elongate by a prescribed amount ΔL. Their

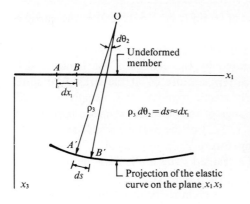

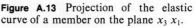

Figure A.13 Projection of the elastic curve of a member on the plane $x_3 \, x_1$.

ends are then fastened to those of the concrete beam and the external forces are removed. However, the bars cannot assume their undeformed configuration because they are restrained by the concrete beam. The resulting internal actions in the steel bars and the concrete beam may be established by regarding the bars as being subjected to a uniform axial component of initial strain equal to

$$e_{11}^I = -\frac{\Delta L}{L}$$

On the basis of the foregoing discussion, when a body made of an isotropic linearly elastic material is subjected to external actions in an environment of varying temperature, its stress-strain relations are

$$e_{11} = \frac{1}{E} \left[\tau_{11} - \nu(\tau_{22} + \tau_{33}) \right] + \alpha(T - T_0) + e_{11}^I$$

$$e_{22} = \frac{1}{E} \left[\tau_{22} - \nu(\tau_{11} + \tau_{33}) \right] + \alpha(T - T_0) + e_{22}^I \qquad \text{(A.49)}$$

$$e_{33} = \frac{1}{E} \left[\tau_{33} - \nu(\tau_{11} + \tau_{22}) \right] + \alpha(T - T_0) + e_{33}^I$$

$$e_{12} = \frac{\tau_{12}}{2G} \qquad e_{13} = \frac{\tau_{13}}{2G} \qquad e_{23} = \frac{\tau_{23}}{2G}$$

or

$$\tau_{11} = \frac{E}{(1 + \nu)(1 - 2\nu)} \left[(1 - \nu)e_{11}^a + \nu(e_{22}^a + e_{33}^a) \right] - \beta(T - T_0)$$

$$\tau_{22} = \frac{E}{(1 + \nu)(1 - 2\nu)} [(1 - \nu)e_{22}^a + \nu(e_{11}^a + e_{33}^a)] - \beta(T - T_0)$$

$$\tau_{33} = \frac{E}{(1 + \nu)(1 - 2\nu)} [(1 - \nu)e_{33}^a + \nu(e_{11}^a + e_{22}^a)] - \beta(T - T_0)$$

$$\tau_{12} = 2Ge_{12} \qquad \tau_{13} = 2Ge_{13} \qquad \tau_{23} = 2Ge_{23} \qquad \text{(A.50)}$$

where e_{ii}^a $(i = 1, 2, 3)$ = total strain e_{ii} minus initial strain e_{ii}^I
$\qquad\qquad T$ = temperature at present state
$\qquad\qquad T_0$ = temperature at the reference stress-free state
$\qquad\qquad \alpha$ = coefficient of linear thermal expansion

and β is given by

$$\beta = \frac{E\alpha}{1 - 2\nu} \qquad \text{(A.51)}$$

The coefficient of linear thermal expansion represents a material property. It may be taken as constant for moderate changes of temperature. At a given temperature, the coefficient of linear thermal expansion changes when the magnitude and the character of the stress alters. However, this change is negligible for elastic states of stress.

Consider a body subject to a general deformation process. The first law of thermodynamics (principle of conservation of energy) states that

$$Q = \Delta E - W \qquad \text{(A.52)}$$

where Q = total heat energy absorbed by body during process of deformation
$\qquad \Delta E$ = change of internal energy of body during process of deformation
$\qquad W$ = work of external forces (body forces and surface actions) during process of deformation

When a body made of an elastic material is subjected to a deformation process in an environment of constant temperature, considerable heat energy could flow in or out of the body, which increases or decreases its internal energy. Thus, the principle of conservation of energy cannot be invoked to prove that the total strain energy in a body made of an elastic material, subjected to a deformation process in an environment of constant temperature, is equal to the work of external forces $[W = (U_s)_T]$. As shown in Sec. A.5.2, this is a result of the definition of $(U_s)_T$. From relation (A.52), it can be seen that the change of the internal energy in an elastic body subjected to deformation in an environment of constant temperature is not equal to its strain energy.

A.6 Equations of Equilibrium for a Segment of Infinitesimal Length of a Straight Member

Consider a straight member of a framed structure, subjected to external concentrated forces $\mathbf{P}^{(i)}$ ($i = 1, 2, \ldots, N$), concentrated moments $\mathbf{M}^{(i)}$ ($i = 1, 2, \ldots, M$) and to external distributed forces $\mathbf{q}(x_1)$ and moments $\mathbf{m}(x_1)$ given in units of force or moment per unit length (see Fig. A.14). Consider a segment of this member of length Δx_1, shown in Fig. A.15. This segment is loaded only by distributed forces $\mathbf{q}(x_1)$ and moments $\mathbf{m}(x_1)$ (the latter are not shown in Fig. A.15 to avoid cluttering it). The internal actions at the end faces of the segment under consideration are assumed positive, as shown in Fig. A.15. Since the segment under consideration is in equilibrium, we have

$$\Sigma F_1 = 0 = -N + q_1\,\Delta x_1 + N + \Delta N$$

$$\Sigma F_2 = 0 = -Q_2 + q_2\,\Delta x_1 + Q_2 + \Delta Q_2$$

$$\Sigma F_3 = 0 = -Q_3 + q_3\,\Delta x_1 + Q_3 + \Delta Q_3$$

$$\Sigma M_1 = 0 = -M_1 + m_1\,\Delta x_1 + M_1 + \Delta M_1 \tag{A.53}$$

$$\Sigma M_2 = 0 = -M_2 + m_2\,\Delta x_1 + M_2 + \Delta M_2 + q_3\frac{(\Delta x_1)^2}{2} - Q_3\,\Delta x_1$$

$$\Sigma M_3 = 0 = -M_3 + m_3\,\Delta x_1 + M_3 + \Delta M_3 - q_2\frac{(\Delta x_1)^2}{2} + Q_2\,\Delta x_1$$

In the limit as $\Delta x_1 \to 0$, the above relations reduce to:

$$q_1 = -\frac{dN}{dx_1} \tag{A.54}$$

$$q_2 = -\frac{dQ_2}{dx_1} \tag{A.55}$$

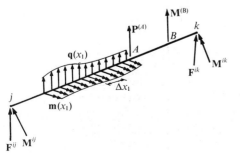

Figure A.14 Member of a framed structure.

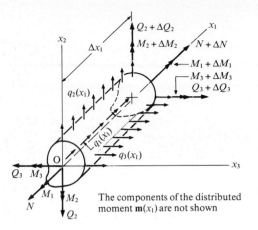

The components of the distributed moment $\mathbf{m}(x_1)$ are not shown

Figure A.15 Segment of a member.

$$q_3 = -\frac{dQ_3}{dx_1} \tag{A.56}$$

$$m_1 = -\frac{dM_1}{dx_1} \tag{A.57}$$

$$Q_2 = -\frac{dM_3}{dx_1} - m_3 \tag{A.58}$$

$$Q_3 = \frac{dM_2}{dx_1} + m_2 \tag{A.59}$$

These relations must be satisfied at any cross section of a member in equilibrium, where the derivatives of the internal actions with respect to x_1 exist. For instance, as shown in Fig. A.16, at a cross section where a concentrated force $P_3^{(A)}$ is applied, the shearing force $Q_3(x_1)$ is a discontinuous function of $x_1[Q_3(a^+) = Q_3(a^-) - P_3^{(A)}]$. Consequently, at this cross section, the derivative of $Q_3(x_1)$ with respect to x_1 does not exist and relation (A.56) is meaningless. Similarly, at the cross section where a concentrated moment $M_2^{(B)}$ is applied, the internal moment M_2 is a discontinuous function of $x_1[M_2(b^+) = M_2(b^-) - M_2^{(B)}]$. Therefore at this cross section relation (A.59) is also meaningless.

Notice that in the mechanics of materials theories presented in this appendix, the internal actions acting on any element of infinitesimal length dx_1 of a member must satisfy the conditions of equilibrium (A.54) to (A.59). This however does not ensure that the conditions of equilibrium (A.27) will be satisfied by the components of stress τ_{ij} $(i, j = 1, 2, 3)$ acting on a particle of the member. In fact the components of stress obtained on the basis of the mechanics of materials theories often do not satisfy the equations of equilibrium (A.27).

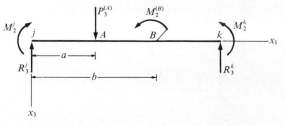

(a) Member subjected to a concentrated force and a
concentrated moment

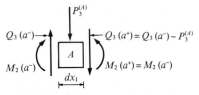

(b) Free-body diagram of a segment of
the member at point A

(c) Free-body diagram of a segment
of the member at point B

Figure A.16 Discontinuity in the internal shearing force and in the bending moment.

A.7 Relations between the Components of Internal Action and the Components of Displacement of Straight Members

In this section, we limit our attention to members subjected to axial deformation and to bending without twisting in an environment of constant temperature. In order that a member be subjected to bending without twisting, the external moments acting on it should not have a torsional component. Moreover, the plane of the external forces acting on the member must pass through the shear center of its cross section (see Fig. A.17). The location of the shear center depends only on the geometry of the cross section of the member (see Sec. A.10).

As discussed in Sec. 1.2, in this text we are concerned with the behavior of idealized structures composed of line members made of idealized materials (isotropic linearly elastic) subjected to external loads. The fact that two of the dimensions of line members are small as compared to their third dimension, their length, allows us to make certain assumptions as to the geometry of their deformed configuration and as to the distribution of their components of stress. The assumptions as to the geometry of the deformed configuration of line members subjected to axial deformation and to bending without twisting (due to Bernoulli) are:

1. Plane sections normal to the axis of a line member, prior to deformation, can be considered plane subsequent to deformation; that is, the warping of the cross sections of a line member is assumed negligible.

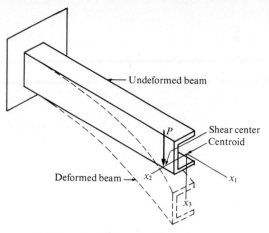

(a) The external force passes through
    the shear center

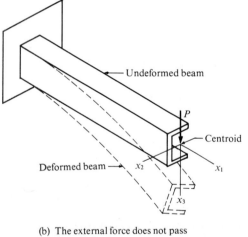

(b) The external force does not pass
    through the shear center

Figure A.17   Bending of a cantilever channel.

2. Plane sections normal to the axis of a line member before deformation can
   be considered normal to its deformed axis subsequent to deformation.

The assumptions as to the distribution of stress in line members are

$$\tau_{22} = \tau_{33} = \tau_{23} = 0 \tag{A.60}$$

These assumptions reduce the relations between the components of stress and
strain (A.46) or (A.47) for line members made of isotropic, linearly elastic

materials to the following:

$$\tau_{11} = Ee_{11} \tag{A.61a}$$

$$e_{22} = e_{33} = -\nu e_{11} \tag{A.61b}$$

$$\tau_{12} = 2Ge_{12} \tag{A.61c}$$

$$\tau_{13} = 2Ge_{13} \tag{A.61d}$$

Actually, when a line member is subjected to an axial force and/or to end bending moments, plane sections normal to its axis remain plane, subsequent to deformation and normal to its deformed axis. However, when a line member is subjected to external transverse forces, its cross sections warp and do not remain normal to its deformed axis. Nevertheless, when the length of a line member is considerably larger than its other dimensions, the warping of its cross sections and the change of the angle between it axis and its cross sections due to its deformation do not appreciably affect the components of displacement and the normal components of strain of its points.

The theory based on the first assumption only is referred to as the *Timoshenko theory of beams*, while the theory based on both assumptions is referred to as the *classical theory of beams*.

A.7.1 The Timoshenko theory of beams

On the basis of the first geometric (Bernoulli) assumption, the component of displacement $\hat{u}_1$ of any point (x_1, x_2, x_3) of a cross section of a straight line member may be expressed as

$$\hat{u}_1(x_1, x_2, x_3) = u_1(x_1) - x_2\theta_3(x_1) + x_3\theta_2(x_1) \tag{A.62}$$

where $u_1(x_1)$ is the axial component of translation of the centroid of the cross section; $\theta_2(x_1)$ and $\theta_3(x_1)$ are the components of rotation of the cross section about x_2 and the x_3 axes, respectively (see Fig. A.18). Substituting relation (A.62) into (A.12), we obtain the following expressions for the components of strain e_{11}, e_{12}, and e_{13}:

$$e_{11} = \frac{\partial \hat{u}_1}{\partial x_1} = \frac{du_1}{dx_1} - x_2 \frac{d\theta_3}{dx_1} + x_3 \frac{d\theta_2}{dx_1} \tag{A.63a}$$

$$2e_{12} = \frac{\partial \hat{u}_1}{\partial x_2} + \frac{\partial \hat{u}_2}{\partial x_1} = -\theta_3 + \frac{du_2}{dx_1} \tag{A.63b}$$

$$2e_{13} = \frac{\partial \hat{u}_1}{\partial x_3} + \frac{\partial \hat{u}_3}{\partial x_1} = \theta_2 + \frac{du_3}{dx_1} \tag{A.63c}$$

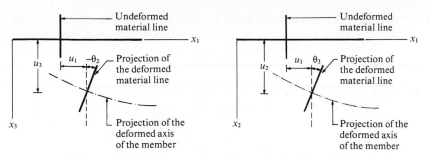

Figure A.18 Deformation of a line member subjected to bending without twisting.

The components of strain $e_{22} = \partial \hat{u}_2 / \partial x_2$ or $e_{33} = \partial \hat{u}_3 / \partial x_3$ of a particle are obtained from relation (A.61b). The right-hand sides of relations (A.63b) and (A.63c) have been obtained by assuming that

$$\hat{u}_2 = u_2(x_1) + f_2(x_2, x_3)$$
$$\hat{u}_3 = u_3(x_1) + f_3(x_2, x_3)$$

Consequently,

$$\frac{\partial \hat{u}_2}{\partial x_1} = \frac{du_2}{dx_1} \qquad \frac{\partial \hat{u}_3}{\partial x_1} = \frac{du_3}{dx_1}$$

Relations (A.63b) and (A.63c) indicate that the shearing components of strain e_{12} and e_{13} do not vary with x_2 and x_3. This, however, is not in agreement with experimental results or with results obtained on the basis of more exact theories. Moreover, the shearing components of stress obtained on the basis of formula (A.104) are not constant at the points of a cross section. In the Timoshenko theory of beams, it is assumed that the values of the shearing components of strain obtained on the basis of relation (A.63) approximate satisfactorily the maximum values of the actual shearing components of strain which occur at the centroid of a cross section. The average values of the shearing components of strain acting on a cross section are obtained by multiplying the values given by relations (A.63) by correction factors. That is,

$$(2e_{12})_{av} = \frac{1}{\lambda_2}\left(-\theta_3 + \frac{du_2}{dx_1}\right) \qquad (2e_{13})_{av} = \frac{1}{\lambda_3}\left(\theta_2 + \frac{du_3}{dx_1}\right) \qquad \text{(A.64)}$$

The correction factors λ_2 and λ_3 depend primarily on the geometry of the cross section of the member.† For instance, the value of the coefficient λ_3 for a mem-

---

† An interesting review as well as a better approach for finding λ_2 and λ_3 are presented in papers by Cowper.[3,4]

ber of rectangular cross section of depth h subjected to transverse forces in the x_1x_3 plane may be established by noting that the distribution of the shearing components of stress τ_{13} is parabolic, vanishes at $x_3 = \pm h/2$, and is maximum at $x_3 = 0$ (see Example 1 of Sec. A.9). Hence,

$$e_{13} = \frac{\tau_{13}}{2G} = (e_{13})_{\text{max}}\left(1 - \frac{4x_3^2}{h^2}\right) \qquad \text{(A.65)}$$

The average value of e_{13} for all particles of a rectangular cross section is equal to

$$(e_{13})_{\text{av}} = \frac{(e_{13})_{\text{max}}}{h}\int_{-h/2}^{h/2}\left(1 - \frac{4x_3^2}{h^2}\right)dx_3 = \frac{2}{3}(e_{13})_{\text{max}} \qquad \text{(A.66)}$$

Thus, the correction factor λ_3 for a rectangular cross section is

$$\lambda_3 = \frac{(e_{13})_{\text{max}}}{(e_{13})_{\text{av}}} = 1.5 \qquad \text{(A.67)}$$

For a circular cross section λ_2 or λ_3 is equal to $\frac{4}{3}$. For an I beam, λ_2 or λ_3 is approximately equal to A/A_w, where A_w is the area of the web of the beam. As discussed in Sec. 5.13, the effect of shear deformation on the components of displacement of the points of a structure obtained on the basis of the Timoshenko theory of beams is not very accurate. This indicates that the assumption that the shearing components of strain obtained on the basis of relation (A.63) satisfactorily approximate their maximum values is not accurate.

By substituting relation (A.63a) and relations (A.64) into (A.61) and the resulting expressions into (A.24) and (A.25) and taking into account that the local axes x_2 and x_3 are principal and centroidal, we obtain

$$(a)\quad N = EA\frac{du_1}{dx_1} \qquad (b)\quad M_2 = EI_2\frac{d\theta_2}{dx_1} \qquad (c)\quad M_3 = EI_3\frac{d\theta_3}{dx_1}$$

$$(d)\quad Q_2 = \frac{GA}{\lambda_2}\left(-\theta_3 + \frac{du_2}{dx_1}\right) \qquad (e)\quad Q_3 = \frac{GA}{\lambda_3}\left(\theta_2 + \frac{du_3}{dx_1}\right) \qquad \text{(A.68)}$$

where I_2 and I_3 are the moments of inertia of the cross section of the member with respect to the x_2 or x_3 axis, respectively.

If the local x_2 and x_3 axes of a member are centroidal but not principal, we get

$$(a)\quad N = EA\frac{du_1}{dx_1} \qquad (b)\quad M_2 = EI_2\frac{d\theta_2}{dx_1} - EI_{23}\frac{d\theta_3}{dx_1}$$

(c) $M_3 = EI_3 \dfrac{d\theta_3}{dx_1} - EI_{23} \dfrac{d\theta_2}{dx_1}$ (d) $Q_2 = \dfrac{GA}{\lambda_2}\left(-\theta_3 + \dfrac{du_2}{dx_1}\right)$

(e) $Q_3 = \dfrac{GA}{\lambda_3}\left(\theta_2 + \dfrac{du_3}{dx_1}\right)$

$$(A.69)$$

These relations may be rewritten as

(a) $N = EA \dfrac{du_1}{dx_1}$ (b) $I_3 M_2 + I_{23} M_3 = E(I_2 I_3 - I_{23}^2)\dfrac{d\theta_2}{dx_1}$

(c) $I_2 M_3 + I_{23} M_2 = E(I_2 I_3 - I_{23}^2)\dfrac{d\theta_3}{dx_1}$

(d) $Q_2 = \dfrac{GA}{\lambda_2}\left(-\theta_3 + \dfrac{du_2}{dx_1}\right)$ (e) $Q_3 = \dfrac{GA}{\lambda_3}\left(\theta_2 + \dfrac{du_3}{dx_1}\right)$

$$(A.70)$$

where I_{23} is the product of inertia of the cross section of the member with respect to the x_2 and x_3 axes.

A.7.2   The classical theory of beams

In the classical theory of beams, the second of the Bernoulli assumptions is employed in order to express the components of rotation θ_2 and θ_3 in terms of the components of translation u_2 and u_3. This assumption implies that the components of rotation θ_2 and θ_3 are equal to the slope of the projection of the deformed axis (elastic curve) of the member on the $x_1 x_3$ and $x_2 x_1$ plane, respectively. Thus, for the classical theory of beams, we have

$$\theta_2(x_1) \approx -\frac{du_3}{dx_1} \qquad \theta_3(x_1) \approx \frac{du_2}{dx_1} \tag{A.71}$$

Relations (A.71) may be verified by referring to Fig. A.19 ($\theta_2 \approx \tan \theta_2 = -\tan(-\theta_2) = -du_3/dx_1$). Substituting relations (A.71) into (A.62), we obtain the following relation for the axial component of displacement of a point of a line member subjected to bending without twisting.

$$\hat{u}_1(x_1, x_2, x_3) = u_1(x_1) - x_3 \frac{du_3}{dx_1} - x_2 \frac{du_2}{dx_1} \tag{A.72}$$

This relation characterizes the deformed configuration of a cross section of a member in the classical theory of beams. Substituting relation (A.72) in the

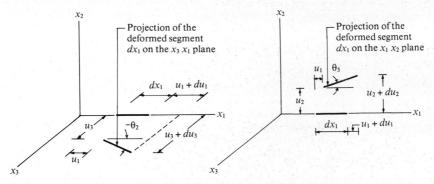

Figure A.19 Relations between the angles of rotation and the slopes.

first of relations (A.12), we get

$$e_{11} = \frac{\partial \hat{u}_1}{\partial x_1} = \frac{du_1}{dx_1} - x_3 \frac{d^2 u_3}{dx_1^2} - x_2 \frac{d^2 u_2}{dx_1^2} \tag{A.73}$$

The components of strain e_{22} and e_{33} are obtained by substituting relation (A.73) into (A.61b). Moreover the shearing components of strain in the classical theory of beams are established from the shearing components of stress on the basis of relations (A.61c) and (A.61d). The latter are obtained from relation (A.104). Substituting relation (A.73) into the first of relations (A.61) and the resulting expression into (A.23a), (A.24b), and (A.24c), and taking into account that the x_2 and x_3 axes are principal ($I_{23} = 0$) centroidal axes, we obtain, for the classical theory of beams, the following relations between the internal actions acting on a cross section of a member and the components of translation of this cross section:

$$N = EA \frac{du_1}{dx_1} \tag{A.74a}$$

$$M_2 = -EI_2 \frac{d^2 u_3}{dx_1^2} \tag{A.74b}$$

$$M_3 = EI_3 \frac{d^2 u_2}{dx_1^2} \tag{A.74c}$$

If the local axes x_2 and x_3 of a member are centroidal but not principal, we get

$$(a) \quad N = EA \frac{du_1}{dx_1} \qquad (b) \quad M_2 = -EI_2 \frac{d^2 u_3}{dx_1^2} - EI_{23} \frac{d^2 u_2}{dx_1^2} \tag{A.75}$$

$$(c) \quad M_3 = EI_3 \frac{d^2 u_2}{dx_1^2} + EI_{23} \frac{d^2 u_3}{dx_1^2}$$

These relations may be rewritten as

$$(a) \quad N = EA \frac{du_1}{dx_1} \qquad (b) \quad I_3 M_2 + I_{23} M_3 = -E(I_2 I_3 - I_{23}^2) \frac{d^2 u_3}{dx_1^2}$$

$$(c) \quad I_2 M_3 + I_{23} M_2 = E(I_2 I_3 - I_{23}^2) \frac{d^2 u_2}{dx_1^2}$$

$$(A.76)$$

Notice that in this case, when a member is subjected to loading which produces only an M_2 or an M_3 component of moment, the component of translation of a point of its axis in the direction x_2 or x_3, respectively, does not vanish.

A.8 Computation of the Components of Displacement of Points of Straight Members

In this section we describe two methods for computing the components of displacement of a member of a framed structure subjected to axial deformation and to bending without twisting. The first method can be employed only for statically determinate structures, while the second method can be employed for both statically determinate and statically indeterminate structures. However the second method involves differential equations of higher order than those used in the first method; consequently it requires lengthier calculations (see Examples 1 and 2 at the end of this section).

Method I. As illustrated in the examples of Chap. 3, the internal actions in any member of a statically determinate structure can be established as functions of the coordinate x_1 by considering the equilibrium of parts of the structure. The axial component $N(x_1)$ of the internal force in a member may be substituted into Eq. (A.68a) or (A.74a) and the resulting equation may be integrated once to yield the axial component of translation of the cross sections of the member as a function of its axial coordinate. The constant of integration is evaluated from the given condition on the axial component of translation [for example, $u_1(0) = 0$]. This condition ensures that the member does not translate as rigid body in the direction of its axis (see Example 1 at the end of this section). Furthermore, the bending components M_2 and M_3 of the internal moment and the shearing components Q_2 and Q_3 of the internal force in a member may be substituted into Eqs. (A.68b) to (A.68e) and the resulting equations may be integrated to yield the transverse components of translation and rotation of the cross sections of the member as functions of its axial coordinate. The transverse components of translation u_2 and u_3 include the effect of shear deformation. The constants of integration are evaluated from the given conditions on the transverse components of translation and rotation. These conditions ensure that the member does not translate in the direction of the x_2 and

x_3 axes as a rigid body and does not rotate about these axes as a rigid body (see Example 2 at the end of this section).

For members whose length is considerably larger than the dimensions of their cross section, the effect of shear deformation on the transverse components of translation is very small and it is disregarded. In this case, the transverse components of translation u_2 and u_3 are established by substituting the bending components of the moment into Eqs. (A.74b) and (A.74c) and integrating the resulting equations. The components of rotation θ_2 and θ_3 are then obtained by substituting the resulting transverse components of translation into relations (A.71). The constants of integration are evaluated from the given conditions on the transverse components of translation and on the components of rotation.

Method II.   In this method, the internal actions are eliminated from relations (A.74) using the relations between the internal actions of a member and the external actions acting on it [see relations (A.54) to (A.59)]. That is, differentiating relation (A.74a) and using relation (A.54) we obtain

$$\frac{d}{dx_1}\left(EA\frac{du_1}{dx_1}\right) = -q_1 \qquad (A.77a)$$

Moreover, differentiating relation (A.74b) twice and using relations (A.59) and (A.56), we get

$$\frac{d^2}{dx_1^2}\left(EI_2\frac{d^2u_3}{dx_1^2}\right) = -\frac{d^2M_2}{dx_1^2} = -\frac{dQ_3}{dx_1} + \frac{dm_2}{dx_1} = q_3 + \frac{dm_2}{dx_1} \qquad (A.77b)$$

Furthermore, differentiating relation (A.74c) twice and using relations (A.58) and (A.55), we obtain

$$\frac{d^2}{dx_1^2}\left(EI_3\frac{d^2u_2}{dx_1^2}\right) = \frac{d^2M_3}{dx_1^2} = -\frac{dQ_2}{dx_1} - \frac{dm_3}{dx_1} = q_2 - \frac{dm_3}{dx_1} \qquad (A.77c)$$

where q_1, q_2, q_3 and m_1, m_2, m_3 are the components of the external forces and moments, respectively, which are distributed along the length of the member.

For given external actions, Eq. (A.77a) can be integrated to yield the axial component of translation while Eqs. (A.77b) and (A.77c) can be integrated to yield the transverse components of translation and rotation of the cross sections of a member of a structure as functions of its axial coordinate. The constants of integration are evaluated using the specified conditions at the ends of the member (see Examples 1, 2, and 4 at the end of this section).

In general, no matter which method is used to establish the components of translation of the cross sections of a member of a space framed structure subjected to axial deformation and to bending without twisting in an environment of constant temperature, the loading on the member must be given and the value of one quantity of each of the following pairs of quantities must be specified at each end of the member:

$$u_1 \text{ or } N \qquad u_2 \text{ or } Q_2 \qquad u_3 \text{ or } Q_3$$
$$\theta_2 \text{ or } M_2 \qquad \theta_3 \text{ or } M_3$$

(A.78)

Moreover, in order to establish the components of translation and rotation of the cross sections of a member of a planar structure, the loading on the member must be given and the value of one quantity of each of the following pairs of quantities must be specified at each end of the member.

$$u_1 \text{ or } N \qquad u_3 \text{ or } Q_3$$

(A.79)

$$\theta_2 \text{ or } M_2$$

When the components of translation and rotation of the cross sections of a member are established by integrating Eqs. (A.68) (Timoshenko theory of beams) or (A.74) (classical theory of beams), the components of external actions specified at the ends of the member are employed in expressing the components of internal actions as functions of the axial coordinate of the member. The components of displacements specified at the ends of the member (kinematic or essential boundary conditions) are used in evaluating the constants of integration. However, when the components of translation and rotation of the cross sections of a member are established by integrating relations (A.77), all the boundary conditions of the member are used in evaluating the constants of integration (see Examples 2 and 4 at the end of this section).

In the following examples, the components of translation and rotation of straight bars and beams are computed by integrating the corresponding relations (A.74), (A.68), or (A.77).

Example 1. Compute the total elongation of the bar shown in Fig. a, subjected to an axial centroidal force P_1 at its end k and to an axial centroidal force q_1 uniformly distributed along its length and given in units of force per unit of length. The bar is made of an isotropic, linearly elastic material with modulus of elasticity E. The cross-sectional area of the bar is denoted by A.

Figure a Geometry and loading of the bar.

solution This is a statically determinate structure. Consequently its component of displacement $u_1(x_1)$ can be established either from relation (A.74a) or (A.77a) using the following boundary conditions.

$$u_1(0) = 0$$

(a)

$$N(L) = EA \frac{du_1}{dx_1}\bigg|_{x_1 = L} = P_1 \tag{b}$$

Case 1: Solution on the Basis of Relation (A.74a)

Referring to the free-body diagram of the portion of the bar shown in Fig. b, the axial force at any point x_1 is equal to

$$N(x_1) = P_1 + q_1(L - x_1) \tag{c}$$

Substituting the above relation into (A.74a) and integrating, we obtain

$$u_1(x_1) = \int du_1 = \int \frac{P_1 + q_1(L - x_1)}{EA} \, dx_1 = \frac{P_1 x_1}{EA} - \frac{q_1(L - x_1)^2}{2EA} + C \tag{d}$$

Notice that in obtaining relation (c), the boundary condition (b) has been employed. Consequently, the axial component of displacement $u_1(x_1)$ given by relation (d) satisfies the boundary condition (b). The constant of integration C in relation (d) is evaluated by requiring that $u_1(x_1)$ satisfy the boundary condition (a). Thus,

$$C = \frac{q_1 L^2}{2EA}$$

Substituting the value of C into relation (d), we obtain

$$u_1(x_1) = \frac{1}{EA}\left(P_1 x_1 - \frac{q_1 x_1^2}{2} + q_1 L x_1 \right) \tag{e}$$

The total elongation of the bar is equal to

$$u_1(L) = \frac{L}{EA}\left(P_1 + \frac{q_1 L}{2} \right) \tag{f}$$

If the bar were subjected only to the axial centroidal force P_1, relation (f) reduces to

$$u_1(L) = \frac{L P_1}{EA} \tag{g}$$

Case 2: Solution to the Basis of Relation (A.77a)

In the above solution, we have established the internal force $N(x_1)$ by considering the equilibrium of a portion of the bar, and we have substituted it into Eq. (A.74a). The resulting first-order differential equation was then integrated and the constant of integration was evaluated from the kinematic (essential) boundary condition $[u_1(0) = 0]$. In the sequel, we establish the component of translation $u_1(x_1)$ of the bar of Fig. a

$N = P_1 + q_1 (L - x_1)$ q_1 P_1

$\overline{ 2 }$

$\longmapsto L - x_1 \longmapsto$

Figure b Free-body diagram of a portion of the bar.

using the second-order differential equation (A.77 a). That is,

$$\frac{d}{dx_1}\left(EA\frac{du_1}{dx_1}\right) = \frac{dN}{dx_1} = -q_1 \qquad\text{(h)}$$

For q_1 = constant, integrating relation (h) twice we get

$$EAu_1 = -\frac{q_1 x_1^2}{2} + C_1 x_1 + C_2 \qquad\text{(i)}$$

The constants C_1 and C_2 are evaluated by requiring that the component of translation $u_1(x_1)$ satisfy boundary conditions (a) and (b). That is

$$u_1(0) = 0 \quad\rightarrow\quad C_2 = 0 \qquad\text{(j)}$$

$$N(L) = EA\left.\frac{du_1}{dx_1}\right|_{x_1=L} = -q_1 L + C_1 = P_1$$

or $$C_1 = P_1 + q_1 L \qquad\text{(k)}$$

Substituting relations (j) and (k) into (i), we obtain

$$EAu_1 = P_1 x_1 + q_1 L x_1 - \frac{q_1 x_1^2}{2}$$

Example 2.   Consider a cantilever beam subjected at its free end to a transverse force P and a moment $M^{(2)}$, as shown in Fig. a. The beam is made of an isotropic, linearly elastic material of modulus of elasticity E. The beam has a constant cross section whose moment of inertia about the x_2 axis is denoted by I_2.

Compute the translation and rotation of the free end of the cantilever beam using:

1. The classicial theory of beams.
2. The Timoshenko theory of beams.

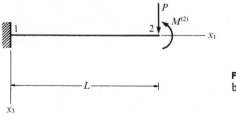

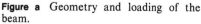

Figure a Geometry and loading of the beam.

solution   The components of translation u_3 and the component of rotation θ_2 of this beam must satisfy the following boundary conditions

$$u_3(0) = 0 \qquad\text{(a)}$$

$$\theta_2(0) = -\left.\frac{du_3}{dx_1}\right|_{x_1=0} = 0 \tag{b}$$

$$Q_3(L) = -\left.\frac{d}{dx_1}\left(EI_2\frac{d^2u_3}{dx_1^2}\right)\right|_{x_1=L} = P \tag{c}$$

$$M_2(L) = -\left.EI_2\frac{d^2u_3}{dx_1^2}\right|_{x_1=L} = M^{(2)} \tag{d}$$

Referring to the free-body diagram of the portion of the beam shown in Fig. b, the shearing force and the moment acting at any cross section of the beam is equal to

$$Q_3 = P \qquad M_2 = M^{(2)} - P(L - x_1) \tag{e}$$

Part 1. Solution on the Basis of the Classical Theory of Beams

CASE 1: SOLUTION ON THE BASIS OF EQ. (A.74b).   Substituting the second of relations (e) into (A.74b), we get

$$\frac{d^2u_3}{dx_1^2} = -\frac{M_2}{EI_2} = -\frac{M^{(2)}}{EI_2} + \frac{P}{EI_2}(L - x_1) \tag{f}$$

Integrating the differential equation (f) twice, we obtain

$$\frac{du_3}{dx_1} = -\frac{M^{(2)}x_1}{EI_2} + \frac{P(L - x_1)^2}{2EI_2} + C_1$$

$$u_3(x_1) = -\frac{M^{(2)}x_1^2}{2EI_2} + \frac{P(L - x_1)^3}{6EI_2} + C_1x_1 + C_2 \tag{g}$$

Notice that in obtaining relations (e), the boundary conditions (c) and (d) have been employed. That is, the shearing force $Q_3(x_1)$ and the moment $M_2(x_1)$ given by Eqs. (e) satisfy the boundary conditions (c) and (d). Moreover, notice that the first of relations (e) is obtained from the second, on the basis of relation (A.59). That is,

$$Q_3(L) = \left.\frac{dM_2}{dx_1}\right|_{x_1=L} = P \tag{h}$$

Thus, the second of relations (e) satisfies the boundary condition (d) directly and the boundary (c) through relation (h). Consequently, the component of translation $u_3(x_1)$ given by relation (g) satisfies the boundary conditions (c) and (d). The constants of integration C_1 and C_2 in relation (g) are evaluated by requiring that the component

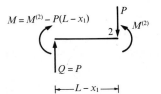

$$M = M^{(2)} - P(L - x_1)$$

$$Q = P$$

$$\longmapsto L - x_1 \longrightarrow$$

Figure b   Free-body diagram of a portion of the beam.

of translation $u_3(x_1)$ satisfy boundary conditions (a) and (b). Noting that $\theta_2 = -du_3/dx_1$ and substituting relations (g) into (a) and (b), we obtain

$$C_1 = \frac{PL^2}{2EI_2} \qquad C_2 = -\frac{PL^3}{6EI_2}$$

thus

$$\theta_2 = -\frac{du_3}{dx_1} = \frac{x_1}{EI_2}\left[M^{(2)} - \frac{P}{2}(2L - x_1)\right]$$

$$u_3 = -\frac{M^{(2)}x_1^2}{2EI_2} + \frac{P}{6EI_2}\left[(L - x_1)^3 + 3L^2x_1 - L^3\right]$$

$$= \frac{x_1^2}{2EI_2}\left[-M^{(2)} + \frac{P(3L - x_1)}{3}\right] \tag{i}$$

The rotation and the transverse component of translation at the free end of the beam $(x_1 = L)$ are equal to

$$\theta_2 = \frac{L}{EI_2}\left(M^{(2)} - \frac{PL}{2}\right) \qquad u_3 = \frac{L^2}{EI_2}\left(-\frac{M^{(2)}}{2} + \frac{PL}{3}\right) \tag{j}$$

CASE 2: SOLUTION ON THE BASIS OF EQ. (A.77b). In what follows, we establish the transverse component of translation (deflection) of the beam under consideration using the fourth-order differential equation (A.77b). That is,

$$\frac{d^2}{dx_1^2}\left(EI_2\frac{d^2u_3}{dx_1^2}\right) = q_3 \tag{k}$$

Integrating relation (k) for $q_3 = 0$, we have

$$EI_2\frac{du_3}{dx_1} = \frac{C_1x_1^2}{2} + C_2x_1 + C_3 \tag{l}$$

and

$$EI_2u_3 = \frac{C_1x_1^3}{6} + \frac{C_2x_1^2}{2} + C_3x_1 + C_4 \tag{m}$$

The constants C_1, C_2, C_3, and C_4 are evaluated from the boundary conditions (a) to (d). Hence, substituting relations (l) and (m) into the boundary conditions (a) and (b), we obtain

$$u_3(0) = 0 \quad \rightarrow \quad C_4 = 0 \qquad \theta_2(0) = -\frac{du_3}{dx_1}\bigg|_{x_1=0} = 0 \quad \rightarrow \quad C_3 = 0 \tag{n}$$

Substituting relations (l) and (m) into the boundary conditions (c) and (d), we get

$$2C_1L + C_2 = -M^{(2)} \qquad -2C_1 = P$$

Thus,

$$C_1 = -\frac{P}{2} \qquad C_2 = -M^{(2)} + P \tag{o}$$

Substituting relations (n) and (o) into (m) we obtain

$$EI_2 u_3 = -\frac{M^{(2)} x_1^2}{2} + \frac{P}{6}(-x_1^3 + 3x_1^2 L) \tag{p}$$

Part 2. Solution on the Basis of the Timoshenko Theory of Beams

Substituting relations (e) into (A.68c) and (A.68d), we get

$$\frac{d\theta_2}{dx_1} = \frac{M^{(2)} - P(L - x_1)}{EI_2} \tag{q}$$

$$\theta_2 + \frac{du_3}{dx_1} = \frac{P\lambda_3}{GA} \tag{r}$$

Integrating relation (q), we obtain

$$\theta_2 = \frac{M^{(2)} x_1}{EI_2} + \frac{P(L - x_1)^2}{2EI_2} + C_1 \tag{s}$$

Substituting relation (s) into (r) and integrating, we get

$$u_3 = -\frac{M^{(2)} x_1^2}{2EI_2} + \frac{P(L - x_1)^3}{6EI_2} - C_1 x_1 + \frac{\lambda_3 P x_1}{GA} + C_2 \tag{t}$$

The integration constants C_1 and C_2 are obtained by requiring that the solution (t) satisfy the boundary conditions (a) and (b). Substituting relation (s) and (t) into (a) and (b), we get

$$C_1 = \frac{P}{2EI_2} \qquad C_2 = -\frac{PL^3}{6EI_2} \tag{u}$$

Using the values of the constants (u), relation (t) becomes

$$u_3 = -\frac{M^{(2)} x_1^2}{2EI_2} + \frac{P}{6EI_2}(-x_1^3 + 3x_1^2 L) + \frac{\lambda_3 P x_1}{GA} \tag{v}$$

The last term in the above relation represents the effect of shear deformation. The constant λ_3 depends on the geometry of the cross section of the beam. For beams of rectangular cross section, $\lambda_3 = 1.5$. Thus, denoting by h the depth of the cross section $(I_2/A = h^2/12)$ and assuming that $G = E/2.7$, the deflection $u_3^{(2)}$ of the end of the cantilever beam of Fig. a is equal to

$$u_3^{(2)} = \frac{L^3}{2EI_2}\left[-\frac{M^{(2)}}{L} + \frac{2P}{3} + \frac{1.5(2.7)Ph^2}{6L^2} \right] \tag{w}$$

Usually, in practice, beams have a very small h/L ratio ($h/L < 1/10$). Consequently, referring to relation (w), we may conclude that the influence of the shear deformation of the beam of Fig. a on its deflection is negligible if its h/L ratio is small.

Example 3. Establish the equations of the elastic curve of the simply supported beam, loaded as shown in Fig. a. The beam is made of an isotropic, linearly elastic material with modulus of elasticity E and has a constant cross section of moment of inertia I_2.

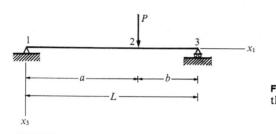

Figure a Geometry and loading of the beam.

solution   This is a statically determinate structure; thus it is more convenient to establish the equations of its elastic curve using relation (A.74b) than relation (A.77b).

The reaction of support 1 may be obtained by referrring to Fig. b and setting the sum of moments about point 3 equal to zero. Thus,

$$R^{(1)} = \frac{Pb}{L} \tag{a}$$

The moment at any section of the beam is given by

$$M_2 = \frac{Pbx_1}{L} \qquad 0 \le x_1 \le a$$
$$\tag{b}$$
$$M_2 = \frac{Pbx_1}{L} - P(x_1 - a) \qquad a \le x_1 \le L$$

Substituting relations (b) into Eq. (A.74b) and integrating twice, we have

$$0 \le x_1 \le a \qquad\qquad a \le x_1 \le L$$

$$EI_2 \frac{d^2u_3}{dx_1^2} = -M_2 = \frac{-Pbx_1}{L} \qquad\qquad Ei_2 \frac{d^2u_3}{dx_1^2} = -M_2 = \frac{-Pbx_1}{L} + P(x_1 - a)$$

$$EI_2 \frac{du_3}{dx_1} = \frac{-Pbx_1^2}{2L} + C_1 \qquad\qquad EI_2 \frac{du_3}{dx_1} = \frac{-Pbx_1^2}{2L} + \frac{P(x_1 - a)^2}{2} + C_2$$

$$EI_2 u_3 = \frac{-Pbx_1^3}{6L} + C_1 x_1 + C_3 \qquad\qquad EI_2 u_3 = \frac{-Pbx_1^3}{6L} + \frac{P(x_1 - a)^3}{6}$$

$$+ C_2 x_1 + C_4 \tag{c}$$

The deflection u_3 must satisfy the following conditions at the ends of the beam.

$$u_3(0) = 0 \qquad u_3(L) = 0$$

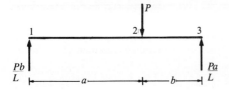

Figure b Free-body diagram of the beam.

Substituting relations (c) in the above, we obtain

$$C_3 = 0 \qquad -\frac{PL^2b}{6} + \frac{P(L-a)^3}{6} + C_2L + C_4 = 0 \qquad \text{(d)}$$

Finally, the continuity of the beam at point 2 imposes the requirement that at $x_1 = a$ the deflection and slope of the elastic curve just to the right of the force are equal to the deflection and the slope, respectively, just to the left of the force. Consequently, referring to relations (c), we get

$$C_1 = C_2 \qquad C_4 = C_3 = 0 \qquad \text{(e)}$$

Substituting relations (e) into the second of relations (d), we have

$$C_1 = C_2 = \frac{P}{L}\left[-\frac{(L-a)^3}{6} + \frac{bL^2}{6}\right] \qquad \text{(f)}$$

Using relations (e) and (f), the expressions for the deflection $u_3(x_1)$ in (c), become

$$EI_2u_3 = \frac{-Pbx_1^3}{6L} + \frac{Px_1}{L}\left[\frac{-(L-a)^3}{6} + \frac{L^2b}{6}\right] \qquad 0 \le x_1 \le a$$

$$EI_2u_3 = \frac{-Pbx_1^3}{6L} + \frac{Px_1}{L}\left[\frac{-(L-a)^3}{6} + \frac{L^2b}{6}\right] + \frac{P(x_1-a)^3}{6} \qquad a \le x_1 \le L$$
$$\text{(g)}$$

The point at which the maximum deflection occurs may be obtained by setting $du_3/dx_1 = 0$. Thus, from relations (g) we get

$$-\frac{Pbx_1^2}{2L} + \frac{P}{L}\left[\frac{-(L-a)^3}{6} + \frac{L^2b}{6}\right] = 0 \qquad 0 \le x_1 \le a$$

$$-\frac{Pbx_1^2}{2L} + \frac{P}{L}\left[\frac{-(L-a)^3}{6} + \frac{L^2b}{6}\right] + \frac{P}{2}(x_1-a)^2 = 0 \qquad a \le x_1 \le L$$
$$\text{(h)}$$

If the first of Eqs. (h) yields a positive value of x_1 which is less than a, then the maximum deflection occurs at that value of x_1. However, if the first of Eqs. (h) yields values of x_1 that are either negative or greater than a, then the maximum deflection occurs in the region $a \le x_1 \le L$ and its location must be obtained from the second of Eqs. (h).

Example 4. Derive the equation of the elastic curve of the fixed-end beam, loaded as shown in Fig. a. The beam is made of an isotropic, linearly elastic material with modulus of elasticity E and has a constant cross section of moment of inertia I_2.

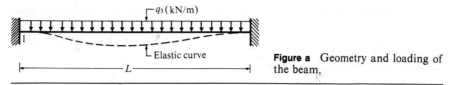

Elastic curve

Figure a Geometry and loading of the beam.

solution This is a statically indeterminate beam; consequently, the distribution of its moments cannot be established only from the equations of equilibrium for parts of the beam. For this reason, the deflection u_3 of the beam cannot be established by integrating relation (A.74b) but, rather, by integrating relation (A.77b). Thus

$$\frac{d^2}{dx_1^2}\left(EI_2\frac{d^2u_3}{dx_1^2}\right) = q_3 \tag{a}$$

Moreover, the deflection u_3 must satisfy the following boundary conditions.

$$u_3(0) = 0 \qquad \left.\frac{du_3}{dx_1}\right|_{x_1=0} = 0$$

$$u_3(L) = 0 \qquad \left.\frac{du_3}{dx_1}\right|_{x_1=L} = 0 \tag{b}$$

Integrating relation (a), we obtain

$$EI_2\frac{du_3}{dx_1} = \frac{q_3x_1^3}{6} + \frac{C_1x_1^2}{2} + C_2x_1 + C_3 \tag{c}$$

$$EI_2u_3 = \frac{q_3x_1^4}{24} + \frac{C_1x_1^3}{6} + \frac{C_2\,x_1^2}{2} + C_3x_1 + C_4 \tag{d}$$

The constants C_1, C_2, C_3, and C_4 are evaluated from the boundary conditions (b). Hence, substituting relations (c) and (d) into the first two of the boundary conditions (b), we obtain

$$C_3 = 0 \qquad C_4 = 0 \tag{e}$$

Substituting relations (c) and (d) into the last two of the boundary conditions (b), we get

$$\frac{q_3L^4}{24} + \frac{C_1L^3}{6} + \frac{C_2L^2}{2} = 0 \qquad \frac{q_3L^3}{6} + \frac{C_1L^2}{2} + C_2L = 0 \tag{f}$$

Solving relations (f), we obtain

$$C_1 = -\frac{q_3L}{2} \qquad C_2 = \frac{q_3L^2}{12} \tag{g}$$

Substituting relations (e) and (g) into Eq. (d), we have

$$u_3 = \frac{q_3 x_1^2 (L - x_1)^2}{EI_2} \tag{h}$$

A.9 Components of Stress in Straight Members Subjected to Bending without Twisting

The value of the normal component of stress at a point of a cross section of a member depends on the values of the axial force N and the components M_2 and M_3 of the moment acting on this cross section. Substituting relation (A.73) into the first of relations (A.61) and using relations (A.74), we obtain the following well-known expression for the normal component of stress at a point (x_2, x_3) of a cross section of a member when the local axes x_2 and x_3 are principal and centroidal:

$$\tau_{11} = E e_{11} = E \left[\frac{du_1}{dx_1} - x_3 \frac{d^2 u_3}{dx_1^2} - x_2 \frac{d^2 u_2}{dx_1^2} \right]$$

$$= \frac{N}{A} + \frac{M_2 x_3}{I_2} - \frac{M_3 x_2}{I_3} \tag{A.80}$$

If we chose a pair of mutually perpendicular centroidal axes which are not principal as local axes x_2 and x_3 of a member, the formula for the normal component of stress τ_{11} is

$$\tau_{11}(x_1, x_2, x_3) = \frac{M_2(x_1) I_3 + M_3(x_1) I_{23}}{I_2 I_3 - I_{32}^2} x_3 - \frac{M_3(x_1) I_2 + M_2(x_1) I_{23}}{I_2 I_3 - I_{32}^2} x_2 \tag{A.81}$$

where I_2, I_3, and I_{23} are the moments and the product of inertia with respect to the axes x_2 and x_3. Moreover, $M_2(x_1)$ and $M_3(x_1)$ are the components of moment in the direction of the axes x_2 and x_3.

The value of the shearing component of stress τ_{12} or τ_{13} at a point of a cross section of a member depends on the value of the shearing force Q_2 or Q_3, respectively, and the torsional component M_1 of the moment acting on this cross section. In the sequel, we derive the relation between the shearing component of stress τ_{12} or τ_{13} at a point (x_2, x_3) of a cross section of a member subjected to bending without twisting and the shearing force Q_2 and Q_3 acting on this cross section.

Consider an infinitesimal segment of length dx_1 of a member (see Fig. A.20).

The components of shearing force and bending moment acting on the left-hand face of this segment of the member are denoted by Q_2, Q_3, M_2, and M_3, while the components of shearing force and bending moment acting on its right-hand face are denoted by $Q_2 + dQ_2$, $Q_3 + dQ_3$, $M_2 + dM_2$, and $M_3 + dM_3$. Referring to relations (A.80), the normal component of stress acting on the left-hand face of the segment under consideration is equal to

$$\tau_{11} = \frac{N}{A} + \frac{M_2 x_3}{I_2} - \frac{M_3 x_2}{I_3}$$

while the normal component of stress acting on the right-hand face of the segment under consideration is equal to

$$\tau_{11} + d\tau_{11} = \frac{N}{A} + \frac{(M_2 + dM_2)x_3}{I_2} - \frac{(M_3 + dM_3)x_2}{I_3}$$

Suppose that the segment under consideration is cut into two parts by a plane $ABCD$ which is normal to the unit vector $\mathbf{n} = n_2\mathbf{i}_2 + n_3\mathbf{i}_3$. Let us consider the equilibrium of part $ABCDEF$ shown in Fig. A.20d. The resultant force H_1 of

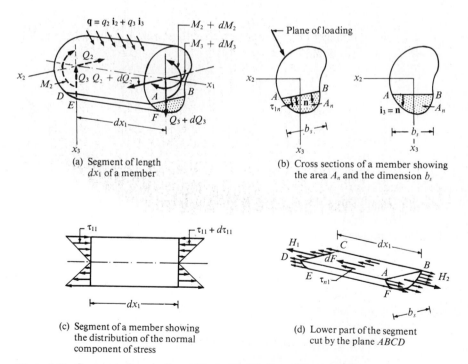

(a) Segment of length dx_1 of a member

(b) Cross sections of a member showing the area A_n and the dimension b_s

(c) Segment of a member showing the distribution of the normal component of stress

(d) Lower part of the segment cut by the plane $ABCD$

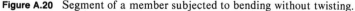

Figure A.20   Segment of a member subjected to bending without twisting.

the normal component of stress acting on face *CDE* of part *ABCDEF* is given by

$$H_1 = \iint_{A_n} \tau_{11} \, dA = \frac{NA_n}{A} + \frac{M_2}{I_2} \iint_{A_n} x_3 \, dA - \frac{M_3}{I_3} \iint_{A_n} x_2 \, dA$$

$$(A.82)$$

while the resultant force H_2 of the normal component of stress on face *ABF* of part *ABCDEF* is equal to

$$H_2 = \iint_{A_n} (\tau_{11} + d\tau_{11}) \, dA = \frac{NA_n}{A} + \frac{(M_2 + dM_2)}{I_2} \iint_{A_n} x_3 \, dA$$

$$- \frac{(M_3 + dM_3)}{I_3} \iint_{A_n} x_2 \, dA \quad (A.83)$$

The equilibrium of the part *ABCDEF* requires that a force $d\mathbf{F} = -dF\mathbf{i}_1$ must exist on its surface *ABCD* given by

$$dF = H_2 - H_1 = \frac{dM_2}{I_2} \iint_{A_n} x_3 \, dA - \frac{dM_3}{I_3} \iint_{A_n} x_2 \, dA \quad (A.84)$$

The force per unit length p_{1n} exerted on the surface *ABCD* is referred to as the *shear flow*, and it is equal to

$$p_{1n} = \frac{dF}{dx_1} = \frac{dM_2}{dx_1} \frac{1}{I_2} \iint_{A_n} x_3 \, dA - \frac{dM_3}{dx_1} \frac{1}{I_3} \iint_{A_n} x_2 \, dA$$

$$= \frac{Q_3}{I_2} \iint_{A_n} x_3 \, dA + \frac{Q_2}{I_3} \iint_{A_n} x_2 \, dA$$

$$(A.85)$$

In general, the distribution of the shearing component of stress τ_{1n} on the surface *ABCD* is not known. In the theories of beams, we do not have a rational means of establishing the distribution of this stress for members of any cross section. However, we can compute the average value of the distribution of the shearing component of stress along line *AB* (see Fig. A.20). That is,

$$\tau_{1n} = \frac{1}{b_s} \frac{dF}{dx_1} \quad (A.86)$$

Substituting relation (A.85) into (A.86), we obtain

$$\tau_{1n} = \frac{b_s Q_3}{I_2} \iint_{A_n} x_3 \, dA + \frac{b_s Q_2}{I_3} \iint_{A_n} x_2 \, dA = \frac{Q_3 A_n \bar{x}_{3n}}{b_s I_2} + \frac{Q_2 A_n \bar{x}_{2n}}{b_s I_3}$$

$$(A.87)$$

The integral $\iint_{A_n} x_3\, dA$ or $\iint_{A_n} x_2\, dA$ is the first moment of the area A_n about the x_2 or the x_3 axis, respectively. It is equal to the product of the area A_n and the distance $\bar{x}_{3n}$ or $\bar{x}_{2n}$ of its centroid from the x_2 or x_3 axis, respectively. If the plane of the external actions is parallel to the x_3 axis ($Q_2 = 0$), relation (A.87) reduces to

$$\tau_{1n} = \frac{Q_3}{b_s I_2} \int \int_{A_n} x_3\, dA = \frac{Q_3 A_n \bar{x}_{3n}}{b_s I_2} \tag{A.88}$$

A positive value of τ_{1n} indicates that its sense is that of the unit vector **n**; that is, toward the area A_n (see Fig. A.20b).

If we choose a pair of mutually perpendicular centroidal axes which are not principal axes as local axes x_2 and x_3 of a member, the formula for the shearing component of stress τ_{1n} is

$$\tau_{1n} = \frac{Q_3 I_3 - Q_2 I_{32}}{b_s(I_2 I_3 - I_{23}^2)} \int \int_{A_n} x_3\, dA + \frac{Q_2 I_2 - Q_3 I_{32}}{b_s(I_2 I_3 - I_{23}^2)} \int \int_{A_n} x_2\, dA \tag{A.89}$$

where I_1, I_2, and I_{23} are the moments and product of inertia with respect to the chosen set of axes. Moreover, Q_2 and Q_3 are the shearing components of internal force in the direction of the chosen axes.

Relations (A.87) to (A.89) give the average value of the distribution of the shearing component of stress τ_{1n} along the distance b_s which is normal to the unit vector **n**. However, for certain members subjected to certain loading, we know a priori that the shearing component of stress τ_{1n} does not vary very much along the direction normal to a unit vector **n**. For these members, relation (A.87) or (A.88) gives a satisfactory approximation of the actual shearing component of stress τ_{1n}. For instance, relations (A.87) or (A.88) can be employed to obtain a satisfactory approximation of the following.

1. The shearing component of stress τ_{13} acting on particles of members having, with respect to the x_3 axis, solid symmetrical cross sections (see Fig. A.21a and b) when they are subjected to forces acting in the $x_1 x_3$ plane parallel to the x_3 axis. The shearing component of stress τ_{13} in such members has been computed on the basis of the theory of elasticity,[†] and it was established that its variation along lines parallel to the x_2 axis is small when the ratio b_s/h is less than $1/2$.

After the shearing component of stress τ_{13} is computed, using relation (A.88), the shearing component of stress τ_{12} can be established by noting that the total shearing stress τ at a particle of the free surface of a member must

---

[†] The theory of elasticity is considerably more inclusive than the mechanics of materials presented in this text.

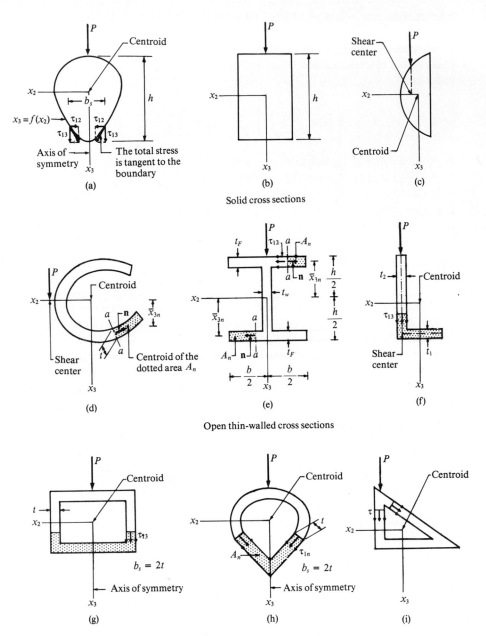

Figure A.21 Types of cross sections.

be tangent to the boundary of the cross section on which it acts. Thus, referring to Fig. A.21a at any point of the surface of the member, we have

$$\tau_{12} = \frac{\tau_{13}}{(df/dx_2)} \tag{A.90}$$

Moreover, since the component of stress τ_{12} must be symmetric about the x_3 axis, it must vanish on the x_3 axis. Assuming that the variation of the component of stress τ_{12} in the x_2 direction is linear, we have

$$\tau_{12} = \frac{2|x_2|\tau_{13}}{b_s(df/dx_2)} \tag{A.91}$$

Inasmuch as df/dx_1 is usually large, the value of τ_{12} at a point is, in general, smaller that the value of τ_{13} at that point.

2. The total shearing stress τ_{1n} at particles of members having open, thin-walled cross sections (see Fig. A.21d to f)† when they are subjected to bending without twisting. The total shearing stress acting on a cross section of a member at particles of its free surface is tangent to the boundary of the cross section. Moreover, since the thickness of the cross section is small, the variation of the total shearing stress along the normal to the boundary is small. For instance, consider a member having a cross section shown in Fig. A.21d or e subjected to external forces acting through the shear center of its cross section in the direction of its x_3 axis. The total shearing stress τ_{1n} acting on any particle of line aa of the cross section is obtained using relation (A.88) with $b_s = t$ or t_F.

3. The total shearing stress in members having closed, thin-walled cross sections with an axis of symmetry (see Fig. A.21g and h) when they are subjected to transverse forces acting along their axis of symmetry. In this case $b_s = 2t$.

Relation (A.87) or (A.88) may not give a satisfactory approximation of the components of shearing stress in members having solid, unsymmetric cross sections or solid cross sections with one axis of symmetry if the external forces to which they are subjected do not act along the axis of symmetry (see Fig. A.21c). In this case, relations (A.87) or (A.88) give the average value of the distribution of the shearing component of stress τ_{1n} along the distance b_s. Moreover, relations (A.87) or (A.88) do not give accurate results when used in computing the total shearing stress in members having closed, unsymmetric, thin-walled cross sections (see Fig. A.21i) or closed, thin-walled cross sections with one axis of symmetry if the external forces to which they are subjected do not act along this axis of symmetry. In this case the total shearing stress at the

---

† Notice that the x_2 and x_3 axes in Fig. A.21f, g, and h are not principal axes. Thus, relation (A.89) must be employed to compute the shearing components of stress τ_{1n}.

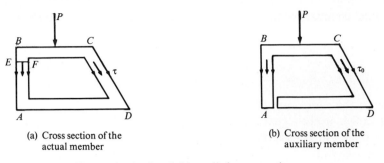

(a) Cross section of the
actual member

(b) Cross section of the
auxiliary member

Figure A.22 Unsymmetric closed thin-walled cross section.

particles of each thin wall is parallel to its boundary and varies very little in the direction normal to it. However we do not know a priori how the distributions of the total shearing stress in each thin wall are related. For example, if we are interested in establishing the total shearing stress at the particles of line *EF* (see Fig. A.22*a*), we do not know a priori the location of the line on the thin wall *CD* whose particles are subjected to equal total shearing stress. Thus we can not establish the boundary of the area A_n.

An approximate value of the shearing stress in members having unsymmetric, closed, thin-walled cross sections may be established by considering an auxiliary member made from the actual member by cutting it by a plane parallel to its axis and normal to its boundary (see Fig. A.22*b*). The total shearing stress $\tau_0(s)$ acting on a cross section of the auxiliary member is obtained using relation (A.87) or (A.88). It is assumed that the total shearing stress acting on a particle of a cross section of the actual member is related to that at the corresponding particle of the auxiliary member by the following relation.

$$\tau = \tau_0(s) + \frac{q}{t} \qquad (A.92)$$

where *s* is the coordinate of the particle measured counterclockwise from a reference point along the center line of the wall of the cross section of the member; *t* is the thickness of the cross section at the point where the shearing stress is desired, and *q* is a constant given by

$$q = -\frac{\oint \tau_0 \, ds/t}{\oint ds/t} \qquad (A.93)$$

The integration in relation (A.93) is performed around the entire closed perimeter of the cross section *ABCDA*. Relation (A.93) is obtained by requiring that the relative displacement of the two cut ends of the auxiliary member vanishes.

In the sequel, we present two examples. In the first example, we compute the distribution of the shearing stress in a beam of rectangular cross section. Moreover, we discuss the results, and we compare them with those obtained on the basis of the theory of elasticity. In the second example, we compute the distribution of the shearing stress in an *I* beam.

Example 1. Compute the distribution of shearing stress in a member of a structure having a rectangular cross section and loaded by forces acting in the $x_1 x_3$ plane and parallel to the x_3 axis.

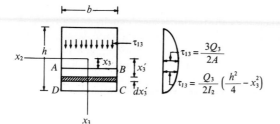

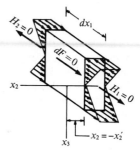

Figure a Distribution of the shearing component of stress τ_{13}.

Figure b Part of a segment of a member cut by a plane normal to the x_2 axis.

---

solution In this case, it is assumed that the shearing component of stress τ_{13} is constant along the width of the member. Hence, the shearing component of stress τ_{13} at any particle of line *AB* (see Fig. a) can be established using relation (A.88). That is

$$\tau_{13} = \frac{Q_3}{bI_2} \iint_{A_n} x_3' \, dA = \frac{Q_3}{I_2} \int_{x_3}^{h/2} x_3' \, dx_3' = \frac{Q_3}{2I_2} \left(\frac{h^2}{4} - x_3^2 \right) \tag{a}$$

where A_n is the area of the portion *ABCD* of the cross section of the member.

The above relation indicates that the distribution of the shearing component of stress τ_{13} along the x_3 axis is parabolic (see Fig. a). It is zero at $x_3 = \pm \frac{1}{2} h$ and assumes its maximum value at $x_3 = 0$. That is

$$(\tau_{13})_{\max} = \frac{3Q_3}{2A} \tag{b}$$

where A is the area of the cross section of the member.

The shearing component of stress τ_{12}, obtained on the basis of relation (A.88), is zero. This may be readily seen by considering the part of the segment of the member, cut by a plane normal to the x_2 axis, shown in Fig. b. The resultant forces H_1 and H_2

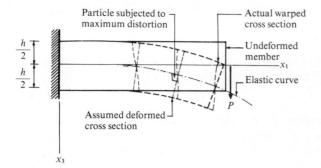

Figure c Deformation of a cantilever beam subjected to a transverse force.

acting on the faces of this part vanish. Consequently, the resultant shearing force dF acting on the plane $x_2 = -x_2'$ must vanish.

The shearing stress τ_{13} induces shearing deformation whose magnitude is a function of x_3. As shown in Fig. c, as a result of shear deformation, the cross section of the member warps. The nature of warping becomes apparent by noting that $\tau_{13} = 0$ at $x_3 = \pm\frac{1}{2}h$. Consequently, the cross section is not distorted at $x_3 = \pm\frac{1}{2}h$, whereas the distortion of the cross section is maximum (τ_{13} is maximum) at $x_3 = 0$. For members whose depth is small as compared to their length, the warping is negligible. Therefore, for such members, the assumption that plane sections remain plane, subsequent to deformation, is a very satisfactory approximation.

The distribution of the shearing components of stress in cantilever beams of constant rectangular cross section has been obtained on the basis of the theory of elasticity. For values of the ratio b/h less than $\frac{1}{2}$, the shearing component of stress τ_{12} has a very small magnitude, while the shearing component of stress τ_{13} is almost uniformly distributed along the x_2 axis. Consequently, for this range of the ratio b/h, the results obtained on the basis of Eq. (a) are in satisfactory agreement with the results obtained on the basis of the theory of elasticity (see Table a). However, for larger values of the ratio b/h, the solution obtained on the basis of the theory of elasticity yields values for the component of stress τ_{13} which vary considerably with x_2 (see Table a). Consequently, in this case, the results obtained on the basis of Eq. (a) represent only the average value of the shearing stress τ_{13}. At the centroid of a cross section of a member, the magnitude of τ_{13} obtained on the basis of the theory of elasticity is less than that given by relation (a); whereas, at $x_2 = \pm b/2$, the magnitude of τ_{13}, obtained on the basis of the theory of elasticity, is greater than that given by relation (a) (see Table a).

TABLE a   Shearing Stress Coefficients for Members of Rectangular Cross Section[†]

b/h	$\frac{1}{2}$	1	2	4
K_1[‡]	0.983	0.940	0.856	0.805
K_2[‡]	1.033	1.126	1.396	1.988

† See Timoshenko and Goodier.[5]
‡ For definitions of K_1 and K_2, see Fig. d.

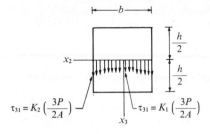

Figure d Distribution of shearing stress at x_3 = 0 obtained on the basis of the theory of elasticity. (*Timoshenko and Goodier*[5])

$$\tau_{31} = K_2 \left(\frac{3P}{2A}\right)$$

$$\tau_{31} = K_1 \left(\frac{3P}{2A}\right)$$

Example 2. Compute the distribution of the shearing components of stress for a member of a structure having an I cross section, subjected to external forces in the $x_1 x_3$ plane parallel to the x_3 axis (see Fig. a).

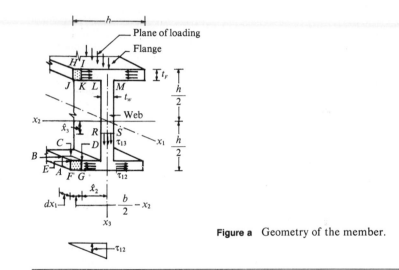

Figure a Geometry of the member.

---

solution The member shown in Fig. a is more suitable for bending about its x_2 axis than a member of rectangular cross section because the former has a smaller Ah/I_2 ratio than the latter.† Notice that the ability of the member of Fig. a to resist bending

---

† Consider two members of equal length and weight made of the same material (consequently, of equal cross-sectional area A). However, the ratio I_2/h of the moment of inertia about the x_2 axis to the height of the cross section of member 1 is larger than that of member 2. From Eq. (A.80) it can be deduced that the moment which must be applied at a cross section of member 1 in order to produce a certain maximum value of the normal component of stress τ_{11} is greater than the moment which must be applied at a cross section of member 2 in order to produce the same maximum value of the normal component of stress. Thus, member 1 is more suitable for bending than member 2.

On the basis of the foregoing, it is apparent that it is desirable that the cross sections of members subjected to bending about the x_2 axis have a ratio Ah/I_2 which is as small as possible.

increases as the depth of its web increases. However, there are considerations beyond the scope of this text which limit both the depth of the web and the width of the flanges of members having an I cross section.

The shearing component of stress τ_{13} in the flanges of the member under consideration generally is small and may be neglected. Moreover, in the flanges, the component of stress τ_{13} varies considerably with x_2 and thus it cannot be computed using relations (A.88). This becomes apparent by noting that at $x_3 = \pm(\frac{1}{2}h - t_F)$ and $x_2 > \frac{1}{2}t_w$ the shearing component of stress τ_{13} must be zero because this stress must be equal to the shearing component of stress τ_{31} acting on the free surface of the flanges. However, across the junction LM ($x_2 < \frac{1}{2}t_w$), the shearing component of stress τ_{13} is different than zero. This indicates that at the points of the flange close to $x_3 = \pm(\frac{1}{2}h - t_F)$, the distribution of the shearing component of stress τ_{13} along the x_2 axis is not uniform.

The values of the shearing component of stress τ_{13} on the web, obtained on the basis of relation (A.88), are a satisfactory approximation of its actual values because on the web the shearing component of stress τ_{13} varies negligibly in the direction of the x_2 axis. In order to establish the shearing component of stress τ_{13} at a point of the web of the member of Fig. a [say on the line $RS(x_3 = \hat{x}_3)$], we compute the first moment of the area of the portion of the cross section of the member below line RS. Referring Fig. a, this area consists of two rectangles, the flange of area bt_F and the portion of the web below line RS of area $t_w(\frac{1}{2}h - t_F - \hat{x}_3)$. The distance of the centroid of the flange from the x_2 axis is equal to $\frac{1}{2}(h - t_F)$ while the distance of the centroid of the portion of the web below line RS is equal to $\frac{1}{2}(h - t_F + \hat{x}_3)$. Thus, the first moment of the area of the portion of the cross section of the member below line RS is equal to

$$\iint_{A_n} x_3\, dA = bt_F\left(\frac{h}{2} - \frac{t_F}{2}\right) + \frac{t_w}{2}\left(\frac{h}{2} - t_F - \hat{x}_3\right)\left(\frac{h}{2} - t_F + \hat{x}_3\right) \tag{a}$$

Substituting relation (a) into (A.88) we get

$$\tau_{13} = \frac{Q_3}{2I_2 t_w}\left\{ bt_F(h - t_F) + t_w\left[\left(\frac{h}{2} - t_F\right)^2 - \hat{x}_3^2\right]\right\} \tag{b}$$

That is, as shown in Fig. b, the distribution of the shearing component of stress τ_{13} on the web is parabolic. The maximum value of τ_{13} is at $x_3 = 0$. For the I beams used in practice, the shearing component of stress τ_{13} on the web accounts for 90 to 98% of the shearing forces Q_3.

In order to establish the shearing component of stress τ_{12} at any point of the bottom flange of the member of Fig. a, consider a segment $ABCDEFG$ of length dx_1 and width $(\frac{1}{2}b - \hat{x}_2)$ (see shaded area in Fig. a). The equilibrium of this segment requires that a force dF acts on plane CDG (see Fig. c). As discussed previously, the shearing com-

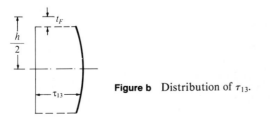

Figure b   Distribution of τ_{13}.

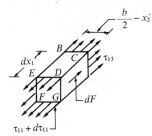

Figure c   Segment of the bottom flange.

ponent of stress τ_{12} acting on the flanges of the member of Fig. a can be considered approximately as being uniformly distributed over the thickness t_F of each flange. Thus, the shearing component of stress τ_{12} acting on the bottom flange is

$$\tau_{12} = \frac{Q_3}{I_2 t_F} \int \int_{A_n} x_3 \, dA \tag{c}$$

where A_n is the area *EDGF*. Referring to Fig. a, Eq. (c) may be rewritten as

$$\tau_{12} = \frac{Q_3(\frac{1}{2}b - \hat{x}_2)(h - t_F)}{2I_2} \tag{d}$$

τ_{12} is positive; consequently, its sense is that of the unit vector **n**. That is, referring to Fig. a, its sense is from line *DG* to *EF*.

The shearing component of stress τ_{12} at any point of line $x_2 = \hat{x}_2$ of the top flange is equal to

$$\tau_{12} = \frac{Q_3(\frac{1}{2}b - \hat{x}_2)[-(h - t_F)]}{2I_2} \tag{e}$$

In this case, τ_{12} is negative; consequently, its sense is opposite to that of the unit vector **n**; that is, referring to Fig. a, we see that its sense is from line *HJ* to *IK*. The distribution of the shearing component of stress τ_{12} is shown in Fig. a.

A.10   Location of the Shear Center of Thin-Walled Open Sections

In the previous section, we have established a formula for the shearing component of stress in members subjected to bending without twisting. In order that a member be subjected to bending without twisting, the external moments acting on it should not have a torsional component. Moreover, the plane of the external forces acting on it must pass through the shear center of the cross section of the member. The shear center of cross sections having an axis of symmetry is located on the axis of symmetry. Thus, the shear center of cross sections having two axes of symmetry is the intersection of these axes; that is, the shear center of such cross sections coincides with their centroid.

When the external forces acting on a member are located in a plane which does not pass through the shear center of its cross section, they can be replaced

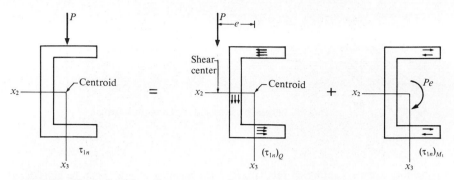

Figure A.23 Channel subjected to transverse forces whose plane does not pass through the shear center of its cross section.

by a statically equivalent system of external forces located in a plane which contains the shear center and a torsional moment (see Fig. A.23). The shearing component of stress $(\tau_{1n})_Q$ due to the external forces which are located in a plane passing through the shear center can be established using relation (A.87). The shearing component of stress $(\tau_{1n})_{M_1}$ due to the torsional moment can be established as discussed in Sec. A.11. Their sum is the shearing component of stress in the member subjected to the given forces. That is

$$\tau_{1n} = (\tau_{1n})_Q + (\tau_{1n})_{M_1} \tag{A.94}$$

The shear center of a cross section of a member can be located by considering a segment of the member and setting equal to zero the sum of the components along the x_1 axis of the moments of the external forces acting on this segment and of the shearing components of stress acting on the end cross sections of this segment. In the example which follows, we locate the shear center of a channel.

Example 1. Locate the shear center of the thin channel shown in Fig. a.

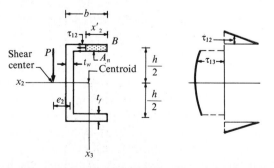

Figure a Geometry of the cross section of the channel and distribution of the shearing components of stress.

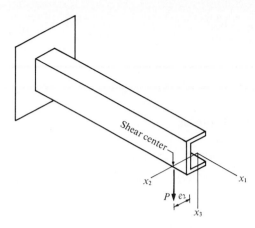

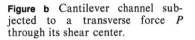

Figure b Cantilever channel subjected to a transverse force P through its shear center.

solution   Referring to Fig. a, the shear center of a thin channel is located on the x_2 axis because it is an axis of symmetry. To establish the distance e_2 of the shear center from the center line of the web of the channel, we assume that the channel is a cantilever beam subjected to a force P (see Fig. b). The component of stress τ_{13} on the flanges of the channel generally is small; thus its effect upon the location of the shearing force Q acting on a cross section of the channel may be disregarded. That is, the shearing force Q is assumed to act through the center line of the web of the channel. The force P induces a component of shearing stress τ_{12} on the flanges [see Eq. (A.87)] equal to

$$\tau_{12} = \frac{QhA_n}{2I_2 t_f} = \frac{Qhx_2'}{2I_2} \tag{a}$$

As shown in Fig. a, the coordinate x_2' is measured from the edge B of the flange of the channel. Using the above relations, we see that the resultant horizontal force F_2 on each flange, due to the component of stress τ_{12}, is equal to

$$F_2 = \int_0^b \tau_{12} t_f \, dx_2' = \frac{Q t_f h b^2}{4I_2} \tag{b}$$

Considering the equilibrium of the segment of the channel whose free-body diagram is shown in Fig. c, we have

$$\Sigma \overline{F}_v = 0 \qquad Q = P$$

$$\Sigma M_1 = 0 \qquad Pe_2 = F_2 h = \frac{P t_f h^2 b^2}{4I_2} \tag{c}$$

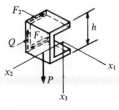

Figure c Free-body diagram of a segment of the cantilever beam of Fig. b.

thus
$$e_2 = \frac{t_f h^2 b^2}{4 I_2}$$
(d)

The location of the shear center is shown in Figs. a and b.

A.11 Components of Stress and Angle of Twist of Prismatic Straight Members Subjected to External Torsional Moments

When a prismatic straight member is subjected to two equal and opposite external torsional moments at its ends, its axis does not elongate or bend. However, a centroidal straight line normal to the axis of the member prior to deformation does not always remain straight subsequent to deformation. It can become a curve whose projection on a plane normal to the axis of the member is a centroidal straight line which rotated by an angle θ_1 about the axis of the member subsequent to deformation. Every centroidal line of a plane normal to the axis of a member prior to deformation rotates by the same angle θ_1 about this axis subsequent to deformation. Thus, plane sections normal to the axis of a prismatic straight member subjected to two equal and opposite external torsional moments at its ends do not remain plane after deformation; they rotate about the axis of the member, and they warp. Only plane sections normal to the axis of prismatic straight members of circular cross section remain plane subsequent to deformation; that is, they do not warp.

When all the cross sections of a member are free to warp, the change of length of the longitudinal fibers of the member is negligible; consequently, the normal component of stress acting on the cross sections of the member is also negligible. In practice, however, one or more cross sections of a member are usually restrained from warping. In this case, the change of length of the longitudinal fibers of members having thin, open cross sections may not be negligible (see Fig. A.24). Consequently, the normal component of stress acting at particles located at or near the cross sections which are restrained from warping and away from the axis of the member may be large and must be taken into account. Moreover, the effect of restraining a cross section of members having thin, open cross sections on the angle of rotation of their cross sections about their axis may not be negligible. For members of other cross sections (hollow-closed or full), the effect of restraining the warping of a cross section on the values of the normal component of stress and on the angle of rotation about their axis is small and it is neglected. The effect of the warping of the members of a structure is taken into account only in computing their torsional constant.

Consider a cylindrical (prismatic) member, subjected at its ends to two equal and opposite torsional moments of magnitude M_1 and assume that it is restrained from rotating as a rigid body in a way that all its cross sections are

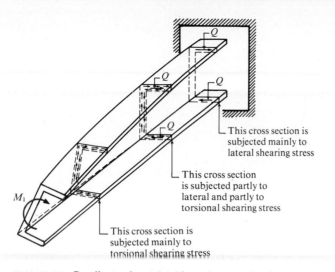

Figure A.24 Cantilever channel subjected to a torsional moment at its free end.

free to warp. Assume that this is accomplished by fixing a very small region around the centroid of the cross section of the member at $x_1 = 0$. It can be shown that the components of displacement $\hat{u}_1$, $\hat{u}_2$, and $\hat{u}_3$ of the particles of the member have the following form:†

$$\hat{u}_1 = \gamma f_1(x_2, x_3) \qquad \hat{u}_2 = -\gamma x_1 x_3 \qquad \hat{u}_3 = \gamma x_1 x_2 \qquad \text{(A.95)}$$

The component of displacement $\hat{u}_1(x_2, x_3)$ represents the warping of the cross section of the member and depends on its geometry. When prismatic members with a circular cross section are subjected at their ends to torsional moments, they do not warp; consequently, for these members $\hat{u}_1$ vanishes. However, when prismatic members with a noncircular cross section are subjected at their ends to torsional moments, they warp. Notice that on the basis of relations (A.95) the normal components of strain e_{11}, e_{22}, and e_{33} vanish.

Consider the material straight line *OP*, shown in Fig. A.25, which is normal to the axis of a prismatic member prior to deformation. When the member is subjected at its ends to two equal and opposite torsional moments, point *P* moves to *P′* whose projection on the plane normal to the axis of the member is denoted in Fig. A.25 by *P″*. Referring to Fig. A.25 and using relations (A.95), we have

$$\tan \psi = \frac{-u_2}{u_3} = \frac{x_3}{x_2} = \tan \phi \qquad \text{(A.96)}$$

---

† See Timoshenko and Goodier (Ref. 5, p. 293).

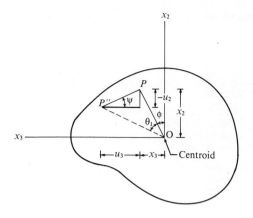

Figure A.25 Cross section of a prismatic member twisted by an angle θ_1.

Consequently, the angles ψ and ϕ are equal; hence, line PP'' is perpendicular to OP. Thus, point P is displaced in the direction perpendicular to OP, and the radial component of displacement vanishes.

Referring to Fig. A.25, on the basis of relations (A.95), it is apparent that the angle of rotation of the material line OP due to the deformation of the member is equal to

$$\theta_1 \approx \tan \theta_1 = \frac{PP''}{OP} = \frac{|u_2|}{\sin \psi (OP)} = \frac{|u_2|}{\sin \phi (OP)} = \frac{|u_2|}{x_3} = \gamma x_1 \quad (A.97)$$

Thus, if relations (A.95) are valid, the angle θ_1 is independent of the coordinates x_2, x_3. Consequently, all material lines through the centroid of a cross section of the member rotate by the same angle θ_1, which is referred to as the *angle of twist of the cross section.* From relation (A.97), we have

$$\frac{d\theta_1}{dx_1} = \gamma = \text{constant} \quad (A.98)$$

Hence, the constant γ is the rate of change of the angle of twist along the axis of the cylindrical member, and it is referred to as *the angle of twist per unit length.*

The component of strain e_{12} and e_{13} in a line member may be obtained by substituting relations (A.95) into the strain displacement relations (A.12). Thus,

$$e_{13} = \frac{\gamma}{2} \left(\frac{\partial f_1}{\partial x_3} + x_2 \right) \qquad e_{12} = \frac{\gamma}{2} \left(\frac{\partial f_1}{\partial x_2} - x_3 \right) \quad (A.99)$$

Substituting the above relations into (A.61) and the resulting expressions into (A.24a), we obtain

$$M_1 = \int \int_A (\tau_{13}x_2 - \tau_{12}x_3) \, dA = 2G \int \int_A (e_{13}x_2 - e_{12}x_3) \, dA$$

$$= G\gamma \left[\int \int_A \left(\frac{\partial f_1}{\partial x_3} x_2 - \frac{\partial f_1}{\partial x_2} x_3 \right) dA + \int \int_A (x_2^2 + x_3^2) \, dA \right]$$

$$= G\gamma \left[\int \int_A \left(\frac{\partial f_1}{\partial x_3} x_2 - \frac{\partial f_1}{\partial x_2} x_3 \right) dA + J \right]$$

$$(A.100)$$

where J is the polar moment of inertia of the cross section of the member. Relation (A.100) can be rewritten as

$$\frac{d\theta_1}{dx_1} = \gamma = \frac{M_1}{KG} \qquad (A.101)$$

where K is referred to as the *torsional constant* and is dependent upon the geometry of the cross section of the member. In order to compute the value of the torsional constant for a prismatic member of noncircular cross section, the warping $\hat{u}_1(x_2, x_3) = \gamma f_1(x_2, x_3)$ corresponding to this cross section must be computed.†

Relation (A.101) may be integrated to give

$$\theta_1 = \frac{M_1 L}{KG} \qquad (A.102)$$

When prismatic members are subjected to a distribution of external torsional moments along their length, the warping of their cross sections varies along their length. That is, the twist per unit length γ is a function of x_1. This implies that the cross sections of the member cannot warp freely because they are partially restrained from warping by the neighboring cross sections which warp by a different amount. In this case, the normal component of stress in members having thin, open sections may not be negligible.

Consider a prismatic member subjected to a distribution of external torsional moments $m_1(x_1)$. Differentiating relations (A.101) and using relation (A.57), we obtain

$$KG \frac{d^2\theta_1}{dx_1^2} = \frac{dM_1}{dx_1} = -m_1 \qquad (A.103)$$

---

† For a detailed analysis of the torsion of cylindrical members, see Ref. 5, p. 291.

This relation may be integrated to yield the rotation θ_1 of a member subjected to a distribution of external torsional moments. The constants of integration are evaluated from the boundary conditions. In general either the angle of twist θ_1 or the torsional moment M_1 must be specified at each end of a member. For instance, the boundary conditions for a member which is fixed at $x_1 = 0$ and free at $x_1 = L$ are

$$\theta_1(0) = 0 \qquad M_1(L) = \text{the given torsional moment}$$

A.11.1   Members of circular cross section

Prismatic members having a circular (hollow or full) cross section do not warp ($\hat{u}_1 = 0$); consequently, for such members, relation (A.100) reduces to

$$\gamma = \frac{M_1}{JG} \tag{A.104}$$

Substituting relation (A.104) into (A.99) and noting that in this case $\hat{u}_1 = 0$ ($f_1 = 0$), we obtain

$$e_{12} = -\frac{M_1 x_3}{2GJ} \qquad e_{13} = \frac{M_1 x_2}{2GJ} \tag{A.105}$$

Thus, the components of stress τ_{12} and τ_{13} in prismatic members having a circular cross section subjected at their ends to two equal and opposite torsional moments of magnitude M_1 are equal to

$$\tau_{12} = -\frac{M_1 x_3}{J} \qquad \tau_{13} = \frac{M_1 x_2}{J} \tag{A.106}$$

A.11.2   Members of rectangular cross section

For prismatic members with rectangular cross section of dimensions h and t, the torsional constant K is expressed as

$$K = Cht^3 \tag{A.107}$$

The values of C obtained from available† analytical solutions are given in Table A.1. The parameter C may also be obtained from the following formula[6]

$$C = \frac{1}{3} - 0.21 \frac{t}{h}\left(1 - \frac{t^4}{12h^4}\right) \tag{A.108}$$

---

† For a detailed analysis of the torsion of cylindrical members, see Ref. 5, p. 312.

TABLE A.1   Values of the Parameter C

h/t	1	1.2	1.5	2.0	2.5	3.0	4.0	5.0	10.0	∞
C	0.1406	0.166	0.196	0.229	0.248	0.263	0.281	0.291	0.312	0.333

Referring to Table A.1, for members having thin (t/h small) rectangular cross sections, Eq. (A.101) may be expressed approximately as

$$\gamma = \frac{3M_1}{Gt^3 h} \tag{A.109}$$

The stress distribution on a cross section of a prismatic member having a thin, rectangular cross section (see Fig. A.26) subjected to torsional moments at its ends, may be approximated† by the following relations:

$$\tau_{12} = 0 \qquad \tau_{13} = \frac{6M_1 x_2}{t^3 h} \tag{A.110}$$

Notice that the stress distribution given by relation (A.110) does not satisfy the first of relations (A.24) because the component of stress τ_{12} is not negligible near the short edges of the thin rectangular cross section ($x_3 = \pm\frac{1}{2}h$). Because of its relatively large distance from the centroid of the cross section, this component of stress contributes half of the resultant moment M_1.

---

† See Ref. 5, p. 308.

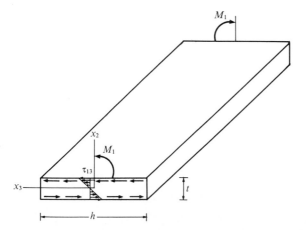

Figure A.26 Cylindrical member having a thin rectangular cross section subjected to end torsional moments.

A.11.3 Members of open, thin-walled cross section

Relation (A.109) may be applied to prismatic members having open, thin-walled cross sections and consisting of n prismatic parts of thin, rectangular, or curved cross sections connected together (see Fig. A.27). When such members are subjected at their ends to equal and opposite torsional moments, each of their parts twists by the same amount γ. Thus, using Eq. (A.109), we see that the torsional moment resisted by the ith part is equal to

$$M_1^{(i)} = G\gamma \frac{t_i^3 h_i}{3} \tag{A.111}$$

Consequently, the torsional moment applied to the ends of a cylindrical member having open cross sections and consisting of n thin-walled parts of rectangular or curved cross sections connected together (see Fig. A.27) can be expressed as

$$M_1 = G\gamma \sum_{i=1}^{n} \frac{t_i^3 h_i}{3} = G\gamma K \tag{A.112}$$

The torsional constant K of four cross sections of engineering interest are given in Fig. A.27.

Consider a prismatic member of open cross section, consisting of n prismatic parts of thin, rectangular cross section or thin, curved cross section (see Fig. A.27) connected together. When this member is subjected at its ends to equal and opposite torsional moments M_1, the shearing stress acting on the cross section of part p may be approximated as

$$\tau_{13}^{(p)} = \frac{6M_1 x_2^{(p)}}{\displaystyle\sum_{i=1}^{n} t_i^3 h_i} = \frac{2M_1 x_2^{(p)}}{K} \tag{A.113}$$

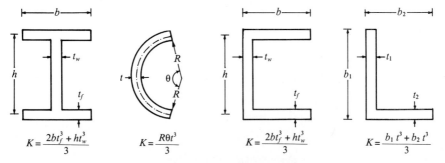

Figure A.27   Torsional constant of open thin-walled cross sections.

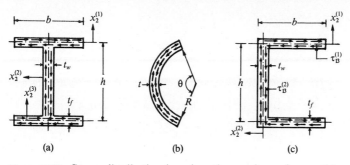

Figure A.28 Stress distribution in prismatic members of open thin-walled cross section subjected to the end torsional moments.

where $x_2^{(p)}$ is the thickness coordinate measured from the center line of the thin, rectangular, or curved cross section of part p of the member; t_i and h_i are the width and height, respectively, of the ith part of the member. The distribution of the shearing component of stress on three cross sections of engineering interest is shown in Fig. A.28.

Notice that members having open, thin-walled cross sections generally are not very suitable for carrying a torsional moment. This becomes apparent by referring to Fig. A.26 and observing that in this case the maximum value of the shearing component of stress τ_{13} obtained from relation (A.110) is equal to

$$(\tau_{13})_{\max} = \frac{3M_1}{t^2 h} \tag{A.114}$$

Since the thickness t is small, the maximum shearing component of stress could be large even for small values of the applied torque.

A.11.4   Members of closed, thin-walled cross section

The torsional constant of prismatic members of closed, thin-walled cross section is given by the following formula.†

$$K = 4A_m^2 \bigg/ \int_0^S \frac{ds}{t} \tag{A.115}$$

where A_m is the area enclosed by the median line of the cross section, S is the length of the median line, and t is the thickness of the cross section, which can be variable.

The shearing stress τ in prismatic members of closed, thin-walled cross sec-

---

† See Gere and Timoshenko.[7]

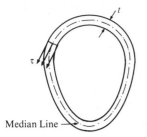

Median Line

Figure A.29 Closed thin-walled cross section.

tions subjected to torsional moments varies negligibly along their thickness. Taking this into account, we can show that the shearing stress is equal to

$$\tau = \frac{M_1}{2tA_m} \tag{A.116}$$

A.12 Components of Stress and Strain in Straight Members Subjected to Bending and Twisting

Consider a straight-line member made of an isotropic, linearly elastic material, and assume we know a priori that the variation of the shearing component of stress τ_{1n} is negligible along the direction of the unit vector **s**. The shearing component of stress τ_{1n} acts in the direction of the unit vector **n** (see Fig. A.20b), which is normal to the unit vector **s**. As discussed in Sec. A.9, when such members are subjected to bending without twisting, the shearing component of stress τ_{1n} is computed using relation (A.87). Thus, referring to relations (A.87) and (A.80), the components of stress at any particle of a member subjected to bending and twisting are equal to

$$\tau_{11} = \frac{N}{A} + \frac{M_2 x_3}{I_2} - \frac{M_3 x_2}{I_3} \qquad \tau_{1n} = \frac{Q_2 Z_3}{I_3 b_s} + \frac{Q_3 Z_2}{I_2 b_s} + (\tau_{1n})_{M_1}$$

$$\tag{A.117}$$

$$\tau_{1s} = (\tau_{1s})_Q + (\tau_{1s})_{M_1} \qquad \tau_{22} = \tau_{33} = \tau_{23} = 0$$

where
$$Z_3 = \iint_{A_n} x_3 \, dA \qquad Z_2 = \iint_{A_n} x_2 \, dA \tag{A.118}$$

For the member under consideration, I_2 and I_3 are the moments of inertia of the cross section about its principal centroidal axes x_2 and x_3, respectively, and A is the area of the cross section of the member. Referring to Fig. A.20b, A_n is the area of the shaded portion of the cross section of the member, while b_s is the dimension of its cross section, parallel to the unit vector **s** at the point where the component of stress is established (see Fig. A.20b). The terms

$(\tau_{1n})_{M_1}$ and $(\tau_{1s})_{M_1}$ represent the part of the components of the shearing stress τ_{1n} or τ_{1s}, respectively, due to the twisting moment M_1. The term $(\tau_{1s})_Q$ represents the part of the shearing component of stress τ_{1s} which is due to the transverse forces. As discussed in Sec. A.9, the shearing component of stress $(\tau_{1s})_Q$ is negligible at the points of thin-walled members having open cross sections. Moreover, the shearing component of stress $(\tau_{1s})_Q$ is negligible at the particles of thin-walled members having closed cross sections.

As is evident from relations (A.95), when a member is free to warp, the twisting moment M_1 does not induce a normal component of strain e_{11} and, consequently, a normal component of stress τ_{11}.

Referring to relations (A.49) (with $\tau_{22} = \tau_{33} = \tau_{23} = 0$) and (A.117), we see that the components of strain at any particle of a member subjected to bending and twisting are

$$e_{11} = \frac{N}{EA} + \frac{M_2 x_3}{EI_2} - \frac{M_3 x_2}{EI_3} + \alpha\, \Delta T(x_1, x_2, x_3) + e_{11}^I$$

$$e_{1n} = \frac{Q_2 E_3}{2GI_3 b_s} + \frac{Q_3 Z_2}{2GI_2 b_s} + (e_{1n})_{M_1} \qquad\qquad (A.119)$$

$$e_{1s} = (e_{1s})_Q + (e_{1s})_{M_1}$$

The terms $(e_{1n})_{M_1}$ or $(e_{1s})_{M_1}$ represent the part of the shearing components of strain e_{1n} or e_{1s}, respectively, due to the twisting moment. The term $(e_{1s})_Q$ represents the part of the shearing component of strain (e_{1s}) due to the transverse forces. $\Delta T(x_1, x_2, x_3)$ is the change of temperature from the uniform temperature T_0 at the stress-free state. α is the coefficient of linear expansion of the material from which the member is made. e_{11}^I is the axial component of initial strain at a particle due to lack of fit of the member (see Sec. A.5.3).

A.13 Problems

1 to 7. Compute the equations of the elastic curve and the value of the maximum deflection of the structure loaded as shown in Fig. A.P1. Repeat for Figs. A.P2 to A.P7.

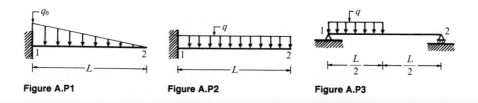

Figure A.P1   **Figure A.P2**   **Figure A.P3**

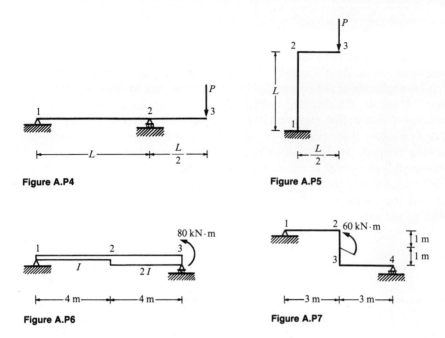

Figure A.P4

Figure A.P5

Figure A.P6

Figure A.P7

8. A steel beam has a T cross section as shown in Fig. A.P8. The axial force N, the shearing force Q_3, and the bending moment M_2 at a cross section of this beam are equal to 100 kN, 200 kN, and 450 kN·m, respectively. Determine:
   (*a*) The shearing component of stress τ_{13} at the points of lines *aa*, *bb*, and *cc*.
   (*b*) The resultant normal force on the 60-mm flange.
   (*c*) The shearing component of stress τ_{12} at the points of line *aa*.

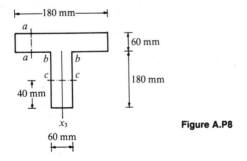

Figure A.P8

9. A cantilever beam made of an angle is loaded as shown in Fig. A.P9. Compute the normal component of stress acting at point *A* of the cross section of the beam adjacent to the wall.

10. A cantilever beam made of a channel is loaded, as shown in Fig. A.P10. Compute the normal and shearing components of stress acting at point *A* of the cross section of the beam adjacent to the wall.

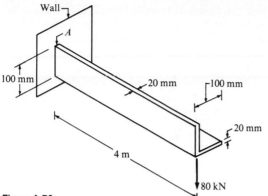

Figure A.P9

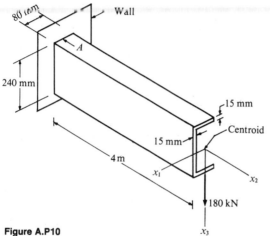

Figure A.P10

11. A cantilever steel beam having the triangular cross section shown in Fig. A.P11 is subjected to a force of 200 kN. Find the total shearing stress at point A. Compute the maximum total shearing stress acting on the cross section.

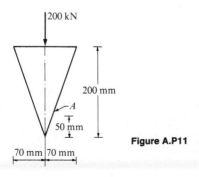

Figure A.P11

12. A cantilever steel beam having the closed, thin-walled cross section shown in Fig. A.P12 is subjected to a force of 200 kN. The thickness of the cross section is $t = 5$m. Find the maximum total shearing stress acting on the cross section.

13. A cantilever steel beam having the closed, thin-walled cross section shown in Fig. A.P13 is subjected to a force of 200 kN. Find the maximum total shearing stress acting on the cross section.

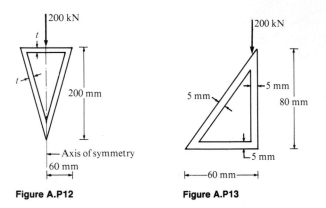

Figure A.P12   Figure A.P13

14. A structure made of steel (modulus of elasticity $E = 210{,}000$ MPa, Poisson's ratio $\nu = 0.3$) is loaded in a constant temperature environment. The components of displacement of the particles of a member of this structure, referred to a rectangular system of axes (x_1, x_2, x_3), are equal to

$$\hat{u}_1 = 0.002x_1 + 0.001x_2x_1 \qquad \hat{u}_2 = 0.004x_1^2 \qquad \hat{u}_3 = 0$$

Compute the components of stress at the particles of the member, referred to the same system of axes, as the given components of displacement.

15. A steel structure ($E = 210{,}000$ MPa, $\nu = 0.3$) is loaded in a constant temperature environment. The components of stress at a particle of this structure, referred to a rectangular system of axes (x_1, x_2, x_3), are

$$\tau_{11} = 120 \text{ MPa} \qquad \tau_{12} = 60 \text{ MPa}$$

$$\tau_{22} = 160 \text{ MPa} \qquad \tau_{33} = \tau_{13} = \tau_{23} = 0$$

Compute the components of strain of this particle, referred to the same system of axes as the given components of stress.

16. A structure made of aluminum ($E = 69{,}000$ MPa, $\nu = 0.3$) is loaded in a constant temperature environment. The components of stress at a particle of this structure, referred to a rectangular system of axes (x_1, x_2, x_3), are

$$\tau_{11} = -160 \text{ MPa} \qquad \tau_{12} = 80 \text{ MPa}$$

$$\tau_{22} = 160 \text{ MPa} \qquad \tau_{23} = 40 \text{ MPa}$$

$$\tau_{33} = 60 \text{ MPa} \qquad \tau_{13} = 0$$

Compute the components of strain of this particle, referred to the same system of axes as the given components of stress.

17. A structure made of steel ($E = 210,000$ MPa, $\nu = 0.3$) is loaded in a constant temperature environment. The components of strain of a particle of this structure, referred to a rectangular system of axes, are

$$e_{11} = 0.012 \qquad e_{22} = 0.006$$

$$e_{12} = 0.004 \qquad e_{23} = e_{13} = e_{33} = 0$$

Compute the components of stress at this particle, referred to the same system of axes as the given components of strain.

18. A structure made of aluminum ($E = 69,000$ MPa, $\nu = 0.3$) is loaded in a constant temperature environment. The components of strain of a particle of this structure, referred to a rectangular system of axes, are

$$e_{11} = -0.008 \qquad e_{12} = 0.004$$

$$e_{22} = 0.012 \qquad e_{23} = -0.006$$

$$e_{33} = -0.004 \qquad e_{13} = -0.002$$

Compute the components of stress at this particle, referred to the same system of axes as the given components of strain.

19. The components of strain e_{11}, e_{22}, and e_{12} of a particle at the surface of a steel beam ($E = 210,000$ MPa, $\nu = 0.3$) were measured with a strain rosette as

$$e_{11} = 0.012 \qquad e_{22} = 0.006 \qquad e_{12} = -0.004$$

Compute the components of stress at this particle, referred to the same system of rectangular axes as the measured components of strain.

20. A steel structure ($E = 210,000$ MPa, $\nu = 0.3$, $\alpha = 10^{-5}\,°\text{C}$) is subjected to external actions while its temperature changes from $T_0 = 5°\text{C}$ to $T = 30°\text{C}$. The components of stress at a particle of this structure, referred to a rectangular system of axes, are

$$\tau_{11} = -160 \text{ MPa} \qquad \tau_{12} = 80 \text{ MPa}$$

$$\tau_{22} = 160 \text{ MPa} \qquad \tau_{23} = 40 \text{ MPa}$$

$$\tau_{33} = 60 \text{ MPa} \qquad \tau_{13} = 0$$

Compute the components of strain of this particle, referred to the same system of axes as the components of stress.

21. An aluminum structure ($E = 69{,}000$, $\nu = 0.3$, $\alpha = 2.3 \times 10^{-5}{}^\circ\text{C}$) is subjected to external actions while its temperature changes from $T_0 = -5\,^\circ\text{C}$ to $T = 35\,^\circ\text{C}$. The components of strain of a particle of this structure, referred to a rectangular system of axes, are

$$e_{11} = -0.008 \qquad e_{12} = \quad 0.004$$

$$e_{22} = \quad 0.012 \qquad e_{22} = -0.006$$

$$e_{33} = -0.004 \qquad e_{13} = -0.002$$

Compute the components of stress of this particle, referred to the same system of axes as the given components of strain.

22. A steel structure ($E = 210{,}000$ MPa, $\nu = 0.3$, $\alpha = 10^{-5}{}^\circ\text{C}$) is subjected to external actions while its temperature changes from $T_0 = 5\,^\circ\text{C}$ to $T = 30\,^\circ\text{C}$. The components of stress at a particle of this structure, referred to a rectangular system of axes x_1, x_2, and x_3, are

$$\tau_{11} = 160 \text{ MPa} \qquad \tau_{12} = 80 \text{ MPa}$$

$$\tau_{22} = 160 \text{ MPa} \qquad \tau_{23} = 40 \text{ MPa}$$

$$\tau_{33} = \quad 60 \text{ MPa} \qquad \tau_{13} = 0$$

The direction cosines of the system of axes x_1, x_2, and x_3 with respect to the system of axes x_1', x_2', and x_3' are

$$\lambda_{11} = \quad 1/\sqrt{3} \qquad \lambda_{21} = \quad 1/\sqrt{3} \qquad \lambda_{31} = 1/\sqrt{3}$$

$$\lambda_{12} = \quad 0 \qquad \lambda_{22} = -1/\sqrt{2} \qquad \lambda_{32} = 1/\sqrt{2}$$

$$\lambda_{13} = -2/\sqrt{6} \qquad \lambda_{23} = \quad 1/\sqrt{6} \qquad \lambda_{33} = 1/\sqrt{6}$$

Compute the components of strain referred to the system of axes x_1', x_2', and x_3'.

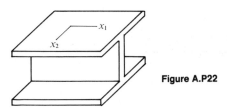

Figure A.P22

23. A circular steel bar ($E = 210{,}000$ MPa, $\nu = 0.3$) of diameter 20 mm is subjected to an axial force resulting in a state of uniaxial stress $\tau_{11} = 100$ MPa $\tau_{22} = \tau_{33} = \tau_{12} = \tau_{23} = 0$. Compute the change of its diameter due to its deformation.

REFERENCES

1. V. V. Novozhilov, *Foundations of the Nonlinear Theory of Elasticity,* Gaylock Press, Rochester, N.Y., 1953.

2. E. P. Popov, *Mechanics of Materials,* 2d ed., Prentice-Hall, Englewood Cliffs, N.J., 1976, p. 264.

3. G. R. Cowper, "The Shear Coefficient in Timoshenko's Beam Theory," *Journal of Applied Mechanics,* 335 (June 1966).

4. G. R. Cowper, "On the Accuracy of Timoshenko's Beam Theory," *Engineering Mechanics Division, ASCE,* 1947 (1968).

5. S. Timoshenko and N. J. Goodier, *Theory of Elasticity,* McGraw-Hill, New York, 1951, p. 326.

6. R. J. Roark, *Formula for Stress and Strain,* McGraw-Hill, New York, 1943.

7. J. M. Gere and S. P. Timoshenko, *Mechanics of Materials,* 2d ed., Brooks Cole Engineering Division, Wadsworth Inc., Belmont, Calif., 1984, p. 161.

Examples of Structures Which Cannot Be Analyzed on the Basis of the Theory of Small Deformation

In this appendix, we present two examples of structures which cannot be analyzed on the basis of the theory of small deformation.

Example 1. Study the response of a simply supported beam column subjected to an axial compressive force having an eccentricity as shown in Fig. a.

solution If the deformation of the beam column is within the range of validity of the theory of small deformations, the effect of the change of its geometry on the beam

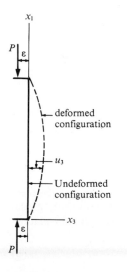

Figure a Beam column subjected to an axial compressive force.

column's internal actions, due to its deformation, is negligible. Thus, the moment at any cross section of the beam column can be approximated by

$$M_2 = P\varepsilon \tag{a}$$

Moreover, as shown in Sec. A.7, for a beam made of an isotropic, linearly elastic material with a modulus of elasticity E, the following linear relation is valid between the moment at a cross section and the component of translation u_3 of this cross section.

$$M_2 = -EI_2 \frac{d^2 u_3}{dx_1^2} \tag{b}$$

I_2 is the moment of inertia of the cross section of the beam column with respect to the x_2 axis. Substituting relation (b) into (a), we get

$$EI_2 \frac{d^2 u_3}{dx_1^2} + P\varepsilon = 0 \tag{c}$$

The solution of this equation is

$$u_3 = -\frac{P\varepsilon x_1^2}{2EI_2} + Ax_1 + B \tag{d}$$

where the constants A and B are evaluated from the boundary conditions of the simply supported beam column. That is

$$u_3 = 0 \quad \text{at } x_1 = 0$$
$$u_3 = 0 \quad \text{at } x_1 = L \tag{e}$$

Thus
$$B = 0 \quad A = \frac{P\varepsilon L}{2EI_2}$$

Consequently, the solution of the differential equation (b) becomes

$$u_3 = \frac{P\varepsilon x_1}{2EI_2}(L - x_1) \tag{f}$$

When the eccentricity ε is different than zero, the component of translation u_3 of the beam column increases as the force increases. However, for values of the force which induce a component of translation u_3 at the middle point of the beam column whose magnitude approaches the value of ε, solution (f) is not valid. In this case the effect on the internal moment of the change of geometry of the beam column, due to its deformation, cannot be neglected. Consequently, referring to Fig. b, the moment at any point of the beam column should be taken equal to

$$M_2 = P(\varepsilon + u_3) \tag{g}$$

Notice that in relation (g), the component of translation u_3 is a function of the force P; consequently the moment is not a linear function of the force P. Substituting relation (b) into (g) and differentiating, we get the following linear differential equation

$$\frac{d^4 u_3}{dx_1^4} + k^2 \frac{d^2 u_3}{dx_1^2} = 0 \tag{h}$$

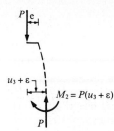

$M_2 = P(u_3 + \varepsilon)$ **Figure b** Free-body diagram of a portion of the beam column.

where
$$k^2 = \frac{P}{EI_2} \qquad \text{(i)}$$

Equation (h) is valid only within the range of validity of relation (b), that is, only for values of P and ε which cause small values of the component of translation u_3. The solution of Eq. (h) is

$$u_3 = A \sin kx_1 + B \cos kx_1 + Cx_1 + D \qquad \text{(j)}$$

where A, B, C, and D are constants of integration which are evaluated from the following nonhomogeneous conditions at the ends of the beam column.

$(a) \quad u_3(0) = 0 \qquad\qquad (b) \quad u_3(L) = 0$

$$(c) \quad \frac{d^2 u_3}{dx_1^2}\bigg|_{x_1=0} = -\frac{P\varepsilon}{EI_2} \qquad (d) \quad \frac{d^2 u_3}{dx_1^2}\bigg|_{x_1=L} = -\frac{P\varepsilon}{EI_2} \qquad \text{(k)}$$

Substituting relation (j) into (k), we get

$(a) \quad B + D = 0 \qquad (b) \quad A \sin kL + B \cos kL + CL + D = 0$

$(c) \quad B = \varepsilon \qquad\qquad\quad (d) \quad A \sin kL + B \cos kL = \varepsilon \qquad \text{(l)}$

This is a set of nonhomogeneous algebraic equations from which the constants A, B, C, and D can be obtained. Thus, we have

$$B = -D = \varepsilon \qquad C = 0$$
$$A = \frac{(1 - \cos kL)\varepsilon}{\sin kL} \qquad\qquad\qquad \text{(m)}$$

Substitution of the value of the constants into solution (j), yields

$$u_3 = \varepsilon \left[\frac{(1 - \cos kL)\sin kx_1}{\sin kL} - 1 \right] \qquad \text{(n)}$$

In the above result, the relation between the component of translation u_3 and the force P is not linear. Notice that when $\varepsilon \neq 0$, the denominator of the first term in the bracket of relation (n) vanishes ($\sin kL = 0$) for certain values of the axial force P, referred to as its critical values (P_{cr}). The critical values of P are obtained from the following equation.

$$\sin kL = 0 \qquad \text{(o)}$$

Consequently, $kL = n\pi \qquad n = 1, 2, 3, \cdots \qquad \text{(p)}$

Substituting relation (i) in the above, we get

$$P_{cr} = \frac{n^2\pi^2 EI_2}{L^2} \qquad n = 1, 2, 3, \cdots \tag{q}$$

On the basis of Eq. (n), the component of translation u_3 of the beam column is a single-valued function of the axial force. That is, the beam column does not reach a condition of unstable equilibrium. When the force P approaches one of its critical values, the component of translation u_3 of the beam column becomes very large. However, as we mentioned previously, the differential equation (h) is valid only for relatively small translations u_3. Thus, for values of the force P approaching the critical values (q), the solution (n) does not give the actual translation of the beam column. Nevertheless, as shown in Fig. c, for values of the axial force P that are close to its lowest critical value, the rate of increase of the component of translation u_3 with P becomes very large. Thus, the beam column must be designed so that the value of the anticipated axial force is smaller than the critical force of the beam column. Hence, the critical force of a beam column is an important design parameter.

In order to establish the component of translation u_3 of the beam column for values of P approaching the critical load at buckling (that is, for relatively large values of the translation u_3), a nonlinear theory must be employed. In this theory, the measure of the deformation is not the strain tensor $[e]$ whose components are defined by relation (1.13), but another tensor $[\varepsilon]$. The relations between the components of the tensor $[\varepsilon]$ and the components of the displacement vector $\hat{u}$ include nonlinear terms. Consequently, the relation between the moment and the component of translation u_3 is not linear. For a beam column made of an isotropic, linearly elastic material ($\tau_{11} = E\varepsilon_{11}$), we have

$$M_2 = \int\int_A x_3\tau_{11}\, dA = \int\int_A x_3 E\varepsilon_{11}\, dA = L(u_3) \tag{r}$$

where L is a nonlinear differential operator. For example, a nonlinear relation between the moment and the component of translation u_3 which has been employed in the solution of buckling problems is

$$M_2 = -EI_2 \frac{d^2u_3}{dx_1^2}\left[1 + \left(\frac{du_3}{dx_1}\right)^2\right]^{-3/2} \tag{s}$$

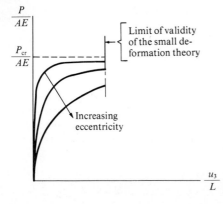

Limit of validity of the small deformation theory

Increasing eccentricity

Figure c Relation between the axial force and the transverse component of translation of the beam column.

Substituting relation (s) into (g), we obtain the following nonlinear differential equation[†] for the component of translation u_3.

$$EI_2 \frac{d^2 u_3}{dx_1^2} \left[1 + \left(\frac{du_3}{dx_1} \right)^2 \right]^{-3/2} + P(u_3 + \varepsilon) = 0 \qquad \text{(t)}$$

The solution of this equation is cumbersome and will not be discussed herein.

When the load acts through the centroid of the cross section of the beam column ($\varepsilon = 0$), the boundary conditions (k) become homogeneous; thus, Eqs. (1) reduce to the following set of homogeneous, linear algebraic equations.

$$B = D = 0 \qquad A \sin kL + CL = 0 \qquad A \sin kL = 0 \qquad \text{(u)}$$

These equations have a solution other than the trivial $A = B = C = D = 0$ ($u_3 = 0$) only when the determinant of the coefficients of the unknown constants (A, B, and C) vanishes. Thus, Eqs. (u) have a nontrivial solution when the values of the load parameter k satisfy the following relation.

$$\sin kL = 0 \qquad \text{(v)}$$

hence
$$kL = n\pi \qquad n = 1, 2, 3, \ldots \qquad \text{(w)}$$

Referring to relation (i), the values of the load corresponding to the above values of k are

$$P_{cr} = \frac{n^2 \pi^2 EI_2}{L^2} \qquad n = 1, 2, 3, \ldots \qquad \text{(x)}$$

Substituting results (u) and (w) into relation (j), we obtain the following expression for the component of translation of the beam column subjected to a compressive force passing through the centroid of its cross sections.

$$u_3 = A \sin \frac{n\pi x_1}{L} \qquad \text{(y)}$$

Thus, we have established that an ideally straight beam column loaded by a concentrated force passing through its centroid ($\varepsilon = 0$) has a position of equilibrium other than the original straight line ($u_3 = 0$) when the force reaches the critical values given by relation (x). That is, in this case, the beam column reaches a state of unstable equilibrium. Notice that as a result of using the linear relation (b) between the moment and the component of translation u_3, relation (y) gives only the shape of the deformed beam column. The amplitude of its translation has not been established. That is, the coefficient A in relation (y) is an unspecified constant.

Example 2.   Consider the structure shown in Fig. a, consisting of two identical slender members of constant cross section, pinned together at one end, while the other end is pinned to rigid supports. The structure is loaded by a concentrated vertical force at joint 2. Compute the displacement of joint 2 as a function of the force P and of the

[†] For $\varepsilon = 0$ see Timoshenko.[1]

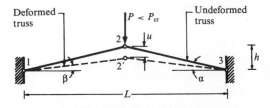

Figure a Geometry and loading of the structure.

angle α. The latter specifies the geometry of the undeformed structure. Use both the theory of small deformations and a nonlinear theory for which the difference between the undeformed and the deformed cross-sectional areas of the members of the structure is very small and can be disregarded. However, the effect of the rotation of the members of the structure is taken into account in computing their internal forces.

In order to compute the displacement of the particles of a structure, we must specify the material from which its members are made, in addition to their geometry. In this example, we assume that the material from which the members of the structure are made is isotropic and linearly elastic. That is, the relation between the axial component of stress and the change of length per unit length (unit elongation) of a particle of the structure is linear. Consequently, since the members of the structure are in a state of uniaxial stress ($\tau_{11} \neq 0$, $\tau_{22} = \tau_{33} = \tau_{12} = \tau_{23} = \tau_{13} = 0$), we have

$$\tau_{11} = EE_{11} \tag{a}$$

where E is the modulus of elasticity of the material from which the members of the structure are made and E_{11} is the change of length per unit length (unit elongation) of a particle of a member of the structure in the direction of its axis.

solution   The undeformed and deformed configurations of the structure are shown in Fig. a. Joint 2 is displaced to point 2'. The change of length per unit length of a particle of a member of the structure in the direction of its axis is equal to

$$E_{11} = \frac{L/\cos \beta - L/\cos \alpha}{L/\cos \alpha} = \frac{\cos \alpha}{\cos \beta} - 1 \tag{b}$$

Moreover, referring to Fig. a, from geometric considerations, we obtain

$$\cos \alpha = \frac{L}{\sqrt{L^2 + h^2}} \qquad \sin \alpha = \frac{h}{\sqrt{L^2 + h^2}}$$

$$\sin \beta = \frac{h - u}{\sqrt{L^2 + (h - u)^2}} \qquad \cos \beta = \frac{L}{\sqrt{L^2 + (h - u)^2}} \tag{c}$$

Substituting the first and fourth of the above relations into (b), we obtain the following relation between the unit elongation E_{11} and the displacement u in the nonlinear theory.

$$E_{11} = \sqrt{\frac{L^2 + (h - u)^2}{L^2 + h^2}} - 1 = \sqrt{1 + \frac{u^2 - 2hu}{L^2 + h^2}} - 1 \tag{d}$$

As becomes apparent later, in order to disregard the effect of the change of the geometry of the structure, due to its deformation, when considering the equilibrium of joint 2, the displacement u must be very small compared to h. Consequently, the term $2hu/(L^2 + h^2) = (2u/h)/(L^2/h^2 + 1)$ must be small compared to unity. Moreover, the term $u^2/(L^2 + h^2)$ has a lower order of magnitude than the term $2hu/(L^2 + h^2)$ and can be disregarded when compared to the latter. Thus, for a deformation within the range of validity of the theory of small deformations, using the binomial expansion formula, relation (d) may be approximated as

$$E_{11} = \left(1 - \frac{2hu}{L^2 + h^2}\right)^{1/2} - 1 = -\frac{hu}{L^2 + h^2} = e_{11} \tag{e}$$

The free-body diagrams of joint 2 are shown in Figs. b and c. In Fig. b, the change of the angle α due to the deformation of the structure has been disregarded, while it is included in Fig. c.

We first analyze the structure on the basis of the theory of small deformation. Referring to Fig. b, from the equilibrium of joint 2, we have

$$\Sigma \bar{F}_v = 0 = 2F \sin \alpha - P$$

Using the second of relations (c), in the above relation we get

$$F = \frac{P}{2 \sin \alpha} = -\frac{P\sqrt{L^2 + h^2}}{2h} \tag{f}$$

moreover

$$e_{11} = \frac{\tau_{11}}{E} = \frac{F}{EA} = -\frac{P\sqrt{L^2 + h^2}}{2hEA} \tag{g}$$

where A is the undeformed cross-sectional area of the members of the structure, which differs negligibly from their deformed cross-sectional area, for deformations within the range of validity of the small deformations theory. Thus, substituting relation (g) into (e), and noting that $h = L \tan \alpha$, we obtain the following expression for the displacement of joint 2:

$$u = P\frac{(L^2 + h^2)^{3/2}}{2EAh^2} = \frac{PL}{2EA}\frac{(1 + \tan^2 \alpha)^{3/2}}{\tan^2 \alpha} \tag{h}$$

It is apparent that this is a linear relation between the applied force and the resulting displacement. Notice that for a given force, the displacement increases as the angle α decreases. When the angle α is zero, that is, when both members of the structure are horizontal, the displacement u and the internal force F in the members of the structure become infinite. In this case, the structure is geometrically unstable.

We now proceed to analyze the structure using the nonlinear theory. In this theory,

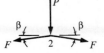

Figure b Free-body diagram of joint 2 disregarding the change of angle α due to the deformation.

Figure c Free-body diagram of joint 2 including the change of the angle α due to the deformation.

as was previously mentioned, the change of the area of the cross sections of the members of the structure is disregarded, as compared to their original area. Thus

$$\tau_{11} = \frac{F}{A} \tag{i}$$

However, in computing the internal forces in the two members of the structure, the effect of the change of the angle α, due to the deformation, is retained. Consequently referring to Fig. c, from the equilibrium of joint 2, we have:

$$\Sigma \overline{F}_v = 0 = 2F \sin \beta + P$$

or
$$F = -\frac{P}{2 \sin \beta} \tag{j}$$

Substituting relation (j) into (i) and the resulting relation into (a), we obtain

$$E_{11} = \frac{F}{EA} = -\frac{P}{2EA \sin \beta} \tag{k}$$

Substituting relations (d) and (c) into (k), we get

$$\frac{P}{EA} = -\frac{2(h - u)}{\sqrt{L^2 + (h - u)^2}} \left[\frac{\sqrt{L^2 + (h - u)^2}}{\sqrt{L^2 + h^2}} - 1 \right]$$

$$= -\frac{2(h - u)}{\sqrt{L^2 + (h - u)^2}} \left[\sqrt{1 + \frac{u^2}{L^2 + h^2} - \frac{2hu}{L^2 + h^2}} - 1 \right] \tag{l}$$

This is the nonlinear, force-displacement relation obtained on the basis of the nonlinear theory. It is plotted in Fig. d for $h/L = 0.01$, and $h/L = 0$. Referring to this figure, it can be seen that for values of h/L different than zero, the displacement of point 2 of the structure increases monotonically as the force increases from zero. For

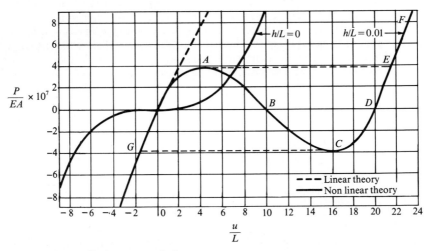

Figure d Force displacement relations.

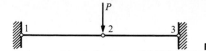

Figure e Kinematically unstable structure.

small values of the force, the results obtained on the basis of the linear theory (h) are in agreement with those obtained on the basis of the nonlinear theory (l). However, as the force approaches its critical value at point A, the linear theory does not predict the response of the structure. When the force reaches its critical value at point A ($P_{cr}/EA \approx 3.85 \times 10^{-7}$), the structure reaches a state of unstable equilibrium and jumps (buckles) to its stable configuration at point E (see Fig. 1.45b). For values of the force above the critical value, the displacement of point 2 of the structure increases monotonically as the force increases. The structure does not reach another state of unstable equilibrium until its members yield. If the force decreases after reaching its value at point F, the displacement decreases monotonically until the force vanishes at point D. The configuration of the truss at point D is shown in Fig. 1.45d. If the structure is subsequently subjected to an upward force, the displacement of point 2 increases monotonically until the force reaches its critical value at point C (see Fig. 1.45f) when the structure jumps (buckles) to its stable configuration at point G.

When h/L is equal to zero (see Fig. e), the structure is kinematically unstable. As soon as a transverse force is applied to this structure, its members move instantaneously without deforming. Moreover, as can be seen from Fig. d, for small values of the force, the rate of increase of the deformation is very large. However, the structure does not necessarily fail. In fact, as the force increases, its ability to withstand it increases. For instance, suppose that the structure of Fig. e is made of American steel A36 [yield stress† $\tau_{11} = 36{,}000$ lb/in² (248 MPa); modulus of elasticity $E = 31 \times 10^6$ lb/in² (210,000 MPa)]. Moreover, suppose that this structure is subjected to the transverse force required to induce yielding of its members. For this value of the force, the unit elongation ($E_{11} = \tau_{11}/E$) in the members of the structure is equal to 1.181 $\times 10^{-3}$. The displacement of point 2 of the structure may be obtained from relation (d) as

$$\left. \frac{u}{L} \right|_{yielding} = \sqrt{E_{11}^2 + 2E_{11}} = 4.8615 \times 10^{-2}$$

Substituting this result into relation (l), the force required to produce yielding of the members of the structure is equal to

$$\left. \frac{P}{AE} \right|_{yielding} = 2 \frac{u}{L} \left[1 - \frac{1}{\sqrt{1 + (u^2/L^2)}} \right] = 1.147 \times 10^{-4}$$

Thus, the phenomena illustrated in Fig. d occur for values of the force P/AE considerably below that which causes the members of the structure to yield.

REFERENCE

1. S. Timoshenko, *Theory of Elastic Stability*, McGraw-Hill, New York, 1936, p. 69.

---

† The yielding stress of American high-strength steel (A242) (A441) could reach as high as 50,000 lb/in² (344 MPa).

Fixed-End Moments

C.1 Notation

A Cross-sectional area.

E Modulus of elasticity.

G Shear modulus.

h Depth of member.

I Moment of inertia with respect to the x_3 axis.

K Torsional constant.

T_0 Temperature during construction.

δT_2 $T_2^{(+)} - T_2^{(-)}$. It is assumed that $\delta T_2 > 0$.

δT_c Change of temperature at the centroid of a cross section. It is assumed that $\Delta T_c > 0$.

$T^{(+)}, T^{(-)}$ Temperature at the points of a cross section of an member where the positive and negative x_2 axes, respectively, intersect its perimeter.

α Coefficient of linear expansion.

TABLE C.1   Members Fixed at Both Ends

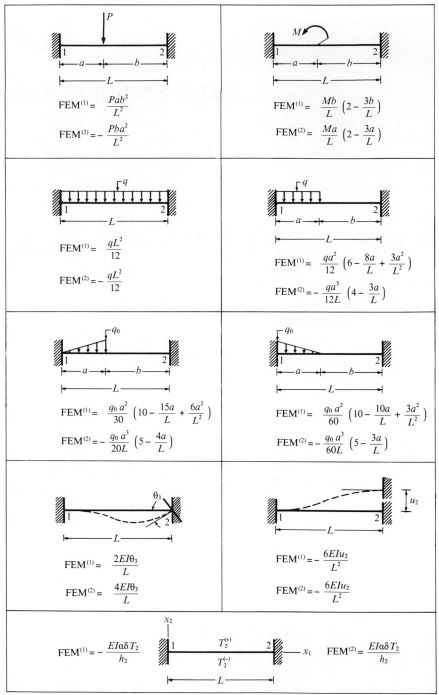

$$\text{FEM}^{(1)} = \frac{Pab^2}{L^2}$$

$$\text{FEM}^{(2)} = -\frac{Pba^2}{L^2}$$

$$\text{FEM}^{(1)} = \frac{Mb}{L}\left(2 - \frac{3b}{L}\right)$$

$$\text{FEM}^{(2)} = \frac{Ma}{L}\left(2 - \frac{3a}{L}\right)$$

$$\text{FEM}^{(1)} = \frac{qL^2}{12}$$

$$\text{FEM}^{(2)} = -\frac{qL^2}{12}$$

$$\text{FEM}^{(1)} = \frac{qa^2}{12}\left(6 - \frac{8a}{L} + \frac{3a^2}{L^2}\right)$$

$$\text{FEM}^{(2)} = -\frac{qa^3}{12L}\left(4 - \frac{3a}{L}\right)$$

$$\text{FEM}^{(1)} = \frac{q_0\, a^2}{30}\left(10 - \frac{15a}{L} + \frac{6a^2}{L^2}\right)$$

$$\text{FEM}^{(2)} = -\frac{q_0\, a^3}{20L}\left(5 - \frac{4a}{L}\right)$$

$$\text{FEM}^{(1)} = \frac{q_0\, a^2}{60}\left(10 - \frac{10a}{L} + \frac{3a^2}{L^2}\right)$$

$$\text{FEM}^{(2)} = -\frac{q_0\, a^3}{60L}\left(5 - \frac{3a}{L}\right)$$

$$\text{FEM}^{(1)} = \frac{2EI\theta_3}{L}$$

$$\text{FEM}^{(2)} = \frac{4EI\theta_3}{L}$$

$$\text{FEM}^{(1)} = -\frac{6EIu_2}{L^2}$$

$$\text{FEM}^{(2)} = -\frac{6EIu_2}{L^2}$$

$$\text{FEM}^{(1)} = -\frac{EI\alpha\delta T_2}{h_2} \qquad \text{FEM}^{(2)} = \frac{EI\alpha\delta T_2}{h_2}$$

TABLE C.2 Members Fixed at One End and Pinned at the Other End

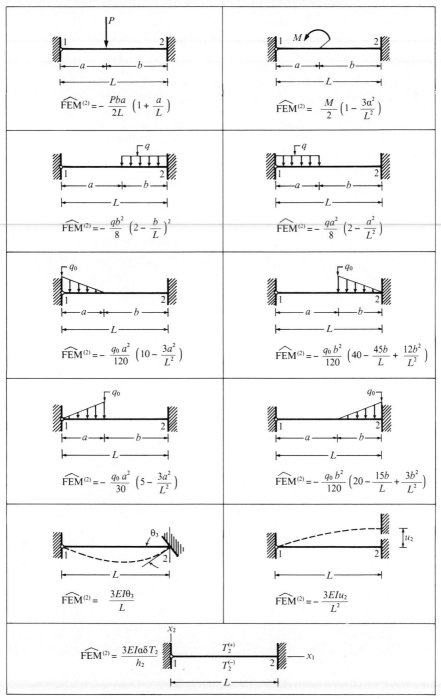

$$\widehat{\text{FEM}}^{(2)} = -\frac{Pba}{2L}\left(1+\frac{a}{L}\right)$$

$$\widehat{\text{FEM}}^{(2)} = \frac{M}{2}\left(1-\frac{3a^2}{L^2}\right)$$

$$\widehat{\text{FEM}}^{(2)} = -\frac{qb^2}{8}\left(2-\frac{b}{L}\right)^2$$

$$\widehat{\text{FEM}}^{(2)} = -\frac{qa^2}{8}\left(2-\frac{a^2}{L^2}\right)$$

$$\widehat{\text{FEM}}^{(2)} = -\frac{q_0 a^2}{120}\left(10-\frac{3a^2}{L^2}\right)$$

$$\widehat{\text{FEM}}^{(2)} = -\frac{q_0 b^2}{120}\left(40-\frac{45b}{L}+\frac{12b^2}{L^2}\right)$$

$$\widehat{\text{FEM}}^{(2)} = -\frac{q_0 a^2}{30}\left(5-\frac{3a^2}{L^2}\right)$$

$$\widehat{\text{FEM}}^{(2)} = -\frac{q_0 b^2}{120}\left(20-\frac{15b}{L}+\frac{3b^2}{L^2}\right)$$

$$\widehat{\text{FEM}}^{(2)} = \frac{3EI\theta_3}{L}$$

$$\widehat{\text{FEM}}^{(2)} = -\frac{3EIu_2}{L^2}$$

$$\widehat{\text{FEM}}^{(2)} = \frac{3EI\alpha\delta T_2}{h_2}$$

Photograph and Brief Description of a Lift Bridge

This graceful lift bridge carries two lanes of traffic and a pedestrian walk over the U.S. Canal in the city of Kaukauna, Wisconsin. The lift span consists of welded plate girders. The four towers and the struts are welded steel sections. The lifting machinery has been tucked into the longitudinal and transverse welded box struts at the top of the towers (see Fig. C.2). This protects the machinery from the weather and, moreover, keeps it out of sight. There are access doors and ladders in each tower and an access hatch in the bottom of the longitudinal strut.

Figure C.1 Lift bridge in the city of Kaukauna, Wis. *(Courtesy of Owen Ayres & Associates, Inc., Consulting Engineers, Eau Claire, Wis.)*

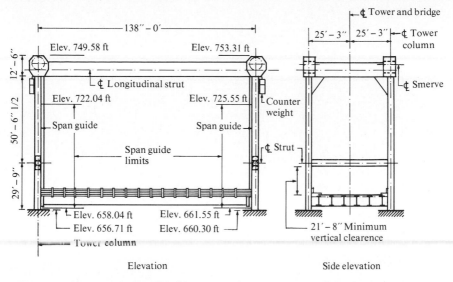

Figure C.2 Elevation of the lift bridge in the city of Kaukauna, Wis. *(Courtesy of Owen Ayres & Associates, Consulting Engineers, Eau Claire, Wis.)*

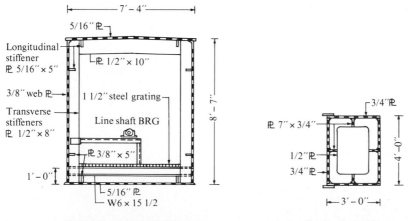

Section through longitudal strut   Section through tower

Figure C.3 Details of the lift bridge in the city of Kaukauna, Wis. *(Courtesy of Owen Ayres & Associates, Consulting Engineers, Eau Claire, Wis.)*

Answers to Problems

Chapter 1

1. (a) 3d degree; (b) 5th degree; (c) 3d degree; (d) 1st degree;
   (e) 9th degree; (f) 15th degree; (g) 5th degree; (h) 7th degree; (i) 5th degree;
   (j) 5th degree; (k) 2d degree; (l) 8th degree;
   (m) 6th degree; (n) 3d degree; (o) 6th degree;
   (p) 12th degree when subjected to the given loading;
   (q) 3d degree when subjected to the given loading;
   (r) 12th degree; (s) 12th degree; (t) 9th degree;
   (u) 3d degree when subjected to the given loading;
   (v) 24th degree; (w) 1st degree.

Chapter 3

1. $R_h^{(1)} = -40$ kN; $R_v^{(1)} = 0$; $R_v^{(5)} = 60$ kN; $N^{(1)} = 0$; $N^{(2)} = 40$-kN tension;
$N^{(3)} = 60$-kN tension; $N^{(4)} = 40$-kN tension; $N^{(5)} = 40$-kN comp.;
$N^{(6)} = 50$-kN tension; $N^{(7)} = 30$-kN comp.

2. $R_h^{(1)} = -80$ kN; $R_v^{(1)} = 57.14$ kN; $R_v^{(4)} = 102.86$ kN; $N^{(1)} = 80.81$-kN comp.;
$N^{(2)} = N^{(4)} = 137.15$-kN tension; $N^{(3)} = 60$-kN tension; $N^{(5)} = 171.43$-kN comp.

3. $R_h^{(1)} = 30$ kN; $R_v^{(1)} = 40$ kN; $R_h^{(3)} = -30$ kN; $N^{(1)} = 50$-kN comp;
$N^{(2)} = 30$-kN comp.

4. $R_h^{(1)} = -120$ kN; $R_v^{(1)} = -120$ kN; $R_v^{(6)} = 120$ kN; $N^{(1)} = 40$-kN tension;
$N^{(2)} = 0$; $N^{(3)} = 60$-kN comp.; $N^{(4)} = 72.11$-kN tension; $N^{(5)} = 40$-kN comp.;
$N^{(6)} = 120$-kN comp.; $N^{(7)} = 0$; $N^{(8)} = 120$-kN comp.; $N^{(9)} = 0$.

5. $R_h^{(1)} = 80$ kN; $R_h^{(2)} = -80$ kN; $R_v^{(1)} = -60$ kN; $N^{(1)} = 0$;
$N^{(2)} = 80$-kN comp.; $N^{(3)} = 100$-kN tension; $N^{(4)} = 0$; $N^{(5)} = 100$-kN tension;
$N^{(6)} = -80$-kN comp.

6. $R_v^{(1)} = 34.44$ kN; $R_h^{(1)} = -40$ kN; $R_v^{(7)} = 65.56$ kN; $N^{(1)} = 43.05$-kN comp.;
$N^{(2)} = 65.83$-kN tension; $N^{(3)} = 18.05$-kN tension; $N^{(4)} = 18.05$-kN comp.;
$N^{(5)} = 76.6$-kN comp.; $N^{(6)} = 87.49$-kN comp.; $N^{(7)} = 31.09$-kN comp.;
$N^{(8)} = 31.95$-kN tension; $N^{(9)} = 68.31$-kN comp.; $N^{(19)} = 56.83$-kN tension;
$N^{(11)} = 81.95$-kN comp.

7. For vertical loading the truss has an axis of symmetry and the loading is symmetric about this axis.

$R_v^{(1)} = R_v^{(8)} = 60$ kN; $N^{(1)} = N^{(13)} = 80$-kN tension;
$N^{(2)} = N^{(12)} = 100$-kN comp.; $N^{(3)} = N^{(11)} = 0$; $N^{(4)} = N^{(8)} = 66.67$-kN comp.;
$N^{(5)} = N^{(9)} = 33.33$-kN comp.; $N^{(6)} = N^{(10)} = 80$-kN tension;
$N^{(7)} = 40$-kN tension.

8. $R_h^{(1)} = -40$ kN; $R_v^{(1)} = 0$; $R_v^{(6)} = 60$ kN; $N^{(1)} = 40$-kN tension; $N^{(2)} = 0$;
$N^{(3)} = 40$-kN comp.; $N^{(4)} = 36.03$-kN comp.; $N^{(5)} = 17.75$-kN tension;
$N^{(6)} = 25.75$-kN comp.; $N^{(7)} = 43.75$-kN comp.; $N^{(8)} = N^{(9)} = 0$.

9. $R_h^{(1)} = -20$ kN; $R_v^{(1)} = 15$ kN; $R_v^{(8)} = 25$ kN; $N^{(1)} = 27.04$-kN comp.;
$N^{(2)} = 42.5$-kN tension; $N^{(3)} = 7.5$-kN tension; $N^{(4)} = 50$-kN tension;
$N^{(5)} = 23.72$-kN comp.; $N^{(6)} = 10.61$-kN comp.; $N^{(7)} = 0$; $N^{(8)} = 50$-kN comp.;
$N^{(9)} = 39.53$-kN comp.; $N^{(10)} = 17.678$-kN comp.; $N^{(11)} = 12.5$-kN tension;
$N^{(12)} = 37.5$-kN tension; $N^{(13)} = 45.07$-kN comp.

10. $R_h^{(1)} = 120$ kN; $R_v^{(1)} = 160$ kN; $R_h^{(2)} = -120$ kN; $R_v^{(2)} = -90$ kN;
$N^{(1)} = N^{(5)} = N^{(9)} = N^{(10)} = 200$-kN comp.;
$N^{(2)} = N^{(4)} = N^{(6)} = N^{(8)} = N^{(11)} = 0$; $N^{(3)} = N^{(7)} = N^{(12)} = 150$-kN tension.

11. $R_h^{(1)} = 0$; $R_v^{(1)} = -120$ kN; $R_h^{(6)} = 0$; $R_v^{(6)} = 170$ kN; $N^{(1)} = 0$;
$N^{(2)} = 170$-kN comp.; $N^{(3)} = 0$; $N^{(4)} = 170$-kN comp.; $N^{(5)} = N^{(6)} = 0$;
$N^{(7)} = 93.33$-kN comp.; $N^{(8)} = 200$-kN comp.; $N^{(9)} = 160$-kN tension; $N^{(10)} = 0$;
$N^{(11)} = 83.33$-kN comp.; $N^{(12)} = 160$-kN tension; $N^{(13)} = 0$;
$N^{(14)} = 83.33$-kN tension; $N^{(15)} = 26.66$-kN tension; $N^{(16)} = 93.33$-kN comp.;
$N^{(17)} = 30$-kN comp.; $N^{(18)} = 26.66$-kN tension; $N^{(19)} = 33.33$-kN comp.

12. For vertical loading the tress has an axis of symmetry, and the loading is symmetric about this axis.

$R_h^{(1)} = 0$; $R_v^{(1)} = R_v^{(16)} = 75$ kN; $N^{(1)} = N^{(27)} = 90.14$-kN comp.;
$N^{(2)} = N^{(28)} = 50$-kN tension; $N^{(3)} = N^{(25)} = 30$-kN tension;
$N^{(4)} = N^{(26)} = 75$-kN tension; $N^{(5)} = N^{(21)} = 50$-kN comp.;
$N^{(6)} = N^{(22)} = 37.5$-kN comp.; $N^{(7)} = N^{(23)} = 37.5$-kN tension;
$N^{(8)} = N^{(24)} = 50$-kN tension; $N^{(9)} = N^{(19)} = 7.5$-kN tension;
$N^{(10)} = N^{(20)} = 22.5$-kN comp.; $N^{(11)} = N^{(15)} = 80$-kN comp.;
$N^{(12)} = N^{(16)} = 12.5$-kN comp.; $N^{(13)} = N^{(17)} = 12.5$-kN tension;
$N^{(14)} = N^{(18)} = 80$-kN tension. $N^{(8-9)} = 10$-kN tension.

13. The truss has an axis of symmetry.

$R_v^{(1)} = -75$ kN; $R_v^{(6)} = 270$ kN; $N^{(1)} = 93.75$-kN tension;
$N^{(2)} = 56.25$-kN comp.; $N^{(3)} = 30$-kN tension; $N^{(4)} = 56.25$-kN comp.;

$N^{(5)} = 131.25$-kN comp.; $N^{(6)} = 0$; $N^{(7)} = 135$-kN tension; $N^{(8)} = 190.92$-kN comp.

14. $N^{(1)} = 56.57$-kN tension; $N^{(2)} = 240$-kN comp.

15. $N^{(1)} = 32.05$-kN comp.; $N^{(2)} = 66.67$-kN comp.

16. $N^{(1)} = 80$-kN comp.; $N^{(2)} = 120$-kN comp.; $N^{(3)} = 70.59$-kN comp.; $N^{(4)} = 117.65$-kN comp.

17. $R_v^{(7)} = 320$ kN; $R_v^{(9)} = -320$ kN; $R_h^{(17)} = -40$ kN; $R_v^{(17)} = 80$ kN: $N^{(1)} = 60$-kN comp.; $N^{(2)} = 66.67$-kN comp.; $N^{(3)} = 33.33$-kN comp.; $N^{(4)} = 100$-kN tension.

18. $R_h^{(1)} = 32$ kN; $R_v^{(1)} = 16$ kN; $R_h^{(21)} = -32$ kN; $R_v^{(21)} = 24$ kN; $N^{(1)} = 106.6$-kN comp.; $N^{(2)} = 80$-kN tension; $N^{(3)} = 96$-kN comp.

19. $R_v^{(1)} = -80$ kN; $R_v^{(4)} = 160$ kN; $R_h^{(4)} = 0$; $R_v^{(15)} = 20$ kN; $N^{(1)} = 0$; $N^{(2)} = 113.14$-kN comp.; $N^{(3)} = 28.24$-kN tension.

20.

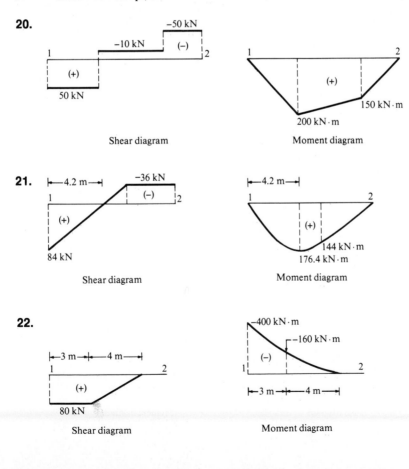

Shear diagram   Moment diagram

21.

Shear diagram   Moment diagram

22.

Shear diagram   Moment diagram

23.

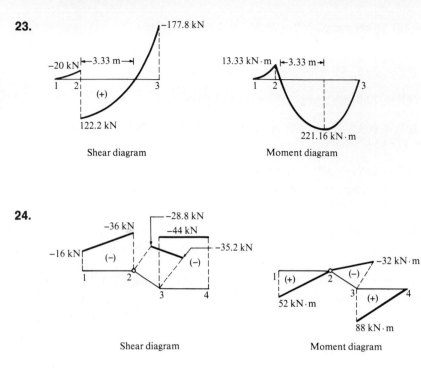

Shear diagram

Moment diagram

24.

Shear diagram

Moment diagram

25.

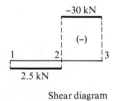

Shear diagram

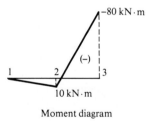

Moment diagram

26.

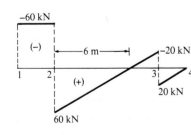

Shear diagram

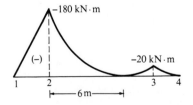

Moment diagram

27.

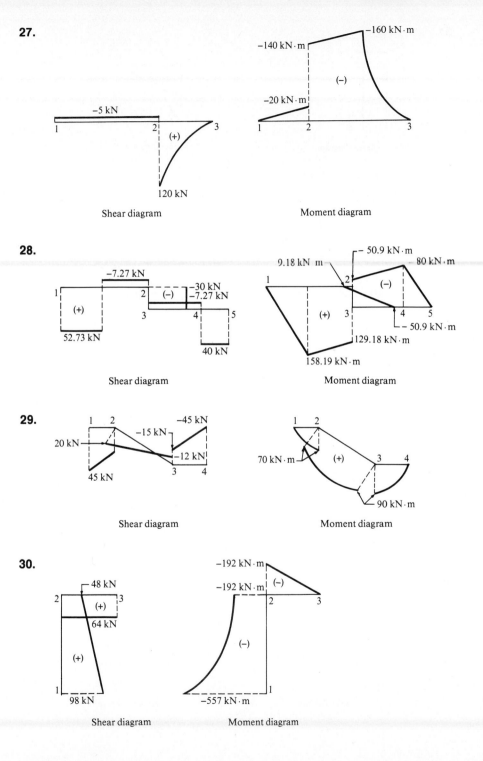

Shear diagram   Moment diagram

28.

Shear diagram   Moment diagram

29.

Shear diagram   Moment diagram

30.

Shear diagram   Moment diagram

31.

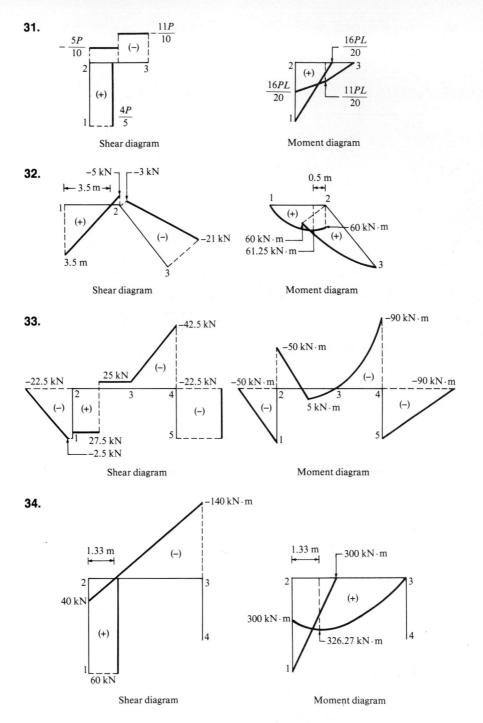

Shear diagram

Moment diagram

32.

Shear diagram

Moment diagram

33.

Shear diagram

Moment diagram

34.

Shear diagram

Moment diagram

35.

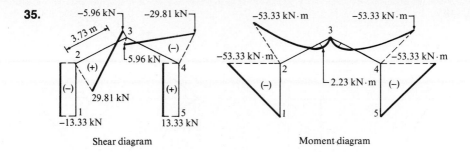

Shear diagram   Moment diagram

36.

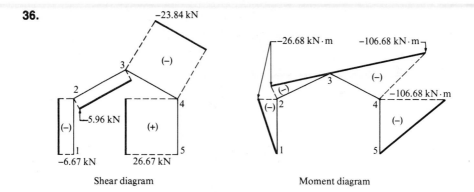

Shear diagram   Moment diagram

37.

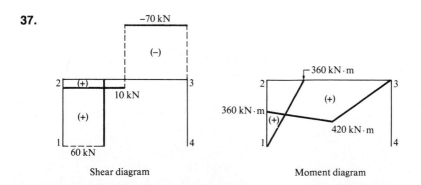

Shear diagram   Moment diagram

38.

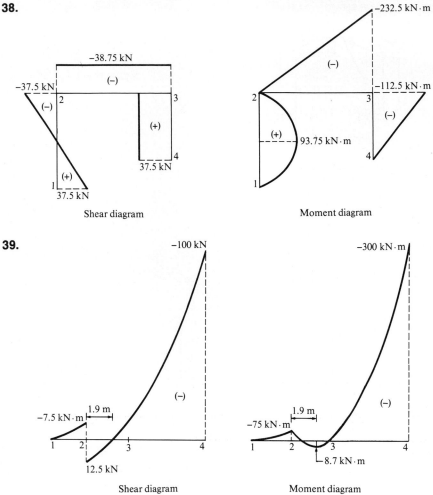

Shear diagram

Moment diagram

39.

Shear diagram

Moment diagram

40.

$R_v^{(1)} = 15$ kN $R_v^{(4)} = -50$ kN $R_h^{(4)} = -300$ kN $R_v^{(7)} = 550$ kN $R^{(8)} = 375$ kN ↘

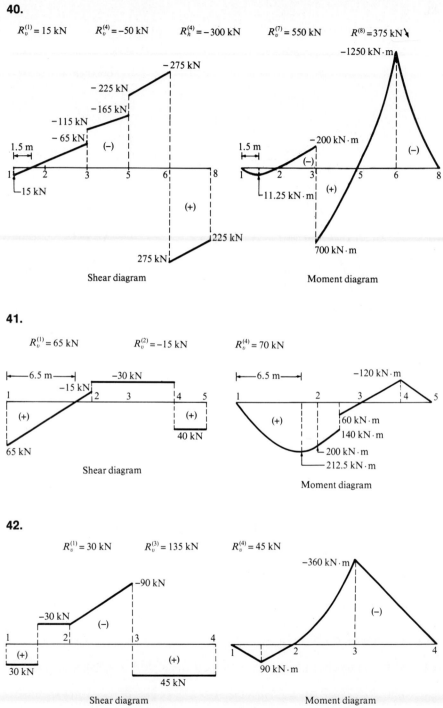

Shear diagram   Moment diagram

41.

$R_v^{(1)} = 65$ kN $R_v^{(2)} = -15$ kN $R_v^{(4)} = 70$ kN

Shear diagram   Moment diagram

42.

$R_v^{(1)} = 30$ kN $R_v^{(3)} = 135$ kN $R_v^{(4)} = 45$ kN

Shear diagram   Moment diagram

43.

$R_v^{(4)} = 128.82$ kN $R_h^{(1)} = 211.25$ kN $R_v^{(10)} = 171.18$ kN $R_h^{(10)} = -211.25$ kN

$N^{(1} = 3.39$- kN tension $N^{(2)} = 250.65$-kN comp. $N^{(3)} = 156.40$- kN comp.

$N^{(4)} = 118.92$-kN comp.

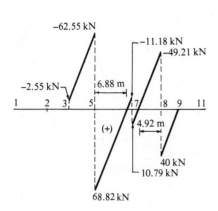

Shear diagram

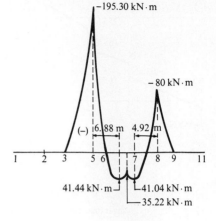

Moment diagram

44.

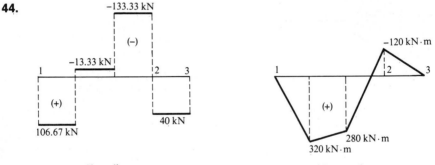

Shear diagram   Moment diagram

45. $R_v^{(1)} = 80$ kN; $R_h^{(1)} = 40$ kN; $R_v^{(3)} = 100$ kN; $R_h^{(3)} = -50$ kN.

46. $R_v^{(1)} = 45.58$ kN; $R_h^{(1)} = 46.0$ kN; $R_v^{(3)} = 54.42$ kN; $R_h^{(3)} = -66.0$ kN.

47. $T_{max} = 484.66$ kN.

49. $N^{(1)} = 11.04$-kN tension; $N^{(2)} = 13.5$-kN comp.; $N^{(3)} = 3.92$-kN tension.

50. $\bar{R}_2^{(1)} = 5.55$ kN; $\bar{R}_3^{(1)} = -20.0$ kN; $\bar{R}_2^{(2)} = 5.55$ kN; $\bar{R}_3^{(2)} = -20$ kN;
$\bar{R}_1^{(3)} = 0$; $\bar{R}_2^{(4)} = 24.45$ kN; $N^{(1)} = 3.33$-kN tension; $N^{(2)} = 24.45$-kN tension;

$N^{(3)} = 14.68$-kN tension; $N^{(4)} = 24.45$-kN tension; $N^{(5)} = N^{(6)} = 7.86$-kN comp.; $N^{(7)} = N^{(8)} = 37.55$-kN comp.

52. $\overline{R}_2^{(1)} = -39.98$ kN; $\overline{R}_3^{(1)} = 7.70$ kN; $\overline{R}_1^{(2)} = -41.75$; $\overline{R}_2^{(2)} = 40.02$ kN; $\overline{R}_3^{(2)} = 7.70$ kN; $\overline{R}_2^{(3)} = -0.033$ kN; $N^{12} = -13.32$ kN; $N^{23} = 0$; $N^{13} = 0$; $N^{45} = 40.0$ kN; $N^{56} = 0$; $N^{46} = 0.033$ kN; $N^{14} = 42.89$ kN; $N^{25} = 0$; $N^{36} = 0.035$ kN; $N^{16} = -0.04$ kN; $N^{24} = -48.68$ kN; $N^{25} = 0$.

53. $\overline{R}_2^{(1)} = -0.75P$; $\overline{R}_3^{(1)} = 0.249P$; $\overline{R}_2^{(2)} = 0.75P$; $\overline{R}_3^{(2)} = -0.249P$; $\overline{R}_1^{(3)} = -P$; $\overline{R}_2^{(3)} = 0$; $N^{(1)} = 0$; $N^{(2)} = -0.768P$; $N^{(3)} = 0.768P$; $N^{(4)} = 0.962P$; $N^{(5)} = 0.962P$; $N^{(6)} = 0$.

54. $\overline{R}_1^{(1)} = 0$; $\overline{R}_2^{(1)} = 0$; $\overline{R}_3^{(1)} = -36$ kN; $\overline{M}_1 = -240.7$ kN·m; $\overline{M}_2 = 121.9$ kN·m; $M_1^{1k} = 240.7$ kN·m; $M_2^{1k} = -78.7$ kN·m; $M_1^{2k} = -78.7$ kN·m; $M_2^{2k} = 183.1$ kN·m.

55. $\overline{R}_1^{(1)} = 40$ kN; $\overline{R}_2^{(1)} = \overline{R}_3^{(1)} = 0$; $\overline{M}_1^{(1)} = -120$ kN·m; $\overline{M}_2^{(1)} = 20$ kN·m; $\overline{M}_3^{(1)} = -80$ kN·m; $N^{(1)} = 37.6$-kN comp.; $M_1^{1k} = 109.29$ kN·m; $M_2^{1k} = -39.63$ kN·m; $M_3^{1k} = -58.13$ kN·m; $M_1^{2k} = -120$ kN·m; $M_2^{2k} = -20$ kN·m; $M_3^{2k} = 0$.

Chapter 4

1.

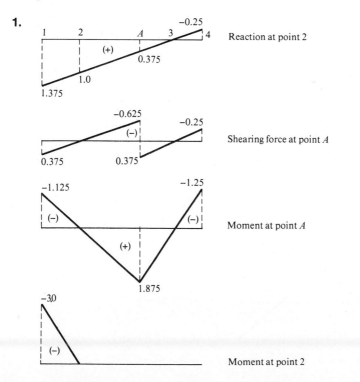

2.

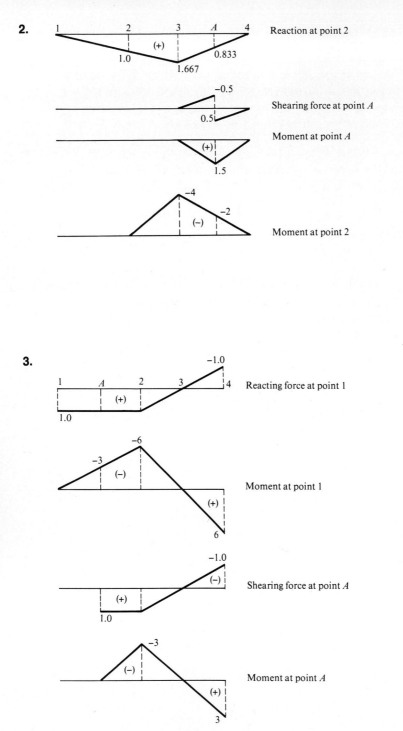

Reaction at point 2

Shearing force at point A

Moment at point A

Moment at point 2

3.

Reacting force at point 1

Moment at point 1

Shearing force at point A

Moment at point A

4.

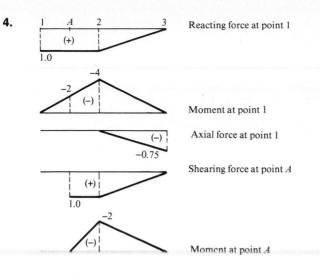

Reacting force at point 1

Moment at point 1

Axial force at point 1

Shearing force at point A

Moment at point A

5.

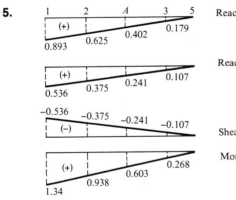

Reaction at point 1

Reaction at point 3

Shearing force at point A

Moment at point A

6.

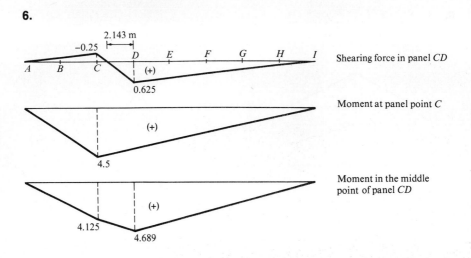

2.143 m

−0.25

A   B   C   D   E   F   G   H   I

(+)

0.625

Shearing force in panel *CD*

(+)

4.5

Moment at panel point *C*

(+)

4.125

4.689

Moment in the middle
point of panel *CD*

7.

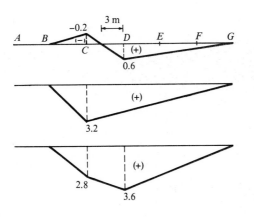

3 m

−0.2

A   B   C   D   E   F   G

(−)

(+)

0.6

Shearing force at panel CD

(+)

3.2

Moment at panel point *C*

(+)

2.8

3.6

Moment in the middle
point at panel *CD*

8.

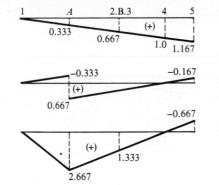

Reaction at point 4

Shearing force at point A

Moment at point A

No shearing force at point B

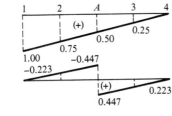

Moment at point B

9.

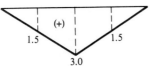

Reaction at point 1

Shearing force at point A

Moment at point A

10.

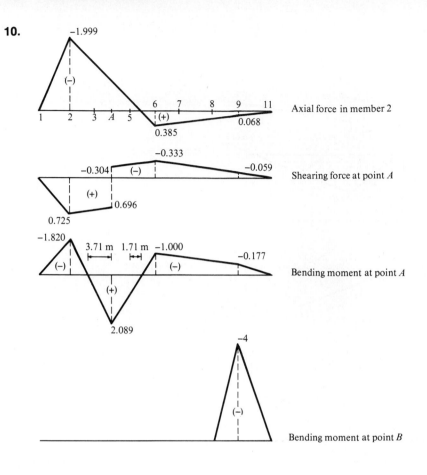

Axial force in member 2

Shearing force at point A

Bending moment at point A

Bending moment at point B

11.

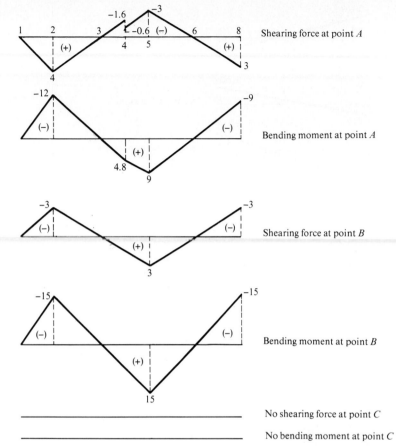

Shearing force at point A

Bending moment at point A

Shearing force at point B

Bending moment at point B

No shearing force at point C

No bending moment at point C

12.

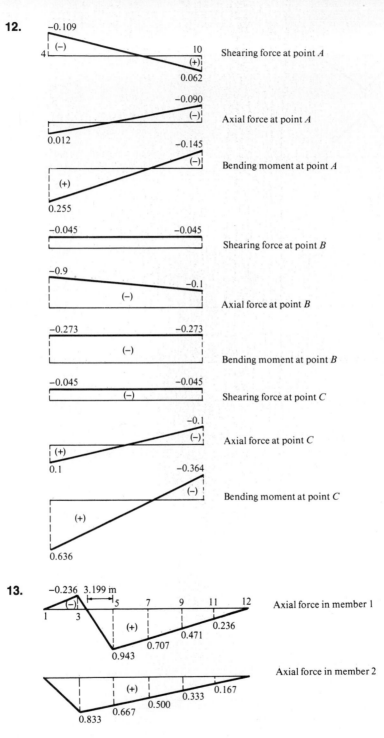

Shearing force at point A

Axial force at point A

Bending moment at point A

Shearing force at point B

Axial force at point B

Bending moment at point B

Shearing force at point C

Axial force at point C

Bending moment at point C

13.

Axial force in member 1

Axial force in member 2

14.

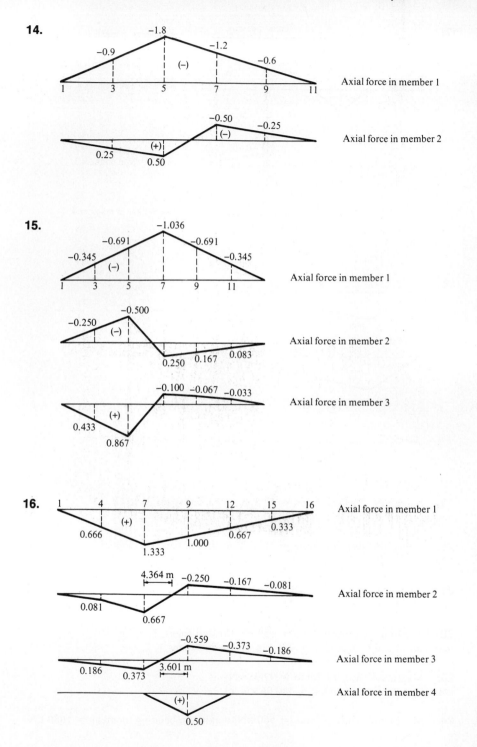

Axial force in member 1

Axial force in member 2

15.

Axial force in member 1

Axial force in member 2

Axial force in member 3

16.

Axial force in member 1

Axial force in member 2

Axial force in member 3

Axial force in member 4

17.

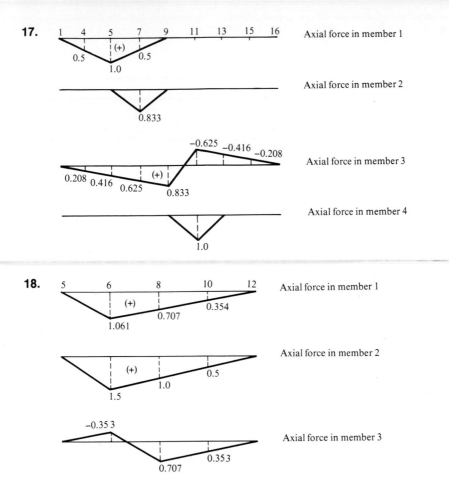

Axial force in member 1

Axial force in member 2

Axial force in member 3

Axial force in member 4

19.   Maximum shearing force = 62.50 kN (loaded spans 2, 3, 4); maximum bending moment = 228.75 kN·m (loaded spans 1, 2, 3).

20.   Maximum shearing force = 45 kN; maximum bending moment = 180 kN·m.

21.   Maximum shearing force = 150 kN; maximum bending moment = 360 kN·m.

22.   Maximum shearing force = 150 kN (loaded spans, 1, 2, 3); maximum bending moment = 260 kN·m (loaded spans 1, 2, 3).

23.   Maximum shearing force = 93.38 kN; maximum bending moment = 480.06 kN·m.

24.   Maximum shearing force = 560 kN; maximum bending moment = 1680 kN.

25. Maximum shearing force = 159 kN;
maximum bending moment = 1088 kN·m.

26. Maximum shearing force = 185.716 kN;
maximum bending moment = 1800 kN·m.

27. Maximum axial force = 362.7 kN; maximum shearing force = 179.22 kN;
maximum bending moment = 140.67 kN·m.

28. Maximum shearing force at pt. A = 640 kN (loaded spans 1, 2, 3);
maximum bending moment at pt. A = 1920 kN·m (loaded spans 1, 2, 3);
maximum shearing force at pt. B = 480 kN (loaded spans 1, 2, 3);
maximum bending moment at pt. B = 2400 kN·m (loaded spans 1, 2, 3).

29. Maximum axial force in member 1 = 147.08 kN;
maximum axial force in member 2 = 149.98.

30. Maximum axial force in member 1 = 378 kN;
maximum axial force in member 2 = 67.5 kN.

31. Maximum axial force in member 1 = 241.74 kN;
maximum axial force in member 2 = 69.98 kN;
maximum axial force in member 3 = 76.28 kN.

32. Maximum axial force in member 1 = 319.74 kN;
maximum axial force in member 2 = 79.58 kN;
maximum axial force in member 3 = 93.91 kN.

33. Maximum axial force in member 1 = 140 kN;
maximum axial force in member 2 = 83.3 kN;
maximum axial force in member 3 = 126.08 kN.

34. Maximum bending moment at pt. 2 = 76.0 kN·m;
maximum bending moment at pt. A = 27.5 kN·m.

35. Maximum bending moment at pt. 2 = 50 kN·m;
maximum bending moment at pt. A = 35.16 kN·m.

36. Maximum bending moment at pt. 1 = 113.33 kN·m;
maximum bending moment at pt. A = 55 kN·m.

37. Maximum axial force in member 1 = 18.37 kN;
maximum axial force in member 2 = 16.31 kN.

38. Maximum axial force in member 1 = 35.4 kN;
maximum axial force in member 2 = 9.67 kN.

39. Maximum axial force in member 1 = 26.27 kN;
maximum axial force in member 2 = 12.93 kN;
maximum axial force in member 3 = 10.92 kN;
maximum axial force in member 4 = 9.58 kN.

40. Maximum bending moment at pt. 2 = 600 kN·m;
maximum shearing force at pt. A = 168.75 kN;
maximum bending moment at pt. A = 306.25 kN.

41. Maximum bending moment at pt. 2 = 966.67 kN·m;
maximum shearing force at pt. A = 108.33 kN;
maximum bending moment at pt. A = 325 kN·m.

42. Maximum axial force in member 1 = 418.45 kN;
maximum axial force in member 2 = 387.35 kN.

43. Maximum axial force in member 1 = 855.00 kN;
maximum axial force in member 2 = 208.35 kN.

44. Maximum axial force in member 1 = 655.10 kN;
maximum axial force in member 2 = 273.75 kN;
maximum axial force in member 3 = 254.66 kN;
maximum axial force in member 4 = 179.17 kn.

45. Maximum bending moment. Under load 3 = 1633.60 kN·m.

46.

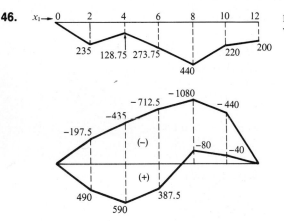

Envelope of the maximum absolute values of the shearing force

Envelope of the maximum values of the bending moment

Chapter 5

1. $u_v^{(3)}$ = 1.13 mm upward.

2. $u_v^{(3)}$ = 0.494 mm downward.

3. $u_v^{(5)}$ = 2.180 mm downward; $\Delta\theta$ = 0.112 rad clockwise.

4. $u_v^{(4)} = 1.92$ mm downward; $u_h^{(4)} = 0.693$ mm to the right.

5. $u_h^{(4)} = 1.00$ mm to the right.

6. $u_v^{(4)} = 1.21$ mm downward.

7. $u_v^{(4)} = 2.22$ mm upward.

8. $u_v^{(5)} = 0.046$ mm downward.

9. $u_v^{(3)} = 8.33$ mm downward.

10. $u_v^{(4)} = 10.00$ mm upward.

11. $u_v^{(5)} = 8.33$ mm downward.

12. $u_v^{(2)} = 10.00$ mm downward.

13. $u_v^{(2)} = 5.00$ mm downward.

14. $u_v^{(2)} = 5.39$ mm downward.

15. $\Delta L^{(2)} = 4.50$ mm; $\Delta L^{(4)} = 1.875$ mm.

16. $u_v^{(11)} = 21.60$ mm downward.

17. $u_h^{(4)} = 0.515$ mm to the right.

18. $u^{(4)} = 3.34$ mm.

19. $u_v^{(9)} = 3.923$ mm downward.

20. $u^{(2)} = 0.750$ mm.

21. $u_v^{(2)} = 3.03$ mm downward; 2.26%; $\theta^{(2)} = 6.35 \times 10^{-4}$ rad clockwise; 0%.

22. $u_v^{(2)} = 10.06$ mm downward; 0.8%; $\theta^{(2)} = 1.39 \times 10^{-3}$ rad clockwise; 0%.

23. $u_v^{(4)} = 0.122$ mm upward; $\theta^{(4)} = 325 \times 10^{-4}$ rad clockwise.

24. $u_v^{(2)} = 9.06$ mm downward; $\theta^{(2)} = 152 \times 10^{-2}$ rad clockwise.

25. $u_v^{(4)} = 2.32$ mm downward.

26. $u_v^{(5)} = 0.0795$ mm upward; $u_h^{(5)} = 90$ mm to the right.

27. $u_v^{(3)} = 11.19$ mm downward.

28. $u_v^{(3)} = 75.75$ mm downward.

29. $u_h^{(4)} = 6.65$ mm to the right; 0.43%.

30. $u_h^{(2,3)} = 21.48$ mm to the right.

31. $u_v^{(3)} = 4.47$ mm downward; $\theta^{(2,3)} = 0.001288$ rad clockwise; $\Delta\theta^{(4)} = 0.002576$ rad.

32. $u_h^{(2)} = 2.64$ mm to the right; 2%; $u_h^{(4)} = 9.04$ mm to the right, 0.144%.

33. $u_h^{(1)} = 6.64$ mm to the left.

34. $u_v^{(3)} = 0.344$ mm upward; $\theta^{(2,3)} = 0.000728$ rad clockwise; $\Delta\theta^{(2)} = 0.001484$ rad.

35. $u_v^{(5)} = 50.5$ mm downward; $\theta^{(3)} = 0.00966$ rad counterclockwise; $\Delta\theta^{(3)} = 0.03091$ rad.

36. $u_v^{(2)} = 122.88$ mm downward; $\theta^{(2,3)} = 0.04444$ rad; $\Delta\theta^{(3)} = 0.04270$ rad.

39. $u_h^{(4)} = 333.68$ mm to the right.

40. $u_v^{(3)} = 27.16$ mm downward

41. $u_h^{(2)} = 6.64$ mm to the right; $u_v^{(3)} = 13.28$ mm up; $u_h^{(4)} = 6.64$ mm to the left.

42. $u_h^{(1)} = 10.52$ mm.

43. $u_v^{(4)} = 5$ mm upward; $\theta^{(2,3)} = 0.001250$ rad counterclockwise; $\Delta\theta^{(3)} = 0.00250$ rad.

44. $u_v^{(3)} = 7.5$ mm downward.

45. $u_v^{(3)} = 5$ mm upward.

46. $u^{(2)} = 0$; $u^{(3)} = 7.14$ mm; $u^{(4)} = 20.45$ mm.

47. $u_v^{(5)} = 9.23$ mm downward; $u_h^{(5)} = 3.59$ mm in the direction $\overrightarrow{34}$.

49. Approximate value: $u_v^{(2)} = 21.4$ mm downward.

50. Approximate value: $u_v^{(2)} = 11.10$ mm downward.

51. Approximate value: $u_v^{(2)} = 18.75$ mm downward.

52. See Problem 2.

53. $u_h^{(4)} = 2.60$ mm to the right.

54. See Problem 17.

55. $u_v^{(2)} = 0.429$ mm downward.

56. See Problem 25.

57. See Problem 26.

58. See Problem 31.

59. $u_v^{(3)} = 52.90$ mm downward.

Chapter 6

1. $u_v^{(2)} = 90.6$ mm downward; $\theta^{(2)} = 0.0194$ rad clockwise.

2. $u_v^{(2)} = 305.2$ mm downward; $\theta^{(2)} = 0.0426$ rad clockwise.

3. $u_v^{(4)} = 2.3$ mm downward.

4. $u_v^{(m)} = 36.0$ mm downward.

5. $u_v^{(m)} = 22.0$ mm downward.

6. $u_v^{(2)} = 11.2$ mm downward; $\theta^{(2)} = 0.0019$ rad clockwise.

7. $u_v^{(4)} = 0.242$ mm upward.

8. $u_v^{(5)} = 50.5$ mm downward.

9. $u_v^{(2)} = 3.17$ mm downward; $\theta^{(2)} = 0.00095$ rad clockwise.

Chapter 7

1. $N^{(1)} = 16.12$-kN tension; $N^{(2)} = 33.77$-kN tension; $N^{(3)} = 23.80$-kN comp.; $N^{(4)} = 26.87$-kN comp.; $N^{(5)} = 16.12$-kN tension.

2. $N^{(1)} = 26.67$-kN tension; $N^{(2)} = 72.11$-kN tension; $N^{(3)} = 72.11$-kN comp.; $N^{(4)} = 40.00$-kN comp.; $N^{(5)} = 0$; $N^{(6)} = 48.1$-kN tension; $N^{(7)} = 0$; $N^{(8)} = 26.67$-kN comp.

3. $N^{(1)} = 48.89$-kN tension; $N^{(2)} = 36.11$-kN comp.; $N^{(3)} = 10.43$-kN comp.; $N^{(4)} = 38.35$-kN comp.; $N^{(5)} = 13.18$-kN comp.; $N^{(6)} = 17.38$-kN tension; $N^{(7)} = 35.0$-kN tension; $N^{(8)} = 30.55$-kN comp.; $N^{(9)} = 7.91$-kN tension; $N^{(10)} = 24.44$-kN tension.

4. $N^{(1)} = 25.15$-kN tension; $N^{(2)} = 41.92$-kN comp.; $N^{(3)} = 41.92$-kN tension; $N^{(4)} = 50.31$-kN comp.; $N^{(5)} = 45.46$-kN tension; $N^{(6)} = 8.1$-kN tension; $N^{(7)} = 8.1$-kN comp.; $N^{(8)} = 40.62$-kN comp.; $N^{(9)} = 5.77$-kN tension; $N^{(10)} = 58.1$-kN tension; $N^{(11)} = 58.1$-kN comp.; $N^{(12)} = 29.1$-kN tension; $N^{(13)} = 6.81$-kN comp.; $N^{(14)} = 37.11$-kN comp.; $N^{(15)} = 37.11$-kN tension; $N^{(16)} = 15.46$-kN comp.; $N^{(17)} = 7.73$-kN tension; $N^{(18)} = 12.9$-kN tension; $N^{(19)} = 12.9$-kN comp.

5. $R^{(2)} = 2.5P$; $M^{(1)} = 0$.

6. $R^{(1)} = 0.9687P$; $R^{(2)} = 1.0625P$; $R^{(3)} = 0.9688P$.

7. $R_v^{(1)} = 36.25$ kN; $R_h^{(1)} = 0$; $R_v^{(2)} = 65.50$ kN; $R_v^{(3)} = 118.75$ kN; $M^{1k} = -15.0$ kN·m.

8. See Problem 6.

9. See Problem 5.

10. $R_v^{(1)} = 2.07$ N; $R_v^{(2)} = 47.20$ kN; $R_v^{(3)} = 10.73$ kN.

11. $R_v^{(1)} = 23.33$ kN; $R_h^{(1)} = 44.94$ kN; $R_v^{(16)} = 76.67$ kN; $R_h^{(16)} = 35.10$ kN; $N^{(1)} = 13.97$-kN tension; $N^{(2)} = 46.96$-kN tension; $N^{(3)} = 78.15$-kN comp.; $N^{(4)} = 46.42$-kN tension; $N^{(5)} = 64.81$-kN comp.; $N^{(6)} = 28.52$-kN tension; $N^{(7)} = 9.25$-kN tension; $N^{(8)} = 99.17$-kN comp.

12. $R_v^{(1)} = 76.40$ kN; $R_h^{(1)} = 235.20$ kN; $M(1)$; $M^{(1)} = 58.41$ kN·m; $R_v^{(2)} = 176.4$ kN; $R_h^{(2)} = 235.20$ kN; $M^{(3)} = 294$ kN·m.

13. $R_v^{(1)} = 7.28$ kN; $R_h^{(1)} = 80$ kN; $M^{(1)} = 121.83$ kN·m; $R_v^{(3)} = 32.72$ kN; $M^{1k} = 38.17$ kN·m.

14. $R_v^{(1)} = 34.87$ kN; $R_h^{(1)} = 19.46$ kN; $R_v^{(4)} = 34.87$ kN; $R_h^{(4)} = 40.54$ kN; $M^{1k} = 116.76$ kN·m; $M^{2k} = -162.16$ kN·m.

15. $R_v^{(1)} = 86.15$ kN; $R_h^{(1)} = 0$; $R_v^{(4)} = 73.85$ kN; $M^{1j} = -49.23$ kN·m; $M^{1k} = 49.23$ kN·m.

16. $R_v^{(1)} = 49$ kN; $R_h^{(1)} = 0$; $M^{1k} = 3.28$ kN·m; $N^{1j} = 68.72$-kN comp.; $N^{(3)} = 72.65$-kN tension; $N^{(4)} = 47.13$-kN comp.

17. $R_v^{(1)} = 2.91$ kN; $R_h^{(1)} = 0$; $M^{(1)} = 34.95$ kN·m; $R_v^{(3)} = 2.91$ kN; $M^{1k} = 34.95$ kN·m.

18. $R_v^{(1)} = 5.11$ kN; $R_h^{(1)} = 0$; $R_v^{(2)} = 8.94$ kN; $R_v^{(3)} = 3.83$ kN; $M^{1k} = 61.29$ kN·m.

19. $R_v^{(1)} = 1.95$ kN; $R_h^{(1)} = 6.66$ kN; $M^{(1)} = 8.93$ kN·m; $R_v^{(4)} = 1.95$ kN; $R_h^{(4)} = 6.66$ kN; $M^{1k} = -35.6$ kN·m.

20. $R_v^{(1)} = 0.286$ kN; $R_h^{(1)} = 0$; $M^{(1)} = 3.43$ kN·m; $M^{1k} = 3.43$ kN·m.

21. $R_v^{(1)} = 1.16$ kN; $R_h^{(1)} = 0$; $R_v^{(2)} = 2.03$ kN; $R_v^{(3)} = 0.87$ kN;
$M^{1k} = -13.86$ kN·m.

22. $R_v^{(1)} = 1.43$ kN; $R_h^{(1)} = 2.02$ kN; $M^{(1)} = 13.46$ kN·m; $R_v^{(4)} = 1.43$ kN;
$R_h^{(4)} = 2.02$ kN; $M^{1k} = 5.37$ kN·m; $M^{2k} = -6.07$ kN·m.

23. (a) $R_v^{(1)} = 100.91$ kN; $R_h^{(1)} = 0$; $M^{(1)} = 167.27$ kN·m; $R_v^{(3)} = 59.09$ kN;
  $M^{1k} = 76.36$ kN·m; $N^{2j} = 12.55$-kN tension; $N^{2k} = 35.46$-kN comp.
  (b) $R_v^{(1)} = 3.88$ kN; $R_h^{(1)} = 0$; $M^{(1)} = 31.06$ kN·m; $R_v^{(3)} = 3.88$ kN;
  $M^{1k} = 15.53$ kN·m; $N^{2j} = 2.23$-kN comp.
  (c) $R_v^{(1)} = 7.33$ kN; $R_h^{(1)} = 0$; $M^{(1)} = 58.67$ kN·m; $R_v^{(3)} = 7.33$ kN;
  $M^{1k} = 29.34$ kN·m; $N^{2j} = 4.40$-kN comp.

24. (a) $R_v^{(1)} = 57.50$ kN; $R_h^{(1)} = 76.47$ kN; $R_v^{(5)} = 102.50$ kN;
  $R_h^{(5)} = 43.53$ kN; $M^{1k} = 98.82$ kN·m; $M^{2k} = -22.37$ kN·m;
  $M^{3k} = -261.18$ kN·m;
  $N^{2j} = 69.32$-kN comp.; $N^{2k} = 21.32$-kN comp.;
  $N^{3j} = 43.82$-kN comp.; $N^{3k} = 96.32$-kN comp.
  (b) Very small internal actions.
  (c) $R_v^{(1)} = 0$; $R_h^{(1)} = 4.50$ kN; $R_v^{(5)} = 0$; $R_h^{(5)} = 4.50$ kN;
  $M^{1k} = 26.97$ kN·m; $M^{2k} = 53.95$ kN·m; $M^{3k} = 26.97$ kN·m;
  $N^{2j} = N^{3j} = 3.60$-kN tension.

26. (a) $R_v^{(1)} = 18.28$ kN; $R_h^{(1)} = 0$; $M^{(1)} = 3.99$ kN·m; $R_v^{(3)} = 21.72$ kN;
  $M^{1k} = -3.99$ kN·m.
  (b) $R_v^{(1)} = 7.78$ kN; $R_h^{(1)} = 0$; $M^{(1)} = 27.24$ kN·m; $R_v^{(3)} = 7.78$ kN;
  $M^{1k} = 27.24$ kN·m.
  (c) $R_v^{(1)} = 11.73$ kN; $R_h^{(1)} = 0$; $M^{(1)} = 41.05$ kN·m; $R_v^{(3)} = 11.73$ kN;
  $M^{1k} = 41.05$ kN·m.

27. (a) $R_v^{(1)} = 115.00$ kN; $R_h^{(1)} = 10.16$ kN; $M^{(1)} = 59.04$ kN·m;
  $R_v^{(8)} = 125.00$ kN; $R_h^{(8)} = 90.16$ kN; $M^{(8)} = 200.96$ kN·m;
  $M^{1k} = -300.00$ kN·m; $M^{3k} = -160.32$ kN·m;
  $M^{4k} = -180.32$ kN·m; $M^{6k} = -340.00$ kN·m;
  $N^{2j} = 70.16$-kN comp.; $N^{3j} = 35.00$-kN comp.;
  $N^{4j} = 90.16$-kN comp.; $N^{5j} = 45.00$-kN comp.; $N^{6j} = 90.16$-kN comp.
  (b) Very small internal actions.
  (c) $R_v^{(1)} = 21.35$ kN; $R_h^{(1)} = 26.23$ kN; $M^{(1)} = 29.27$ kN·m;
  $R_v^{(8)} = 21.35$ kN; $R_h^{(8)} = 26.23$ kN; $M^{(8)} = 71.98$ kN·m;
  $M^{1k} = -128.13$ kN·m; $M^{3k} = -52.47$ kN·m;
  $M^{4k} = -137.88$ kN·m;
  $M^{6k} = -85.42$ kN·m; $N^{2j} = 26.23$-kN comp.; $N^{3j} = 21.35$-kN tension;
  $N^{4j} = 26.23$-kN comp.; $N^{5j} = 21.35$-kN comp.; $N^{6j} = 26.23$-kN comp.

28. (a) $R_v^{(1)} = 27.50$ kN; $R_h^{(1)} = 38.54$ kN; $R_v^{(6)} = 57.50$ kN; $R_h^{(6)} = 21.46$ kN;

$M^{2k} = 56.88$ kN·m; $M^{3k} = 163.12$ kN·m; $M^{4k} = -98.75$ kN·m;
$N^{2j} = 66.67$-kN comp.; $N^{3j} = 21.46$-kN comp.; $N^{4j} = 27.5$-kN comp.

(*b*) Very small internal actions.

(*c*) $R_v^{(1)} = 0$; $R_h^{(1)} = 5.11$ kN; $R_v^{(6)} = 0$; $R_h^{(8)} = 5.11$ kN;
$M^{2k} = 45.99$ kN·m; $M^{3k} = 45.99$ kN·m; $M^{4k} = 30.66$ kN·m;
$N^{2j} = 4.08$-kN tension; $N^{3j} = 5.11$-kN tension; $N^{4j} = 0$.

29. $R_v^{(1)} = 28.03$ kN; $M^{(1)} = 82.13$ kN·m; $R_v^{(2)} = 97.72$ kN; $R_v^{(3)} = 42.25$ kN.

30. $R_v^{(1)} = 28.56$ kN; $R_v^{(2)} = 98.65$ kN; $R_v^{(2)} = 11.49$ kN; $R_v^{(4)} = 21.03$ kN.

31. $R_v^{(1)} = 10.81$ kN; $R_v^{(2)} = 62.58$ kN; $R_v^{(4)} = 37.50$ kN; $R_v^{(5)} = 13.07$ kN;
$R_v^{(6)} = 2.18$ kN.

32. $R_v^{(1)} = 33.60$ kN; $M^{(1)} = 96.00$ kN; $R_v^{(2)} = 6.40$ kN.

33. For $L = 6$ m: $R_v^{(1)} = R_v^{(2)} = 60$ kN; $M^{(1)} = 67.5$ kN·m.

34. Approximate answers: $R_v^{(1)} = 38.85$ kN; $M^{(1)} = 91.92$ kN·m;
$R_v^{(2)} = 21.15$ kN.

35. Approximate answers: $R_v^{(1)} = 35.99$ kN; $M^{(1)} = 51.95$ kN·m;
$R_v^{(2)} = 4.01$ kN.

36. Approximate answers: $R_v^{(1)} = 75.35$ kN; $M^{(1)} = 128.38$ kN·m;
$R_v^{(2)} = 44.65$ kN; $M^{(2)} = 80.28$ kN·m.

37. $R_v^{(1)} = 5.17$ kN; $R_h^{(1)} = 114.53$ kN; $M^{(1)} = 252.22$ kN·m; $R_v^{(4)} = 65.17$ kN;
$R_h^{(4)} = 25.47$ kN.

38. $R_v^{(1)} = 6.53$ kN; $R_h^{(1)} = 97.5$ kN; $M^{(1)} = 182.29$ kN·m; $R_v^{(4)} = 53.47$ kN;
$R_h^{(4)} = 42.5$ kN; $M^{(4)} = 112.47$ kN·m.

39. $R_v^{(1)} = 124.14$ kN; $R_h^{(1)} = 9.90$ kN; $R_v^{(5)} = 39.31$ kN; $R_v^{(6)} = 96.55$ kN;
$R_h^{(6)} = 9.90$ kN; $M^{3k} = -90.51$ kN·m; $M^{3k} = -57.39$ kN·m.

40. $R_v^{(1)} = 4.92$ kN; $R_h^{(1)} = 28.35$ kN; $R_v^{(4)} = 20.16$ kN; $R_h^{(4)} = 22.35$ kN;
$R_v^{(6)} = 14.92$ kN; $R_h^{(6)} = 9.30$ kN.

41. $R_v^{(1)} = 12.5$ kN; $R_h^{(1)} = 40$ kN; $R_v^{(8)} = 27.5$ kN; $N^{(1)} = 56.67$-kN tension;
$N^{(2)} = 20.83$-kN comp.; $N^{(3)} = 10.17$-kN tension; $N^{(4)} = 59.78$-kN comp.;
$N^{(5)} = 3.89$-kN tension; $N^{(6)} = 16.95$-kN comp.; $N^{(7)} = 70.22$-kN tension;
$N^{(8)} = 17.22$-kN comp.; $N^{(9)} = 21.0$-kN comp.; $N^{(10)} = 56.53$-kN comp.;
$N^{(11)} = 53.47$-kN tension; $N^{(12)} = 24.83$-kN tension; $N^{(13)} = 12.60$-kN tension;
$N^{(14)} = 45.83$-kN comp.; $N^{(15)} = 36.67$-kN tension.

42. $R_v^{(1)} = 83.98$ kN; $R_h^{(1)} = 89.10$ kN; $R_v^{(4)} = 14.53$ kN; $R_h^{(4)} = 10.90$ kN;

$N^{(1)}$ = 9.73-kN tension; $N^{(2)}$ = 3.18-kN comp.; $N^{(3)}$ = 24.50-kN tension;
$N^{(4)}$ = 17.72-kN comp.; $N^{(5)}$ = 20.17-kN tension; $N^{(6)}$ = 12.91-kN comp.;
$N^{(7)}$ = 15.50-kN comp.; $N^{(8)}$ = 98.51-kN comp.; $N^{(9)}$ = 115.99-kN tension;
$N^{(10)}$ = 18.16-kN tension.

43. $R_v^{(1)}$ = 135.86 kN; $R_h^{(1)}$ = 266.05 kN; $M^{(1)}$ = 217.97 kN·m;
$R_v^{(4)}$ = 135.86 kN; $R_h^{(4)}$ = 183.95 kN; $M^{(4)}$ = 235.04 kN·m; M^{1k} = 202.53 kN;
M^{2k} = −205.07 kN.m; N^{1j} = 265.7-kN tension; N^{2j} = 183.97-kN comp.;
N^{3j} = 197.35-kN comp.

44. $R_v^{(1)}$ = 130.77 kN; $R_h^{(1)}$ = 115.97 kN; $R_v^{(4)}$ = 820.77 kN; $R_h^{(4)}$ = 115.97 kN;
$M^{(4)}$ = 270.38 kN·m; $R_v^{(7)}$ = 88.46 kN; $N^{(1)}$ = 426.26-kN tension;
$N^{(2)}$ = 461.88-kN tension.

45. $R_v^{(1)}$ = $R_v^{(6)}$ = 71.43 kN; $R_h^{(1)}$ = −$R_h^{(6)}$ = 15.24 kN;
$M^{(1)}$ = −$M^{(6)}$ = 30.48 kN·m; $R_v^{(4)}$ = 177.14 kN; $R_h^{(4)}$ = 0; $M^{(4)}$ = 0;
M^{2k} = −129.52 kN·m.

46. $R_v^{(1)}$ = −$R_v^{(6)}$ = 12.64 kN; $R_h^{(1)}$ = $R_h^{(6)}$ = 23.97 kN;
$M^{(1)}$ = $M^{(6)}$ = 87.20 kN·m; $R_v^{(4)}$ = 0; $R_h^{(4)}$ = 32.06 kN; $M^{(4)}$ = 103.37 kN.

47. $R_v^{(1)}$ = $R_v^{(5)}$ = 80 kN; $R_h^{(1)}$ = $R_h^{(5)}$ = 4.88 kN; tension in tie = 45.97 kN;
M^{2k} = −16.97 kN·m.

48. $R_v^{(1)}$ = $R_v^{(5)}$ = 80 kN; $R_h^{(1)}$ = −$R_h^{(5)}$ = 23.38 kN;
$M^{(1)}$ = −$M^{(4)}$ = 51.9 kN·m; M^{2k} = 24.82 kN·m.

49. $R_v^{(1)}$ = $R_v^{(4)}$ = 100 kN; $R_h^{(1)}$ = −$R_h^{(1)}$ = 55.19 kN;
$M^{(1)}$ = −$M^{(4)}$ = 114.19 kN·m; M^{1k} = −22.17 kN·m.

50. $R_v^{(1)}$ = −$R_v^{(4)}$ = 17.74 kN; $R_h^{(1)}$ = $R_h^{(4)}$ = 30.0 kN;
$M^{(1)}$ = $M^{(4)}$ = 31.33 kN·m; M^{1k} = 35.47 kN·m; N^{1k} = 32.19-kN tension;
N^{2k} = 30.00-kN comp.

51.   −$R_v^{(1)}$ = $R_v^{(3)}$ = 40 kN; $R_h^{(1)}$ = $R_h^{(3)}$ = 30 kN; $M^{(1)}$ = $M^{(3)}$ = 0.

52. $R_v^{(1)}$ = −$R_v^{(8)}$ = 18.01 kN; $R_h^{(1)}$ = $R_h^{(8)}$ = 30 kN;
$M^{(1)}$ = $M^{(8)}$ = 113.90 kN·m; M^{2k} = −5.96 kN·m; M^{4j} = 54.04 kN·m.

53. $R_v^{(1)}$ = −$R_v^{(4)}$ = 45 kN; $R_h^{(1)}$ = $R_h^{(4)}$ = 0; M^{1j} = −9.0 kN·m;
M^{1k} = −9.0 kN·m.

54. $R_v^{(1)}$ = $R_v^{(4)}$ = 30 kN; $R_h^{(1)}$ = $R_h^{(4)}$ = 0; $M^{(1)}$ = −$M^{(4)}$ = 6.14 kN·m.

57. $R_v^{(1)}$ = 20; $R_h^{(1)}$ = 0; $N^{(1)}$ = 26.67-kN tension; $N^{(2)}$ = 33.33-kN comp.;
$N^{(3)}$ = 9.85-kN tension; $N^{(4)}$ = 40.20-kN comp.; $N^{(5)}$ = 16.42-kN comp.;
$N^{(7)}$ = 39.80-kN tension; $N^{(8)}$ = 20.30-kN comp.

58. From the figure: $L_1 = L_3 = 3.30$ m, $L_2 = L_4 = 2.45$ m;
uniform load at top $= 35.1$ kN/m;
uniform load at bottom $= 54.7$ kN/m; side load at top $= 28.9$ kN/m;
side load at bottom $= 70.55$ kN/m; $M^{2j} = 38.91$ kN·m; $M^{3j} = -16.10$ kN·m.

59. See Appendix C.

60. (*a*) $R_v^{(1)} = 82.74$ kN; $R_h^{(1)} = 3.70$ kN; $M^{(1)} = 90.28$ kN·m;
$R_v^{(4)} = 33.27$ kN; $R_h^{(4)} = 64.30$ kN; $M^{(4)} = 120.26$ kN·m;
$M^{1k} = 22.70$ kN·m;
$M^{2k} = -56.94$ kN·m; $M^{3j} = 136.94$ kN·m;
$N^{1j} = 63.97$-kN comp.; $N^{2j} = 64.30$-kN comp..

(*b*) $R_v^{(1)} = 80.089$ kN; $R_h^{(1)} = 1.001$ kN; $M^{(1)} = 23.32$ kN·m;
$R_v^{(4)} = 0.089$ kN; $R_h^{(4)} = 1,001$ kN; $M^{(4)} = 24.13$ kN·m;
$M^{1k} = -19.59$ kN·m; $M^{2k} = -20.13$ kN·m; $N^{1j} = 0.672$-kN tension;
$N^{2j} = 1.001$-kN tension.

61. $\bar{R}_3^{(1)} = P/2$; bending moment $\bar{M}_1^{(1)} = 0.5PL$;
torsional moment $\bar{M}_2^{(1)} = 0.109 \ PL$.

67. $\bar{R}_v^{(1)} = 31.74$ kN; bending moment: $\bar{M}_2^{(1)} = -126.96$ kN·m;
torsional moment: $\bar{M}_1^{(1)} = -72.40$ kN·m.

68. $\bar{R}_v^{(1)} = 10$ kN; bending moment: $\bar{M}_2^{(1)} = -38.61$ kN·m;
torsional moment: $\bar{M}_1^{(1)} = 13.61$ kN·m.

69. See Problem 5.

70. See Problem 6.

71. See Problem 10.

72. See Problem 2.

73. See Problem 3.

74. See Problem 13.

75. See Problem 14.

76. See Problem 16.

77. See Problem 23.

78. See Problem 37.

79. See Problem 29.

80. See Problem 30.

81. See Problem 31.

83. $R_v^{(1)} = 37.07$ kN; $R_h^{(1)} = 2.07$ kN; $M^{(1)} = 1.77$ kN·m; $R_v^{(4)} = 37.07$ kN; $R_h^{(4)} = 2.07$ kN; $M^{(4)} = 16.48$ kN·m.

84. $R_v^{(1)} = 7.81$ kN; $R_h^{(1)} = 37.5$ kN; $R_v^{(4)} = 7.81$ kN·m; $M^{1j} = -12.13$ kN·m; $M^{2j} = 3.99$ kN·m; $M^{3j} = -6.50$ kN·m; $M^{2k} = 18.63$ kN·m, $N^{1j} = -2.62$-kN tension; $N^{2j} = 6.05$-kN comp.; $N^{3j} = 2.62$-kN comp.; $N^{4j} = 6.05$-kN tension.

85. $u_h^{(2)} = 0.102$ mm.

86. $u_h^{(4)} = 3.5$ mm.

87. $u_v^{(3)} = 2PL^3/3EI$; $\theta^{(3)} = 5PL^2/6EI$.

88. $u_v^{(4)} = 6.31$ mm; $\theta^{(4)} = 0.0024$ rad.

89. $u_v^{(4)} = 11.5$ mm; $\theta^{(4)} = 0.0044$ rad.

90. $u_v^{(4)} = 102.16$ mm; $\theta^{(4)} = 0.0235$ rad.

91. $u_h^{(3)} = 25.4$ mm.

92. $u_h^{(3)} = 11.4$ mm.

94. $u_h^{(3)} = 6.46$ mm.

95. $u_h^{(3)} = 4.13$ mm.

96. Rotational stiffness of end 2 is $9EI_c/L$; carry-over factor $= \frac{1}{3}$.

97. Horizontal component of force in cable $= 2.70\ L_p$; $R_v^{(1)} = 3.06qL_p$; moment at midspan $= 7.1\ qL_p$.

Chapter 8

3. Approximate values: $k^{ij} = 32,639$ kN·m; $C^{ij} = 0.3508$; $k^{ik} = 16,385$ kN·m; $C^{ik} = 0.6988$.

4. Approximate values: $k^{ij} = 22,531.66$ kN·m; $C^{ij} = 0.3923$; $k^{ik} = 11,156.74$ kN·m; $C^{ik} = 0.7923$.

5. Approximate values: $k^{ij} = 1.008 \times 10^7$ kN·m; $C^{ij} = 0.44381$; $k^{ik} = 6.264 \times 10^6$ kN·m; $C^{ik} = 0.71417$.

6. Approximate values: $k^{ij} = k^{ik} = 5.887613 \times 10^6$ kN·m; $C^{ij} = C^{ik} = 0.61715$.

7. Approximate values: $k^{ij} = k^{ik} = 23{,}969.664$; $C^{ij} = C^{ik} = 0.44933$.

8. $R_v^{(1)} = 28.25$ kN; $M^{(1)} = 20.95$ kN·m; $R_v^{(2)} = 46.64$ kN; $R_v^{(3)} = 33.11$ kN; $M^{(3)} = 68.29$ kN·m; $M^{1k} = -43.43$ kN·m.

9. $R_v^{(1)} = 7.24$ kN; $M^{(1)} = 5.52$ kN·m; $R_v^{(2)} = 191.86$ kN; $R_v^{(3)} = 80.90$ kN; $M^{1k} = -191.03$ kN·m.

10. $R_v^{(1)} = 20.00$ kN; $R_v^{(2)} = 100$ kN; $R_v^{(3)} = 100$ kN; $R_v^{(4)} = 20.00$ kN; $M^{1k} = -80$ kN·m.

11. $R_v^{(1)} = 100.83$ kN; $M^{(1)} = 162.22$ kN·m; $R_v^{(2)} = 48.80$ kN; $R_v^{(3)} = 70.37$ kN; $M^{1k} = 4.44$ kN·m.

12. $R_v^{(1)} = 45.99$ kN; $M^{(1)} = 139.28$ kN·m; $R_v^{(2)} = 144.33$ kN; $R_v^{(3)} = 89.67$ kN; $M^{1k} = -103.32$ kN·m.

13. $R_v^{(1)} = 118.18$ kN; $M^{(1)} = 246.34$ kN·m; $R_v^{(2)} = 26.89$ kN; $R_v^{(3)} = 74.93$ kN; $M^{1k} = 59.12$ kN·m.

14. $R_v^{(1)} = 42.84$ kN; $R_h^{(1)} = 34.32$ kN; $R_v^{(3)} = 175.94$ kN; $R_h^{(3)} = 25.68$ kN; $M^{(3)} = 21.36$ kN·m; $R_v^{(4)} = 61.22$ kN; $M^{1k} = -102.95$ kN·m; $M^{2j} = 47.30$ kN·m.

15. $R_v^{(1)} = 22.37$ kN; $R_h^{(1)} = 6.28$ kN; $R_v^{(3)} = 40.08$ kN; $R_h^{(3)} = 7.23$ kN; $M^{(3)} = 120.48$ kN·m; $R_v^{(5)} = 2.93$ kN; $R_h^{(5)} = 0.95$ kN; $M^{(5)} = 1.59$ kN·m; $R_v^{(6)} = 0.475$ kN.

16. $R_v^{(1)} = 74.83$ kN; $R_h^{(1)} = 27.86$ kN; $M^{(1)} = 37.14$ kN·m; $R_v^{(4)} = 239.68$ kN; $R_h^{(4)} = 7.16$ kN; $M^{(4)} = 9.55$ kN·m; $R_v^{(5)} = 78.22$ kN; $R_h^{(5)} = 1.18$ kN; $M^{(5)} = 101.92$ kN·m; $R_v^{(7)} = 87.27$ kN; $R_h^{(7)} = 21.88$ kN; $M^{(7)} = 126.08$ kN·m; $M^{2j} = -74.30$ kN·m; $M^{2k} = -115.68$ kN·m; $M^{5j} = -19.61$ kN·m; $M^{5k} = -67.89$ kN·m; $M^{3j} = 19.10$ kN·m.

17. $R_v^{(1)} = 40$ kN; $R_h^{(1)} = 0$; $R_v^{(4)} = 40$ kN; $M^{1j} = -2.54$ kN·m; $M^{1k} = -20.31$ kN·m.

18. $R_v^{(1)} = 60$ kN; $R_h^{(1)} = 43.63$ kN; $M^{(1)} = 43.62$ kN·m.

19. $R_v^{(1)} = 16.67$ kN; $M^{(1)} = 50.00$ kN·m; $R_v^{(3)} = 43.33$ kN; $M^{(3)} = 80.00$ kN·m.

20. $R_v^{(1)} = 52.44$ kN; $M^{(1)} = 127.09$ kN·m; $R_v^{(2)} = 7.56$ kN.

21. $R_v^{(1)} = 59.22$ kN; $R_h^{(1)} = 0.937$ kN; $M^{(1)} = 72.40$ kN·m; $R_v^{(4)} = 20.78$ kN; $R_h^{(4)} = 0.937$ kN; $M^{(4)} = 7.74$ kN·m; $M^{(2)} = -31.36$ kN·m; $M^{2k} = 3.99$ kN·m.

22. $R_v^{(1)} = 34.49$ kN; $R_h^{(1)} = 2.81$ kN; $M^{(1)} = 36.95$ kN·m; $R_v^{(4)} = 34.49$ kN;

$R_h^{(4)} = 57.19$ kN; $M^{(4)} = 141.03$ kN; $M^{2j} = 59.45$ kN·m; $M^{2k} = 12.28$ kN·m; $N^{2j} = 66.57$-kN comp.

23. $R_v^{(1)} = 76.46$ kN; $R_h^{(1)} = 29.99$ kN; $M^{(1)} = 113.89$ kN·m; $R_v^{(5)} = 83.54$ kN; $R_h^{(5)} = 29.99$ kN; $M^{(5)} = 25.56$ kN; $M^{1k} = -6.07$ kN·m; $M^{2k} = 49.78$ kN·m; $M^{3k} = -34.42$ kN·m; $N^{2j} = 69.87$-kN comp.; $N^{3j} = 26.12$-kN comp.

24. $R_v^{(1)} = 73.81$ kN; $R_h^{(1)} = 10.90$ kN; $M^{(1)} = 48.89$ kN: $R_v^{(5)} = 128.64$ kN; $R_h^{(5)} = 22.70$ kN; $M^{(5)} = 70.20$ kN·m; $R_v^{(7)} = 37.56$ kN; $R_h^{(7)} = 16.41$ kN; $M^{(7)} = 63.91$ kN·m; $M^{2j} = 21.63$ kN·m; $M^{3j} = 71.18$ kN·m; $M^{3k} = 7.03$ kN·m; $M^{5j} = 4.92$ kN·m; $M^{5k} = 14.68$ kN·m.

25. $R_v^{(1)} = 5.59$ kN; $R_h^{(1)} = 13.71$ kN; $M^{(1)} = 35.83$ kN·m; $R_h^{(3)} = 3.11$ kN; $R_v^{(4)} = 5.59$ kN; $R_h^{(4)} = 10.58$ kN; $M^{(4)} = 1.37$ kN·m; $M^{2j} = 7.41$ kN·m; $M^{2k} = 40.94$ kN·m; $N^{1j} = 3.75$-kN tension.

26. $R_v^{(1)} = 25.66$ kN; $R_h^{(1)} = 11.99$ kN; $M^{(1)} = 33.72$ kN·m; $R_v^{(4)} = 54.35$ kN; $R_h^{(4)} = 71.99$ kN; $M^{(4)} = 75.88$ kN·m; $N^{1j} = 27.72$-kN comp.; $N^{2j} = 11.99$-kN comp.; $N^{3k} = 86.68$-kN comp.; $M^{2j} = 62.70$ kN·m; $M^{2k} = -49.05$ kN·m;

27. Approximate results: $R_v^{(1)} = 72.5$ kN; $R_v^{(2)} = 133.75$ kN; $R_v^{(3)} = 26.25$ kN; $M^{(3)} = 52.5$ kN·m; $M^{2j} = 105.00$ kN·m.

28. Approximate results. The length of member 2 is taken as 7 m. $R_v^{(1)} = 99.65$ kN; $R_v^{(3)} = 406.36$ kN; $R_h^{(3)} = 6.05$ kN; $M^{(3)} = 13.82$ kN·m; $R_v^{(5)} = 214.00$ kN; $R_h^{(5)} = 0$; $M^{(5)} = 0$; $M^{3j} = -531.98$ kN·m; $M^{2j} = -28.51$ kN·m.

29. $R_v^{(1)} = 7.05$ kN; $R_v^{(3)} = 52.58$ kN; $R_h^{(3)} = 57.35$ kN; $M^{(3)} = 443.87$ kN·m; $R_v^{(6)} = 5.41$ kN; $R_h^{(6)} = 45.13$ kN; $M^{(6)} = 416.07$ kN·m; $R_v^{(7)} = 0.49$ kN; $R_h^{(7)} = 102.47$ kN.

30. See Problem 8.

31. See Problem 9.

32. See Problem 10.

33. See Problem 11.

34. See Problem 14.

35. See Problem 15.

36. See Problem 16.

37. See Problem 17.

38. See Problem 18.

39. (*a*) $R_v^{(1)} = 0$; $R_h^{(1)} = 33.03$ kN; $M^{(1)} = 8.19$ kN·m; $N^{4j} = 16.91$-kN comp.;
$N^{5j} = 46.06$-kN comp.; $M^{1k} = -5.89$ kN·m; $M^{2k} = -2.00$ kN·m.
(*b*) $R_v^{(1)} = 0$; $R_h^{(1)} = 32.92$ kN; $M^{(1)} = 8.14$ kN·m; $N^{4j} = 17.28$-kN comp.;
$N^{5j} = 45.80$-kN comp.; $M^{1k} = -5.01$ kN·m; $M^{2k} = -2.31$ kN·m;
$M^{4j} = 0.31$ kN·m; $M^{5j} = 0.37$ kN·m.

40. $R_v^{(1)} = 27.89$ kN; $R_v^{(3)} = 156.02$ kN; $R_h^{(3)} = 6.49$ kN; $M^{(3)} = 17.29$ kN·m;
$R_v^{(6)} = 143.41$ kN; $R_h^{(6)} = 3.23$ kN; $M^{(6)} = 8.61$ kN·m; $R_v^{(8)} = 103.44$ kN;
$R_h^{(9)} = 2.92$ kN; $R_h^{(8)} = 0.34$ kN; $M^{(8)} = 0.90$ kN·m; $R_v^{(9)} = 29.23$ kN;
$M^{1k} = -221.05$ kN·m; $M^{2j} = 34.59$ kN·m; $M^{5j} = -17.21$ kN·m;
$M^{6j} = -175.85$ kN·m; $M^{6k} = -87.94$ kN·m; $M^{7j} = -1.79$ kN·m.

41. $R_v^{(1)} = 300$ kN; $R_h^{(1)} = 293.51$ kN; $M^{(1)} = 640.03$ kN·m; $R_v^{(4)} = 0$;
$R_h^{(4)} = 0$; $M^{(4)} = 0$; $M^{1k} = -827.53$ kN·m; $M^{2k} = 772.47$ kN·m; $M^{3j} = 0$;
$M^{5j} = -401.57$ kN·m.

42. $R_v^{(1)} = 80$ kN; $M^{1j} = 42.07$ kN·m; $M^{1k} = -52.74$ kN·m;
$M^{2k} = 15.26$ kN·m; $M^{3j} = 0$; $M^{4k} = 49.93$ kN·m; $N^{1j} = 57.00$-kN comp.;
$N^{2j} = 31.60$-kN comp.; $N^{3j} = 46.00$-kN comp.; $N^{4j} = 31.60$ kN.

43. Approximate answers: $R_v^{(1)} = 100.35$ kN; $R_h^{(1)} = 96.2$ kN·m;
$M^{(1)} = 11.05$ kN·m; $R_v^{(3)} = 199.65$ kN; $R_h^{(3)} = 0.36$ kN; $M^{(3)} = 0.65$ kN·m;
$M^{2k} = 12.64$ kN·m; $M^{2j} = 0.78$ kN·m; $M^{3j} = 11.86$ kN·m.

44. See Problem 21.

45. See Problem 22.

46. See Problem 23.

47. See Problem 24.

48. See Problem 25.

49. See Problem 26.

50. See Problem 27.

51. See Problem 28.

52. $R_v^{(1)} = 365.31$ kN; $R_h^{(1)} = 15.62$ kN; $M^{(1)} = 155.37$ kN·m;
$R_v^{(5)} = 120.69$ kN; $R_h^{(5)} = 356.14$ kN; $M^{(5)} = 1533$ kN·m.

53. $R_v^{(1)} = 19.92$ kN; $R_h^{(1)} = 33.67$ kN; $M^{(1)} = 118.68$ kN·m; $R_v^{(5)} = 2.63$ kN;

$R_h^{(5)} = 32.78$ kN; $M^{(5)} = 99.90$ kN·m; $R_v^{(7)} = 22.55$ kN; $R_h^{(5)} = 33.56$ kN; $M^{(7)} = 100.67$ kN·m.

54. $R_v^{(1)} = 17.84$ kN; $R_h^{(1)} = 28.21$ kN; $M^{(1)} = 97.95$ kN·m; $R_v^{(5)} = 23.72$ kN; $R_h^{(5)} = 12.26$ kN; $M^{(5)} = 53.61$ kN·m; $R_v^{(7)} = 41.56$ kN; $R_h^{(7)} = 51.53$ kN; $M^{(7)} = 100.85$ kN·m.

55. $R_v^{(1)} = 17.73$ kN; $R_h^{(1)} = 27.85$ kN; $M^{(1)} = 96.56$ kN·m; $R_v^{(5)} = 22.43$ kN; $R_h^{(5)} = 8.94$ kN; $M^{(5)} = 47.96$ kN·m; $R_v^{(6)} = 6.50$ kN: $R_v^{(7)} = 40.14$ kN; $R_h^{(7)} = 56.71$ kN; $M^{(7)} = 95.71$ kN·m.

56. $R_v^{(1)} = 422.88$ kN; $R_h^{(1)} = 23.02$ kN; $M^{(1)} = 28.77$ kN·m; $R_v^{(8)} = 977.12$ kN; $R_h^{(8)} = 23.02$ kN; $M^{(8)} = 57.60$ kN·m; $M^{1k} = -86.34$ kN·m; $M^{2j} = 100.22$ kN·m; $M^{3j} = -117.69$ kN·m; $M^{4j} = -186.56$ kN·m; $M^{3k} = -543.70$ kN·m; $M^{4k} = -531.72$ kN·m; $M^{6j} = -369.99$ kN·m; $M^{8j} = -339.95$ kN·m.

57. $R_v^{(1)} = 15.13$ kN; $R_h^{(1)} = 47.71$ kN; $M^{(1)} = 92.62$ kN·m; $R_v^{(6)} = 1.29$ kN; $R_h^{(6)} = 28.82$ kN; $M^{(6)} = 81.98$ kN·m; $R_v^{(9)} = 16.42$ kN; $R_h^{(9)} = 23.47$ kN; $M^{(9)} = 73.05$ kN·m; $M^{1k} = 20.91$ kN·m; $M^{2j} = -29.06$ kN; $M^{2k} = 12.34$ kN·m; $M^{3k} = -15.12$ kN·m; $M^{5k} = -43.58$ kN·m; $M^{4k} = 24.21$ kN·m; $M^{6j} = 62.14$ kN·m; $M^{7k} = -20.87$ kN·m; $M^{8k} = -48.33$ kN·m.

Chapter 9

6. **Ordinates of Influence Lines**

x_1, m	$R_v^{(1)}$	$M^{(2)}$	Q_A	M_A
1	0.848	−0.219	0.0219	−0.131
2	0.698	−0.417	0.0417	−0.250
3	0.553	−0.573	0.0573	−0.344
4	0.417	−0.667	0.0667	−0.400
5	0.290	−0.677	0.0677	−0.406
6	0.177	−0.583	0.0583	−0.350
7	0.0794	−0.365	0.0365	−0.219
9	−0.0594	−0.475	−0.0525	0.315
10	−0.100	−0.800	−0.120	0.720
11	−0.124	−0.992	−0.201	1.210
12	−0.133	−1.070	−0.293	1.760
13	−0.130	−1.040	0.604	1.380
14	−0.117	−0.933	0.493	1.040
15	−0.0948	−0.758	0.376	0.745
16	−0.0667	−0.533	0.253	0.480
17	−0.0344	−0.275	0.128	0.235
19	0.0347	0.278	−0.128	−0.233
20	0.0694	0.556	−0.256	−0.467
21	0.104	0.833	−0.383	−0.700

7. Ordinates of Influence Lines

x_1, m	$R_v^{(1)}$	$R_v^{(2)}$	$R_v^{(4)}$	$R_v^{(5)}$	$M^{(5)}$	$M^{(2)}$	Q_A	M_A
1	0.804	0.226	−0.052	0.022	0.0591	−0.177	−0.00222	0.0296
2	0.613	0.441	−0.094	0.041	0.102	−0.324	−0.0405	0.0541
3	0.432	0.636	−0.120	0.052	0.138	−0.410	−0.0513	0.0684
4	0.266	0.802	−0.118	0.050	0.135	−0.405	−0.0507	0.0676
5	0.120	0.927	−0.081	0.034	0.093	−0.279	−0.0348	0.0465
7	−0.092	1.017	0.132	−0.057	0.150	−0.549	0.0564	−0.0752
8	−0.158	0.982	0.308	−0.132	0.351	−0.946	0.132	−0.176
9	−0.204	0.908	0.518	−0.222	0.593	−1.220	0.222	−0.296
10	−0.234	0.802	0.756	−0.324	0.865	−1.410	0.324	−0.431
11	−0.255	0.676	1.013	−0.434	1.160	−1.530	0.434	−0.579
12	−0.270	0.540	1.277	−0.547	1.460	−1.620	0.547	−0.730
13	−0.190	0.380	1.230	−0.420	1.120	−1.140	0.420	−0.560
14	−0.115	0.230	1.174	−0.289	0.770	−0.689	0.289	−0.385
15	−0.050	0.100	1.100	−0.150	0.400	−0.299	0.150	−0.200
17	0.031	−0.062	0.868	0.163	−0.430	0.186	−0.163	0.223
18	0.046	−0.092	0.713	0.333	−0.846	0.274	0.333	0.486
19	0.048	−0.095	0.547	0.500	−1.190	0.285	−0.500	0.808
20	0.040	−0.081	0.384	0.657	−1.420	0.243	−0.657	1.210
21	0.029	−0.057	0.234	0.794	−1.470	0.171	0.206	0.710
22	0.015	−0.030	0.112	0.903	−1.280	0.0912	0.0973	0.329
23	0.005	−0.009	0.030	0.974	−0.811	0.0266	0.0258	0.0854

8. Ordinates of Influence Lines

x_1, m	$R_v^{(1)}$	$R_v^{(2)}$	$R_v^{(3)}$	$M^{(3)}$	$M^{(2)}$	Q_A	M_A
1	0.843	0.195	−0.038	0.127	−0.254	0.038	−0.102
2	0.690	0.383	−0.073	0.242	−0.484	0.073	−0.194
3	0.542	0.558	−0.100	0.333	−0.665	0.100	−0.266
4	0.403	0.713	−0.116	0.387	−0.774	0.116	−0.310
5	0.277	0.841	−0.118	0.393	−0.786	0.118	−0.315
6	0.165	0.937	−0.102	0.339	−0.677	0.102	−0.271
7	0.072	0.992	−0.064	0.212	−0.423	0.064	−0.169
9	−0.049	0.958	0.091	−0.299	−0.392	−0.091	0.245
10	−0.077	0.874	0.203	−0.650	−0.619	−0.203	0.568
11	−0.089	0.759	0.330	−1.010	−0.711	−0.330	0.969
12	−0.087	0.624	0.463	−1.330	−0.697	−0.463	1.450
13	−0.076	0.479	0.597	−1.570	−0.605	0.403	1.010
14	−0.058	0.336	0.722	−1.690	−0.465	0.278	0.646
15	−0.038	0.205	0.833	−1.630	−0.305	0.167	0.364
16	−0.02	0.099	0.921	−1.360	−0.155	0.079	0.162
17	−0.005	0.026	0.979	−0.833	−0.044	0.021	0.041

9   Ordinates of Influence Lines

x_1, m	$R_v^{(1)}$	$R_v^{(3)}$	$R_h^{(3)}$	$Q^{(2)}$	$M^{(2)}$
2	0.733	0.298	0.081	0.031	−0.185
4	0.483	0.571	0.141	0.054	−0.323
6	0.266	0.795	0.162	0.062	−0.369
8	0.097	0.947	0.121	0.046	−0.277
12	−0.050	0.941	0.140	0.890	−1.342
14	−0.085	0.797	0.237	0.712	−2.273
16	−0.113	0.613	0.312	0.500	−3.000
18	−0.065	0.353	0.180	0.288	−1.727
20	−0.025	0.134	0.069	0.110	−0.658
24	0.010	0.057	0.029	−0.046	0.279
26	0.014	0.075	0.038	−0.062	0.369
28	0.012	0.066	0.034	−0.054	0.323
30	0.007	0.038	0.019	−0.031	0.185

Index

ABOUT THE AUTHOR

Anthony E. Armenàkas is currently a Professor of Aerospace
Engineering at the Polytechnic University of New York. He
has more than 35 years' experience teaching undergraduate
and graduate courses in structural analysis and applied
mechanics. He has taught at the University of the City of
New York, the Cooper Union for the Advancement of
Science and Art, the University of Florida, and the National
Technical University of Athens. In addition, he has acted as
a consultant to many companies and government agencies,
both in the U.S. and abroad, on problems of structural analysis, stress analysis, and the design of complex structures and
machines. Doctor Armenàkas received his B.S. in civil engineering from the Georgia Institute of Technology, his M.S. in
civil engineering from the Illinois Institute of Technology, and
his Ph.D. in applied mechanics from Columbia University.
He has authored numerous articles and technical papers on
subjects related to structures.